Questioning Gender

Questioning Gender

A Sociological Exploration

Second Edition

Robyn Ryle

Hanover College

Los Angeles | London | New Delhi
Singapore | Washington DC

Los Angeles | London | New Delhi
Singapore | Washington DC

FOR INFORMATION:

SAGE Publications, Inc.
2455 Teller Road
Thousand Oaks, California 91320
E-mail: order@sagepub.com

SAGE Publications Ltd.
1 Oliver's Yard
55 City Road
London EC1Y 1SP
United Kingdom

SAGE Publications India Pvt. Ltd.
B 1/I 1 Mohan Cooperative Industrial Area
Mathura Road, New Delhi 110 044
India

SAGE Publications Asia-Pacific Pte. Ltd.
3 Church Street
#10-04 Samsung Hub
Singapore 049483

Acquisitions Editor: Jeff Lasser
Editorial Assistant: Nick Pachelli
Production Editor: Libby Larson
Copy Editor: Sheree Van Vreede
Typesetter: C&M Digitals (P) Ltd.
Proofreader: Sally Jaskold
Indexer: Kathy Paparchontis
Cover Designer: Edgar Abarca
Marketing Manager: Erica DeLuca

Printed in the United States of America

Library of Congress Cataloging-in-Publication Data

Ryle, Robyn.
Questioning gender : a sociological exploration / Robyn Ryle, Hanover College. — Second Edition.

pages cm
Includes bibliographical references and index.

ISBN 978-1-4522-7586-4 (pbk.)
ISBN 978-1-4522-7584-0 (web pdf)

1. Sex role. 2. Sex (Psychology)--Social aspects.
3. Sex differences (Psychology)—Social aspects.
4. Gender identity. 5. Sex discrimination. I. Title.

HQ1075.R95 2014
305.3—dc23 2013040640

This book is printed on acid-free paper.

SFI label applies to text stock

14 15 16 17 18 10 9 8 7 6 5 4 3 2

Contents

PART II HOW ARE OUR LIVES FILLED WITH GENDER? 107

This section of the text focuses on everyday aspects of gender, using an interactional, micro level approach to issues. In this part of the book, students will begin to see the ways in which gender matters in their day-to-day lives and how that impact is socially constructed both historically and globally.

10 How Does Gender Affect What You Watch, What You Read, and What You Play? The Gender of Media and Popular Culture 409

11 How Does Gender Help Determine Who Has Power and Who Doesn't? The Gender of Politics and Power 459

Specific Areas of Interest

INDIVIDUAL LEVEL

Chapter 2

Pp. 30–33 "Sex Roles" to "Interactionist Theories"

Chapter 3

Pp. 64–72 "Psychological Approaches to Gender" to "Anthropological Approaches to Gender"

Chapter 4

Pp. 118–128 "Social Learning Theory" to "The Early Years: Primary Socialization Into Gender"

Chapter 9

Pp. 389–390 "Socialization as an Explanation for Sex Segregation" to "Human Capital Theory"

Pp. 390–394 "Human Capital Theory" to "Gendered Organizations"

INTERACTIONIST LEVEL

Chapter 2

Pp.22–43 "Interactionist Theories" to "Institutional or Structural Approaches"

Chapter 4

Pp. 138–143 "Gender Camp: The Importance of Peer Groups" to "Learning Gender Never Ends: Secondary Socialization"

INSTITUTIONAL LEVEL

Chapter 2

Pp. 43–48 "Institutional or Structural Approaches" to "Intersectional Feminist Theory"

Chapter 9

Pp. 394–396 "Gendered Organizations" to "Transmen at Work"

Chapter 10

Pp. 412–420 "Behind the Scenes: The Gender of Media Organizations" to "Media Power Theory: We're All Sheep"

Chapter 11

INTERSECTIONAL

Chapter 2

Chapter 4

Chapter 6

Chapter 7

Chapter 8

Chapter 9

Chapter 10

MASCULINITY

Chapter 3

Chapter 5

Chapter 7

Chapter 8

Chapter 9

Chapter 10

Chapter 11

GLOBAL

Chapter 3

Chapter 4

Chapter 5

Chapter 6

Chapter 7

Chapter 8

Chapter 9

Pp. 360–361 "What Is Work?" to "Measuring the World's Work"

Pp. 361–365 "Measuring the World's Work" to "A Man's Job: Masculinity and Work"

Chapter 10

Pp. 438–444 "Soap Operas, *Telenovelas*, and Feminism" to "Masculinity and Video Games: Learning the Three Rs"

Chapter 11

Pp. 478–481 "Violent Intersections: The Gender of Human Trafficking" to "Gender Rights and Human Rights?"

Pp. 482–484 "Hijab and Ethnocentrism" to "Institutional Power: Nations and Gender"

Preface to the Second Edition

NEW TO THE SECOND EDITION

The following features are new to the 2nd edition.

Mix and Match

The 2nd edition includes more subheadings that make it easier for instructors to mix and match certain topics and sections. This allows instructors and students to customize the textbook to their own interests and needs. Using the subheadings and the Mix and Match table, instructors can emphasize certain topics or theoretical approaches within chapters while choosing not to cover other topics. For example, in a class with an emphasis on masculinity, instructors might have students read those specific sections relating to masculinity while ignoring other topics. Clearer and more frequent subheadings make it easier for instructors to mix and match topics and approaches.

Glossary

A list of all the terms with definitions gathered in one place at the end of the textbook make it easier for students and instructors to locate and use key terms.

Updated Cultural Artifact Boxes

Cultural artifact boxes have been brought up to date with more contemporary material.

Updated Material on Transnational Work and Families

Transnationalism is an increasingly important topic for study in all areas of social life. Chapters 8 and 9 include more material on how transnationalism affects the gender of work and the gender of families.

Updated Section on Intersectionality

The section in Chapter 2 on intersectionality has been updated to include evolving debates about the nature of an intersectional approach to gender.

ABOUT QUESTIONING GENDER

Questioning Gender: A Sociological Exploration is an un-textbook. If textbooks are presumed to be the place you go to get all the answers, this book is an un-textbook in that our goal is to raise as many questions as we can about gender. We explore some possible answers to those questions, but this is an un-textbook in that the main goal is to start a conversation that helps you to question some of the prevailing assumptions you might have about gender. This book is just the start of that conversation, and it is designed to serve as a resource for discussion inside and outside of the classroom. *Questioning Gender* is a book based on the premise that a good conversation about gender helps you to connect all the complicated scholarship that has been conducted on gender to a thorough investigation of the role of gender in your own life, and for that reason, you'll find this book packed to the brim with questions. Each chapter title is a question, there are question boxes inserted in each chapter, questions at the end of the cultural artifacts that help you think about the prevalence of gender in our everyday lives, and big questions to help you make connections at the end of each chapter. This un-textbook uses a wide range of theories from within and without sociology, assuming that theories are useful for the ways in which they can suggest new questions or focus the questions you already have. Then for a wide range of topics related to gender, including socialization, sexuality, friendship and dating, bodies, marriage and families, work, and media and politics, we use a historical and cross-cultural perspective to question the things we might think we know about gender. In this book, we'll unpack many of the truths we take for granted about our social lives related to gender, making basic concepts like sex, marriage, love, and friendship into moving targets with many potential meanings depending on who you are, and when and where you happened to be born. We place the experiences of people who are usually at the margin of gender conversations (gay, lesbian, bisexual, and transgendered; women and men of color; women and men of the global South; and poor and working-class women and men) at the center of our conversations because their experiences throw open the door on a whole new set of questions that need to be asked about gender. As I tell the students in my own gender course, if you finish this book with more questions about gender than you started with at the beginning, our goal will have been accomplished and you'll be well on your way to a lifetime conversation about what gender is and what it means in your own life.

The primary course this text is aimed at is sociology of gender. This text would be ideal for upper-class undergraduate students (juniors and seniors), although it is cast at a level that would make it accessible to lower-class undergraduates, as well. The text is firmly grounded within a sociological approach to gender, with a focus on sociological theories related to gender and research within social science disciplines. However, it is impossible to discuss gender in a contemporary context without also addressing theories from outside sociology, such as feminist theories and queer theory. Because of this interdisciplinary approach, *Questioning Gender* would be appropriate for introductory courses in women's studies and gender studies as well. This text is best suited for courses that seek to use the social construction of gender in a global and historical perspective to challenge students' preconceptions about gender and to demonstrate how gender as a system creates and reinforces inequality.

UNIQUE APPROACHES

There are several unique approaches in *Questioning Gender* that set it apart from other gender textbooks. First, *Questioning Gender* takes a **global approach** to gender. In an increasingly global world, it is difficult to justify an approach to gender issues that focuses solely on the United States or the developed world. Examining gender in a global context also helps to demonstrate the social construction of gender and the persistence of gender inequality around the world. For some of the same reasons, *Questioning Gender* also uses an **intersectional approach**. Since women of color first brought attention to the ways in which gender intersects with race and ethnicity, those who study gender have become increasingly concerned with how to discuss gender while grounding it firmly within the complex web of identities such as race, class, sexuality, disability, religious background, and so on. Gender does not exist in a vacuum, and an intersectional approach helps to demonstrate that there is no "normal" experience of what it means to be gendered. *Questioning Gender* is also unique in incorporating **the perspective of gays, lesbians, bisexuals and transgendered, and queer theory** throughout the textbook. As it is the goal of many instructors to help students understand the social construction of gender, focusing on transgender concerns helps to blur the boundaries between male and female, masculine and feminine, and homosexual and heterosexual previously perceived as rigid. Queer theory questions all categories of difference, and the experiences and perspectives of transgender individuals provide a vantage point for students to see beyond our dimorphic gender constructions. *Questioning Gender* also assumes that questions related to gender must be answered through a consideration of both women and men and, therefore, includes the growing scholarship on **the study of men and masculinity**. All of these unique approaches to gender in *Questioning Gender* help to create a textbook that resists the tendency to normalize certain understandings of gender while marginalizing others. For example, many textbooks in gender segregate discussions of gender in global perspective in one chapter or in certain sections of the chapter. In this book, global experiences of gender are incorporated throughout the text, decentering the idea of a Western, predominant idea of what it means to be a man or a woman. In addition, rather than segregating a discussion of transgender issues in a chapter on sexuality or biology, *Questioning Gender* integrates these concerns throughout in discussions about a wide variety of topics including socialization, work, and dating.

The final distinctive approach in *Questioning Gender* is **the integration of theoretical perspectives throughout the text**. Many textbooks cover theory in the first few chapters and then move on to a topical focus on gender. In this book, theory is covered in early chapters, but then it is discussed throughout the book as each theory applies to different topics. This use of theory throughout the text is highlighted for students and instructors through "Theory Alerts!" This approach reminds students of the importance of theory to our understanding of gender, and it models for them how different theories might be applied to different topics related to gender. Other key features of the text, as defined in the next section, such as Cultural Artifacts, question boxes, and Big Questions, all incorporate theory. In this way, *Questioning Gender* seeks to put theory at the center of discussions of gender rather than at the peripheries.

KEY FEATURES OF THE TEXT

Question Boxes

Distributed throughout each chapter in *Questioning Gender* are question boxes. Question boxes? contain discussion questions that push students to explore or test further the theories and concepts being discussed. Question boxes appear in the text integrated with the material to which they relate and are specific to that material. Question boxes serve as starting points for students and instructors in beginning their discussions and explorations of gender in their own lives. So, for example, in a discussion of research on heteronormativity in high schools in Chapter 5, a question box asks students, "Can you think of other examples of heteronormativity on your college or university campus, or other ways in which practices and values reinforce the idea that heterosexuality is normal and right?" Question boxes can be used by students to enhance their experience of reading the text or by instructors in the classroom to generate discussion or serve as the prompts for free-writes, short response papers, or journals.

Cultural Artifacts

Also distributed throughout the text are Cultural Artifact boxes. The premise of using cultural artifacts is adapted from an assignment I use in my own classroom in which students have to bring in some artifact related to gender to share with the class once over the course of the semester. These gender artifacts are anything they observe that is related to gender and have included television commercials, YouTube clips, magazine articles, print advertisements, comic strips, viral e-mails, conversations, sports equipment (men and women's basketballs), and personal hygiene products (men and women's shampoo, deodorant, razors). Cultural Artifact boxes attempt to accomplish the same goal as this class assignment in encouraging students to begin to see gender all around them in their daily lives. Cultural Artifact boxes include references to films, television shows, music, websites, and video games as they help to illustrate key concepts discussed in the text. This feature helps students to make connections between sociological theories and concepts related to gender and the everyday world around them; this allows them to consider how gender matters on a day-to-day basis in their lives.

Big Questions

At the end of each chapter are Big Questions. Big Questions accomplish one of three different goals. First, Big Questions get students to apply concepts and theories to a broader set of issues or questions than those discussed in that particular chapter. These questions often take the form of "what if" questions, and they ask students to expand in directions just suggested at in the chapter. On a smaller scale, some Big Questions ask students to think about integrating concepts and ideas within the chapter. An example might be looking at how two different theories discussed in the chapter fit together, or how a theory applied to one particular topic in the chapter might work when applied to a different topic. Finally, some Big Questions encourage students to make connections between concepts, theories, or topics discussed in that particular chapter with those discussed in other chapters in the book. All of these different types of questions encourage students to think of the "big picture" as it relates to gender and address some of the larger themes identified in the book. They encourage students and instructors to use the book interactively by making connections between different chapters in the book for themselves, as well as expanding beyond the topics covered in the book itself. Big Questions might be used as the basis for longer, critical thinking essays; essay examination questions; or small group discussions or debate exercises.

Gender Exercises

Also at the end of each chapter are Gender Exercises, which are more interactive than Big Questions and encourage students to embark on their own projects of gender inquiry. Gender Exercises suggest ways in which students could use social science techniques like interviews, observation, surveys, and existing statistics to investigate gender for themselves. They also introduce exercises that help students apply theory to a particular situation from their own lives or popular culture to understand that theory better. In Chapter 7 on gender and bodies, a Gender Exercise asks students to visit a local art museum or look at images of art online and think about how male and female bodies are depicted. Gender exercises could be used by students and instructors as the basis for research projects or research exercises. These exercises help encourage students to become active learners and to engage in social science research techniques.

Terms and Suggested Readings

Within each chapter in *Questioning Gender,* important terms and concepts are highlighted and included in a list of terms at the end of each chapter. This feature helps students and instructors to organize information presented in the chapter with the ultimate goal of learning how to apply these terms and concepts to other gender issues and situations. Each chapter in *Questioning Gender* also includes a list of suggested readings organized by different topics covered in the chapters. This can serve as a resource for students and instructors interested in finding out more about some of the issues discussed in the chapters and as a good starting place for student research papers.

ORGANIZATION OF THE TEXT

Part I: What are the important questions to ask about gender?

These first three chapters set up the basic foundations for an exploration of gender. They introduce the main goals in learning about gender, the basic theories that help to understand gender, and the ways in which those theories will be used throughout the text. Chapter 1 introduces and defines basic concepts in the exploration of gender and discusses why the study of gender is a worthwhile pursuit. Chapter 2 explores the feminist background of many gender theories and outlines sociological theories of gender. In Chapter 3, we explore gender theories from disciplines outside of sociology, including psychology, anthropology, queer theory, development theory, and eco-feminist theories of gender.

Part II: How do we experience gender in our day-to-day lives?

This section of *Questioning Gender* focuses on everyday aspects of gender through a more interactional, micro-level approach to issues. In this part of the book, students will begin to consider the ways in which gender matters in their day-to-day lives, and how that impact is socially constructed historically and globally. In Chapter 4, we explore questions related to socialization and theories that explain how we learn to be gendered. The gender of sexuality is explored in Chapter 5 where we look at the complicated ways in which sexuality and gender intersect. Chapter 6 explores the gender of friendship in dating, including the different ways in which attraction works on a global scale and how the gender of friendship has changed over time. In Chapter 7, we look at the gender of bodies, including issues of body image and health.

Part III: How is gender an important part of the way our society works?

This portion of *Questioning Gender* moves toward a focus on how gender permeates various institutions in society. Working at the institutional, macro level, these chapters are more concerned with how gender operates as a system of power and reinforces inequality. In Chapter 8, we examine the important intersections between gender, marriage, and families, taking a historical look at how marriage as an institution has changed over time and how this has affected ideas about gender. Chapter 9 looks at how the institution of the workplace has gendered implications, including a consideration of sex segregation and the gender wage gap. The unique intersections between gender and the media as an institution are examined in Chapter 10. Finally, Chapter 11 explores gender in the realm of states and governments through a consideration of the politics of gender.

A NOTE ON LANGUAGE

Like many other authors, I chose to capitalize "Black" but not "white" in this textbook. Being Black in America constitutes an ethnic group, similar in its sense of belonging and group cohesion to Irish Americans, Arab Americans, and many other groups. Being white in the United States generally does not carry with it this sense of ethnic identity and belonging, although ethnic identities such as Polish, Italian, or German might. Although both categories are socially constructed and have no underlying scientific basis, I chose to mark this distinction in how the categories are experienced.

ACKNOWLEDGMENTS

Writing and revising a textbook are activities that truly take a village, in the sense of a supportive community of individuals. Although I was quite happily teaching a successful sociology of gender course, the thought of writing a textbook based on that class would never have occurred to me had my colleague, Keith Roberts, not suggested it to an editor at

SAGE who gave me the opportunity to try my hand at a proposal. I owe thanks to the rest of my colleagues in the Department of Sociology and Anthropology at Hanover College for supporting me in the extra effort required in writing a textbook, and to the Faculty Development Committee at Hanover for granting me two leaves to work on the book. Kelly Joyce and Heather Loehr of the Hanover College library provided invaluable research expertise in finding statistics related to gender. Jessica Hoover and Jessica Caldwell spent many hours with JSTOR and interlibrary loan gathering research articles. Fahima Eissar helped assemble the glossary for the 2nd edition. My dear friends and fellow co-conspirators from graduate school at Indiana University—Katy Hadley, Jeni Loftus, Sandi Nenga, and Carla Shirley—may not have read any drafts of the chapters, but they surely helped to keep me sane in the life of academia during the long process of writing the book.

Many thanks go to the team of editors at SAGE for having faith in me and for helping in the long and arduous task of bringing this textbook to life, including David Repetto, Diane McDaniel, Lauren Johnson, Jeff Lasser, and Erica DeLuca, as well as copy editor Sheree Van Vreede. I thank both my parents for teaching me important lessons about gender throughout my life. I could not have completed this book without the support, love, encouragement, and patience of my husband, Jeff, and my stepdaughter, Grace. Thank you, Jeff, for continually and repeatedly assuring me that I could, in fact, do this. Thank you, Grace, for putting up with frequent states of distraction.

My final thanks are to all the students at Hanover College over the years who served as guinea pigs in the long process of figuring out how to teach a good course on the sociology of gender. I learned how to teach about gender from learning with you and could not have had the courage to put one word on the page without the memories of our shared laughter and exploration, and I could not have sustained myself over the long haul without the hope that I could pass along to others some of the joy of learning together we shared.

PART I

What Are the Important Questions to Ask About Gender?

CHAPTER 1

What Is Gender and Why Should We Care About It?

Introducing Gender

What do students expect when they sit down on the first day in a class on sociology and gender? What notions do they already have about what gender is and how it matters? How important do they believe gender is to their own lives? What are the stories they tell themselves about gender? These are questions I often ask *myself* about gender, and let me just say at the outset, I certainly don't claim to have all the answers. Learning about gender inevitably involves learning about yourself and your own life. When we combine gender and sociology, it also involves learning about the importance of social forces as they relate to gender. When you begin to think seriously about the impact of gender on your life and the larger world, it becomes difficult not to see gender everywhere you look. You can decide for yourself whether that's a good or a bad thing, but I believe that being aware of the ways in which gender permeates our lives can give us invaluable insight into the world.

SWIMMING WITH THE FISHES: LEARNING TO SEE GENDER

Part of what we will be doing throughout this book is learning to see gender in the world around us. That might seem like a fairly stupid goal at first; most of us believe that we can see gender in the world. Imagine yourself standing on a busy sidewalk in a city somewhere. As the people walk by, we imagine we could identify the gender of most of the people with a fairly high degree of certainty. Every now and then someone might walk by who gave us pause, but by and large, we believe that we're pretty good at seeing gender—in the sense of identifying the gender of those around us. In this chapter, we will begin to cast some doubt on whether what you're seeing when people walk by on the street is really gender, and we will, therefore, begin to question whether we really see gender as well as we might think. But we'll also begin to extend exactly what we mean by *seeing gender*. Seeing gender the way we'll talk about it involves more than just identifying the gender of the people around you. It means beginning to reveal the invisible ways in which gender works, which are not always apparent on the surface of our lives. This is what Judith Lorber (1994) means when she describes gender as like water to fish. For many people who study this topic, gender is the substance that's all around us and inside us but that we largely take for granted most of the time. Learning to see gender means developing a special kind of vision.

Gender and the Social Construction of Reality

One particularly useful tool that sociologists have to help us develop this kind of gender vision is called *the social construction of reality*. If you have taken sociology classes before, there's a strong chance you've heard about the social construction of reality and about how a wide range of phenomena can be seen as socially constructed. The social construction of reality describes the historical process by which our experiences of the world become put into categories and treated as real things (Roy, 2001). What does this mean in relation to gender? Probably the best way to help us understand the social construction of reality as it relates to gender is to use the Thomas principle, from Dorothy and W. I. Thomas, which states, "If people define situations as real, they are real in their consequences" (Thomas & Thomas, 1928, p. 572). Take categories of race as an example. Anthropologists and biologists have definitively shown that there is no biological basis for what we think of as racial categories. Extensive research using DNA from people all over the planet demonstrates that among the small amount of genetic variation that exists in humans as a species, most exists within the groups we commonly refer to as *races*.

What this means is that any two people within what we think of as a racial category (white or Caucasian, for example) are as likely to be genetically different from each other as they are from someone in what we think of as a different racial category (Hispanics or Latinos, for example). If you consider yourself Latino, you are no more likely to be genetically similar to someone else who considers herself Latino than you are to be genetically similar to someone who considers herself Asian. Genetic variation among humans simply does not map along racial lines in part because the genes that influence skin color, hair form, and facial features (the traits we generally associate with race) evolved much later in our human history than other characteristics like intelligence, athletic ability, or musical ability (Adelman, 2003).

How does this help us understand the social construction of reality? Over time, people began to believe that things like skin color, hair form, and facial features meant something more than they actually do. The categories of both white and Black were created over a long historical time period and were used to justify a system of exploitation and oppression. Today, many people believe that these categories are real things, based in some underlying biological reality. But historically, the definitions of who is white, Black, Latino, Asian, or Native American in the United States have constantly shifted; since the first year the United States conducted a census in 1790, the way in which race is measured has changed every single year. Race is just one example of how, through a historical process, people's experiences of the world (seeing different skin colors) were put into categories (Black, white, Asian, Latino, Native American) and treated as real things.

The Thomas principle points us to the important part of this equation, socially speaking. Scientists can demonstrate for you that race has no underlying biological reality, but that does not mean race does not have a real impact on the lives of people in the United States and around the world. As we will discuss throughout this book, glaring inequalities exist along racial lines. But these inequalities are not a result of any biological explanations but of the ways in which our belief in racial categories as something real makes them into something that has real consequences. Because many people in the United States believe that the categories of Asian, Latino, and Black are real, those categories have actual effects on our lives. When someone turns to an Asian American student and asks for help with math homework, drawing on the stereotype that all Asians are good at math, knowing that the category "Asian American" is socially constructed doesn't really do much good. In the social world (which is the world we all live in), what we believe matters, and the things we treat as real become, for all intents and purposes, real.

Once we understand the ways in which the world around us is socially constructed, we can begin to untangle the complex ways in which our understanding of the world is shaped by our particular social constructions. Understanding the social construction of gender helps us to see the water in which we're swimming, as well as to understand why we didn't notice it in the first place. There are two specific methods for helping us to see the social constructions around us that we'll be using in this book. The first, which we've already considered in our discussion of race, is a historical approach. Part of what makes things seem real is the sense that this is how they always have been and how they always will be. A historical approach helps us see that this is very often not the case. The "real" things we call racial categories were constructed very differently in the United States in 1790 (free whites, other free persons, slaves), in 1890 (white; Black; mulatto, or half-Black; quadroon, or one-fourth Black; octoroon, or one-eighth Black; Japanese, Indian), and in 1990 (white, Black or African American, American Indian and Alaskan Native, Asian, Native Hawaiian and other Pacific Islander, and some other race). If people thought of mulatto (being half-white and half-Black) as a real racial category in 1890, what does that say about the reality of our current racial categories?

The second approach that helps us reveal social constructions is cross-cultural. If something that you very much believe is real in your particular culture is perceived as ridiculously impossible in another culture, what exactly is real? Cross-culturally, racial categories become even more complicated. *Black* in Great Britain can refer to people of African origin but also to Indian and Pakistani immigrants, who in the United States, would officially (by the Census rules) be considered Asian. In South Africa, Blacks are the native Africans, but Coloureds are those who are of some mixed African ancestry, whereas Indians under the system of apartheid (legalized racial segregation and reduced rights for minority groups) were called Blacks, Asians, or Indian. In many Latin American countries, members of the same family can be categorized in different racial groups based on their skin color; this means that your sibling may be considered white while you are considered Black. Many people around the world find that their racial category changes as they move from place to place and country to country. What does this say about the reality of these categories? Throughout this book, we will be using both the historical approach and the cross-cultural approach to help us see the ways in which gender is also socially constructed.

SEX OR GENDER: WHAT'S THE DIFFERENCE?

By now you might have noticed an important relationship between biology and our perceptions of reality; race is perceived as something that's real because many people believe in its biological truth. This is true in many parts of the world where science forms a dominant way of thinking about the world. If something is rooted in our biology, it can be empirically and objectively observed and must, therefore, be real. Later, we'll explore the way in which even our trust in science is gendered and has gender implications.

Biological explanations of gender differences are often called on to establish their reality, and this brings us to our first point of vocabulary, the difference between *sex* and *gender.* When I ask students in my sociology of gender class early in the semester what the difference is between sex and gender, they usually agree that sex describes the biological differences between women and men while gender pertains to the social differences. Although not everyone follows this general usage, this is a fairly accurate description of how social scientists employ these terms. Sex describes the biological differences between people we call males and people we call females; gender is the social meanings that are layered onto those differences. This neat division leaves sex up to those concerned with biology and gender up to those interested in the social world. For much of the history of the study of gender in the Anglo-European world, this was a standard way of understanding sex and gender, and it was often called a biosocial approach.

A Biosocial Approach

A *biosocial approach* to the study of gender acknowledges that much of what we experience regarding gender is socially constructed. There are differences in how gender as a social category is constructed across historical time periods and across different places. However, from the biosocial perspective, there are real limits to that social construction in the form of the biological reality of male and female bodies. One way to understand this perspective is to say that biosocialists believe in sexual dimorphism. Sexual dimorphism is the claim that sex marks a distinction between two physically and genetically discrete categories of people. If you subscribe to sexual dimorphism, you believe that we can use certain characteristics to sort people objectively into two categories called male and female. Discrete here means that you can only be one or the other (male or female) and not both at the same time. Biosocialists do not believe that our sex is the only thing that determines how we interact with the world; gender is constructed onto the differences we call sex. But they do believe that there are two kinds of people in the world—females and males.

At this point, you might be thinking to yourself, well, duh! Who doesn't believe that there are two types of people in the world, males and females? Even a child knows that there are girls and there are boys and that's sex, right? Or is it? If you stood on a busy street, watching people walk by, would you be using sex or gender to categorize people? Genitalia are one of the physical criteria that we use to sort people into sex categories. Can you see people's genitalia when they're walking down the street? Perhaps there are some streets where people are walking around naked or with their genitalia exposed but generally not.

What can we see of sex in our everyday lives? We can see the shape of people's bodies and their faces. If someone has breasts, does that mean the person is a woman? Is that a criterion we could use? Yes, except that some men have breasts (or there would have been no need for Kramer, a character on the hit sitcom *Seinfeld,* to invent the "bro," a bra for men) and some women do not. What about facial hair? We can certainly see facial hair. And yet, again, some men do not have facial hair, and some women do (think of the bearded lady). Adam's apples? Again, not all men have them and not all women lack them. On average, the females should be slightly shorter than the males, but again, we will also encounter very short males and very tall females. What is it you're really seeing as you stand on this street?

MYTHS OF GENDER: MEN ARE TALLER THAN WOMEN

One way to think about the strong social constructionist position is to consider how biological differences can be influenced by social reality. This helps us see how the social can influence the biological. One biological difference most people would identify between females and males is that males are, on average, taller than females. This is true across human cultures, although it's important to note that there's a great deal of overlap. The tallest male, according to the *Guinness Book of World Records,* was 8 ft. 11 in., whereas the world's tallest female was 8 ft. 1 ¾ in.—a difference of only 9 ¼ inches. What's particularly interesting about sex differences in height from a strong social constructionist perspective is that they change over time. The average height differences between females and males have been shrinking in the last 100 years. How do we explain this change in biology? The best possible answer has to do with changes in nutrition brought about by shifting cultural beliefs. Adult height is greatly influenced by nutrition as a child; this is part of why we see average height variations in people from different parts of the world. If you benefit from good nutrition as a child, you're more likely to reach your maximum height potential. How does that explain changing sex differences in height? One theory is that cultural norms across many societies that once clearly favored male children over female children have changed. In the past, one expression of this preference for male children was to direct more of the family's resources toward male children, and one of those resources was food. Male children received better nutrition than their sisters and therefore grew taller. As these cultural norms have changed, the nutritional lives of female and male children have evened out, and the height differences between male and female adults have lessened over time. Average height differences between males and females have not completely disappeared, but this historical story demonstrates how our biological makeup is often conditioned by cultural beliefs.

Fine, you might say, but if they were naked on the street, then of course, you would be able to sort everyone into categories of male and female. Some people would have penises and some people would have vaginas. No one would have both, right? The answer is that, in fact, some people *would* have both penises and vaginas. People born with congenital adrenal hyperplasia (CAH) have XX chromosomes, but masculinization of the genitalia. As infants, these babies would have what appear to be a penis as well as a vagina. Individuals born with androgen insensitivity syndrome (AIS) have XY chromosomes, but feminized genitalia, which often means they have a vagina, but also testes (Fausto-Sterling, 2000). As we will read in Chapter 4, these infants are unlikely to survive into adulthood with their ambiguous genitalia intact because doctors in the United States usually perform surgery on intersex infants to create a consistent sex category for them. Thus, the infant with CAH is likely to have the penis removed, whereas the infant with AIS is likely to be raised as a girl despite the presence of testes and the development of masculine secondary sex characteristic at puberty.

Gender scholars identify individuals who have any of a wide number of conditions that cause ambiguity in regard to sex category as *intersexed*. Intersex conditions are estimated to occur at a rate of 1.7% of all births, making intersexuality more common than albinism (being born albino or lacking skin pigmentation). In a city of 300,000 people, 5,100 of them would have varying stages of intersexual development, meaning that there's a chance someone walking down your city sidewalk might in fact have both a penis and a vagina. Now, how would you decide their sex?

OK, so you say, we might not be able to see sex with the naked eye. But we live in the 21st century, and there are other ways to determine sex. There are internal sex organs. Today, medical doctors generally use the presence or absence of a penis to determine initially sex in infants. In the past, doctors went by internal organs and emphasized the presence or absence of a uterus because without a uterus, a woman could not reproduce. But conditions of intersexuality deal with both internal and external genitalia. Some intersexual individuals have both an ovary and a testis, one on each side of their body. In other individuals, the ovary and testis grow together, forming an ovo-testis. The presence or absence of internal sex organs is then also an imperfect method of determining sex category.

What about hormones, then? What we call sex hormones are not differentiated in children before they reach puberty, and post-puberty, there is also a wide variation in the presence and absence of sex hormones. Individuals with androgen insensitivity syndrome have testosterone in their bodies, but they cannot metabolize it and, therefore, develop breasts at puberty. Is the presence of testosterone, then, a good measure of who's male and who's female? What about genetics? As we learn in high-school biology, males have XY sex chromosomes while females have XX sex chromosomes. Does modern genetic testing provide a definitive answer to sex category? No. Those born with Turner syndrome lack a second sex chromosome, making them XO, whereas those with Klinefelter's syndrome have two X chromosomes and a Y (XXY). How should we identify the sex of someone who is XXY or XO?

To the complications that intersexed conditions bring to the idea of sex category, you might also add the complexities of the transgender community. Transgender is a broad label that includes a wide variety of people who seek to change, cross, or go beyond culturally defined gender categories (Ferber, Holcomb, & Wentling, 2008). Transgender can include individuals who seek surgery to change their underlying anatomy but also can include individuals who wish to live as a different gender than that which lines up with their biological sex category. Children as young as four and five years old express that their biological sex category does not match their internal sense of who they are. Transgender individuals have existed throughout human history and across very different cultures. How do we fit this reality into a biosocial approach, where sex category is supposed to provide a real limit on the ways in which gender is expressed?

From a biosocial perspective, we might say that, sure, these variations exist. But most people can be fit into two categories based on some set of agreed upon criteria. Sexual dimorphism is not a claim but actual reality. There are two kinds of people in the world, males and females. Even the intersexed can surely be fit into one of these categories with a little tweaking.

A Strong Social Constructionist Perspective

For a second perspective, what we'll call strong social constructionism, there is no tweaking possible. What intersexuality teaches us is that sex itself is socially constructed. In contemporary U.S. society, we have looked at what is a much more complex biological reality and have decided that there are two types of people—when it's really much more complicated than that. There are no criteria that will consistently and reliably sort people into the categories of male and female. This is true of physical as well as behavioral characteristics. For strong social constructionists, this means that the categories themselves just don't work.

Think of a behavioral trait that is generally associated with one sex or the other (crying, fighting, looking pretty, nurturing, etc.). Can you think of examples of females who engage in the behaviors generally associated with males or of males who engage in the behaviors generally associated with females?

Strong social constructionists, then, believe sexual dimorphism is a claim and not the truth, whereas biosocialists believe sexual dimorphism is the truth and not a claim. Another way to help understand the differences between these two perspectives is to consider how they understand the relationship between sex and gender. From the biosocial perspective, we could logically propose that sex by and large precedes and therefore causes gender. Babies are born and assigned a sex category at birth and, based on that assignment, are raised into an appropriate gender. Although biosocialists would acknowledge that there are many variations in the particular form that gender takes across cultures and time periods (because gender is socially constructed), they would also point out that there are limits to the degree to which gender can be socially constructed. Those limits are caused by the biological reality of something called sex.

For example, biosocialists might acknowledge that the particular arrangements for child rearing vary a great deal across cultures and historical time periods, but they would also point out that females have almost always been the sex primarily involved in child rearing. Biosocialists might argue that this is a result of the biological reality that females carry and give birth to children. Although the specific arrangements vary, the biological reality of women as the ones who get pregnant and nurse children places real limits on the possibility of those arrangements. Sex, then, is a causal factor dictating how gender gets expressed.

From the strong social constructionist perspective, the causal arrow is reversed. Gender, in the form of the social meanings we attach to sex categories, causes us to believe there are real categories out there called "female" and "male." Sex itself is socially constructed, and therefore, it is culture that dictates how we understand sex. For many of us who have grown up in a culture founded on the truth of sexual dimorphism, this is initially a hard perspective to understand. But strong social constructionists, following the two approaches we outlined earlier, point to the ways in which sex categories have, in fact, varied across time and place. Today, in Anglo-European societies, we believe there are two sexes. But as we will discuss in Chapter 5, the ancient Greeks believed in a one-sex model (Roy, 2001). Females were not a completely different sex than males, but they were an inferior version of males in a hierarchy that included the gods and other kinds of people (slaves, dwarves, eunuchs, etc.). This particular gender system (a set of cultural beliefs) shaped the ways in which the Greeks saw biological reality. Remember, in ancient Greece, autopsies were performed on dead bodies, and the Greeks therefore knew the differences in the organs of males and females. How did they explain these differences if they believed in only one sex? The Greeks saw the vagina and the penis as the same organ; in women this organ was internal (vagina), whereas in men it was external (penis). Similarly, the Greeks saw ovaries and testes as the same organ in males and females. The same biological reality was used to justify a completely different understanding of sex.

Strong social constructionists also point to a wide range of cultures that have third sex categories, or a space within their particular conceptualization for people who are considered neither male nor female. These include the hijra in India, the berdache in Native American cultures, the kathoey of Thailand, and the sworn virgins of the Balkans (Nanda, 2000). If different societies construct different kinds of sex categories, not all of which are based on a dimorphic system (only males and females), then surely sex itself is socially constructed. Gender, in the form of cultural meanings, therefore produces our notions of sex, rather than the other way around.

The debate between biosocial and strong social constructionist perspectives is ongoing, but it has important implications for how we think about the relationship between sex and gender. From the strong social constructionist perspective, we're always talking about gender because there really is no such thing as sex. This does not mean that biology does not exist and that people don't have bodies. It also doesn't mean that people don't have differences in their genitalia, DNA, sex hormones, and other biological realities. But for strong social constructionists, these biological differences do not line up with the categories we have created and labeled sex, and the claim that they do is false.

From a biosocial perspective, it's important to specify when we mean sex and when we're actually referring to gender. Scholars from both perspectives will use the terms, but keep in mind that, depending on your perspective, the terms can mean different things. From a strong social constructionist perspective, using sex infers a *belief* in two kinds of people, male and female. From a biosocial perspective, sex indicates the *existence* of two different kinds of people, male and female. Throughout the book, we'll refer back to these two perspectives in explaining different ways of discussing gender. But regardless of the particular perspective, this is a sociology of gender textbook, and therefore, we are more interested in questions of gender than in questions pertaining to sex.

Which of these two perspectives makes more sense to you? Which is easier to understand? Which do you think is the more commonly held perspective?

A WORD ABOUT BIOLOGY AND STRONG SOCIAL CONSTRUCTIONISM

The idea that sex categories are socially constructed—that there are *not* two kinds of distinguishable male and female bodies in the world—is a hard idea to swallow for some people. Students are often left wondering whether biology exists at all from the strong social constructionist perspective. Are these theorists arguing that there are no such things as penises, vaginas, testes, hormones, or chromosomes? Are they saying that we don't have physical, biological bodies at all?

Most theorists who argue from the strong social constructionist point of view would say that, yes, of course we have physical bodies. The problem is that the particular categories we use to describe those physical bodies don't fully convey that reality. In fact, many would argue that the diversity in our physical bodies is greater than our categories would lead us to believe. They might go so far as to argue that our belief in how bodies *should* be gets in the way of our perceiving the reality of the way bodies *actually are.* Because we believe that everyone should have a penis or a vagina, we tend to ignore the repeated cases of people who have both. Because we believe that your biological sex category should match up with the gender you express, we stigmatize transgender people who violate these norms.

In other words, strong social constructionists believe that our social ideas about what sex categories should look like gets in the way of us seeing what the actual biological reality is. Can we ever perceive that biological reality free from the particular set of blinders that our cultural beliefs give us? Maybe not, but we can at least begin to think about exactly what those cultural beliefs are.

ESSENTIALISM AND CONSTRUCTIONISM

In addition to the distinction between a biosocial and a strong social constructionist approach, gender scholars make distinctions between essentialist and constructionist approaches. One of the best metaphors I've encountered for understanding these two perspectives involves baseball, but feel free to substitute the sport of your choice (Fish, 1996).

In baseball, the umpire is responsible for identifying pitches as strikes or balls. The essentialist pitcher would describe what he does by saying there's such a thing as a strike and a ball and he calls them like he sees them. (I use *he* because most umpires in the Major League in the United States are, in fact, men.) This view implies that there are real, objectively identifiable things out there in the world called strikes and balls, and the job of the umpire is merely to identify them. With the proper criteria, any number of different umpires would be able to identify the same pitches as strikes or balls because strikes and balls *really* exist. They have an "essence," which is what gives essentialism its name.

The constructionist umpire, on the other hand, would say something like, "Some are pitches and some are strikes and some are balls, but they're nothing until I call them." In this perspective, the umpire *determines* what is a strike and what is a ball, and his determination is what makes them so. There is no essence to what makes a strike or ball—just the uniquely subjective perspective of human beings. We see reality only through our own constructions of the world, and therefore, what might be a ball to one umpire is likely to be a strike to another. If you've ever played baseball, you might have your own opinion on whether most umpires are constructionists or essentialists.

How does this apply to sex and gender? An essentialist perspective on sex and gender is similar to a biosocial perspective in that essentialists believe there is an identifiable essence that makes people male or female, man or woman. But essentialism is somewhat different in that it doesn't necessarily have to be grounded in biology. Someone who believes there are real people called males and females in the world because God created them that way is an essentialist. Some would argue that those who follow the psychoanalytic approach outlined in Chapter 4 are essentialists in that gender becomes something deeply ingrained in our personalities and, therefore, is difficult to change. Essentialists tend to see sex and gender as timeless and unchangeable, part of the very essences of our being. Constructionists by and large see sex and gender as things that society has made. This means that sex and gender have changed over time and can be changed in the future.

Some Notes About Vocabulary

Language is an important component that shapes our social construction of reality, and so it matters for our conversations about gender. You may have already noticed that I used "herself" rather than "himself" earlier when discussing racial categories. Historically, masculine pronouns were used in much writing because "men" was perceived as a universal category. So "mankind," at least in theory, refers to both women and men. In this book, I use masculine and feminine pronouns interchangeably.

In English, *sex* can mean both the biological categories of males and females as well as engaging in some kind of sexual act (a subject we will return to in Chapter 5). Because of this confusion, it's sometimes easier to use the term *sex category* to distinguish between these two different meanings of the word. You will see these two terms used interchangeably throughout this book.

WHY STUDY GENDER?

This question—why study gender—brings us back to some of the questions with which we began this introduction. You may have your own reasons that bring you to this book, but the general answer to this question is that gender matters. Perhaps you've already noticed the ways in which gender matters in your own life. In this book, we'll push that understanding even farther by raising questions about what gender is and how it operates. Our journey can be summarized with three main goals. First, we'll be building an understanding of the ways in which gender is socially constructed in a global and historical context. Remember that the two approaches to revealing most clearly the ways in which gender is socially constructed are to examine how gender varies across time and place. In this textbook, we'll look at both of these dimensions. A global examination of gender is especially important in today's society where events around the world are brought closer and closer by technologies like the Internet. When we focus only on the gender experiences of those in a certain part of the world (usually those parts of the world described as the West or as Anglo-European society), we tend to normalize those experiences, making what it means to be gendered outside of these areas seem strange or weird or wrong. One fundamental truth about gender that those who study this topic have arrived at is that gender varies a great deal based on where, when, and who you are. A global perspective helps us to see that what is actually normal is gender variation, rather than one particular way of being gendered.

This nicely summarizes our second goal in this textbook, which is to debunk any ideas about what is normal and abnormal in regard to gender. We do this through looking at gender globally and by placing the experiences of people of color, gays, lesbians, transgenders, and working class people at the center rather than at the margins of our inquiries. Looking at what it means to be a gay man or a Middle Eastern woman should not teach us what it means to be different from some unspoken norm (straight, white women and men), but it should help us reveal the unique lessons to be learned about gender in the experiences of many different kinds of people. Because of this goal, the particular language we use to talk about different experiences and places in the world is important. The term *Western* assumes a certain actual geographic centering but also an economic and social one. Societies are Western if you see Europe as the center of the world, and this terminology derives from colonial philosophy. We will generally use the term "Western" or "Anglo-European" when we are referring to cultural phenomenon. But you will also hear terminology like *developing* and *developed* or *global North* and *global South,* which reflect different ways of understanding global divisions (which we will discuss in Chapter 3).

We will be emphasizing the social construction of gender, but remember from the Thomas principle that just because something is socially constructed does not mean it does not have real consequences in the lives of people. Gender may be socially constructed, but it is also a system of inequality; understanding gender in this light is the third goal we will be pursuing. What we believe about gender has real consequences for the lives of women and men around the world. Gender distributes power to people. As we will explore, it may cut short the lives of many men, cause some women to live with the fear of physical assault, and influence how much pleasure you experience in your sexual life, how much money you make, and how much leisure time you have. When we begin to see gender around us, we will also begin to see the ways in which gender sometimes works to help some and hurt others. But we will argue that as a system that distributes privilege, in the end, gender can negatively affect everyone at some point in their lives. For many people who study gender, the answer to the question, why study gender, is that understanding gender is the first step toward deciding what needs to be changed and then taking action.

Figuring out what aspects of gender as a social system need to be changed and how to accomplish those changes is part of what we will be exploring throughout this book.

There are a lot of questions about gender to be asked and a lot of different answers to be explored. Many answers contradict each other. During one semester of my sociology of gender course, a student complained after class that his head hurt—not because of a hangover or too much yelling, but because the class was making him think too much. Can you think too much about gender? Perhaps. Sometimes students are frustrated by the lack of ready answers when it comes to gender, but asking questions seems to be the first step in finding out something that's truly meaningful to you, and that is what we will be seeking to do in this book.

WORKS CITED

Adelman, L. (Director). (2003). *Race: The power of an illusion* [Motion Picture]. United States: California News Reel.

Fausto-Sterling, A. (2000). *Sexing the body.* New York, NY: Basic Books.

Ferber, A. L., Holcomb, K., & Wentling, T. (2008). *Sex, gender and sexuality: The new basics.* Oxford, England: Oxford University Press.

Fish, S. (1996, May 21). Professor Soka's bad joke. *New York Times,* pp. A22–23.

Lorber, J. (1994). *Paradoxes of gender.* New Haven, CT: Yale University Press.

Nanda, S. (2000). *Gender diversity: Crosscultural variations.* Prospect Heights, IL: Waveland Press.

Roy, W. G. (2001). *Making societies.* Thousand Oaks, CA: Pine Forge Press.

Thomas, W. I., & Thomas, D. S. (1928). *The child in America: Behavior problems and programs.* New York, NY: Knopf.

CHAPTER 2

What's the "Sociology" in the Sociology of Gender?

Understanding Sociology and Gender

WHY DO YOU NEED THEORY TO UNDERSTAND GENDER?

Students hear the word *theory,* even as it relates to something as interesting as the study of gender, and have one of several possible reactions. They may become nervous and uneasy at the difficult prospect of trying to grasp seemingly unreachable, complex theoretical concepts. They may they roll their eyes and sigh loudly while muttering, "What does theory matter to me?" A few students may anticipate with great enthusiasm the intellectual endeavor that learning to understand and apply theory involves. But I suspect those students may be in the minority, and they might cause some more eye rolling and sighing on the part of some of their classmates. Given the difficulty of learning theory and the anxiety it may cause, why is it important to understand theories about gender?

Learning and understanding gender theories becomes a little less scary when you realize that everyone already has some working theory about gender and the way it operates in the world. From the first moment someone said to you as a small child, "That's what girls do and not what boys do," you probably began to develop your own explanation for why that was so. As we get older, our theories about how gender works become more sophisticated. They may be grounded in a sense that we act in gendered ways because of our biology or because that's what everyone around us seems to expect. We may have a working explanation of gender for certain situations (family life and intimate relationships) and a different set of explanations for other contexts (work and school life).

As you learn the theoretical approaches various people have developed in relation to gender, there's a very strong possibility that some of them will sound a bit familiar and that some will make more or less sense to you. This is likely because some theories match more or less closely the working theory you've already developed about how gender works.

Regardless of the specific content of your own theories about gender, we all have a general sense of what gender is and how it works, and at a very basic level, this is what a theory is—a set of statements and propositions that seeks to explain or predict a particular aspect of social life (Newman, 2004).

Three Reasons to Learn Gender Theories

If we all already have our own theories about gender, then why is it still important to learn those that have been developed by other people? Why aren't our own personal theories good enough? There are three answers to that question. One answer is that although we all have our own theories about gender, we may never have had the opportunity or inclination to test those theories in a meaningful and rigorous way. You can test your individual theory of gender against your own experiences, but as we've already discussed in Chapter 1, your own experiences are likely to be very different from those of people in other parts of the world and with other identities. For example, your theory may work very well at explaining why many married, working women—perhaps including your own mother—still do much of the household labor in American families, but can it help explain how gay or lesbian families divide household labor? Your theory may work for some situations in your own life, but not for others.

Most of the theories we'll be discussing in the next two chapters were developed by people who have more time, opportunity, and inclination to develop their theories and to test their usefulness in a variety of ways, including conducting social research. Ideally, that research tests these theories in a variety of settings and situations, making the explanatory power of the theory that much greater. Throughout the book, you'll see these theories applied to specific situations to explain a wide variety of behaviors.

A second reason theory is important is that it helps us to test the explanatory "wings," so to speak, of our own way of understanding gender. You may have very strong beliefs about your own particular theory of gender. But your ability to defend that belief depends on being able to demonstrate why your beliefs are right and others are wrong. Learning theory forces you to consider seriously the strengths and weaknesses of your own way of thinking. This happens through gaining a thorough and workable understanding of how other theories work. Why? Because to demonstrate that another theory is wrong, you have to have a pretty good understanding of what it says and how it works. You may read about radical feminism in this chapter and not at all agree with the way in which radical feminists understand gender. It may drastically contradict your own way of understanding gender. But developing your own explanation for why radical feminism is wrong requires that you further develop your own way of understanding gender in response to their ideas. In other words, it's not enough to simply say someone else's explanation of gender inequality is wrong; you must be able first to demonstrate *how* they're wrong and then to demonstrate how *your* explanation is better. If you think of your own way of understanding gender as a set of wings you've constructed for yourself to navigate through life, learning other theories about gender is like putting those wings through a series of test flights to see whether they really work.

The final reason it's important to learn theories about gender has to do with our own ability to see and understand the world accurately. Throughout this book, we'll be coming back to these issues, but they are especially important to our understanding of the social world: Can we trust our own vision of the world? Is what we see true or real, and what does it mean to say something is *real,* anyway? Do the beliefs you may already have about gender influence what you see and feel? For example, psychologists identify **confirmation bias** as our tendency to look for information that confirms our preexisting beliefs while ignoring information that contradicts those beliefs. If you believe gay men act more feminine, confirmation bias predicts that you will pay special attention to all the gay men you know or see who act more feminine while ignoring both the gay men who *don't* act feminine and the non-gay men you may encounter who *do* act feminine. Confirmation bias suggests that our own working theories of gender can serve as blinders, preventing us from seeing and considering certain gendered phenomena in our lives.

Learning about other theories is a way to remove those blinders through focusing your attention on aspects of gender that you might not otherwise have seen or noticed. Along these lines, many sociologists speak of using theory as a kind of lens through which to see the world. Like binoculars, magnifying lenses, microscopes, telescopes, and 3-D glasses, these different lenses provide us with different views of the world. All three of these reasons suggest that learning theories of gender is important because they help us become better thinkers in general—and especially better thinkers as related to issues of gender in our lives. So let's begin our exploration of theories of gender by looking at feminism and its influence on sociological ways of thinking about gender.

Gender in Sociology Before Feminism

Sociology, like many of the traditional academic disciplines, is a discipline developed primarily by white, upper-class, European, presumably heterosexual men. Early sociologists, such as Auguste Comte, Émile Durkheim, Karl Marx, and Max Weber, developed the foundations of sociological theory as a response to the problems they perceived in their own lifetimes—problems such as industrialization, urbanization, and the spread of capitalism. But as feminists would later point out, their view of the world was inevitably shaped by their own positions as white, largely upper-class, heterosexual (as far as we know or can tell historically), European men. These men assumed that they were describing experiences and concepts that applied to everyone, regardless of their gender, race, class position, or sexuality; but as we will continue to discover in this book, there are problems with assuming any kind of universal experience. Although women such as Mary Wollstonecraft, Harriet Martineau, and later Jane Addams and Jesse Bernard had important influences on sociology and the sociological study of gender, for much of its history, the consideration of gender within sociology was limited.

There are two main types of limitations in the way sociology addressed gender in its early history. The first limitation involves an approach that used **men as a proxy** for all human beings. Within this approach, many important sociological studies included only men as research subjects. The assumption was that either men's and women's experiences were the same or women's experiences were not important to the larger questions in which early sociologists were interested. From either perspective, there was no need to include women in their research. Studying men told sociologists all they needed to know about society because men represented the most important and universal human experience. For example, an early study of morality concluded that women were less morally developed than men, although the study made the conclusion based on research that excluded women as research subjects (Gilligan, 1982).

From our current standpoint, it's not very difficult to see the limitations with this approach. It is no more correct to assume men represent the universal experience than it would be to assume that women do. In addition, women may have often been prohibited from full participation in many spheres of social life, but that does not mean that the lives and experiences of women were not important to understanding how the world worked. In this approach, gender was ignored by assuming that men and women were essentially the same or that women were unimportant.

Are there still ways in your society in which men's experiences are assumed to be universal? Can you think of any specific examples of this tendency? How would our thinking be different if we assumed women's experiences were universal? For example, what if giving birth to children was assumed to be a basic, universal experience?

The second type of limitation in sociology as it related to gender did not literally ignore women, but it uncritically assumed and reinforced a certain gender status quo that confined women to the private sphere of the home. The main area within sociology that included any serious discussion of gender prior to the influence of feminism was sociology of the family (Hamilton, 2007). That this was the case already tells you something about the approach to gender being used. Examining gender through the family emphasizes women's roles within that institution while ignoring women's roles in many other areas of social life. This treatment of gender failed to examine critically prevailing societal notions about gender and was more a description of the status quo than a theoretical analysis of gender as a social concept. Although there were sociologists, such as Mirra Komarovsky, who challenged this status quo even before the advent of feminism, mainstream sociology remained fairly uncritical about gender until the 1970s (Hamilton, 2007).

You might at this point ask why, given how great sociology as a discipline is (assuming you think sociology is great), sociologists themselves were so oblivious to the importance of gender. Here's one possible answer to that question. Like many academic disciplines before the advent of feminism, sociology departments were made up of mostly men, and so they didn't think much about the importance of gender. Before you accept that answer too quickly, take a moment to think fully about the implications of that statement. Is there a particular reason that men would be expected to be less concerned about gender than women? In the way our dimorphic, Western gender system works, you need women to have men, and that means both men and women have a gender. In sociology departments full of *gendered* men, why did that necessarily mean that they didn't think about the importance of gender?

The answer to that question brings us to a consideration of how power and privilege work in society. The men in the sociology departments were in a privileged position as a result of their status as men. **Privilege** is a set of mostly unearned rewards and benefits that come with a given status position in society. Privilege can take the form of actual rewards, such as the privilege of heterosexual couples in the United States to have their intimate relationships recognized and sanctioned by the law. This is not a privilege that homosexual couples have in most places in the United States, although they do have that privilege in countries such as the Netherlands, Belgium, South Africa, Spain, and Canada. That privilege translates into economic benefits in the form of tax relief and health insurance for married couples, as well as into social benefits, such as the sense of legitimacy and approval that comes with participating in the institution of marriage.

Privilege is trickier to identify when it signifies the absence of barriers that exist for less privileged people. When a white, upper-class woman goes to the store to return an item without a receipt, chances are she will not be questioned as to whether she really bought the item at the store or be challenged by the salesperson on her reason for returning the item. Having to convince strangers to trust you is an example of a barrier that is not faced by those in a privileged status to the same degree as less privileged people in society. This kind of privilege has been described as functioning like the wind at your back if you're peddling on a bicycle (Wimsatt, 2001). Privilege, like the wind, makes moving through the world that much easier, and you might assume you're moving so quickly because of your own effort—your pedaling. You probably won't realize how helpful the wind at your back was until you find yourself having to bike *into* the wind. It's difficult to realize what that's like until you've had to do it yourself—or maybe until you've talked to someone else who has had to do it.

Where does that leave us with the men in the field of sociology and their lack of concern about gender? Being mostly white, upper-class men allowed these sociologists to bike with the wind, and as far as they knew, everyone else was doing the same thing. Gender didn't seem very important to them in part because its effect on their lives, although still very important, was also less visible. This is, in fact, another form of privilege. Part of the benefit of being in a privileged status is that you don't have to spend a lot of time thinking about it. Do people who live in places that take such things for granted think about how lucky they are to be living in a place where there's access to electricity and clean water? Do people in the United States consider how convenient it is that people around the world know much more about American culture than the average American knows about other cultures? Probably not. Being American or from a place where these things are taken for granted is a privilege, and most people don't spend large chunks of their days thinking about the privileges they *have.*

Male sociologists were privileged by their gender, and that meant they didn't have to spend a lot of time thinking about it. This is one small part of the reason why those male sociologists did not seriously consider gender and an important lesson in the blinding properties of privilege. Privilege, in fact, is one of the reasons we need to be cautious in trusting the reality and objectivity of our own views about gender, and it is therefore another reason learning gender theories can be useful.

What are other examples of statuses that come with privilege? Can you think of specific barriers that don't exist for people with those privileges? How does having privilege affect the way you see and understand the world?

FEMINIST THEORIES AND THEIR INFLUENCE ON SOCIOLOGICAL THINKING ABOUT GENDER

The "F-Word": A Brief Introduction to Feminism and How It's Been Around Longer Than You Might Think

Across history, cultures, and civilizations, when women asserted their power and asked for equality, it has almost always been perceived as dangerous. It is important to remember that the various versions of feminism we will discuss are just one manifestation of a long, global history of questioning the gender status quo and advocating for the rights of women. Women in Kenya organized to fight the effects of colonial governments on their livelihood in 1948, and women in India were involved in working for their own rights along with their country's independence long before it was achieved in 1947 (Basu, 1995). Neither of these groups of women would have described themselves as feminists, though. Feminism in all its many forms assumes certain models of what it means to be a woman, what the goals of women should be relative to their status, and how to go about achieving those goals. But the feminist model, although it continues to expand and adapt to fit the diverse needs of women and men across the world, is, like sociology, a product of Anglo-European thought.

Globally, women define their own interests and goals very differently, and they sometimes perceive feminism as another attempt by the global North to make the rest of the world into their own image (Basu, 1995). Given this history, we should remember that those particular ideologies we label as feminism do not describe the totality of how women think about or organize in their own interests globally, as we will explore throughout this book. For now, because we're focusing on the relationship between feminism and sociological thought, we'll be talking about feminism and women's movements as they developed mainly in the Anglo-European world.

In the 21st century, calling yourself a feminist is probably no more popular than it was in the 1800s. In fact, feminists at times have been branded immoral by the Catholic Church and have been frowned on by some Islamic and Hindu traditions (Basu, 1995). In the United States, the declining rates at which young women have identified themselves as feminist have been used to bolster arguments about the decline of feminism—despite the fact that these same young women also espouse egalitarian values (Peltola, Milkie, & Presser, 2004). According to one survey in the United States, in 1996 only 29% of women self-identified as feminists (McCabe, 2005), but evidence that younger generations of women are truly less "feminist" than previous generations is mixed (Burn, Aboud, & Moyles, 2000; Cowan, Mestlin, & Masek, 1992; Houvaras & Carter, 2008; Huddy, Neely, & LaFay, 2000; Williams & Wittig, 1997). Despite the *word's* declining popularity, feminism is integral to any discussion of gender and especially to a sociological exploration of gender. So what exactly is feminism and why is it seen as so dangerous around the world?

First Wave Feminist Movement

Feminists generally divide their discussion of feminism as a social movement into three different periods, and these are important to understand to give context to the different feminist approaches we will discuss later. The first wave of feminism coincided with suffrage movements in both Europe and the United States (Taylor, Whittier, & Pelak, 2004). This is different from the history of women's movements in most of the developing world simply because nearly all of the men and women in these nations were deprived of the right to vote or govern themselves by colonial powers and so women's suffrage was often connected to suffrage for native peoples more generally. This first phase in the women's movement in the Anglo-European world is specific to the historical context of existing democracies in which male (and in the United States, *white* male) citizens had long ago achieved the right to vote.

The early suffragettes were a diverse group in both their backgrounds and their goals, but many of their efforts focused primarily on enfranchisement, or getting women the right to vote. For some women in the movement, this was because they wished to pursue less gender-specific social reform goals, such as the legal prohibition of alcohol. These women saw getting the vote as the first step in this larger project. Other women of the first wave had more radical goals, including sexual freedom and expanding the roles of middle-class women in the workplace. Gaining an expanded role for women in the workforce was an important goal for white, middle-class women, who largely did *not* do paid work outside the home. Suffragettes such as Sojourner Truth, a former slave who served as a conductor on the Underground Railroad, drew attention to the differing experiences among women of the first wave. As a former slave, Sojourner Truth never had the luxury of *not* working, and the kind of work she did was considered "man's work" by many of the white suffragettes—although she was certainly never paid for it. As would be the case throughout the history of the feminist movement, the way women were positioned in society often led to a different outlook on what were the main problems faced by women in society and how to go about fixing those problems. Regardless of these differences, the first wave movement was successful in gaining the right to vote for women in 1920 in the United States (and in 1928 in England, but not until 1944 in France).

Although there were important links between this first wave of feminism and the second wave that developed during the 1960s, some social movement scholars argue that the years in between characterized a period of social movement abeyance. **Social movement abeyance** is a way to keep the basic ideas of a movement alive during a period of decreased activism, often as a result of increased resistance and hostility to the movement or to a shift in the opportunities that make movements more or less successful. During periods of abeyance, social movements may focus on creating alternative cultures to survive, rather than on directly confronting dominant institutions (Taylor et al., 2004). Social movements turn inward rather than outward during these periods, turning to activities such as small group meetings and consciousness-raising rather than to lobbying governments or engaging in large-scale protests. The idea of social movement abeyance is to keep the core ideas of the movement alive and to maintain a small group of activists who can carry the movement into its next phase.

Based on your own knowledge and sense of the current state of the feminist movement, or movements for women's rights more generally, would you say the movement is currently in a state of abeyance? What evidence would you use to support that conclusion? What do you think it would take to see a resurgence in feminism or in social movements that focus on gender issues more generally?

Second Wave Feminist Movement

When most people in the 21st century think of feminism, their frame of reference is the second wave feminist movement that began in the 1960s in the global North. This movement was part of a larger, global social movement cycle that included independence movements in the developing world as well as the civil rights movement in the United States. A **social movement cycle** is a period of increasing frequency and intensity in social movement activities that spread throughout various parts of society and globally across countries (Snow & Benford, 1992). During a social movement cycle, many connections are made between different social movements, with one social movement either directly or indirectly influencing the evolution of very different and geographically distinct movements. For example, some women who got their initial social movement experience within the civil rights movement in the United States moved on to the women's movement. In the developing world, women worked within nationalist movements to throw off colonial rule and establish democracies. Although they often included women's rights within those larger agendas, these movements were necessarily different because of the historical context of the postcolonial world.

The second wave of feminism, like the first wave, was characterized by diversity in the types of women involved and in the articulation of their goals. To understand this second wave of feminism is to understand that, as with all social movements, there was no one movement, no one group, and often no single, unanimously agreed upon agenda. As we will discuss later, organizations like NOW (National Organization for Women) focused on passing legislation in the United States that would have institutionalized the prevention of gender discrimination into the Constitution in the form of the Equal Rights Amendment (ERA). Consciousness-raising also became an important part of the movement as feminists focused on finding connections between their personal lives and the politics of gender. For many feminists, the development of their own theories of gender were inextricably grounded in an examination of their own personal lives, and this work was just as important, if not more important, than changing institutions such as the government. Charlotte Bunch epitomized this connection when she wrote in 1968, "There is no private domain of a person's life that is not political and no political issue that is not ultimately personal. The old barriers have fallen" (Bordo, 2003). From these examinations of the connections between the personal and the political came a focus on issues such as women's rights in the workplace (including the right to be free from sexual harassment), domestic violence, reproductive rights, and sexual violence.

Like the first wave before it, the second wave contained a great deal of diversity in both ideology and identity. Branches of the movement emerging from more leftist organizations formed the basis of what we call *radical feminism* and *socialist feminism*. As the movement evolved, more feminists drew attention to the problems in attempting to organize around or articulate one experience of being a woman. Lesbian feminists often felt their voices were not heard in the larger movement, and women of color began to feel that many of the goals of the movement centered on the experiences of white, middle-class women. Like all social movements, second wave feminism struggled with maintaining a common purpose in the face of these conflicting experiences and identities.

Third Wave Feminist Movement

Third wave feminism was in many ways a response to the contradictions of second wave feminism. Emerging during the 1980s and 1990s, third wave feminism encompassed a diverse range of theories and orientations among both academics and activists. The voices of women of color, who have been a strong influence in all three waves of the feminist movement, were even stronger in the development of third wave feminism. Women such as Audre Lorde, bell hooks, and Rebecca Walker questioned essentialist tendencies in feminism, or the tendency to assume some universal experience of being a woman. They fought to organize around issues of race and sexual orientation in addition to gender.

Theory Alert!
Essentialist

In her collection of third wave essays, *To Be Real: Telling the Truth and Changing the Face of Feminism,* Rebecca Walker (1995) attempted to articulate a way to be feminist that is inclusive enough to include both men and women, whites and people of color, lesbians, gays and straight people, super models like Veronica Webb, and second wave feminists such as Gloria Steinem. Third wave feminism was influenced by postmodernism, postcolonialism, the work of Michel Foucault, and eventually, queer theory. The third wave is characterized in many ways by coming to terms with and being up front about the many contradictions that always lay beneath the surface of feminism as it developed. For example, postcolonialism influenced feminism in the third wave by drawing attention to the ways in which women had ignored the experiences of women outside of the Anglo-European world. This examination raised questions about whether women could claim one global movement, or whether the interests and goals of women in the global North and global South were so different and opposed as to make any umbrella movement impossible. Third wave feminism, rather than ignoring or suppressing these types of questions, embraced them as crucial to the next phase of achieving gender equality.

Liberal Feminism

Like many gender theories, feminists start with the assumption that gender inequality exists; then they go about explaining that inequality in a variety of different ways. **Liberal feminism** posits that inequality between men and women is rooted in the way existing institutions such as the government treat men and women. When these institutions limit the opportunities for women to compete with men in economic and political arenas, they create inequality. Why should women and men be provided with equal rights? Liberal feminism grounds these claims in a set of basic rights that all humans are entitled to in modern societies. Thus, liberal feminism is part of a larger social movement **master frame of equal rights** that includes the early civil rights movement and some, but not all, versions of the modern women's movement and the gay rights movements (Snow & Benford, 1992). A master frame is a method of interpreting the world that identifies a particular problem, suggests a particular cause for that problem, and proposes a way to resolve the particular problem. The equal rights master frame assumes that diverse groups of individuals in society, such as African Americans, women, and gays and lesbians, are entitled to the same rights as everyone else in society because we are all fundamentally the same. Thus, liberal feminists base their arguments for equality on the *similarities* between men and women; because we are all basically the same, we all deserve the same basic rights.

Like other rights-based movements, the methods liberal feminism prescribes for achieving equality are to gain and ensure those basic rights that give women the opportunity to compete on an equal playing field with men. So during both the first and second waves of the feminist movements, liberal feminists worked for the right to an education, to own property, to vote, to be employed, and to be free from discrimination in the workplace. The assumption under liberal feminism is that once these barriers to competition are removed, the experiences, views, and attitudes of men and women will converge, or become increasingly similar.

Liberal feminists often pursue changes in legislation, such as Title IX, which made gender discrimination within any educational institution receiving federal financial assistance illegal. These kinds of changes in government policies based on the language of universal rights have had widespread and substantial effects on the lives of men and women across the globe, from increasing the rights of Islamic women when they divorce their husbands in India to partially accounting for the increased participation in organized sports for women in the United States. Note that both of these examples fall within the frame of equal rights. In India, why should women not be entitled to the same rights to property and compensation after divorce as men? In the United States, why should women be deprived of the same right to participate fully in sports that is provided to men? Although liberal feminism is more strongly associated with the first and second waves of feminism, it is still an important method for understanding gender and a strong basis for organizing movements centered on gender issues.

Radical Feminism

Although threads of liberal feminism run throughout the first and second waves of the feminist movement, radical feminism is primarily associated with the second and third waves. Radical feminism in its many forms and incarnations starts with the premise that women and men are fundamentally different (Taylor et al., 2004). **Radical feminism** locates this difference in a variety of sources, from their treatment as a "sex class" (Beauvoir, 1952) to early socialization patterns based on women's dominant role in raising children (Chodorow, 1978). Gender is a fundamental aspect of the way society functions, according to a radical feminist perspective, and serves as an integral tool for distributing power and resources among people and groups. Inequality is not solely a result of women being denied the opportunity to compete, but it is built into all aspects of society; gender affects our social and personal lives, as well as our political and economic institutions. From a radical feminist perspective, men directly benefit from the subordination of women, and men's superior position in many societies is premised on ensuring women's inferiority.

How does this basic understanding of the roots of inequality affect the solutions and actions pursued by radical feminists to decrease gender inequality? For radical feminists, fundamental changes to the basic structure of society are necessary to bring about gender equality. A world in which women are no longer subordinate to men requires revolutionary changes to the current social order, rather than legal changes that give women increased opportunities. Important to this goal of radical restructuring for radical feminists are consciousness-raising activities. These activities are at the core of the popular feminist slogan, "The personal is political." **Consciousness-raising** seeks to help women see the connections between their personal experiences with gender exploitation and a larger sense of the politics and structure of society. For example, consciousness-raising might help a woman to realize that her experience as the victim of sexual harassment is not an isolated, personal incident but a fundamental and inevitable product of the patriarchal way our society is structured; it is not an aberration, but one part of the way in which men maintain control in a society through fear and intimidation. Consciousness-raising helps radical feminists to generate necessary critiques of the way society functions to uncover its gendered implications.

These two feminist approaches just begin to scratch the surface of feminist thinking. As this overview of liberal feminism and radical feminism suggests, it would be a mistake to assume there is one and only one way to be a feminist, just as it would be a mistake to assume there is one and only one way to be a man or a woman. Although these two particular feminist orientations are more important to the development of sociological thought, the wide range of other feminisms is testament to the great diversity of experiences and perspectives that exist in the world surrounding gender. We could easily include a discussion of socialist feminism, Marxist feminism, lesbian feminism, postcolonial feminism, ecofeminism, and postmodern feminism, just to name a few. If we can define feminism in general as an ideology that seeks to combat gender inequality, there are an infinite variety of specific ways people have developed to address that central problem. Given this reality, it is more appropriate to talk about feminisms, rather than any one correct version of feminism or way of being a feminist.

Men and Feminism

At this point you might ask yourself, "Where were the men while all this was happening?" You might ask whether conversations about feminism are at all compatible with conversations about men, and if so, how? You might be nervous, given the considerable amount of time we've just spent talking about feminism and the assumptions many people have about feminism, about whether this book is going to be all about how much men suck.

Let's start with the first question: Where were men during these various waves of feminist movements? The answer is complicated. Some men *were* involved in various places and times in women's movements. In the first wave feminist movement in the United States, men such as abolitionists Frederick Douglass and Henry Ward Beecher (father of Harriet Beecher Stowe, who wrote *Uncle Tom's Cabin*) were involved in working for women's suffrage, along with the abolition of slavery. Many 19th-century activists saw these issues as deeply connected, although the two issues also sometimes led to divisions within the movement. Globally, men and women have often worked side by side in movements for national liberation and establishing democracy. In Kenya, women fought alongside men in the Mau Mau war of 1952 for Kenya's independence from colonial control. Women involved in social movements often argue that the accomplishment of their goals would benefit both men and women in society, as we will see later. Although this may be true, most people involved in women's movements are often still women.

Are conversations about feminism compatible with conversations about men? The answer to this second question is a definite yes. As mentioned, feminists argue in various forms that a society with more gender equality is a society that's good for everyone, women and men. Although in the United States the second wave of the feminist movement was often referred to as *women's liberation,* many women felt that the movement would liberate men as well. What did men need to be liberated from? As we will see throughout this book, although gender as a social system often privileges men, it does not *always* do so, and when it does provide privileges, they often come with a price. Our culture demands that both men and women conform to gender and sexual norms, and men's access to power and privilege is conditional on their conformity to these norms. Part of the goal of feminists is to loosen these restrictions for everyone.

As we discussed, feminism has increasingly become a dirty word for younger generations, and feminists have always been subject to accusations of disliking men. But as you should begin to see, feminism is not about positioning men against women in some kind of epic battle for power and control of the universe, although that might make for an interesting video game. As feminism evolved over time, questions about how to involve men and how feminism mattered for men become increasingly important. As we will see in our discussion of sociological theory that follows, sociologists have become especially concerned with studying gender as it applies to men and women, and masculinity has become a central topic for gender research. Because gender is a system that is always relational (you can't have a category called "women" without a category called "men"), a comprehensive understanding of gender must examine both men and women. So no, this book will not be about how much men suck but about how both women and men are part of the social system called gender.

SOCIOLOGICAL THEORIES OF GENDER

The most important contribution of feminism to sociology as a discipline was to place topics such as gender and sexuality on the agenda of sociologists and to encourage a critical reflection on the place of gender within the larger discipline. This change was signaled in part during the 1970s when academic journals within sociology began publishing issues that focused on gender or featured the works of feminist scholars (Hamilton, 2007). This was the first step to developing theories that sought to explain gender from a specifically sociological perspective.

Gender and the Sociological Imagination

So what exactly is this sociological approach to gender? To answer these questions, we need to develop a basic understanding of what sociology is as a discipline. C. Wright Mills famously said that sociology lies at the intersection of history and biography. He wrote this in his essay on the sociological imagination (Mills, 1959/2000) that the sociological imagination is a good place to start with a brief overview of what makes up a sociological perspective.

What did Mills mean by the intersection of biography and history? Well, we've already talked some about biography. You come to the topic of gender with an already intensely gendered biography. If you grew up somewhere on this planet Earth, among other human beings, you have already been touched deeply by gender. Gender is a part of your biography, and the first contribution sociology makes is to help you literally act like an archaeologist in the rich and fascinating material of your own life to unearth the impact of gender. Archaeology is a good metaphor to use for this process because it requires a careful method and precision. A backhoe won't work—but think of the typical method of archaeologists, sifting through the debris of an archaeological site, carefully cataloguing each minute piece. Sociology seeks to give you the analytical and theoretical tools to turn that kind of detailed examination onto your own life, carefully examining interactions, beliefs, relationships, and decisions you may have made to uncover the influence of gender. And although it's usually interesting to think and talk about ourselves, it can also be unnerving; like in archaeology, you never know what you might uncover.

Understanding your own personal biography isn't necessarily the easy part for Mills, but it's certainly more accessible to us than the second part: history. In his essay on the sociological imagination, Mills (1959/2000) worried about people's ability to understand fully the forces acting on them historically as individuals. He explained this concern by making a division between private troubles and public issues. **Private troubles** are those problems we face that have to do with ourselves and our immediate surroundings, or what Mills called our "milieux" (p. 11). Private troubles are solvable within ourselves individually or within a limited range of the people around us. A private trouble can still be troubling, but the idea is that the power to redress the problem is basically within our own control. Public issues are a different matter.

Public issues exist beyond the individual or her own immediate milieu, and they are located within the larger structures of our societies—such as social institutions. Public issues are the history part of the intersection between biography and history. Understanding public issues requires taking a difficult step back and looking at the big picture of society, asking ourselves what kind of social forces are working on us that are beyond our control. Like the archaeology described earlier, this can be difficult because it involves seeing the ways in which we are sometimes subject to social forces beyond our control. Understanding that your free will is somewhat limited is not an easy thing for many people to reconcile; it diminishes the sense that our choices are ours and ours alone to make.

One way to think about public issues is to ask how you might think about and experience gender differently if you lived in a different historical period. If you lived 50 years ago, how would your experience of gender be different? 100 years ago?

How do these concepts help us develop a sociological imagination as it relates to gender? Here's a real story a student told me based on her experiences interning for a minor league baseball team, which helps demonstrate these differences. Baseball is a sport played primarily by boys and men in the United States, except for a few girls sprinkled in at the little league level when they're young. But by high school, most women are playing softball and men are playing baseball. It's probably safe to say then that baseball is a masculine sport. In the minor league baseball organization for which my student worked, there were two managers—one male and one female—who had roughly the same job. My student observed that the kinds of things the man and woman asked their employees to do, as well as their interactional styles, were about the same. Yet many of her fellow workers, both men and women, complained about the woman, calling her pejorative names, like *bitch*. But they did not complain about the male manager. How do we make sense of this situation?

If we see the differences between how the employees perceived the male and female manager as a private trouble, we would assume that there was just something different about the personality or management style of the two managers. Maybe the female manager just didn't fit in well with this particular organization or group of workers. Maybe she did something that rubbed someone the wrong way. Regardless of the particular explanation, if this is a private trouble, the cause of the problem lies primarily in the specific personality and disposition of the female manager. But what if the differences are better described as a public issue? How would we describe this situation then? Maybe her employees complain about the female manager because she's a woman and it's difficult for them to see women in positions of authority, especially in the world of sports in general and in baseball specifically. Maybe it's hard for the employees to believe that a woman knows enough about baseball to work successfully for a minor league team. Maybe it's hard for them to take orders and direction from a woman. There's even a possibility that her employees suspect that she only got the job because of affirmative action policies and, therefore, she's not really as qualified as her male counterpart. If the female manager is facing a private trouble, she makes adjustments to her management style or talks to her employees about their problems with her. Maybe she takes another job in a different but similar organization.

But what if she's facing a public issue? How does she deal with the cultural norm in her society that says that men are better at jobs involving predominantly male sports like baseball? What does she do about the perceptions that women or minorities get privileged by programs such as affirmative action over white men and are therefore less likely to be competent? What does she do about the difficulties men and women have in the workplace taking orders from women in a position of authority? Taking another job in a similar organization doesn't solve her problem anymore because these problems are not about individuals but about the structure of these types of organizations themselves. These problems probably seem a lot more difficult for one person alone to solve, and that's the nature of a public issue.

This is the core of what Mills was pointing to in his discussion of the sociological imagination. To understand the world around us, we have to understand where our private troubles leave off and public issues begin. It doesn't help us that in societies like the United States, with a strong emphasis on individualism, we are more likely to attribute our problems to private troubles. This tendency to explain behavior by invoking personal dispositions while ignoring the roles of social structure and context (public issues) is called the ***fundamental attribution error*** (Aries, 1996). The sociological imagination as outlined by Mills (1959/2000) seeks to correct this tendency, and although feminists used a different language, this was part of their agenda as well. For too long, many of the problems faced by women had been perceived as merely private troubles. Feminists demanded that these problems begin to be treated as public issues, or as connected to larger social forces and therefore beyond the control of just one woman or man. From a sociological perspective, using the **sociological imagination** to investigate gender means performing the detailed archaeology of our own biography and learning to identify the larger structural forces at work in our lives surrounding issues of gender.

All of the theories we will discuss will serve as tools to help in building your sociological imagination, or in learning to see the connection between your own life and larger social structures. There are several questions to ask yourself and to think about that will help you as you read through these theories. The first question to ask yourself is, *How do these theories define gender?* As we discussed in Chapter 1, there is no complete agreement on exactly how to define gender or sex, or on how the two are related to each other. Some theories adhere to the idea of sexual dimorphism, that there are two types of bodies, male and female. Others largely ignore the question of sexual dimorphism or suggest that biological sex is largely unimportant to a consideration of gender. Although theorists usually do not provide an explicit definition of gender, their theories often imply a certain way of understanding what gender is. Think about what the implied definition of gender is in each of these theories.

This first question is related to the second: *Where does the theory locate gender?* Is gender something that exists inside a person, or it is something that's created through interaction? Is gender inside our heads, or is it something deeply embedded in the major institutions of our society? This question has become increasingly important in the sociological study of gender because theories are often divided into three different categories: individual, interactionist, and institutional. Individual theories locate gender inside individuals in some form or another. Some individual theories might see these internal traits as related to sex and therefore biology. Other theories of socialization emphasize that gender becomes internalized over time as we learn gender. From either approach, an individualist theory understands gender as something located inside the individual, and its influence is realized from the inside out through our individual actions and behaviors. Sex role theory, which we will discuss later, is an individual approach, as are some of the psychological and biological approaches in Chapter 3.

Interactionist approaches locate gender metaphorically in the space between people. Rather than something that resides inside a person, gender is something that is created primarily in the interaction between people. For many interactionist theories, gender therefore does not exist as an internal trait or disposition, but only as a phenomenon that is created in our interactions with other people. Status characteristics theory and doing gender are both examples of interactionist approaches. Finally, institutional approaches draw attention to the way in which large-scale organizations and institutions in society help create and reinforce gender. Like radical feminism, these theories emphasize the way in which gender is woven into all the structures of society. When we are plugged into those institutions and structures, the slots we fill are already gendered. Gender is created by the working of these social structures. Gendered organizations and social network theory are both examples of institutional theories of gender.

The third question to think about for these and any theory of gender is the following: How does the theory explain the connection between gender and inequality? Does the theory address issues of inequality, and what is its explanation for why this inequality exists? How is the explanation of inequality rooted in the way the theory defines and locates gender?

There is one final question to ask about a theory: What are its implications? Theories begin with the abstract, but if they are useful, they should have some practical value as well. When you thoroughly explore the implications of the theory, does it still make sense and does the theory still work? This final question involves some critical thinking as you imaginatively expand the theory. For example, exploring the implications of the theory might mean collecting a series of examples of gendered interactions or situations and thinking about whether the theory explains all those interactions and situations. Can you think of a situation in which the theory just doesn't seem to work? Why does the theory not work in that particular instance? Exploring a theory's implications also means thinking through its practical applications.

This is especially relevant for how the theory explains or treats gender inequality. If you follow the assumptions of the theory, what would be the most important step in reducing gender inequality? For example, we saw with radical feminism that they define gender as deeply embedded in the workings of society. Given this way of thinking about gender, the implications of radical feminism for practical action are that you must make radical transformations to society to reduce gender inequality. As you read through these theories, think about what practical action they imply.

SEX ROLES

Sex role theory is our first example of an individual approach to gender. An **individual approach to gender** assumes gender works from the inside out. This means gender exists as something internal to people and individuals that then affects their participation and actions in the outside world. Individual theories of gender vary in their explanations of exactly how gender comes to be something internal; some use explanations partly based on biology, whereas others place more emphasis on theories of socialization. We will discuss more examples of individual approaches to gender in Chapter 3, in regard to theories from other disciplines, and in Chapter 4, in the context of socialization.

As we discussed in Chapter 1, the language we use to discuss gender is important and reveals a lot about the assumptions we're making about what gender is and how it works. We've already discussed the important distinctions between the terms *gender* and *sex,* so it shouldn't come as a surprise that early works exploring the role of gender in sociology most commonly used the term *sex roles.* The idea of a sex role begins with the idea of a social role more generally. A **social role** is simply a set of expectations that are attached to a particular status or position in society. Statuses can be very general, like man or woman, white or Black, and gay or straight. Or they can be more specific, such as professor of sociology, sophomore biology major, president of the science fiction club, or father of three children. Then different expectations go along with these different statuses. For example, we might expect a sophomore biology major to come to class hung over but not the professor of sociology; it wouldn't be unusual for the president of the science fiction club to spend 10 hours a week playing *World of Warcraft,* but we probably wouldn't expect that from a father of three (unless maybe he was playing with his three children).

A **sex role** is then the set of expectations that are attached to your particular sex category. What are the expectations that people in society have for you based on your status as a man or woman? An easy way to think about this is to think about what kinds of behaviors you might get in trouble for engaging in or thoughts that might seem strange to have as a man or woman in your society. For example, in American society, it might be strange to see an adult man cry, but in popular Bollywood movies produced in India, it is an expectation that the male heroes cry. The very best male actors are those who are the best at shedding some tears. Social roles vary by society, but most cultures impose some set of expectations on individuals based on their assignment into a sex category. Violations of a role are a good way to tease out what exactly the expectations are that go along with that particular role.

Make a list of some of the expectations that go along with being a man or a woman in your society. Do you follow all these expectations? What are the results if or when you don't follow some of the expectations?

Sex roles seem like a very useful and sociological way of thinking about gender because gendered behavior is perceived as a product of the social expectations we have of each other in society. A problem with this way of thinking as it developed involves how these sex roles were described. Sociologists within this model often assumed that sex roles, although partly a result of socialization, were built on and perhaps reinforced by underlying biological differences. Within the particular theoretical approach of the time, sex roles were defined in ways that justified the existing status quo. The flawed logic was that if the existing sex roles seemed to work for society, then that necessarily meant those sex roles were optimal for the functioning of society. As previously discussed, sex roles were linked to social functions, and so the belief common to American society at the time that women were best suited to child care and men best suited to the workplace became an important component of sex roles.

One way functionalist sociologists such as Talcott Parsons explained these differences was in terms of instrumental versus expressive roles (Parsons & Bales, 1955). Men were taught in childhood and throughout their lives to be **instrumental**, or goal and task-oriented, whereas women were taught to be **expressive**, or oriented toward their interactions with other people. This particular division of labor into instrumental and expressive activities was considered functional for society, so women who worked outside the home or men who wanted to stay home and take care of their children created dysfunction for society. The concept of sex roles, rather than challenging the predominant gender ideology of the time, actually reinforced it. You should note at this point that this sociological theory, based on "objective" social science methods, also contained a normative component. Functional theory doesn't just describe the way society is, but it also makes claims to the way society *should* be to function most effectively.

You can imagine how this approach would have seemed problematic for sociologists who were influenced by feminism. It is difficult to imagine how the idea of sex roles as described might be used to combat gender inequality, a primary goal of feminism. During the period when sociologists were dealing with the concept of sex roles, Betty Friedan (1963) wrote her famous book *The Feminine Mystique* and identified what she called "the problem with no name." That problem was the lack of fulfillment experienced by many women (but not all) who did not work outside the home. The 1950s in America witnessed a period when many women were encouraged to become homemakers, a relatively new and rare phenomenon in the long history of gender, as we will discuss in Chapter 7. How could sex role theory help create a justification for these women moving outside of the home and back into the workplace, a goal of many feminists? A theory that supported these sex roles as good for the whole of society seemed unlikely to help serve that purpose.

Beyond this critique, the concept of sex roles was underdeveloped theoretically. It rested on the assumption that sex roles develop through socialization, or the process through which we learn the ways of a particular group. We will discuss socialization in more depth in Chapter 4, but this explanation does not account for the particular content of sex roles in different societies. How do we account for the variations in sex roles we have already discussed? Why is it OK for some men to cry and not for others? The theory also did not fully address how sex roles work in an interaction. Remember, sociologists are interested in what gender is and how it works. How did the concept of sex roles help answer this question? As it developed under functionalism, sex roles were understood as a disposition or trait. This meant sex roles were something that became internalized and that men and women carried around with them, inside their heads. These dispositions then affected how they thought and acted in various situations.

To think about sex roles as dispositions still leaves a series of questions unanswered. Do sex roles act the same in every situation? If social roles are situated identities, as many theorists proposed they were, that means they are put on and taken off as necessary in interaction. You perform the student role in the classroom but probably not when you're hanging out with your friends. A master status, on the other hand, is something that you don't have the power to put on or take off as you choose. **Master statuses** cut across all other identities and situations, and they are the most important status in dictating how people respond to us. Is gender something we put on and take off, or is it more like a master status that is with us all the time? If we're always doing gender, then it doesn't really work as a social role, and this is another weakness in conceiving of gender as a role. The concept of sex roles evolved through the study of families, but are sex roles important outside of the family as well, for instance, in the classroom, the workplace, or in our government institutions?

The concept of sex roles as it was initially articulated left many of these questions unanswered, but the use of social roles to explore how gender works remains a common way of understanding this aspect of society. For example, research on the gender of love, which we will discuss in Chapter 6, uses the concept of instrumental and expressive types of intimacy. Although the particular terminology of sex roles is less popular in today's study of gender, the idea of gender as a social role was an important contribution, one that has been built on by subsequent theorists.

INTERACTIONIST THEORIES

The next two theories move to the level of interaction as the important site for the study of gender. This means that both status characteristics theory and doing gender argue for the importance of interaction to our understanding of gender and see gender as something produced in groups among individuals, rather than as residing within an individual.

Status Characteristics Theory

Status characteristics theory, also sometimes called *expectation states theory,* firmly locates gender in the realm of interaction. From this perspective, gender as an internal disposition is much less of a central focus than the concrete things that we do when we interact with other people. Status characteristics theory doesn't say that our internal traits and identities don't matter at all, but it suggests that interaction is especially important. Why the emphasis on interaction as the important site for the study of gender? Status characteristics theory answers this question with an emphasis on the way we divide up the world into male and female, or our particular method of putting people into sex categories; within our Anglo-European system of sex categorization, roughly half of our population is seen as women and the other half as men (although those two categories are certainly not universal). This means that there are bound to be a lot of people out there in the world who are of the opposite gender, and structurally that increases your chances of interacting with someone of a different gender.

Status characteristics theorists argue that this is somewhat different from other statuses, such as race or age, where demographics do not necessarily dictate interaction. For example, in the United States, approximately 15% of the population is Hispanic and approximately 12% of the population is older than 65. If you are neither Hispanic nor older than 65, there's structurally a good chance you can avoid interacting with either of those groups because there are simply fewer Hispanics or people older than 65 in the population. The chances of you being able to avoid interacting with anyone of the opposite gender in the United States are pretty low.

Status characteristics theorists also argue that interaction is an important site for the study of gender because gender is a social category that crosscuts many other divisions in society. This means that even in a society in which there is very little interaction with people outside of your extended family, your family is still very likely to be made up of both men and women. For example, Purdah is a tradition common in Muslim societies, but also among Hindus in India and in some African countries, where women live in various states of seclusion and isolation within their households and rarely leave their homes. Even women in Purdah interact with men on a daily basis because their household is very likely to have men in it (fathers, husbands, sons, brothers, etc.). So, in part, interaction is important to an understanding of gender because we interact with the other gender a lot.

What exactly happens in interaction that's so important to gender, though? What's going on in regard to gender when you interact with someone? Status characteristics theory begins with the importance of sex categorization. **Sex categorization** is the way we use cues of culturally presumed appearance and behavior to represent physical sex differences that we generally cannot see (Ridgeway & Smith-Lovin, 1999). Compared to many species, sex differences between humans are not very dramatic. In many bird species, males and females are different colors, and many other animal species have considerable size differences between the sexes; it's fairly easy to tell the difference between a male and a female robin, for example.

CULTURAL ARTIFACT 1: RAISING GENDER NEUTRAL CHILDREN

Incorrectly categorizing someone's sex can be upsetting—to the person doing the categorizing as well as to the person being categorized. Infants don't have any of the physical cues we generally use to sex categorize adults, and this is why some parents make sure to provide their babies with clear gender markers. How many baby girls have you seen with pink bows fixed to their hairless heads? Dressing your newborn in clearly feminine or masculine clothing prevents that awkward moment when you ask about someone's little girl only to find out he's really a baby boy. But some parents intentionally prevent people from sex categorizing their children by refusing to reveal the sex of their child. Couples in Sweden, England, and Canada have all stirred controversy by keeping the sex of their child a secret from anyone outside the family, and sometimes even from siblings. One couple kept their son Sasha's sex a secret until the age of five, revealing his sex only when he started primary school (Wilkes, 2012). Like many other parents who have made this decision, Sasha's parents were motivated by a desire to protect their child from the stereotyping that gender brings. But in the media, these parents face harsh criticism suggesting that their decisions to raise gender-neutral children are potentially damaging. Kathy Witterick, who has not disclosed the gender of her child, Storm, explained, "The discussion that emerges [about raising gender neutral children] not only 'outs' people (in their rush to judge, they demonstrate the prevailing views), but also has the effect of helping people examine whether they truly do believe the status quo to be the best that we can do. Is this the best that we can do to grow healthy, happy, kind, well adjusted children?" (Gillies, 2011, para. 15). Because the phenomenon of raising gender-neutral children is relatively new, there's no evidence on the long-term effects. Is imposing sex categorization on children in their best interests?

As humans, we rely mainly on cultural cues like clothing, hair length, movement, gestures, and conversation to differentiate male from female because most of us don't walk around naked or with our distinguishing body parts exposed to plain view. We engage in categorization for all kinds of statuses, like age, race, class background, and sexual orientation. But research shows that we engage in *sex* categorization automatically and, most of the time, unconsciously (Brewer & Lui, 1989; Stangor, Lynch, Duan, & Glass, 1992). In general, the only time we might have to stop and think about sex categorization is when there's some doubt as to the sex category of the person with whom we're interacting. But in most cases, because we put people into sex categories without even really having to think about it, we are never interacting with a gender-neutral person. It is never as simple as the "doctor talks to the patient" but always the "*female* doctor talks to the *male* patient."

As we will see, sex categorization is important to both of the interactionist theories we'll discuss simply because you have to be able to identify the sex of people (or, at least, *believe* that you can identify the sex of someone) before you can argue that sex or gender matters in interaction.

That we engage in this sex categorization is interesting in and of itself, but how and why does it matter to our understanding of gender? That we put people into sex categories automatically doesn't mean much unless those categories themselves mean something. Why is it important that we engage in sex categorization? This is explained by the concepts of expectations and status characteristics, which makes sense given that that is also what the theory is called. Status characteristics theory addresses itself to goal-oriented interactions, so it's especially well suited to explaining settings like the workplace. But many other, less formal types of interactions are also goal oriented. For example, a couple in a car trying to figure out how to get somewhere, a father teaching his daughter how to change a car tire, and a grocery clerk and a customer checking out are all examples of goal-oriented interactions. In goal-oriented interactions, status characteristics theorists argue, we form performance expectations about our fellow interactants, or the people with whom we're interacting.

A **performance expectation** is a guess about how useful your own contributions will be to accomplishing the goal of the interaction, as well as how useful the other people in the group will be (Ridgeway, 1993). In other words, performance expectations are our best guesses about how useful someone is going to be in performing some task, whether that someone is us or another person. Performance expectations usually function more as a hunch than as a conscious, rational calculation we do in any given interaction. So you're probably not consciously aware of the performance expectations you carry into a situation. However, these expectations are important because they influence how we engage in the interaction, how others react to us, and how we interact with everyone else, as we'll see in the example that follows.

If people enter into goal-oriented interactions with some kind of performance expectations, what kind of expectations are they likely to have related to gender? This is where the second important concept, status characteristics, enters into the picture. A **status characteristic** is some kind of difference that exists between people in society and to which a sense of lesser or greater value and esteem is given (Ridgeway, 1993). That means it's not just about being different in some way, but that people who are different in that particular way are considered better or more worthy. This is important to gender because studies show that gender is a status characteristic and that men are generally deemed better and more competent than women (Ridgeway, 1997).

This belief in the superiority of men constitutes a specific **gender status belief**. Because gender is a status characteristic and because of the existence of gender status beliefs, the performance expectations regarding women in interactions generally put them at a disadvantage relative to men. If, in interactions, we use performance expectations to guess how much someone can help us achieve our goal, and as a society we generally believe women are less competent, status characteristics theory predicts that this will have an important effect on the way men and women interact with each other.

Let's walk through a specific example of how this all might work to help us put the theory together. I had a student in my gender class who worked summers on a construction site for her family, and she shared her experiences with the class. Her story can serve as a useful example for demonstrating status characteristics theory. The first step in how gender becomes important in these interactions is that her fellow workers would sex categorize her as female. Remember they would do this immediately and without much thought, but despite anyone's best attempts, she would never be perceived as just a construction worker but, rather, as a *female* construction worker. Working in construction is a good example of a goal-oriented task, and so we would expect both my student and her fellow workers to form performance expectations in this setting—regardless of whether they were consciously aware of them. Keep in mind, performance expectations are about your own performance as well as about others in the interaction.

Because gender is a status characteristic, and is therefore associated with a diffuse belief in women's inferiority and men's superiority, these performance expectations are going to be negatively impacted by my student's gender. These expectations become especially important because of the power of self-fulfilling prophecies. If my student had some doubts as a woman about her ability to do construction work, this is a good example of a performance expectation. This would affect how she interacted with the other construction workers. If she was very confident in her abilities, she'd be less likely to ask for help or for the opinions of others, and she'd be more likely to speak up about her own opinions and to defend those opinions in the face of disagreement (Ridgeway, 1993).

In other words, if she believes in her own competence about her construction abilities, she's more likely to act and appear like someone who's competent and her fellow workers are more likely to interact with her based on that assumption as well. If, however, gender status beliefs tell both my student and her fellow workers that women are just not as good at construction work, that's likely to create some doubts for her as well as for her coworkers. My student will be more likely to ask for help and less likely to assert herself; she'll look like someone who's generally less competent. This is the nature of a **self-fulfilling prophecy**. It is a statement that comes true solely as a result of the prediction being made. Because gender status beliefs tell us that women are less competent at construction, the power we place in that belief actually makes it come true.

At this point, you might think, well, if your student just had more confidence in herself and her construction abilities, no problem, right? But if her coworkers still hold gender status beliefs, and research tells us it's pretty likely that they do, the self-fulfilling prophecy still applies. They believe she's a less competent member of their team, they treat her as such, and she probably becomes less competent. Why? Her coworkers are not going to cooperate or support her in her job, and all interactions we engage in require some cooperative support. Think about a task you've tried to accomplish without the full cooperation of everyone involved and how difficult that can be. The main point to remember here is that as a result of performance expectations and gender status beliefs, gender has an important effect on the interaction that takes place. In fact, the student in this story did experience lack of cooperation and eventually outright hostility from her male coworkers on the construction site.

Is this true of all kinds and types of interaction? What if, by some strange circumstance, my female student was working at a construction site solely with other women? Status characteristics theory predicts that gender status beliefs will be most important under two conditions. The first condition is that the interaction involves both men and women. Status characteristics theorists argue that gender is simply not as relevant as a status characteristic in same-sex groups. The second condition is that gender is believed to be important to the goal or task on which the group is focused. The first condition, as we already discussed, happens fairly often. You might count up in the course of your day how many exclusively same-sex interactions you have compared with other-sex interactions. There might be variations in the relative balance of same-sex and other-sex interactions, but chances are you have some other-sex interactions.

What about the second condition? How often is gender important to the goal or task of interaction? If we go back to the examples we've already mentioned, is gender relevant to the tasks of navigating in a car, of a father teaching a daughter how to change a tire, or of the customer buying groceries from a clerk, assuming that all these interactions involve people of different sexes? It's probably relatively easy to imagine gender as relevant to the father teaching his daughter how to change a tire, and there are many comedians who play on assumed differences between men and women in following directions and navigating. There are certainly gender status beliefs about men and women's competence in finding their way, so we can imagine gender status beliefs affecting interaction in that imaginary car. What about the grocery clerk and the customer?

Thinking back on your day, or on an average day for you, can you make a list of interactions that fit both of the circumstances of status characteristics theory, or that are both mixed sex interactions and interactions where gender is relevant to the task or goal? Are there interactions that are mixed sex but where gender is not relevant to the task or goal? How many of your interactions fall under these criteria, and what does that say about the theory?

Status characteristics theorists would answer yes; even in what might seem to us like a gender-neutral interaction between say a female customer and a male grocery clerk, gender is still a part of the interaction. Why? First, because as we already noted, sex categorization takes place as part of any interaction. The grocery clerk and the customer have already identified each other as a man and a woman. More important, status characteristics theory argues that gender stereotypes and gender status beliefs are so diffuse that they can be considered relevant to a wide variety of situations and contexts (Ridgeway, 1997). This means that our beliefs about gender are pervasive and broad; they cover a wide range of situations, not just those that we think of as stereotypically gendered. The theory then argues that for a great deal of our interactions, gender is relevant to the task at hand because we live in a society that perceives gender as relevant to a lot of things.

For status characteristics theory, gender is understood largely as an interactional effect of sex categorization and gender status beliefs. Theorists believe this is an especially useful way to understand gender because it helps explain the persistence of gender inequality, especially in settings like the workplace. How does this work? Research shows that when people enter into interactions, they generally assign hierarchies based on resource differences; in other words, if someone in the interaction seems to have more of the important stuff, whether that be money or power, we give them more respect and consider them more competent and powerful (Ridgeway, 1997). If they also have some distinguishing characteristic, like gender, then we associate all those good things with people in that category in general; if men in interactions have more resources, we're more likely to believe that men are more competent, powerful, and worthy of respect. This means that as long as men have a resource advantage in interaction, they will continue to be considered diffusely more competent, playing into existing gender status beliefs. And as those gender status beliefs favor them in interaction, they will continue to be more likely to have a resource advantage over women. Status characteristics theory perceives gender inequality as self-perpetuating, recreating itself through this combination of resources and interaction. Gender status beliefs continue to be reinforced through interactions, and this makes gender inequality an especially difficult reality to reverse.

Doing Gender

The next theoretical perspective we take up shares with status characteristics theory its emphasis on interaction. In fact, the theory of gender developed by Candace West and Don Zimmerman (1987) was one of the first theories to emphasize strongly interaction as the site for the study of gender, a shift that these theorists argue makes interactionist theories more inherently sociological. Doing gender goes farther than status characteristics theory in arguing that gender does not exist outside of interaction, positing that any idea of gender as an internal trait or disposition is an illusion *created* by interaction. From a doing gender perspective, gender is simply a performance, and we're all constantly on stage.

The perspective of doing gender is informed by a particular approach in sociology called ethnomethodology, so you will sometimes see this perspective referred to as an ethnomethodological approach. If you break down the etymology of **ethnomethodology**, a hard enough word to pronounce, it basically means the study of folkways. Ethnomethodologists are interested in uncovering the taken-for-granted rules that govern our social life but lie largely unexamined most of the time. They're interested in how aspects of our social lives that seem objective, real, and universal are actually created in specific situations and contexts (Zimmerman, 1978). A famous example of how early ethnomethodologists engaged in this project is the use of breaching experiments.

In breaching experiments, ethnomethodologists send students out to create a purposeful **breach** in social life, or a disruption that requires some kind of explanation because it does not fit into the particular cultural story being told. In one breaching experiment, students responded to the statement, "How are you?" with a detailed description of her or his current state of being. For example, they might say, "Well, I'm getting over a cold, so my nose is a bit stuffed up, and I had a fight with my best friend, so I'm worried about that. I feel like I did a good job on my biology exam, so at the moment, I guess I'm okay." As with many breaching experiments, people responded with confusion, discomfort, and sometimes annoyance or anger. Answering the question, "How are you?" in this way breaks with a basic kind of faith and trust we have in our fellow interactants that they are aware of and will follow the rules for smooth and successful interaction.

What taken-for-granted assumption does this breaching experiment reveal? One assumption is that when we use the phrase, "How are you?" we're not asking for an answer to that literal question. We use that phrase as a greeting, and appropriate responses include "Fine" or "Good" or maybe "So-so," but not a detailed answer about your physical or mental state. This breaching experiment shows us that we use language in creative ways and in ways that make sense only within a specific social context. If you were not aware of the cultural norm in the United States that treats "How are you?" as a greeting rather than as a question, you might find it a bit confusing.

That's a very short and abbreviated summary of what is involved in an ethnomethodological approach, but it's meant to give you some sense of the backdrop against which doing gender was developed as a theory. As applied to gender, this perspective is interested in uncovering the assumptions that are concealed in the way we think about and live gender. One particularly influential ethnomethodologist, Harold Garfinkel (1967), began the process of applying ethnomethodology to gender when he became interested in the case of a male to female transsexual named Agnes (a pseudonym Garfinkel used to protect the real identity of Agnes). Agnes was raised as a boy but, at the age of 17, adopted a female identity and eventually underwent sex reassignment surgery to become female. Garfinkel, and later West and Zimmerman (1987), used Agnes as a case to begin an examination of how gender works. Why would you pick a transsexual person as a starting place for examining gender? For the doing gender perspective, it has to do with the way Agnes as a case helps in understanding the differences among sex, sex category, and gender.

We've already discussed conflicting definitions of sex in Chapter 1. The perspective of doing gender most closely resembles what we called a strong social constructionist approach. This means that from a doing gender perspective, sex is something that is produced socially rather than something that exists objectively as real biological or genetic differences between females and males. Agnes's desire for sex reassignment surgery to correct her perception that her penis was a "mistake" in need of remedy is itself an act that reaffirms that sex is socially produced. Despite having a penis (an indicator of being male and a man), Agnes felt like a woman. But because we put so much emphasis on sex in our society, Agnes felt that she could only *become* a woman by having surgery to remove her penis. This desire reaffirms the idea that you can only ever be male or female, but never both, and it supports the idea that sex and gender must match up in expected ways (penis = male = masculine).

From a doing gender perspective, **sex assignment**, or putting someone into one or the other sex category, usually at birth, is considered by society to be merely a case of discovering the "facts of the matter" (West & Fenstermaker, 1995). These facts are based on genitalia, chromosome type, and perhaps the presence or absence of various hormones, and we assume these will all line up in the appropriate ways (presence of a penis lines up with XY chromosomes, which lines up with the presence of appropriate amounts of testosterone at puberty). As we discussed in Chapter 1, this is not necessarily true, as the case of intersexed individuals demonstrates, and the criteria for sex assignment can vary across time periods and cultures (Kessler & McKenna, 1978). Nonetheless, most of us in Anglo-European society believe in the existence of two sexes and in our ability to see a world that consists of two sexes, even though the indicators of sex (genitalia, chromosomes, and hormones) are generally hidden from us.

But what about Agnes, who was born as a biological male, but wanted to become a biological female? How does someone unlearn masculinity and learn to pass as a woman? In her everyday life before and after she had sex reassignment surgery, Agnes's desire was to pass easily as belonging to the female sex category. What would be her best strategy for doing so? How does Agnes, or any of us for that matter, successfully pass as belonging to our particular sex category? How could Agnes accomplish her main goal, to preserve her categorization as female and pass the perfunctory test of everyday sex categorization without further scrutiny?

Answering these questions leads us to the important connection between sex categorization and *the accomplishment of gender*. From this theoretical perspective, being categorized as female does not mean everything you do is automatically feminine. If you stop for a minute, you can probably think of many females who do not act in feminine ways and many males who do not act in masculine ways. But acting "unfeminine" does not also make you "unfemale." This means the relationship between sex categorization and gender is more complicated. What Agnes needed was some configuration of behavior that would lead people to see her as femininely gendered and, therefore, to assume that she was also female (West & Zimmerman, 1987). How might Agnes learn to act feminine?

One potential source would be various women's magazines or books on etiquette. These manuals lay out specific ground rules for what makes feminine or, sometimes more specifically, lady-like behavior. Agnes could simply follow these rules, but from the doing gender perspective, there are problems with this solution. Strictly following some set of rules for feminine behavior might get Agnes into trouble because the enactment of gender is deeply situational and contextual. For example, one particular middle-class American norm is that women are generally expected to smile more, often including smiling at people they do not know. A guide to proper, American feminine behavior might tell Agnes to smile often to be accountable as a woman. But the rule would have to be more specific and contextual than that. It might be OK for Agnes to smile at a grocery clerk or waiter, but should she also smile at everyone in the women's restroom? Should she smile at strangers on a busy city street? If Agnes is with her boyfriend, should she smile at other men? In addition, if Agnes were trying to pass as a woman in many cultures outside of the United States, smiling at strange people would not be considered a particularly feminine behavior, and it could get her into trouble. Even in some European countries, a woman who makes eye contact with a strange man, let alone smiles at him, is interpreted as making a pass. To accomplish femininity, should Agnes smile or not? One simple rule—women should smile more than men—won't always work from situation to situation and from context to context. The rules about smiling quickly become infinitely complex and too complicated to be codified into any simple set of rules.

The doing gender perspective argues that rather than a set of rules, what someone like Agnes needs is to make sure her actions are accountable as feminine. Accountability is an important concept from ethnomethodology, and it highlights the importance of the interactional nature of gender. **Accounts** are the descriptions we engage in as social actors to explain to each other the state of affairs, or what we think is going on (Heritage, 1987). They are important because they serve a variety of purposes in interaction; they can help us identify, categorize, explain, or just draw attention to some activity or situation and therefore provide us with some kind of framework for understanding it (West & Zimmerman, 1987). For example, if a professor showed up to teach class dressed as a fairy, students would expect some kind of account (here, an explanation) for this behavior—he lost a bet, it's Halloween, or he's gone crazy. Accounts help us to make sense of the events and interactions that go on around us.

Accountability, then, has an inherently social and interactional quality; it's determined by how people react to each other. What does all that mean? When you tell a story about a chicken crossing the road or about a priest, a rabbi, and an imam, it's really only a joke if it's *accountable* as a joke—or if the people to whom you're telling the story acknowledge it as a joke. Laughing at the story is one kind of account, in that when you laugh, you're identifying the story as a joke. But even if no one laughs, your audience can provide an account that acknowledges your story as a joke. They might roll their eyes, or shake their heads, or groan, but all of those reactions acknowledge that the story you told was, in fact, a joke . . . just not a very good one. Your story is not accountable as a joke if someone stares at you blankly or asks what a story about a chicken has to do with anything. Ethnomethodologists argue that in our interactions, we work to make our actions accountable—even though this is always dependent on the reaction of others. When we tell a joke, even if it's bad, we'd like it to at least be considered a joke by our audience. **Accountability** means that we gear our actions with attention to our specific circumstances so that others will correctly recognize our actions for what they are (Heritage, 1987); we want our joke to be understood as a joke, and we're likely to be embarrassed and chagrined if it is not.

How does that apply back to our particular focus—gender? The doing gender perspective argues that gender is all about rendering your actions accountable as gendered. Sex category, as we discussed previously, is omnirelevant, which is to say that it matters in all situations, all the time. If sex category matters in every situation, then every activity you engage in can be held accountable as a performance of gender, or as being a man or a woman. Using our example of the joke, this highlights two important aspects of the doing gender perspective. The first is that you don't have to conform to normative ideas of gender in order for your performance to be perceived as accountable. Even a bad joke is still a joke as long as it's considered by your audience as accountable as a joke. Even behavior that doesn't conform to gender norms can be viewed as gendered if your fellow interactants judge it accountable as a performance of gender. This is important because it explains why people can engage in behavior that is not consistent with their gender while being perceived as belonging to their particular sex category. For many people, crying is considered unmanly, but a man who starts crying is generally not suddenly categorized as a female. Even with crying, his overall set of behaviors is probably still considered accountable as a man.

The second important aspect of doing gender that's demonstrated here is that gender is inherently interactional. You can gear your actions to make them accountable, but in the end, accountability is a product of social interaction. Try as you may to tell something that you consider a joke, if no one else considers it as such, it's just not really a joke. In the same way, regardless of whether you are or are not trying to portray your actions as gendered, if they are accountable as such, they are in fact an example of doing gender. Gender, from this perspective, is defined not just as your own performance but, rather, as that performance combined with its accountability. Where is gender located? It's in the intersection between what you do and whether others consider those actions to be accountable as gender. Agnes's solution is not to follow some guidebook or rules for how to be a woman; rather, she needs to become practiced at producing a set of behaviors that are held as accountable for a woman in her particular context and culture.

This unique way of thinking about gender is hard to grasp because gender becomes more than the sum product of how we think or feel (dispositions or traits) or even what we do (behaviors). Theorists within the doing gender perspective argue that by moving gender in this way, outside of individuals and into the space of interaction, the result is a uniquely sociological approach to understanding gender: Gender becomes deeply social in nature. In addition, it's important to be able to understand gender in this way because it helps explain the powerfully strong belief most people in society have in the notion of sex as something objectively real. When we produce gender on a day-to-day basis in routine, recurring ways, we also produce the notion that our actions and perceptions of those actions reflect our masculine and feminine natures (West & Zimmerman, 1987). In this sense, the doing gender perspective argues that gender works like a magic act. Our accomplishment of gender confirms our notions that we're merely acting out the "natural" division of human beings into two sexes, and it creates the illusion that such a thing exists. It's a magic act in the sense that this performance creates a belief in something that is really all smoke and mirrors: the existence of something called sex.

How does the doing gender perspective explain the existence of gender inequality? To say that gender is a performance for which we are accountable does not automatically imply anything about inequality between men and women. Theorists working from this perspective explain inequality through the concept of allocation (West & Fenstermaker, 1993). **Allocation** is simply the way decisions get made about who does what, who gets what and who does not, who gets to make plans, and who gets to give orders or take them (West & Fenstermaker, 1993). The doing gender perspective argues that the accountability of gender is more likely to be called into question when issues of allocation are involved. Like status characteristics theory, the doing gender perspective posits that there is a widespread and deeply held belief in our society that women are both different and inferior to men. This shapes the way in which women will be held accountable for gender, especially when issues of allocation are involved.

Allocation can become important in something as simple as a routine conversation, in decisions about who does housework and child care in a family, and in the different expectations for men and women in the workplace. Research in the doing gender perspective demonstrates allocation in simple conversations between white, middle-class men and women, where the particular kind of work to be allocated is shifting topical transitions when one topic runs out of steam in a conversation. The study of how people in a conversation change topics emerges from a fundamental set of questions in the ethnomethodological tradition about the rules of very basic, mundane interactions—including how to have a conversation; we take it for granted that in a conversation, topics shift and change, but ethnomethodologists are interested in the details of how this happens and what it reveals about social life. Studies show that people in a conversation usually change topics collaboratively, but sometimes a topic change is unilateral or initiated by one person alone.

Research on conversations has shown that unilateral topic changes are always initiated by men, and from the doing gender perspective, this is an issue of allocation—in this case, of who has control over what two people will talk about in a conversation. Men accomplish gender in conversation by changing the topic, and this seems to be especially true when women move the conversation toward topics that are not necessarily considered consistent with ideas of masculinity (West & Garcia, 1988). In this small way, men are producing an accountable performance of masculinity. Later, we'll read about how the division of household labor among couples can also serve as a resource for doing gender, as well as the battle over who gets to hold and use the remote control in families. In all these examples, allocation of certain duties and responsibilities explains how gender inequality is accomplished in a doing gender perspective.

INSTITUTIONAL OR STRUCTURAL APPROACHES

The next two theories we will discuss locate gender at the level of institutions or in the social structure of society. If you were thinking about our exploration of theory as a very large zoom camera, we would have started with an extreme close-up on individuals and on what goes on inside their heads. Then we zoomed out a bit to the level of groups and interaction. In this final zoom, we move our lens very far out to encompass large organizations and the way societies as a whole fit together.

Gendered Organizations

In the chronological evolution of theoretical approaches to gender, the shift to a focus on organizations comes relatively late. As you read, radical feminism made the claim that gender is an integral part of societal institutions and structures during the second wave of the feminist movement in Anglo-European society. But systematic analysis of the gendered ways in which these structures were organized was somewhat slow to emerge in sociological analysis. As Joan Acker (1990) pointed out, this analysis requires seeing through what appear to be the gender-neutral practices of organizations to uncover their powerfully gendered nature. This approach, sometimes called *macrostructural,* shifts the focus from individuals or interaction to social aggregates (Dunn, Almquist, & Chafetz, 1993). **Social aggregates** are *composed* of individuals, but they become more than the sum of the individuals within them. This reflects a basic sociological belief that at a certain level, the group, organization, or institution is no longer reducible to a collection of unique individuals.

A macrostructural approach to gender assumes that gender is more than individual traits or interaction between those individuals; as a part of social aggregates, gender is something that takes on a life of its own. Individual-level variables related to gender are a product of larger, structural processes. Within the gendered organizations approach, this means that organizations can create gendered individuals and shape gendered interactions, but gender is working from the top down (from organizations down to the level of individuals and interactions) rather than from the bottom up (from individuals or interactions to organizations).

What exactly does it mean to call an organization gendered? A **gendered organization** is one in which "advantage and disadvantage, exploitation and control, action and emotion, meaning and identity, are patterned through and in terms of a distinction between male and female, masculine and feminine" (Acker, 1990, p. 146). This is another way of saying that the normal functioning of the organization is based on gender and has important gendered consequences. From this perspective, gender is not simply something that you add into your basic understanding of how an organization works; because everything about the organization is gendered, there are no gender neutral processes in this type of organization. Instead, gender is a basic and integral part of how the organization functions.

According to Acker (1990), organizations produce gender through five interrelated processes. First, they create divisions along gender lines, whether in physical location, power, or behaviors. For example, your average gym in the United States has a separate locker room for women and men, as well as areas that are generally considered more or less masculine (the free weight room versus the aerobics room). Second, gendered organizations construct symbols or images that can support or oppose those divisions. There may be inspirational posters in the gym that portray slim women and bulked-up, muscular men, reinforcing the sense that women are at the gym to become thin while men are there to become larger. Third, gendered organizations produce types of interaction that reinforce these divisions and inequality. In the gendered space of the free weight room, men may interact in very different ways than the women in the aerobics rooms or using cardio equipment. The lone woman wandering around in the free weight room may be made to feel out of place or be offered assistance by gym workers, just as the lone man taking an aerobics class may be expected to be a little less capable in that area. Gendered organizations also have an impact on individual identity, the fourth important process. The woman who repeatedly finds herself surrounded by men in the free weight room may feel subtle pressure to stop lifting weights, and this pressure may change her own idea of appropriate exercise for women. Note, for both the third and fourth processes—interaction and individual identity—gendered organizations affect these processes from the top down, with the organization itself impacting your internal sense of gender as well as how you interact.

The fifth and final process is the way in which gender helps to both create and reinforce social structures. Acker (1990) argued that this happens through **organizational logic**, or the assumptions and practices that underlie an organization. If gyms as organizations are structured in ways that assume basic gender differences, then they are building gender into the organizational structure. To get at the organizational logic of a gym, you might ask why the gym is divided into different areas in the first place. Why do most gyms separate the free weights from the weight machines or from the treadmills, bikes, and stair masters? Is there an underlying gender logic that separates the uses of the gym based on the assumed gender of the typical user (i.e., free weights are for men and stair masters are for women)?

Gendered organization theory has been used most often to explain the gendered dynamics of the workplace. In her research, Joan Acker (1998, 1999) pointed to the ways in which the organizational logic of the workplace is deeply gendered, despite all appearances of gender neutrality. In the workplace, this is especially true in the realm of jobs and hierarchies. According to the organizational logic of the business world, a **job** is "a set of tasks, competencies, and responsibilities represented as a position on an organizational chart" (Acker, 1990, p. 149). This means that the job is a place holder that exists completely separate from the particular person or type of people that might hold that job. The idea of hierarchy is also important to organizational logic and very simply involves the ranking of jobs. The necessity and usefulness of hierarchies are rarely questioned in the business world, and like jobs, hierarchies are based on abstract differentiations rather than on actual human workers.

Both jobs and hierarchies are supposed to be abstract categories, devoid of gender. But how would organizational logic describe the perfect human to hold the generic job? The closest thing to that abstract worker is a male employee whose whole life is centered on his job and who has a wife or some other woman to take care of his own needs as well as those of any family he might have (Acker, 1990). This assumption spills over into the construction of hierarchies as well. The person who can commit the most time and energy to the job should logically be more highly placed in the hierarchy than those who are less able to commit those resources. This again favors the male employee who can often be more committed to his job. Both of these "gender-neutral" components of business logic are demonstrated to be inherently gendered, in that the organizational logic of both jobs and hierarchies is influenced by gender. Subsequent research on gender and organizations has demonstrated how this organizational logic also works in the granting of patents (Bunker Whittington & Laurel, 2008), in high-technology companies in Silicon Valley, California (Baron, Hannan, Hsu, & Kocak, 2007), and even at the United Nations (Keaney-Mischel, 2008).

Homophily: A Social Network Approach to Gender

Like gendered organization theory, a network approach to gender focuses on the importance of social structure. The main focus of this theory is a different type of social aggregate—the social network. Network theory, like status characteristics theory, takes a more general theory from sociology and applies it specifically to explain how gender works. Because of its focus on social structure, network theory conceives of gender in a very different way than either individual or interactional approaches. Network theory is not concerned with the particular attributes of an individual—their gender, race, religious background, age, and so on. In network theory, individual actors are understood only in light of their relationships to other actors. People, according to network theory, are "identically endowed, interchangeable nodes" (Smith-Lovin & McPherson, 1995). It might be odd to think of yourself as a mere interchangeable node, but for network theorists, your individual characteristics are less important than the people you are connected to and how you are connected to them. From this perspective, gender is considered a product of the social relationships in which women and men are embedded; the behaviors or attitudes we perceive as masculine and feminine are really a result of our positions in particular social networks.

To understand this particular way of thinking about gender, we have to understand first some basic concepts about networks. Network theorists conceive of a network in terms of an **ego**, or focal person, and the **alters**, or other people in the network. When they focus on these specific relationships, between an ego and the alters, they are focusing on an **ego network** rather than on the whole, simultaneous web of connections in a society. Ego networks can have three different properties: size, density, and heterogeneity (Smith-Lovin & McPherson, 1995). **Size** is the number of others to whom someone is linked in a network. **Density** measures how interconnected the alters are in an ego network; in a very dense network, many of the people you are connected to are also connected to each other. **Diversity** in a social network means you have contacts with people in multiple spheres of activity, as opposed to all of your alters being very similar to each other. A diverse network is good in the sense that it provides you with information from many distinct sources in your network. Network theorists also describe networks in terms of their structure. A way to think about network structure is to consider what ties are present or absent in your network; what spots are occupied in your network, and where are the holes?

One way of talking about network structure that is particularly important to the discussion of gender is the concept of homophily. *Homophily* is a very basic concept in sociology in general, not just in network theory, and forms one of the very few general truisms of sociology as a science. What can we say that seems to generally be true of people? There seems to be a tendency for their networks to be **homophilous**. That means that "similar nodes are more likely to have a relationship than dissimilar nodes" (Smith-Lovin & McPherson, 1995, p. 228). Remembering that you are one of those interchangeable nodes simply means that you're more likely to be friends or in a relationship with someone who's like you than you are with someone who's not like you. If two men are more likely to discuss baseball with each other than with a woman, then their relationship (discussing baseball) is homophilous.

Homophily works as a concept regardless of whether it's an individual choice, and some research suggests that it's more about the kind of people to whom you're exposed. So saying your networks are homophilous doesn't necessarily mean you don't like people who aren't like you; it's more likely to reflect the fact that our society is structured in ways that make it difficult to form relationships with people who aren't like you. Network theorists argue that homophily has important implications for how gender impacts the social networks of men and women. Their argument is that very small initial differences in the types of networks and experiences of boys and girls lead to men and women occupying very different social worlds (McPherson, Smith-Lovin, & Cook, 2001; Smith-Lovin & McPherson, 1995).

What aspects of your own life might contribute to the formation of homophilous networks? For example, how diverse are the places where you grew up or where you live now? What about the places where you go to school or work?

To understand how this works, we have to go back to the experiences of boys and girls in childhood. At about the time children enter school, they first begin to demonstrate homophily in their play patterns, and in addition, girls demonstrate a tendency to play in smaller groups than boys. Why exactly this is true is subject to debate, but network theorists argue that it has important implications. Research demonstrates that children are generally more likely to add other children of the same sex to their particular friendship groups, or networks, than they are to add children of the other sex. When you compare boys and girls, girls are generally less likely to extend or receive offers of friendship. The end result of these two dynamics is networks among children that are gender-segregated and that are larger and more diverse (heterogeneous) for boys than for girls (McPherson et al., 2001).

Network theorists then point to the important ways in which knowledge is related to networks. Research suggests that our knowledge about the world around us is grounded in the relationships to which we belong. Boys and girls in their gender-segregated networks will gradually come to be exposed to different types of knowledge. The effect of these networks organized around gender is to make what were initially small distinctions between boys and girls seem more and more important. The little boy who spends most of his time talking to other boys about the latest video game will find it difficult to participate in girls' discussions of *High School Musical III*. The little boy's inability to know about *High School Musical III* will begin to seem like a real and important difference between him and little girls. That difference will only be further reinforced by network differences across the life course, but network theorists argue that gender is ultimately the result of the time spent within these gender-segregated groups in childhood. The composition, size, and heterogeneity of the networks boys and girls are in produce gender differences.

These differences in networks continue long after we leave childhood. Research reveals that although men and women generally have networks of similar size, women's networks have fewer ties to nonkin, or to people who are not family (Ibarra, 1992; Marsden, 1987; McPherson et al., 2001; Wellman, 1985). Homophily remains an important characteristic of networks into adulthood, including homophily along the lines of gender. Because women have more ties to kin and neighbors, their networks are also less racially, ethnically, and religiously heterogeneous, but more heterogeneous in regard to age and sex. Certain life-course events can also have important impacts on the characteristics of women's networks. Some research has shown that having children greatly reduces women's cross-sex contacts in part because of their increased responsibility for child rearing (Wellman, 1985).

Some researchers argue that this contraction of women's networks as a result of having children, along with the higher inclusion of kin in their networks, makes women's networks less useful in the realm of work (Campbell, 1988; Ibarra, 1992; McPherson et al., 2001). The flow of information through women's networks is less likely to help them find a job or move up in their occupation than men. This is because kin networks tend to be fairly dense and local; your mother, father, brother, and sister are closely connected and therefore likely to know the same things about the same geographic place. Networks that are less dense have fewer overlapping members and so are more likely to carry different information from different sources. Gender differences also exist in the organizational networks to which women belong because women tend to belong to fewer organizations than men (Booth, 1972), and the organizations they do belong to are more likely to be organized around social and religious activities than around work-related activities (McPherson & Smith-Lovin, 1986; McPherson et al., 2001).

Networks begin in childhood and school years in gendered patterns, and these patterns, once begun, continue throughout our lives according to network theory. These networks are important to the creation of gender because network theorists argue we are a product of the spaces we fill in these networks. The relationships in which we spend most of our time determine the roles we play (McPherson et al., 2001; Stryker, 1987). The nature of women's networks, consisting of more kin and fewer coworkers, means that women will spend more time performing the roles associated with kin rather than with coworker; network theory predicts a woman will spend more time being a mother, daughter, or sister than she will spend being a worker, professional, or boss (Smith-Lovin & McPherson, 1995). A man, with his network composed of more coworkers or organization members, will spend more time performing those nonkin roles. What we experience as gender is really a product of the networks of relationships we find ourselves embedded in, going all the way back to the very first networks we belonged to as children.

The network theory explains inequality because of the general effects a network can have on basic aspects of your life, such as health and wealth. Interestingly, the health of both men and women benefits from ties to women (House, Umberson, & Landis, 1988), which may be why widowers (husbands who have lost their wives) suffer declines in mental and physical health after the loss of their wives. This is good for women's health, as they receive more social support from their kin-heavy networks, but they are also often expected to be the caregivers in these networks (Smith-Lovin & McPherson, 1995). But although women experience health benefits from their social networks, their networks do not work to their benefit economically. Namely, women lack the widely spread, weak ties to nonkin individuals that are so important to finding a job and achieving advancement (Granovetter, 1974). Rosabeth Moss Kanter's (1977) work on women in corporations provided some of the most powerful evidence for the importance of networks for women in the workplace.

According to Kanter's research, even men and women who occupy the same position in a corporation or firm are likely to have different types of networks. The male worker will tend to be closer friends with his male boss and other male clients. He might socialize after work with large numbers of his coworkers, and because of these relationships, clients might be more likely to come to him than to his identical female coworker. His female coworker with the same educational background and experience will have networks composed of fewer coworkers in general, and those coworkers might tend to be the few other women at the firm. Networks serve as a kind of handicap for women in the workplace and make it more difficult for them to achieve the same kind of success as men. Thus, the differences in the networks of women and men explain the inequality that exists between these two groups.

INTERSECTIONAL FEMINIST THEORY

Our discussion of these different theoretical approaches to gender might leave you with a lot of unanswered questions. That's not a bad thing, as a good theory *poses* questions in addition to answering them. There are at least two unanswered questions that sociologists who study gender have focused on most recently in relation to much of the work that's been done in theorizing about gender. The first has to do with how gender theories deal with other categories of difference, such as race, social class, and sexuality. The second, to be discussed in the next section, asks what we do with all these different theories, which seem, at the very least, to contradict each other and to focus our attention on very different sets of questions in our quest to understand what gender is and how it works. How can we use all these diverse theories without becoming impossibly tangled up in their differences and contradictions?

The first question is not a new one, and as we have already discussed, it goes back to the very beginnings of the global women's movements. The question has to do with the tendency of feminists and gender theorists to focus on gender to the exclusion of any other type of identities. Here's something you may or may not have noticed about the theories we've discussed. They all separate gender from any of the other many identities and statuses that we occupy. Gender is considered separately from statuses such as race, ethnicity, age, class, nationality, or sexuality. These theories, as I have described them, assume that gender is something that you can pluck out of the very complicated, messy stuff that is our lives to examine in a pristine state. Does gender really work that way? Can you ever simply be a "man," in interaction, or are you always also raced, classed, and sexualized?

From the very beginning of the feminist movement in the United States, women like Sojourner Truth asked this very same question. Sojourner Truth eloquently described this question in a speech with the phrase "Ain't I a Woman?" Truth compared her own experiences as a slave woman with those of the wealthy, white women organizing for suffrage and wondered how her life of hard physical labor compared with the dominant definition of womanhood as centered on frailty and passivity. Truth's experiences of what it meant to be a woman had very little in common with the experiences of her privileged, white counterparts, so what, after all, did it mean to call all these people women? The question has been asked over time by women of color and working class women like Sojourner Truth, bell hooks, Angela Davis, Patricia Hill Collins, Rosalinda Mendez Gonzalez, and Maxine Baca Zinn, to name a few.

Beginning especially in the second wave of feminism, this question surfaced in relation to sexual identity. Lesbian women often felt pushed to the sidelines during the women's movement by heterosexual women, who sometimes felt that the perception of feminists as lesbians would hurt the overall cause of the movement. In addition, some of the issues of women's movements, such as reproductive rights, were geared more toward the experiences of heterosexual women than lesbian women. Regardless of whether it's asked by people of color, working class people, or gays and lesbians, this question as to whether the categories of "women" and "men" can really describe all the diverse experiences of the people we place into those categories implies two distinct but deeply connected problems with theorizing about gender that need to be solved.

To some extent, the first problem deals with trying to figure out why, if this is something feminists and social scientists have been struggling with since the 19th century, they haven't yet come up with a better way of integrating race, class, sexual orientation, and other identities into their theorizing. The most agreed upon answer to this question is that there is a distinct white, middle-class, and heterosexual bias to much theorizing about gender (West & Fenstermaker, 1995). This bias exists both in the composition of those doing the theorizing and in the specific ways in which they theorize. First, for a variety of reasons, many of those who have the power and resources to theorize about gender and make those theories heard tend to be those in positions of privilege as a result of their race, nationality, class, or sexual orientation. According to bell hooks (1984), this was because the women who are most oppressed by sexism are also least likely to question the system because of their victimization. Larry W. Hurtado (1989) drew attention to the fact that the world of colleges and universities is where most feminist and gender theory is produced. Given that it is still difficult for low-income and minority women and men to gain access to that world, and especially to the positions of power (for example, editor of a scholarly journal that might publish articles on gender), it's not surprising that their viewpoints have generally not been represented in the body of publications that make up our theories of gender.

Finally, many theorists argue that those doing the theorizing about gender have their own set of biases (Collins, 1990; Davis, 1971; Hunt & Zajicek, 2008; Lorde, 1984; Moraga & Anzaldúa, 1981; Rupp, 2006; Zinn, Cannon, Higginbotham, & Dill, 1986). Feminists and social scientists are themselves racist, classist, and homophobic. These biases can take the form of an assumption that everyone else's experiences are just like theirs, what Adrienne Rich (1979) called "white solipsism." Classism, racism, and homophobia thus work when these white, middle-class, heterosexual women assume that their whiteness, middle-class status, and heterosexuality accurately describe the rest of the world (Rich, 1979). None of these specific explanations for why gender theorizing has failed to take account of these other statuses necessarily excludes the possibility that all of these work together to create an overall bias. Together, they help to explain why this question is one with which feminists and social theorists continue to struggle.

The second problem raised by the question of how gender theorists should address other identities is the following: What do we do now? Given that we now have some understanding of some of the dynamics that made this question so difficult to answer in the past, how do we develop a theory that accurately accounts for the diverse experiences of many different types of women and men? This question is connected to a second and equally important question, which we raised previously in our discussion of global feminism. Is it even possible to theorize about gender, or are the diverse experiences of gender so incompatible as to make that grouping together impossible? For example, if women worldwide cannot agree on any one issue that is everywhere and always an important issue for women, does it even make sense to talk about women's movements, or do we always have to be more specific (*middle-class, white, American women's movement* as opposed to *lower caste, Hindu, Indian women's movement*)? Does it make sense to even attempt to write a textbook about something called gender when there's nothing similar about the experience of gender across time and place?

It's an important and difficult question to answer, but don't throw your book down in disgust and move on to something less frustrating yet. Thankfully, gender theorists have done quite a bit of thinking about this question and have developed several different answers to the question of what we do now, or of how to think about gender without ignoring other important categories of difference. As we will discuss next, these conversations about how to go about a more sophisticated and comprehensive analysis of gender continue to evolve.

Multiple Consciousness and Intersecting Models of Intersectionality

Sociologist Patricia Hill Collins (1990), in attempting to formulate a way of thinking about multiple identities, argued that there are two basic premises underlying previous models of intersectionality that must be overcome. The first premise is an underlying belief in dichotomies or oppositions. You can either be Black *or* white, man *or* woman, straight *or* gay, but you cannot be *both* at the same time, and these things exist as opposites of each other. This oppositional, either-or thinking is especially problematic when you add the ideas of oppressed or oppressor and privileged or not privileged. A Hispanic man in the United States is in a position of privilege relative to his gender, but he may be oppressed relative to his ethnicity; his status makes him both oppressed and oppressor, and this violates the underlying assumption of mathematical models that you are either in one category or the other.

The second problematic premise is the idea that these dichotomous differences can be put into some meaningful system of ranking (Collins & Andersen, 1993). The assumption that men are superior to women, whites are superior to Asians, and heterosexuals are superior to homosexuals all assumes that there is some quantitative component to the categories because this ranking implies some kind of logical, numerical order. If you can rank men above women and whites above Asians, then you must also be able to create some ranking of all these identities. But as Collins (1990) pointed out, this assumption breaks down once you begin to think about the real people who occupy these identities. Collins is often asked as a Black woman whether she is more oppressed by her gender or her race, and she describes that what the question essentially asks her to do is "divide myself into little boxes and rank my various statuses" (Collins & Andersen, 1993, p. 71). Her actual experience of oppression is not the either-or described by dichotomous, oppositional thinking, but the both-and that exists when you cannot, as she said, conveniently divide yourself up into neat little boxes of identity.

Make a list of all the identities you occupy that are privileged identities and all the identities you occupy that are not privileged identities. Does your experience of these contradictions feel like the either-or option or the both-and option? How do these various identities interact in your daily life?

One model for thinking about intersecting identities that seeks to overcome the problems of these two premises uses the metaphor of interlocking identities. This perspective is also sometimes called **multiple consciousness**, which describes a way of thinking that develops from a person's position at the center of "intersecting and mutually reliant systems of oppression" (Ward, 2004, p. 83). This perspective seeks to correct for the tendency to perceive one system of domination as more important or fundamental than another. Gender, race, class, and sexual orientation all need to be recognized as distinct social structures that can be experienced by individuals simultaneously. Multiple consciousness does not ask us to divide ourselves up into neat little boxes of gender, race, class, and sexuality, it acknowledges that these identities interact dynamically in any given situation.

The idea of multiple consciousness in the various forms in which it has been articulated has four basic assumptions. The first is that identities such as race, class, gender, and sexual orientation work as social structures in what theorists call a matrix of domination. The **matrix of domination** means that the social structures of race, class, gender, and sexual orientation work with and through each other so that any individual experiences each of these categories differently depending on his or her unique social location (Zinn & Dill, 1996).

This is another way of saying that it is impossible to ever completely separate gender from the other identities we occupy and the unique ways in which those identities intersect.

In addition, the matrix of domination assumes that gender needs the social structures of race, class, and sexual orientation to work, just as race as a social structure needs gender, class, and sexual orientation. For example, the subordination of lower classes works through the social structure of gender. As we will discuss in Chapter 8, the idea of the housewife had important gender implications, but it was also a tool used to enforce class and racial differences, as poor women and women of color were rarely able to choose not to work outside of the home and be housewives in the way upper-class, white women were. Thus, the ideal of the housewife had gender, class, and racial implications, and leaving out any of these social structures results in an incomplete understanding of how and why the ideal of the housewife developed in Anglo-European society.

The second basic assumption of multiple consciousness is that these social structures are interlocking and simultaneous. This is a rejection of the idea of statuses as additive or multiplicative in favor of a model of interlocking circles of experience. At any given moment, a person can be within the circles of race, gender, and sexual orientation and experiencing all of them at the same time. The model is more dynamic in that we can imagine how the movement of one ring might affect the other (West & Fenstermaker, 1995). This model also gets away from the idea that oppressions can or should be ranked and added to each other. In the interlocking model, this is impossible because you cannot separate the categories from each other long enough to do so. The multiple consciousness theory also allows theorists to escape from an either-or ideology. The third basic assumption of this theory is that because of the interlocking and simultaneous nature of categories of difference, they can produce both oppression and privilege. Because of the complex ways in which these identities interact with each other, it is possible to be both-and: oppressed and privileged.

The fourth assumption of multiple consciousness has less to do with how to conceptualize the relationships between these multiple identities and more to do with the implications of this approach. The fourth assumption is that multiple consciousness allows for a fuller examination of how identities like race, class, gender, and sexual orientation work themselves out in the lived experiences of real people. Autobiographies and memoirs have sometimes provided the best accounts of the complex interplay of these identities because they focus on lived experiences. Multiple consciousness theorists argue that this approach is especially helpful in highlighting the complexity of these relationships and dynamics.

Other theorists have built on this idea of interlocking systems of oppression and have taken the basic premise in a slightly different direction. The doing gender theory we discussed previously is used to explain how racial and class identities function in interaction. This means that like gender, both race and class are things that are produced in the dynamic world of interaction through the idea of accountability (West & Fenstermaker, 1995). Like gender, race is something that has no underlying biological reality; anthropologists, biologists, and other scientists have long since demonstrated the emptiness of race as a biological category. Yet people continue to engage in race categorization based on the belief that there is a category called race and that people can easily be placed into such a category. Moreover, the fact that many people are unsettled when others don't act in ways considered appropriate to their race is evidence of the importance of accountability. We might be disturbed at a white person who "acts Black," and for the doing difference perspective, this is evidence that accountability applies to race as well. In regard to class, the prevalence of certain types of uniforms is good evidence of the performative nature of class. Candace West and Sarah Fenstermaker (1995) gave the example of a maid whose employer insisted she wear her uniform when accompanying the family to the beach. Without the uniform, the class background and status of her maid were not immediately obvious to her peers on the beach, and she was not therefore accountable as a maid without the uniform. Perceiving gender, race, and class as situated, ongoing accomplishments makes it easier to imagine how they can be "done" simultaneously. You don't just do your gender in any given interaction; you do your gender, your race, your class, and any other identities that are relevant to the situation and subject to accountability.

Almost 20 years after Kimberle Crenshaw's original formulation of intersectionality in the context of Black women in the legal system, theorists are still exploring exactly what it means to use an intersectional approach. As an analytic framework, intersectionality is a way of thinking about sameness and difference and its relation to power (Cho, Crenshaw, & McCall, 2013, p. 795). An intersectional approach, then, conceives of "categories not as distinct but as always permeated by other categories, fluid and changing, always in the process of creating and being created by dynamics of power" (p. 795). Especially important is the emphasis on how power permates these categories.

PUTTING IT ALL TOGETHER: INTEGRATIVE THEORIES

At the beginning of this chapter, I told you that learning theories of gender is useful as a way to help you make sense of the role of gender in your own life. Now that you've learned about the wide variety of theories about gender, and those just from within the discipline of sociology, you may be feeling a little skeptical about their usefulness. Many of these theories define gender in different and often contradictory ways. They give different explanations for why gender inequality exists, and they focus our attention on very different areas of social life in our study of gender. How, in the middle of all this confusion, can these theories of gender help us make sense of gender in our own lives? What are we to do with all these theories?

Increasingly, gender theorists themselves have come to be a bit frustrated by this confusion and have begun to think about how to bring all these very different theories together into some coherent whole (Connell, 2002; Ferree, Lorber, & Hess, 1999; Lorber, 1994; Risman, 1998, 2004). Developing a more integrative approach to gender should be useful for two reasons (Risman, 2004). First, an integrative theoretical model of gender should help to make sense of the vast array of research findings on gender. The good news is that studying gender is a booming business in the world of colleges and universities, and a lot of research on gender is being produced. But what's the best way to make sense of all this research, especially when much of it assumes very different theoretical perspectives? A more integrative theory could allow for an exploration of these diverse topics in the study of gender while retaining the sense of a common endeavor or body of knowledge being built. In other words, researchers asking very different kinds of questions and motivated by different theoretical orientations could still feel like they're working on the same basic question. Second, a theory that works more toward integration moves away from what Risman (2004) called the "modernist warfare version of science" (p. 434). In this model, much like in war, the goal of gender theorists was often to defeat the other theory in a winner-takes-all battle. You might call this the death match model of theorizing.

In the next chapter, we will discuss the postmodern perspective and how it relates to gender, but a more postmodern approach to theorizing is one that allows room for multiple versions of the "truth" about gender. Postmodernism gives up the modernist idea of one theory, or metanarrative, that will explain gender as it has always worked, everywhere and for all time. If we can accept that no *one* theory has to explain *everything,* then there's more room for perceiving theories as working together. This approach makes integration of theories a much more sensible endeavor, and the end product of this integration should be a more complex account of how gender works in the world.

There are several different versions of these integrative theories out there (hopefully not engaged again in the death match model), but in the spirit of bringing these ideas together, we can talk about two common elements in these approaches. First, most of these theories attempt to integrate the many divergent strands in gender theorizing by perceiving gender as working at several different levels or in several different ways (Collins & Andersen, 1993; Lorber, 1994; Risman, 2004). Luckily for us, these levels line up pretty closely with the three locations for gender we've already begun to outline: individual, interactional, and institutional. Each level implies different types of questions to be asked about gender and different ways in which gender functions. You might think of the integrative approach as something like the way physicists conceive of light. Sometimes light acts as a particle and sometimes light acts as a wave, and those have important implications for the overall behavior of light. Integrative approaches ask us to acknowledge that gender can sometimes be something that exists internally, sometimes something that's produced through interaction, and sometimes something that resides in the structure of organizations and social networks.

This may seem strange, but if physicists are OK with this particular way of theorizing, then why not gender theorists? At the individual level, gender works internally as a set of values or predispositions that influences our behavior. But gender is also something that can be created through interaction and become deeply embedded in the functioning of our institutions and societies. Perceiving gender this way acknowledges that for any gendered phenomenon, there's probably something important going on at all three levels. At the same time, there may be more to learn about certain phenomena at one level or another.

To show how this works, let's pick a specific example of a gendered phenomenon. Research into how men and women use language has demonstrated that, in general, women in the United States tend to apologize more than men, even when they are not directly responsible for whatever event for which they are apologizing (Tannen, 1990). A woman is more likely to apologize for the weather or someone's illness than a man, despite the fact that she did not cause either one. How do we make sense of this research finding? We could go with the battle royal model of how to explain gender phenomena and generate competing explanations of apology behavior from the three different perspectives—individual, interactional, and institutional. Each explanation would sound perfectly legitimate and convincing, but it would still leave us with the problem of what is actually true about why men seem to apologize less than women. From an integrative approach, a better strategy might be to think about how all three levels of gender interact to produce this empirical finding.

From the individual perspective, we might say that the tendency for women to apologize more is part of the social role associated with the status of being a woman in American society. Women learn through the process of gender socialization that apologizing is an important way for women to demonstrate passivity through taking responsibility for things that go wrong. The tendency to apologize more in women and to apologize less in men is considered an internal trait. But how does this predilection for apology *become* internalized? There we might turn to network theory and its emphasis on the way in which very small differences in network structure become intensified throughout the life course. In the homophilous groups of childhood, girls might develop and reinforce this behavior while boys might not. In their networks of kin and neighbors, apologizing is an activity that is reinforced for women. In the world of work and other organizations, apologizing is not reinforced or rewarded and so men become less likely to apologize.

This would be a good explanation of how this works at the institutional or structural level. But exactly how does that process happen? How do these day-to-day interactions produce these gendered behaviors? An interactional theory might be best suited to answering these particular questions. A man doesn't apologize when it starts raining because to do so would not be accountable as masculine within the context of that interaction, according to the doing gender perspective. This repeated need to produce actions that are accountable as gender then reinforces the original behavior, helping the need to apologize or not to feel like an internal trait, and so on. This example demonstrates how we can begin to perceive these different levels at which gender operates as interconnected and complementary, rather than as contradictory and incompatible.

One advantage of integrative theories that the previous example demonstrates is that they help to bridge the gap between structure and agency, or the macro and micro, in sociology. The root of this problem goes back to C. Wright Mills's question of public issues and private troubles. What is the connection between what we do as individuals and the larger social structures that exist in society? We will discuss throughout this book how gender influences a wide variety of decisions we make on a daily basis. An argument strong on the structural or macro end of this spectrum would be that those decisions are always constrained and limited by social structure. If you're a man in the United States, you didn't get up this morning and make a decision about what to wear. You probably made a decision about what to wear drawing from the subset of clothing that is deemed acceptable for your gender, as well as for your class, sexual orientation, and perhaps race. A whole set of clothing options (skirts, halter tops, lacy underwear, a sari, a cravat, etc.) were already effectively off limits for you. So, then, is that really a choice?

An argument on the micro or agency end of the spectrum would argue that you're still a unique individual with the ability to choose the sari or the skirt if you really want and that those choices have important implications for changing the larger structure. The social structures of society, after all, are dependent on the support of millions of individuals like you making decisions about what to wear in the morning. In trying to explain why men apologize less, the individual approach emphasizes the choices we make. Social roles become part of who we are, and so our decisions as individuals are important to the way in which gender works. Interactional theories move to the next level, somewhere on the spectrum between the macro and micro. Yes, we make choices, but those choices are constrained by the need to be accountable; they're constrained by the people around us and the ways in which we gear our actions toward them as an audience. In the institutional approach, you are a product of the network of relationships in which you are embedded. Gender is produced as a result of homophily, a process you as an individual have little control over, and this puts network theory on the macro or structural end of the spectrum. By looking at gender on all of these levels, integrative theories can help us develop a more sophisticated sense of the give-and-take relationship between our individual actions and the social structures that make up gender.

The second common element in integrative approaches is that they are generally also intersectional. In addition to attempting to bring together the different gender theories, they also acknowledge the importance of the ways in which gender overlaps with race, class, sexual orientation, and other identities. There is still some debate as to exactly how intersectionality fits into the larger theoretical agenda of studying gender. Some theorists argue that not all research needs to be intersectional and that there is value in studying various social structures, such as heterosexism, separately from race, class, or gender (Risman, 2004). Others argue that an intersectional approach does not imply that every research project should focus on all identities equally; rather, it is important to acknowledge at all times the many identities at play in whatever group you happen to be researching.

CONCLUSION

In the beginning of this chapter, I assured you that there were many good reasons to do the hard work that mastering gender theories involves. You should now begin to consider how different theories of gender can work as lenses, drawing your attention to different aspects of your own life and assisting with the archaeological examination of gender in your particular biography. Hopefully, you've begun to ask yourself questions that are informed by these theories. Does gender feel like a trait or a disposition, something that's an integral part of your identity and the way you think about yourself? How does gender work differently in the different interactions in which you participate? Are the rules or underlying logic of the organizations to which you belong gendered in nature? As we discussed at the beginning, and as we learned through our discussion of intersectionality and integration, the goal is not to pick a winner among the many theories you'll learn; rather, the point is to use what you learn to develop your own unique way of understanding what gender is and what it means to you.

BIG QUESTIONS

- In this chapter, we discussed individual, interactional, and institutional theories. Which level makes the most sense to you in understanding gender, and why? Which level do you think is best at describing the way in which people in your society generally understand gender? Which level is most difficult to understand, and why?

- We discussed many different theories of gender in this chapter, and we explored only those from a sociological perspective. Why do you think there are so many different theories for explaining gender? Does the wide range of gender theories reflect how difficult gender is as a concept to understand, or does it say more about the group of people who study gender? Is it an advantage or disadvantage to those studying gender that so many different theories exist?
- In Chapter 1, you learned about sexual dimorphism and the different ways in which gender theorists approach this question. How do the theories in this chapter deal with the question of sexual dimorphism? For each theory, does it seem to rest on the idea that there are two different types of bodies, male and female? How important does a belief in objectively real sexes seem to each theory?
- Doing gender as a perspective has been used to explain the categories of social class and race as well as gender. Can any of the other theories in this chapter also be used to explain other categories of difference and inequality, such as race, social class, sexuality, and disability? Do some of the theories seem more flexible than others in being used to explain these other categories? Does this flexibility reflect a strength or weakness of the theory? Should we be able to develop one theory that explains all these different categories? What might that theory look like?
- One central issue raised in this chapter and in Chapter 1 is the diversity of experiences with gender and whether it makes sense to focus our attention on gender when what it means to be gendered is so varied historically and cross-culturally. Is it worthwhile to discuss gender and assume there is something common about that concept across times and places, or are we focusing on a concept (gender) that has no real, consistent meaning?

GENDER EXERCISES

1. For each theory described in this chapter, identify how that perspective would answer the following questions: (a) What is gender? (b) Why does gender inequality exist? (c) What would be the best method for reducing gender inequality? Then think about what your own theory of gender would look like and how it would answer these three fundamental questions. Write a paragraph describing your own theory, and then explain how it could be used to explain a specific example of how gender matters in your own life.
2. Pick one or more of the theories described in this chapter. Then pick a scene from a favorite movie or episode from a television show that involves gender. How could you use this theory to explain the role of gender in that particular scene or episode?
3. Interview several friends, coworkers, or family members about their views on gender. Try to ask questions that help your interview respondent explain their own particular theory of gender. Then think about how the theory outlined by your interview respondent lines up with the theories discussed in this chapter. How does the particular identity of the people you interviewed seem to impact their own particular theory of gender? For example, do older people tend to have different theories of gender than younger people?

Are women's and men's theories different? College students compared with those who are in the workplace?

4. Make a list of the important aspects of what it means to you to be gendered, including how you think, whom you interact with and how, what you wear, what you do or don't do, and so on. For each item on the list, does this aspect of what it means to you to be gendered apply to every culture or time period? In other words, do you think any of the items on your list are universal (true for everyone in that gender category, everywhere, and in every time period), or are they all specific to your particular culture, time period, and social background? Are they true for people in different racial groups, different social classes, different sexualities, or different nationalities?

5. Try typing *feminist* and *feminism* into a news search engine. What kinds of articles and news stories do you find on these topics? Do feminists seem to be engaged in a great deal of activity, and what kinds of issues do feminists and feminist organizations seem to be focusing on? What does this suggest about the current state of feminism? Do you see evidence that feminism is in social movement abeyance, as discussed in the chapter? Do the stories seem to reflect the influence of radical feminism, liberal feminism, or some combination of the two?

6. Interview a group of friends, coworkers, or family on their views about feminism. You might ask them what they think feminism is, whether they consider themselves a feminist, whether they think being a feminist is a good or bad thing, whom they know (personally or not) who is a feminist, and what they know about the history of the feminist movement or current feminist groups. What do their answers tell you about the history of feminism and the current state of feminism? Why do people tend to identify or not identify as feminists? Do there seem to be differences in the way your interview respondents think about feminism based on their age or other factors (education, race, social class, etc.)? How do their responses compare with what you read about feminism in this chapter?

TERMS

confirmation bias

men as a proxy

master frame

privilege

social movement abeyance

social movement cycle

liberal feminism

master frame of equal rights

radical feminism

consciousness-raising

private troubles

public issues

fundamental attribution error

sociological imagination

individual

social role

sex role
instrumental
expressive
master statuses
sex categorization
performance expectation
status characteristics
gender status belief
self-fulfilling prophecy
ethnomethodology
breach
sex assignment
accounts
accountability
allocation
social aggregates
gendered organization
organizational logic
job
ego
alters
ego network
size
density
diversity
homophilous
multiple consciousness
matrix of domination

SUGGESTED READINGS

On Institutional Approaches

Acker, J. (1990). Hierarchies, jobs, bodies: A theory of gendered organization. *Gender & Society, 4* (2), 139–158.

On Intersectionality

Basu, A. (1995). *The challenge of local feminisms: Women's movements in global perspective.* Boulder, CO: Westview Press.
Collins, P. H. (1990). *Black feminist thought.* New York, NY: Routledge.
Collins, P. H., & Andersen, M. L. (1993). *Race, class, and gender.* Belmont, CA: Wadsworth.
Lorber, J. (1994). *Paradoxes of gender.* New Haven, CT: Yale University Press.
Risman, B. J. (2004). Gender as a social structure: Theory wrestling with activism. *Gender & Society, 18* (4), 429–450.

On Third Wave Feminism

Moraga, C., & Anzaldúa, G. (1981). *This bridge called my back: Radical writings by women of color.* New York, NY: Kitchen Table Press.
Walker, R. (1995). *To be real: Telling the truth and changing the face of feminism.* New York, NY: Anchor Books.

On Doing Gender

Kessler, S. J., & McKenna, W. (1978). *Gender: An ethnomethodological approach.* New York, NY: Wiley.

West, C., & Fenstermaker, S. (1993). Power, inequality and the accomplishment of gender: An ethnomethodological view. In P. England (Ed.), *Theory on gender/feminism on theory* (pp. 151–174). New York, NY: Aldine de Gruyter.

West, C., & Fenstermaker, S. (1995). Doing difference. *Gender & Society, 9,* 8–37.

West, C., & Zimmerman, D. H. (1987). Doing gender. *Gender & Society, 1,* 125–151.

WORKS CITED

Acker, J. (1990). Hierarchies, jobs, bodies: A theory of gendered organization. *Gender & Society, 4* (2), 139–158.

Acker, J. (1998). The future of "gender and organizations": Connections and boundaries. *Gender, Work & Organization, 5,* 195–206.

Acker, J. (1999). Gender and organizations. In J. S. Chafetz (Ed.), *Handbook of the sociology of gender* (pp. 177–194). New York, NY: Springer Science & Business Media.

Aries, E. (1996). *Men and women in interaction.* New York, NY: Oxford University Press.

Baron, J. N., Hannan, M. T., Hsu, G., & Kocak, O. (2007). In the company of women. *Work & Occupations, 34* (1), 35–66.

Basu, A. (1995). *The challenge of local feminisms: Women's movements in global perspective.* Boulder, CO: Westview Press.

Beauvoir, S. de. (1952). *The second sex.* New York, NY: Bantam.

Booth, A. (1972). Sex and social participation. *American Sociological Review, 37,* 183–191.

Bordo, S. (2003). *Unbearable weight: Feminism, western culture, and the body.* Berkeley: University of California Press.

Brewer, M., & Lui, L. (1989). The primacy of age and sex in the structure of person categories. *Social Cognition, 7* (3), 262–274.

Bunker Whittington, K., & Laurel, S. D. (2008). Women inventors in context: Disparities in patenting across academia and industry. *Gender & Society, 22,* 194–218.

Burn, S. M., Aboud, R., & Moyles, C. (2000). The relationship between gender social identity and support for feminism. *Sex Roles, 42,* 1081–1089.

Campbell, K. E. (1988). Gender differences in job-related networks. *Work and Occupations, 15,* 179–200.

Cho, S. Crenshaw, K. W., & McCall. L. (2013). Toward a field of intersectionality studies: Theory, application and praxis. *Signs, 38,* 785–810.

Chodorow, N. (1978). *The reproduction of mothering.* Berkeley: University of California Press.

Collins, P. H. (1990). *Black feminist thought.* New York, NY: Routledge.

Collins, P. H., & Andersen, M. L. (1993). *Race, class, and gender.* Belmont, CA: Wadsworth.

Connell, R. (2002). *Gender: Short introductions.* Malden, MA: Blackwell.

Cowan, G., Mestlin, M., & Masek, J. (1992). Predictors of feminist self-labeling. *Sex Roles, 27,* 321–330.

Davis, A. (1971). Reflections on the black woman's role in the community of slaves. *Black Scholar, 3* (4), 2–15.

Dunn, D., Almquist, E. M., & Chafetz, J. S. (1993). Macrostructural perspectives on gender inequality. In P. England (Ed.), *Theory on gender/feminism on theory* (pp. 69–90). New York, NY: Aldine de Gruyter.

Ferree, M. M., Lorber, J., & Hess, B. (1999). *Revisioning gender.* Thousand Oaks, CA: Sage.

Friedan, B. (1963). *The feminine mystique.* New York, NY: W. W. Norton.

Garfinkel, H. (1967). *Studies in ethnomethodology.* Englewood Cliffs, NJ: Prentice-Hall.

Gillies, R. (2011, May 27). *NBC News.* Retrieved April 16, 2013, from Mom defends keeping baby's gender secret: http://www.nbcnews.com/id/43199936/ns/health-childrens_health/t/mom-defends-keeping-babys-gender-secret/#.UW2knKJweSo

Gilligan, C. (1982). *In a different voice: Psychological theory and women's development.* Cambridge, MA: Harvard University Press.

Granovetter, M. (1974). *Getting a job: A study of contacts and careers.* Cambridge, MA: Harvard University Press.

Hamilton, R. (2007). Feminist theories. In C. D. Bryant & D. L. Peck (Eds.), *21st century sociology: A reference handbook* (Vol. 2, pp. 43–53). Thousand Oaks, CA: Sage.

Heritage, J. (1987). *Garfinkel and ethnomethodology.* Cambridge, England: Polity Press.

hooks, b. (1984). *From margin to center.* Boston, MA: South End.

House, J. S., Umberson, D., & Landis, K. R. (1988). Structures and processes of social support. *Annual Review of Sociology, 14,* 293–318.

Houvaras, S., & Carter, J. S. (2008). The F word: College students' definitions of a feminist. *Sociological Forum, 23* (2), 234–256.

Huddy, L., Neely, F. K., & LaFay, M. R. (2000, Fall). The polls—Trends: Support for the women's movement. *Public Opinion Quarterly, 64* (3), 309–350.

Hunt, V. H., & Zajicek, A. M. (2008). Strategic intersectionality & the needs of disadvantaged populations: An intersectional analysis of organizational inclusion and participation. *Race, Gender & Class, 15*(3–4), 180–203.

Hurtado, A. A. (1989). Relating to privilege: Seduction and rejection in the subordination of white women and women of color. *Signs, 14*(4), 833–855.

Ibarra, H. (1992). Homophily and differential returns: Sex differences in network structure and access in an advertising firm. *Administrative Science Quarterly, 37,* 422–447.

Kanter, R. M. (1977). *Men and women of the corporation.* New York, NY: Basic Books.

Keaney-Mischel, C. L. (2008, July 31). *Gender politics at the UN: How far have we come and where are we headed?* Paper presented at the annual meeting of the American Sociological Association, Sheraton Boston and the Boston Marriott Copley Place, Boston, MA. Retrieved June 6, 2010, from http://www.allacademic.com/meta/p243187_index.html

Kessler, S. J., & McKenna, W. (1978). *Gender: An ethnomethodological approach.* New York, NY: Wiley.

Lorber, J. (1994). *Paradoxes of gender.* New Haven, CT: Yale University Press.

Lorde, A. (1984). *Sister outsider.* Trumansburg, NY: Crossing.

Marsden, P. V. (1987). Core discussion networks of Americans. *American Sociological Review, 52* (1), 122–131.

McCabe, J. (2005). What's in a label? The relationship between feminist self-identification and feminist attitudes among U.S. women and men. *Gender & Society, 19* (4), 480–505.

McPherson, J. M., & Smith-Lovin, L. (1986). Sex segregation in voluntary associations. *American Sociological Review, 51,* 61–79.

McPherson, J. M., Smith-Lovin, L., & Cook, J. M. (2001). Birds of a feather: Homophily in social networks. *Annual Review of Sociology, 27,* 415–436.

Mills, C. W. (1959/2000). *The sociological imagination.* Oxford, England: Oxford University Press.

Moraga, C., & Anzaldúa, G. (1981). *This bridge called my back: Radical writings by women of color.* New York, NY: Kitchen Table Press.

Newman, D. M. (2004). *Sociology: Exploring the architecture of everyday life.* Thousand Oaks, CA: Sage.

Parsons, T., & Bales, R. F. (1955). *Family, socialization, and interaction process.* New York, NY: Free Press.

Peltola, P., Milkie, M. A., & Presser, S. (2004). The "feminist" mystique: Feminist identity in three generations of women. *Gender & Society, 18* (1), 122–144.

Rich, A. (1979). Disloyal to civilization: Feminism, racism, gynophobia. In A. Rich (Ed.), *On lies, secrets, and silence* (pp. 275–281). New York, NY: W.W. Norton.

Ridgeway, C. L. (1993). Gender, status, and the social psychology of expectations. In P. England (Ed.), *Theory on gender/feminism on theory* (pp. 175–197). New York, NY: Aldine de Gruyter.

Ridgeway, C. L. (1997). Interaction and the conservation of gender inequality: Considering employment. *American Sociological Review, 62,* 218–235.

Ridgeway, C. L., & Smith-Lovin, L. (1999). The gender system and interaction. *Annual Review of Sociology, 25,* 191–216.

Risman, B. J. (1998). *Gender vertigo: American families in transition.* New Haven, CT: Yale University Press.

Risman, B. J. (2004). Gender as a social structure: Theory wrestling with activism. *Gender & Society, 18* (4), 429–450.

Rupp, L. J. (2006). Is the feminist revolution still missing? Reflections from women's history. *Social Problems, 53,* 466–472.

Smith-Lovin, L., & McPherson, J. M. (1995). You are who you know: A network approach to gender. In P. England (Ed.), *Theory on gender/feminism on theory* (pp. 223–251). New York, NY: Aldine de Gruyter.

Snow, D. A., & Benford, R. D. (1992). Master frames and cycles of protest. In A. D. Morris & C. M. Mueller (Eds.), *Frontiers in social movement theory* (pp. 133–155). New Haven, CT: Yale University Press.

Stangor, C., Lynch, L., Duan, C., & Glass, B. (1992). Categorization of individuals on the basis of multiple social features. *Journal of Personality and Social Psychology, 62,* 207–218.

Stryker, S. (1987). Identity theory: Developments and extensions. In K. Yardley & T. Honess (Eds.), *Self and identity: Psychological perspectives* (pp. 89–103). Chichester, England: Wiley.

Tannen, D. (1990). *You just don't understand: Women and men in conversation.* New York, NY: Morrow.

Taylor, V., Whittier, N., & Pelak, C. F. (2004). The women's movement: Persistence through transformation. In L. Richardson, V. Taylor, & N. Whittier (Eds.), *Feminist frontiers* (pp. 515–531). Boston, MA: McGraw-Hill.

Walker, R. (1995). *To be real: Telling the truth and changing the face of feminism.* New York, NY: Anchor Books.

Ward, J. (2004). "Not all differences are created equal": Multiple jeopardy in a gendered organization. *Gender & Society, 18* (1), 82–102.

Wellman, B. (1985). Domestic work, paid work and net work. In S. Duck & D. Perlman (Eds.), *Understanding personal relationships* (pp. 159–191). London, England: Sage Ltd.

West, C., & Fenstermaker, S. (1993). Power, inequality and the accomplishment of gender: An ethnomethodological view. In P. England (Ed.), *Theory on gender/feminism on theory* (pp. 151–174). New York, NY: Aldine de Gruyter.

West, C., & Fenstermaker, S. (1995). Doing difference. *Gender & Society, 9,* 8–37.

West, C., & Garcia, A. (1988). Conversational shift work: A study of topical transition between women and men. *Social Problems, 35,* 551–575.

West, C., & Zimmerman, D. H. (1987). Doing gender. *Gender & Society, 1,* 125–151.

Wilkes, D. (2012, January 20). *The Daily Mail.* Retrieved April 16, 2013, from Boy or girl? The parents who refused to say for FIVE years finally reveal sex of their "gender-neutral" child: http://www.dailymail.co.uk/news/article-2089474/Beck-Laxton-Kieran-Cooper-reveal-sex-gender-neutral-child-Sasha.html

Williams, R., & Wittig, M. A. (1997). "I'm not a feminist, but . . ." Factors contributing to the discrepancy between pro-feminist orientation and feminist social identity. *Sex Roles, 37,* 885–904.

Wimsatt, W. U. (2001). *Bomb the suburbs.* New York: Soft Skull Press.

Zimmerman, D. H. (1978). Ethnomethodology. *American Sociologist, 13,* 6–15.

Zinn, M. B., Cannon, L. W., Higginbotham, E., & Dill, B. T. (1986). The costs of exclusionary practices on women's studies. *Signs: Journal of Women in Culture and Society, 11,* 290–303.

Zinn, M. B., & Dill, B. T. (1996). Theorizing difference from multiracial feminism. *Feminist Studies, 22* (2), 321–331.

CHAPTER 3

How Do Disciplines Outside of Sociology Study Gender?

Some Additional Theoretical Approaches

Gender is a big concept, as we've just begun to discover, and it has important implications for your life in a wide variety of ways. You might think about gender in Anglo-European society up until the point of the second wave feminist movement as being invisible, the elephant in the room that no one was talking about. Or maybe even something bigger, like a whale, and being ignored in not just one room, but in all rooms, everywhere. With the advent of the feminist movement, a lot of people began to notice the rather large whale in their presence, and you might think about the world since then as an explosion of people thinking about, writing about, and experimenting with gender. This has happened in many different disciplines and areas of life, so the study of gender is an inherently interdisciplinary one. This means it takes not just the perspective of sociology, but many different perspectives to understand fully how gender works.

Many of the women's studies departments that were formed during the second wave of the feminist movement on college and university campuses are composed of more than one discipline, and gender studies departments continue to be interdisciplinary. For my women's studies program as an undergraduate, I took courses from English professors, sociologists, and classicists, just to name a few. Just as we discussed in Chapter 2 that many different theories help to give us a full understanding of what gender is and how it works, different disciplines are also important to getting a full picture of this phenomenon called gender. Although sociology will be the primary framework we'll be using in our discussion of gender, it certainly wasn't the first, or only, discipline on the gender scene. In this chapter, we will give a brief overview of some theoretical approaches from disciplines outside of sociology, paying special attention to those disciplines we'll be drawing on throughout the book. The list of disciplines that could have been included in this discussion is endless. For example, the connections between art and gender, architecture and gender, gender and the practice of medicine, or gender and history are all fascinating subjects I encourage you to explore. However, in this chapter, we will explore approaches in psychology and anthropology, and then we will look at the more recent theoretical areas of queer theory: globalization, development, and the environment.

PSYCHOLOGICAL APPROACHES TO GENDER

In the previous chapter, we discussed how theorists using an interactional approach to gender argue that interactional theories are more sociological in nature because they focus on groups rather than on individuals. This is a broad and general distinction to make between sociology and psychology, two disciplines that frequently overlap. In fact, classes in social psychology—a particular subdiscipline of sociology or of psychology, depending on whom you ask—are as likely to be taught in sociology departments as they are in psychology departments. In general, though, psychologists are more interested in individuals and the processes that go on in the minds and brains of individuals than are sociologists. This is not to say that sociologists are not also interested in these matters or that psychologists are not also interested in more macro phenomena like interaction, organizations, and institutions. But in the interest of teasing out the differences between these two approaches, we can say that psychology leans more in the direction of the individual, whereas sociology leans more strongly in the direction of groups of individuals. For this reason, you can assume that the psychological perspectives we'll discuss here fit within the category of individual approaches, which we discussed in Chapter 2.

Freud

Sigmund Freud is important to the discipline of psychology in general, but he is especially important to the study of gender. As we will see later in our discussion of identification theory, Freud's ideas about gender served as a starting point for some important theories of gender socialization and gender identity that emerged within feminism and postmodern traditions. Freud located the origins of gender early in our childhood development and in the ways in which children unconsciously make sense of anatomical differences. This explains the phrase you might sometimes hear associated with Freud: "Anatomy is destiny." The anatomical differences of boys and girls, and how they make sense of those differences, are fundamental to the formation of gender identity for Freud.

Oedipus complex is the first important concept to Freud's articulation of how men develop a sense of masculine identity. The name comes from a Greek myth in which Oedipus unknowingly kills his father and marries his mother, eventually gouging out his own eyes when he discovers the truth. Luckily, in Freud's model, men do not generally suffer the same fate as Oedipus and lose their eyes, although they do experience similar dynamics. The Oedipal conflict, according to Freud, develops because a young boy's attachment to his mother eventually grows into seeing her as a love object. The boy wants to have sex with his mother and perceives his father as a threat standing in the way of that goal. In addition, the boy sees that women lack a penis and, on an unconscious level, draws the conclusion that penises can be cut off. Boys develop a **castration complex**, a fear that their own fathers will castrate them if they act on their sexual desire toward their mother. The end result of all these dynamics is boys' ultimate rejection of their mother and their sexual desire for her. Men learn to look down on women because they lack a penis and identify with their father; this identification helps them to develop a masculine identity modeled on that of their father. Although this process may sound a bit complicated, for Freud, masculine development is less problematic and generally has a more satisfactory resolution than feminine development.

Women's psychological development is again based in anatomy, and their unconscious dynamics are based in early childhood. A little girl comes to realize at some point in her development that she lacks a penis. She sees that her mother also lacks a penis and, therefore, places the blame for her own lack of a penis on her mother. As a result of these dynamics, women eventually come to envy men's possession of a penis and, like men, disdain their mother and all other women because of their lack of a penis. This is, of course, the explanation for the famous concept of **penis envy**, which has made its way into common cultural usage. Girls and eventually women come to see other women as competitors in the ultimate goal of possessing their own penis; Freud called this the **Electra complex**. Electra was another figure out of Greek myth—a woman who persuaded her brother to murder her mother. Women can resolve the Electra complex when they come to identify with their mother as a symbolic means of gaining access to their father.

For women, the successful resolution of the Electra complex and penis envy is to become resigned to possessing a penis through possessing a man or, perhaps, having a baby. Freud felt that this made women's development much more problematic than that of men, in part because women's superego never develops the strength and independence of men's (Freud, 1933). The **superego** in Freud's theory is the part of our personality responsible for both our own morality and for lining our morality up with societal standards. The end result of women's underdeveloped superego is narcissism, where *being* loved is more important than *loving*. Women have a less well-developed sense of justice and a harder time suppressing their instincts, both because of their inferior superego.

Where would you put Freud on the spectrum between biosocial approaches to gender and strong social constructionist approaches to gender? Why would you put him there?

Freud's theory of gender development might sound a bit strange to our modern ears, and it's probably not at all difficult for you to understand that more recent gender theorists have serious problems with Freud's theory of gender. Women don't fare so well in Freud's framework, and one critique is that Freud assumes men as the norm and then focuses on how women deviate from that norm. His bias is demonstrated in his focus on what men have (a penis) and what women lack. Freud's focus on the penis is the origin of another term that made its way into popular culture: **phallic symbols**, or any abstract representation of the penis. You might argue that Freud works from the assumption that men are better than women and then develops a theory to explain that conclusion. In addition,

Freud locates gender development in the dynamics of early childhood and the unconscious. He assumed these dynamics were universal, that feminine and masculine personalities should follow his predictions everywhere and in all time periods. There is very little evidence that any of the personality traits Freud identified as masculine or feminine are universally "feminine" or "masculine." It is also unclear whether children everywhere identify with the same sex parent in the way Freud predicts.

CULTURAL ARTIFACT 1: GUITARS, PHALLIC SYMBOLS, AND THE GENDER OF MUSICAL INSTRUMENTS

The growing list of controversies over Super Bowl halftime performances has some perhaps not so surprising connections to gender. Before Justin Timberlake revealed Janet Jackson's nipple, pop musician Prince generated his own controversy in creating a rather striking image of a guitar as a phallic symbol. With his shadow projected against a large, billowing curtain, the neck of the pop icon's guitar, shaped like his symbol, emerged from between his legs in an image that Freud would certainly have recognized as phallic. This image merely brings to the forefront suggestions that guitars are, in fact, a kind of phallic symbol. Rock music, with its roots planted firmly in the Black Southern blues tradition in the United States, is a sexualized musical form to begin with. Combined with the image of Elvis Presley, John Lennon, or Chuck Berry with a guitar hanging roughly between their legs, it doesn't take much to see how guitars can be understood as phallic symbols. Does this explain why so few women and girls play guitar? If you participated in band in school, you may have noticed a distinct gendering to who played which musical instrument. In the United States, girls play predominately flutes, violins, and clarinets while boys play drums, trumpets, and trombones (Abeles, 2009). Studies show little changes in these preferences over the past 30 years. A similar study in the United Kingdom focused on musical education provided in schools (Geoghegan, 2008). The most popular instruments among boys were the electric guitar, bass guitar, tuba, kit drums, and trombone. For girls, the top instruments included harp, flute, voice, fife/piccolo, and oboe. Julianne Regan, lead singer and bass guitarist for the group All About Eve said, "It seemed like a freakish thing (bass guitar) for me to be interested in. I was quite popular at school and had a load of friends, but this was just seen as 'one of my little quirks'." Female bass guitarist Donna Dresch, who played with Dinosaur, Jr., noted that the bass guitarist is sometimes considered the "submissive quiet person holding a song together without much notice"(p. 75), so perhaps the bass is appealing because it meshes with stereotypically feminine qualities in this way (Dresch, 1995). On the other hand, Dresch noted that, for her, the bass is "heavy, loud and powerful" (p. 75), the exact opposite of what girls are supposed to be. More sociological reasons suggest that women often play bass in rock bands because the bass is easier to learn in a short amount of time, and women come to play rock music much later than do men. Reagan had to teach herself to play bass guitar, and none of her peers at an all-girl high school were interested in the electric guitar. In one study, women in rock bands first began to play their instrument at the average age of 19, joining their first band at the median age of 21. In contrast, men began to play their instrument at the average age of 13, and they joined their first band at the median age of 15 (Clawson, 1999). Research also suggests that more women play bass because of a vacancy created by the allure of the lead guitar for male rock musicians, leading to a form of occupational segregation by sex within rock music itself. In what ways are other musical instruments gendered, and which of those instruments are phallic? What other gender trends might you notice in the area of rock music?

Not many gender theorists still use Freud's theory of development in the exact form we've laid out here. But many of Freud's ideas continue to have important impacts. Freud focused his theory of gender on early childhood development, and we'll see in the next chapter how subsequent theories of gender socialization also prioritize these early years. Psychoanalytic theories of gender build explicitly on some of the early childhood dynamics identified by Freud. Freudian theory serves as the starting point for some feminist literary theory and is an important aspect of some postmodern theories—including queer theory, which we'll discuss later. Psychoanalysis, the particular therapeutic approach that Freud developed, is still practiced by many therapists; and the idea of the unconscious, which was so central to Freud's theorizing, remains an important concept for us today. Although many dismiss some of the specific dimensions of Freud's theories, his formulations remain a reference point for many gender theorists, and many of his ideas (including the unconscious) have become taken-for-granted ways of thinking about the world.

Sex Differences Research

We fast forward here from Freud's work on gender in the early part of the 20th century to the interest in sex differences research that began during the 1970s in psychology. In the previous chapter, we discussed how the women's movement in the United States and Europe began to influence the academic discipline of sociology. The same was true for psychology during the period of second wave feminism, and this influence was exemplified in a new concern with research on sex differences. **Sex difference research** stands clearly within the individualist approach because its concern is with identifying whether a wide range of traits, dispositions, and behaviors differ significantly between men and women. Sex difference researchers are interested in how gender operates from the inside out in classic individualist fashion.

Entire books have been written that summarize the huge volume of findings on sex differences (Deaux & Major, 1990; Eagly, 1987; Epstein, 1988; Maccoby & Jacklin, 1974); in addition, countless articles have been published. Given that amount of research, it's not surprising that a great many kinds of differences between men and women have, in fact, been recorded. In their seminal book on the subject of sex differences, Eleanor Maccoby and Carol Jacklin (1974) concluded that differences in cognitive abilities between the sexes can be grouped into three domains: mathematical abilities, verbal abilities, and visuospatial abilities. For example, in terms of visuospatial abilities, women are better at decoding facial expressions, whereas men are better at visuospatial tasks that involve aiming, such as throwing a ball (Halpern, 1996).

Years of subsequent research have sometimes confirmed these general findings and sometimes have contradicted them. For example, you may have heard of research demonstrating that men have better mathematical abilities than women. This often-cited difference is based primarily on a study of SAT scores—and so measures men's and women's performance on a standardized test (Benbow & Stanley, 1980, 1983). The study was biased in that it used a sample of 50,000 high-ability students, but can you imagine other problems with using performance on standardized tests to measure mathematical ability? Not surprisingly, the more math classes you take, the better you are likely to do on the math portion of the SAT, and studies have shown that boys are encouraged to take math classes in school to a much greater extent than are girls (Buchmann, DiPrete, & McDaniel, 2008; Caplan & Caplan, 1997). Boys' better performance on the SAT may therefore reflect a social pressure to take more math classes rather than any inherent superiority in mathematical ability over that of girls. Other research has demonstrated a great deal of global variation in men's and women's mathematical abilities; research suggests that although boys perform slightly better than girls in France and Israel, girls and boys perform equally well in England and Japan, whereas girls outperform boys in Iceland, Finland, and Thailand (Baker & Jones, 1993; Yucel, 2007).

To Research or Not to Research?

This very brief examination of one finding should begin to highlight some problems with this body of research and with what conclusions we can or should draw from it. It is important to note here the original goal of many of the researchers involved in sex difference research. Early researchers were motivated by the growing women's movement, and so part of their goal was to challenge many of the unfavorable stereotypes of women that existed at the time (Eagly, 1995). These researchers felt that sex difference research would demonstrate the absence of any essential differences between women and men and that this would help in the goal of reducing inequality. You should notice that this is a goal consistent with the orientation of liberal feminism as we discussed it in Chapter 2. Liberal feminism holds that men and women are essentially the same and any sense of difference is created by laws and other societal structures that prevent women from having access to the same opportunities as men. Sex difference research, then, would demonstrate the lack of any important differences and demonstrate that these differences are created by the structures of society, structures that can be changed.

But this was not exactly the way things went. The most important findings from Maccoby and Jacklin's (1974) book were that there were both fewer differences between women and men than many people had expected and that the magnitude of those differences was relatively small (Hyde, 2005). In other words, their main findings were that the differences weren't that many and weren't that great. And in fact, this is what most of the research into sex differences has consistently demonstrated. For even the most powerful findings regarding sex differences, the differences are described in terms of averages—average SAT scores or average scores on various types of experiments and tests performed by researchers. What does an average difference tell us? Because averages are based on the entire range of all the scores, it means that there's a great deal of overlap for any average difference. A sex difference is considered large when 53% of the scores for any given test overlap. That means that for the very largest sex differences, greater than 53%, or most, of the scores between men and women still overlap. One question posed by many of those who are critical of sex difference research is simply this: Why do we seem to be infusing so much importance into what is a relatively small difference? The size of sex differences is the first problem to consider in the research on sex differences.

We'll come back to the question of why researchers continue to be interested in relatively small differences, but first let's point to another problem with sex differences research. The second problem relates to how sex differences stand up when they are examined across age and ethnicity, as well as across different societies and social contexts (Crawford & Chafin, 1997). For a sex difference truly to stand up as a sex difference, and therefore as rooted in some biological phenomenon, it should be true across many different age groups, racial groups, cultures, and social situations. To address these questions, psychologists developed the technique of **meta-analysis** (Hyde & Linn, 1986). In this technique, researchers use the results from many published studies on sex differences to obtain an average difference based on a larger and more diverse sample size (the combined samples of all published studies).

This technique only solves the underlying problem of sample composition if the underlying samples are diverse, and many times they are not in terms of race, class, or educational experience. For example, any research based on standardized test scores like the SAT will include only those American students planning on attending college, or those able to afford the increasingly expensive fee for taking the test. This can be important because some research suggests that boys with poorer verbal skills tend not to take the SAT, and this may explain why girls outperform boys on the SAT in this area. In addition, some research has demonstrated that the social context of the test or experiment itself matters. For example, some research has demonstrated that women score differently on tests of visuospatial ability based on how much the spatial nature of the task is emphasized to the test-takers (Sharps, Welton, & Price, 1993). If women's knowledge of the nature of the test affects their performance on it, then this suggests that something besides their biological sex is affecting the results. For sex differences to be important, they should be consistent across different racial, class, and national groups, and the evidence that this is so is contradictory and contentious.

One final possible critique of sex difference research is with the conceptualization of sex differences in general. We outlined three different areas of possible differences between women and men: mathematical abilities, visuospatial abilities, and verbal abilities. But how exactly are these conceptualized, and what gender biases might be hidden within these conceptualizations (Galliano, 2003)? The list of abilities included as visuospatial includes reading a map, completing jigsaw puzzles, spatial perception, and the mental rotation of three-dimensional objects (Caplan & Caplan, 1997). Do these different abilities really have that much in common, and do they translate into real-world implications? For example, how does the ability to rotate mentally a three-dimensional object, assuming this is a sex difference for women and men, matter in the course of our day-to-day lives outside the context of an IQ test? In addition, the conceptualization of these concepts often seems to reflect a subtle gender bias. Why should visuospatial skills not include the ability to plan a garden, arrange furniture in a room, or use patterns to make clothing—skills for which women would probably be expected to outperform men (Galliano, 2003)?

This brings us back to the original goal of sex differences research—to demonstrate underlying similarities between women and men—and the direction this body of research actually took. Why is sex difference research such a booming business given the problems we've outlined? There are several different ways to answer this question, and there is much disagreement about whether sex difference research should continue to be pursued at all by social science researchers. The size of sex differences discovered by researchers is generally small, but the structure of both science and public interest means that we focus less attention on the small size of those differences and more attention on the differences themselves (Hyde, 2005). In the world of scientific research, where getting published is important to a successful career, negative findings are never as interesting or publishable as positive findings. Say that, as a psychologist, you hypothesize that there are sex differences in women's and men's ability to play chess (and, yes, there is research out there on chess ability) (Chabris & Glickman, 2006). You put all the time, money, and energy into setting up and running this experiment. Negative findings would mean you find that your hypothesis was wrong, that there are no important sex differences between women and men in their ability to play chess.

How exciting is that? If you've ever done a research project yourself, you may have had this experience. Besides being personally disappointing, negative findings are difficult to publish in academic and scientific journals, and having articles and books published are important ways in which academics and researchers are evaluated in their jobs. The research on sex differences in mathematical abilities we discussed was published in *Science*, a fairly prestigious and widely read magazine. Would that research study have been published had the researchers found no real difference between men and women? Probably not. Sex difference research gets published not just in scholarly journals, but it is very frequently picked up by the popular press as well. For some reason, differences seem to be perceived as more interesting to both fellow scientists and the general public. Thus, there is some pressure to emphasize the existence of differences rather than the relatively small size of those differences.

Why do you think differences seem to be more interesting than similarities when it comes to sex and gender? Can you think of other examples or areas of social life where differences are more interesting than similarities? Does establishing differences necessarily make arguments about inequality easier to make?

The Bottom Line on Sex Differences Research

As you may have gathered by now, there is a great deal of debate in this area as to both the quality of the research on sex differences as well as the value of sex difference research.

Researchers continue to argue that the small average differences that have been demonstrated cannot be ignored, whereas others have suggested a more productive area of research would be to investigate gender similarities, rather than sex differences (Epstein, 1988; Hyde, 2005, 2007). Some researchers point to a fundamental bias in the design of this area of research because it completely ignores any differences that exist within sex categories. Why is there not a research agenda focused on uncovering the many ways in which those labeled as women differ from other people labeled as women? Others argue that discovering these very real differences is important because they can have important implications for education and other types of policies that should take these differences into account. For example, Eagly (1995) argued that if there are important visuospatial differences between women and men, programs designed to train individuals on spatial tasks could take into account these differences. On the other hand, those supporting the importance of gender similarities argue that gender difference research can adversely affect women's performance in the workplace, girls' ability to succeed in math, the social support provided to both boys and girls during the equally difficult period of adolescence, and the health of heterosexual relationships (Hyde, 2005).

Whatever you yourself make of the evidence on sex differences, there are four things that are important to take away from this discussion. First, the task of establishing sex differences beyond any doubt is a complex and difficult project, given the problems we've already discussed. Second, it's a project that's always going to have implications beyond the realm of scientific research. These findings can and do have real implications for gender inequality in places like schools and the workplace (Weil, 2007). Some schools have experimented with math classes for girls only, based in part on many of the findings in sex differences research on girls' and boys' different mathematical abilities, whereas other schools segregate boys and girls for a wide variety of subjects.

The third important thing to take away from our discussion of sex difference research is that it is, for the most part, an example of descriptive research. This means that the research tries to identify whether sex differences exist, but it is not as concerned with explaining the reasons for those differences. You may have already noticed that we've been calling this *sex* differences research, rather than *gender* differences research. As you're learning, that kind of terminology matters. Because this research is descriptive rather than explanatory, it's often unclear whether these differences are thought to be a result of *sex,* and therefore rooted in some biological phenomenon, or a result of *gender,* and therefore something that is grounded in social life. Some sex difference research looks at differences in brain anatomy between women and men, and this research would seem to be specifically about *sex* research, rather than about gender research. But even that distinction is not as clear as it might seem. Cognitive psychologists and other scientists who study the human brain find increasing evidence for its plasticity. This means the actual anatomy of our brains can change as a result of the experiences we have throughout our lives. If women and men have different experiences because they live in a very gendered social world, those differences might very well be written into the physical structure of their brains.

Finally, sex difference research raises an important question we will explore throughout this book: What is the connection between difference and inequality? Many of those who believe sex differences research is not a productive or worthwhile area of inquiry base their arguments in part on the belief that searches for difference are the same as searches for inequality. They point to the ways in which many of the findings of sex differences research seem to demonstrate men's superiority in the skills and abilities most valued by society. Differences, according to these researchers, are almost always translated into a power imbalance (Hollander & Howard, 2000). On this side of the argument, similarities can be drawn between sex difference research and the body of research, most of which was conducted in the 18th and 19th centuries, that sought to find biological evidence for racial differences. Few would argue that these studies of skull size and other features of anatomy were not connected to a larger project of demonstrating the racial inferiority of nonwhite races and, therefore, that difference was not connected to inequality. Others argue that difference does not have to lead to inequality if those differences are equally valued by society (Eagly, 1995). Investigating sex differences does not inevitably lead to a story about female inferiority but to an honest acknowledgment of the fundamental ways in which females and males are different from each other. If we value those differences equally, this does not have to lead to gender inequality. This question continues to be debated in the area of sex differences research, and it is a question we will take up again.

ANTHROPOLOGICAL APPROACHES TO GENDER

Like psychology, anthropology has much in common with sociology as a discipline but several broad differences in its focus and methodological approach. The first point to make is that anthropology as a broad discipline studies many aspects of human behavior that sociologists do not include in their field of study. For example, physical anthropologists study early human evolution, often using methods more akin to biology than to sociology. In a broad sense, this is also an examination into a type of human behavior, but the study of nonhuman species and the fossil record is fairly far removed from what sociologists do. On the other hand, social or cultural anthropology, as opposed to physical anthropology, is more closely connected to sociology. Cultural anthropology and sociology ask some of the same questions about societies and how they work. How do societies interact with their physical environments? How do they meet the material needs of their members or the need for food and shelter? What are the prevailing beliefs and values of a society? The differences are in the relative emphasis each discipline places on these questions and in the way in which they go about answering these questions.

Sociology has historically focused less on the question of how societies interact with their environment, perhaps because sociology has focused more on societies within the developed world. This assumes that societies in the developed world are more buffered from their physical environments by technology, an assumption that is proving to be increasingly suspect. The second difference involves how sociologists and anthropologists go about answering the questions they pose about human behavior. Methodologically, ethnography is the main tool of cultural anthropologists. **Ethnography** involves sustained time spent in the field, living among the group you're studying to develop an insider's perspective on the group's culture and society. Finally, cultural anthropologists are generally engaged in the collection of information across a wide range of societies and cultures to answer questions about how they are similar to and different from each other (Mascia-Lees & Black, 2000). Thus, anthropologists are deeply committed to the cross-cultural study of various aspects of society, including gender.

As a result of this cross-cultural emphasis among cultural anthropologists, they are especially well suited to helping answer certain kinds of questions about gender. Specifically, anthropology as a discipline can help us to explore some connections between sex and gender that we have already touched on. If sex differences are based on some underlying biological characteristic of humans, then we should see evidence of that across many different cultures and societies. For example, anthropologists have found many examples of societies in which women's roles are to provide economically for the family, questioning the modern, Anglo-European assumption that it is the natural place of men to be the "breadwinner" (Mascia-Lees & Black, 2000). If anthropologists discover that there are cultural variations both in what traits and characteristics are associated with men and women and in the categories into which people are placed (male and female, as opposed to male, female, and third sex), then this suggests the idea that sex is, like gender, socially constructed. Throughout our global exploration of gender, anthropological research will be used precisely because of these important cross-cultural contributions to our understanding of gender and sex.

Sex and Temperament in Three Primitive Societies

In any discussion of anthropology and gender, it would be difficult not to at least mention the contributions of Margaret Mead. Mead achieved a great deal of fame outside of the world of anthropology in large part because of the arguments she made about sex and gender—arguments that were fairly radical for the time period. Mead was a student of Franz Boas, one of the founders of anthropology as a discipline in the United States during the early part of the 20th century. In his study of Native American cultures in the United States, Boas was deeply invested in using anthropology to demonstrate the importance of *nurture* over *nature* in the development of many human behaviors (Gordon, 1993). When Mead examined gender cross-culturally, she was partly interested in demonstrating the importance of nurture, or culture, in explaining gender differences. The idea that the most essential beliefs about what makes someone masculine or feminine are cultural and therefore unique to our own culture was in 1935, and still is for many people, a rather difficult idea to swallow.

Mead explored gender differences in several different works, including her study of adolescence in Samoa, *Coming of Age in Samoa* (1928). But probably her most famous work on gender is *Sex and Temperament in Three Primitive Societies* (1935). In this ethnography, Mead compared three groups in New Guinea, an island located north of Australia in the Pacific Ocean. As her title suggests, Mead was particularly interested in how these three groups thought about sex categories (male and female) and the particular traits, characteristics, and behaviors associated with those categories.

The first two groups Mead (1935) studied in New Guinea drew very few distinctions between women and men in terms of personality. Among the Arapesh, women and men were not considered opposites but as equally endowed with the cultural qualities of non-aggression and gentleness that were valued in Arapesh culture as a whole. From Mead and her intended audience's Western perspective, the ideal characteristics for both genders in Arapesh society were what Anglo-European society would call "womanly," namely, being "maternal, gentle, responsive and unaggressive" (p. 161). What's more, according to Mead, among neither the Arapesh nor the second group she studied, the Mundugumor, did the culture conceive of important differences based on sex. The Mundugumor, like the Arapesh, ignored "sex as a basis for the establishment of personality differences" (p. 165), but unlike the Arapesh, the ideal cultural traits for the Mundugumor were what Westerners would consider masculine characteristics. Mead described the Mundugumor as "violent, competitive, aggressively sexed, jealous and ready to see and avenge insult, delighting in display, in action, in fighting" (p. 225). In these two groups, the characteristics that were most valued in society differed, but the importance of sex as a way of characterizing people was relatively unimportant as compared with contemporary Anglo-European culture.

The final group in Mead's (1935) book was the Tchambuli, yet another very different culture living on the island of New Guinea. Mead argued that the Tchambuli were a society concerned foremost with artistic endeavors, but what she found most interesting about them was again the way in which they conceived of sex. Mead argued that although Tchambuli society seemed patriarchal on the surface, women were the ones who truly "dominate[d] the scene" (Mead, 1935, p. 270) through their control of economic arrangements in the society. Among the Tchambuli, the woman was the "dominant, impersonal, managing partner," whereas the man was the "less responsible and emotionally dependent person" (Mead, 1935, p. 279). Mead identified this culture as a virtual reversal of the way sex roles are conceived of in Anglo-European society.

From her examination of all three groups, Mead (1935) concluded the following:

> We may say that many, if not all, of the personality traits which we have called masculine or feminine are as lightly linked to sex as are the clothing, the manners, and the form of head-dress that a society at a given period assigns to either sex. (p. 280)

What is sex in Mead's formulation? It is something resembling a hat you can put on or take off, but in the end, it is a result of what she called social conditioning rather than of any innate, biological, or essential differences between men and women. The kind of "clothes" that are required of each sex vary from culture to culture, demonstrating that these cannot be inherent, biological traits. Thus, Mead made a strong case for the role of nurture in determining sex temperament through the unique processes of what we would call socialization (social conditioning) in each culture. Mead's comparison of these three cultures and the very different ways in which they conceived of sex demonstrated the incredible malleability of human nature under the influence of culture.

Since the publication of Mead's various works on gender, criticism has emerged regarding her methods as an ethnographer and whether her end goal (to show the power of nurture over nature) biased her view of the people she studied in important ways (Feinberg, 1988; Freeman, 1983; Shore, 1981). But Mead's work continues to be significant in that it demonstrates one of the central concerns within the anthropological study of gender. In various ways, anthropologists are concerned with the question of the universality of what we call sex. For Mead and many other anthropologists, this involves the cross-cultural examination of sex as it is currently conceived of in different societies. But a related question asks about the universality of male domination in contemporary societies, as well as in our historical past, both human and prehuman. What does our knowledge about the evolution of humanity and about the social world of our closest primate relatives tell us about the truth or falsehood of sex as a category?

Man the Hunter and Woman the Gatherer?

Margaret Mead considerably predates the influence of second wave feminism on anthropology as a discipline, and you might take her own position as well as the prominence of other women anthropologists such as Elsie Clews Parson, Ruth Benedict, and Zora Neale Hurston as evidence of women's relatively superior status within anthropology compared with other disciplines. However, Mead, despite her popularity, never held a regular academic position, and Ruth Benedict, another student of Franz Boas, was passed over to replace him as chair of the department at Columbia University probably in part because she was a woman (di Leonardo, 1991). In the 1970s, when the second wave of the feminist movement in the global North began to have a serious influence on anthropology, feminists began to reconsider some of the basic assumptions of anthropology as a discipline.

Among those was the idea of Man the Hunter and Woman the Gatherer. This may sound vaguely familiar to you because its influence reached far beyond the bounds of anthropology as a discipline and influenced the common images we have of "Cave Men" as prehistoric peoples. These ideas are connected to a particular model of human evolution that emphasizes the importance of bipedalism (walking upright on two feet), which led to early humans' usage of tools to hunt large game animals (Washburn & Lancaster, 1968). Much that makes current humans unique, according to this theory put forward by Sherwood Washburn and Jane Lancaster (1968)—including elements of our basic social life—can be traced back to evolutionary adaptations necessary to be successful hunters. Gender enters the picture because the *man* in **Man the Hunter** is not meant to represent generically all humans; rather, it represents the assumption that the hunters were, in fact, all men. This model of human evolution assumes that a sexual division of labor emerged very early on in human society, where men hunted while women gathered other types of food. More important, **Women as Gatherers** were also responsible for taking care of infants, and this was part of what prevented them from being hunters. This basic arrangement between a man who hunts and a woman who gathers and cares for offspring explains the evolution of the human family as a unit most efficient for the sharing of food and, therefore, survival of the species as a whole.

CULTURAL ARTIFACT 2: CAVE MEN AND CAVE WOMEN

The image of the Cave Man is an immensely popular one in media representations. Since World War II, at least one movie, cartoon, or television show featuring a Cave Man has been introduced every year (Berman, 1999). A small sampling of movies that involve a Cave Man include *SQuest for Fire* (1981), *The Lost World* (1925), *Teenage Caveman* (1958), and *2001: A Space Odyssey* (1968). Cartoons include *B.C.*, *Alley Oop*, and occasionally *The Far Side*, and from television, *The Flintstones* and *Captain Caveman.* More recently, a *Saturday Night Live* skit that aired in the 1990s featured Phil Hartman as the Unfrozen Caveman Lawyer and an insurance company, Geico, used Cave Men in a series of advertisements to demonstrate how easy it is to save money with its insurance. The image of the Cave Man is easily recognizable and remarkably consistent across his depictions. He's usually standing in front of a cave or in perhaps some other wild setting confronting ferocious animals. He has some primitive tool made of wood or stone and generally designed for violence. He's draped in fur if he's dressed at all, and perhaps most consistently, the Cave Man is always hairy. His hair is long, unkempt, often sticking out all over his head, and growing all over his body (Berman, 1999). We all know a Cave Man when we see one. But what about a Cave Woman? Where is she? Sometimes the answer is that she's cowering beside her Cave Man, with the implication that the primitive club he's carrying isn't just for hunting. Or she's being dragged around by her also significant amount of hair. Cave Women aren't depicted nearly as often as are Cave Men, which says something about the gendered ways in which we imagine early human life. When Cave Men and Cave Women are depicted together, the women are generally perceived as relatively unimportant and completely submissive to their men. What do these popular images of early humans reveal about our own ways of thinking about gender? How do these images line up with anthropological theories of Man the Hunter and Woman the Gatherer? Are the gender relations between Cave Men and Cave Women what we imagine would exist in a pure state of nature, assuming such a thing exists?

Is there evidence that this was in fact the way early human beings organized their lives? Obviously, it's difficult because there are no early humans left around to ask. But you might have noticed that this model of *human* evolution greatly downplays the role of women in the evolution of the species. Washburn and Lancaster (1968) assumed that men's early evolutionary importance in hunting helped to explain the current pleasure experienced by men in warfare and violent sports. Leaving aside the question of whether men, or women, actually enjoy warfare, feminist critics pointed out that this assumes that women are only barely human (Slocum, 1975). If the psychology of the species was set by early hunting experiences, and women did not hunt, then women's supposed lack of desire to hunt and kill makes them somewhat less than human. In addition, feminist critics argued that food sharing observed in primates is not between males and females as predicted in the man-the-hunter model, but between females and their offspring. If we can assume something about the behavior of early humans from observation of our primate relatives, then this sharing is likely to have come before the origins of hunting and is just as likely to have led to the origins of family units as male hunting was. In other words, family units evolved to share the resources gathered by women rather than to distribute the spoils of men's hunts.

Other feminist anthropologists have pointed to more flaws in the man-the-hunter theory, including the flawed logic that led to the conclusion that hunting was central to human evolution. In fact, evidence from most contemporary foraging, or hunter-gatherer, societies indicates that women, as well as men, do some hunting and that the food gathered by women accounts for more than half and sometimes nearly all of what these groups consume (di Leonardo, 1991). Despite the fact that the key tools used by women as gatherers do not survive the fossil record in the way hunting tools do (women's woven baskets as compared with the stone points of spears and arrows), some feminist anthropologists have suggested that the food-sharing behavior of women, rather than hunting, should be considered a key factor driving human evolution (Tanner & Zihlman, 1976).

Which version of human evolution seems to make more sense to you? The importance of hunting or of food sharing? What does each version imply about the essential aspects of our human natures?

Do Men Rule the World? The Universality of Patriarchy

In general, these critiques demonstrate the ways in which anthropologists projected their own assumptions about sex and gender into the human past to develop **androcentric** evolutionary theories, or theories that assumed that what men do is more important than what women do. They also put sex differences in the form of childbearing and the responsibility for it at the center of an argument that also seeks to explain gender as it exists in contemporary society, placing themselves firmly on the nature end of the nature versus nurture debate. Connected to this nature versus nurture debate among anthropologists is a second concern about the universality of male domination. Because anthropologists are engaged in the task of exploring variation across cultures and within our evolutionary past, an intriguing question arises as to whether women have been subordinate to men everywhere in human societies from our very beginnings as a species to our modern world. In other words, has there ever been, in human history, a society that was not patriarchal, or not dominated by those designated male? Before we explore the possible answers, it's useful to think about why this question might be important.

We can understand how this question is important by exploring the implications of some of the possible answers. Let's pretend that the answer to the question of whether there has ever been a human society that was not dominated by men is "no" and that all the human societies that do and have existed are patriarchal. What kind of implications might that answer have? One implication connects us back to our previous discussion about nature versus nurture and the biological underpinnings of sex. If the people we call male have everywhere dominated the people we call female, does this suggest this is the "natural" state of being? Is there something deeply rooted in our past evolution and current

DNA that leads to male domination? This is a difficult and politically loaded question to answer, although many would argue that even if the answer is yes, this does not necessarily mean male domination should continue to be the status quo. Many would argue that biology does not have to be destiny, and if male domination is somehow written into our genes, humans have demonstrated that we do not have to be prisoners to our biology.

Do you know of people or do you yourself believe this perspective, that it is "natural" for males to be dominant over females in all societies? How common do you think this perspective is? Are there examples of other aspects of humans as a species that might be "natural" but that are not important to human society? Do we have the power to change things that are natural?

Others would argue that even if male domination is a universal phenomenon, this does not necessarily imply it is biological, and we will explore these theorists' attempts to locate universal male domination in some universal aspects of human societies and cultures, rather than in human genetics. But this consideration leads us to the second implication of concluding that patriarchy is universal. If we have no models of societies that are not male dominated, how do we go about creating such a society? If it has not happened so far in human history, is it possible at all? As Salvatore Cucchiari (1981) noted in her essay on the origins of gender hierarchy, "A look to the future is the unspoken purpose of the backward looking origins investigation" (p.70). The search for a society that is not dominated by men is also a search for the future possibility of such a society and a road map for how to get there.

Given the weight of these implications, it's not surprising that a lot of debate exists around this topic. The idea of a matriarchal society that existed before the rise of private property was a notion that was popular among Victorian anthropologists and developed most influentially in an essay written by Frederick Engels, who you may also know as co-author of *The Communist Manifesto* with Karl Marx. In his essay, *The Origin of the Family, Private Property, and the State*, Engels (1884/1972) argued that property relations and the corresponding emergence of a class system led to what he called the "world historic defeat of the female sex" (Engels, 1884/1972, p. 120). Namely, in the matriarchal society that preceded the historic defeat of the female sex, property was communal rather than private and group marriages were the predominant form of family life. In this family situation, with more than two people commonly having sex with each other, only the mother can be identified with any certainty as the parent, and therefore, these societies were matrilineal. **Matrilineal** means that both kinship (who's related to whom) and property are passed down based on your mother; being ethnically Jewish is conceived of as matrilineal because you are considered Jewish if only your mother is Jewish but not if only your father is Jewish. Engels assumed that the matrilineal structure of this society gave women power over the household, which was the productive unit. This power made for a matriarchal society.

Subsequent generations of anthropologists questioned Engels and other Victorian notions of a lost matriarchal society that was destroyed by the rise of private property and classes. In part, this was a result of the lack of archaeological evidence in support of the theory (Webster, 1975). But feminist anthropologists during the 1970s revived interest in Engels's work and searched for contemporary examples to support his claims (di Leonardo, 1991). This meant looking for egalitarian, small-scale, matrilineal societies that seemed to be less characterized by male domination.

This more contemporary project had its own set of problems, though. First, using contemporary societies to support the notion that a historical, matriarchal society did in fact exist treats currently existing societies as nothing more than living examples of our historical and prehistorical past (di Leonardo, 1991). It assumes that these societies have somehow survived with no changes while the rest of the world progressed; this assumption is misguided as well as potentially insulting to those groups who are being considered examples of the "less evolved." Also, as more and more ethnographies were conducted focusing on women's status in small-scale matrilineal societies, the evidence did not support many of Engels's assertions (di Leonardo, 1991). Although some women in matrilineal societies do enjoy higher status, the status of some seems to be even worse than in **patrilineal** societies, or those that assume descent through the father and the male line. Within North American Native American populations alone, Seneca and Pueblo women did seem to enjoy relatively higher status, but among the Plains Indians, women held much lower status relative to men.

Nonetheless, some anthropologists continue to investigate the possible existence of a purely matriarchal society. Some argue that it is in fact impossible to say for sure that no such society has ever existed because we may simply lack any archaeological record of a historical matriarchal society or contemporary matriarchal societies as they existed before Western gender notions were spread through colonialism and globalization (Mascia-Lees & Black, 2000). Cucchiari (1981) described a theoretically pregendered society she called "the bisexual horde," in which a loose categorization existed based on divisions between Child Tenders and Foragers. These distinctions were not gender based because this pregendered group had no concept of gender distinctions and, therefore, no distinctions based on sexuality either—because our categories of sexuality are based on the existence of gender or sex categories, or both. Although she suggested some potential archaeological evidence in support of this theory, the question remains open to debate.

When feminist anthropologists argue that patriarchy *is* universal, their next step is to try to uncover what features societies have in common that produce male domination. These theories make the assumption that although patriarchy is universal, that does not necessarily mean it is rooted in our biology. Different theorists have pointed to various features of society that may contain the key to explaining patriarchy. Nancy Chodorow's (1974) theory turned Freudian dynamics around and located male domination in the structure of the family. We will discuss her theory in more depth in the next chapter, but male domination, according to Chodorow, is created by women's universally primary role in caregiving. Sherry Ortner (1974) located patriarchy in the symbolic structure of women's equation with nature and men's equation with culture. This dichotomy means that women, as more natural, are also perceived as "unthinking, inferior, untouched by human creativity" while men are "intelligent and creative, superior and representing humanness" (di Leonardo, 1991, p. 13). This formulation is sometimes used within the environmental movement, and it influenced the development of a specific type of feminism called ecofeminism, which we will examine later in this chapter. Finally, Michelle Rosaldo (1974) argued that the separation of social life into domestic and public domains was the most plausible explanation for male domination, a perspective we will discuss in Chapter 9 when we examine gender and work. This brief overview should demonstrate that all these attempts to formulate an explanation for women's universal subordination to men have had important impacts on thinking about gender in general, even if the ultimate question of the universality of male domination remains unresolved.

Do you believe male domination, or patriarchy, is universal across time and different cultures? What evidence might you use to support your argument? Which of the nonbiological theories for universal male domination seems to make the most sense to you? Can you imagine a world as described by Cucchiari (1981), where neither sex category, nor gender, nor sexuality, as we understand these concepts, exists?

QUEER THEORY

An examination of queer theory as a body of knowledge takes us firmly away from what some would call the pretense of disciplinary boundaries. Are the differences among anthropology, psychology, and sociology really that important? From the perspective of queer theory, the answer is a definite "no." Disciplinary boundaries are a manifestation of various actors making claims to knowledge, and knowledge and power are inextricably bound together. Psychology is often considered closer to the "hard" sciences, such as biology and physics. Physical anthropology is closely aligned with biology and zoology and is, therefore, more scientific than social or cultural anthropology. Within sociology as a discipline, quantitative studies, or research that transforms data into numbers, are considered less biased and more objective than qualitative research, such as ethnographies or in-depth interviews. As queer theory and many of the theoretical traditions from which it draws would argue, these divisions are all about attempting to privilege certain types of knowledge over others, usually by making claims to being more scientific or objective.

If I told you that the status of being a Black man in the United States is a dangerous one, would you be more convinced by statistics on the average life expectancies of different groups (69.2 years for Black males, 75.4 years for white males, 76.1 years for Black females, and 80.5 years for white females) (National Center for Health Statistics, 2005), by an ethnographic account of the daily lives of Black men in an inner-city urban community, or by an autobiography written by a Black man? Each type of knowledge has its own value, but since at least the Enlightenment in Western Europe, knowledge more closely aligned with "pure" science has generally been privileged over other ways of knowing in many parts of the developed world. As many feminists have pointed out, this privileging of different ways of knowing has a definite gender component because objectivity and rationality are traits associated with masculinity much more so than with femininity. Sources both within the global North (like feminism, postmodernism, and fundamental Christianity) and from outside (postcolonialism) have challenged the dominance of science and rationality, and queer theory and its antecedents are examples of that challenge. As noted by William Turner (2000), queer theory begins with a definite suspicion that the regular modes of contemporary intellectual activity, including the idea of disciplinary boundaries, do not serve the interests of queers.

It might be important before we delve into the substance of queer theory to say a bit about the usage of the word *queer*. Queer theorists use this word in ways very different from how it may be used outside of the academic world. For some people, perhaps those not familiar with television shows such as *Queer Eye for the Straight Guy* and *Queer as Folk,* a common reaction to the word *queer* might still be one of shock; *queer* has historically been used as a derogatory term for gay, lesbian, and transgendered individuals and considered similar to words like *faggot*. Why is it now being used to describe a theory about identity as well as in the titles of television shows? As queer theory and sociology seek to teach us, language matters. It is an act of empowerment to take a word that has been used as an insult against a group to which you belong and to claim it for yourself and the group. Thus, you may have heard the phrase, "We're here. We're queer. Get used to it," a slogan originally used by the gay rights organization, Queer Nation. The use of the word *queer* is partly political, a way of refusing and rechanneling its negative connotations. But the word *queer* also fits very well with what we will discuss as part of the ideological and theoretical agenda of queer theory. The literal dictionary meaning of queer is "not usual," "eccentric," or "suspicious." A theory that is queer is therefore a strange or unusual theory, a theory that is different in some important way. It is just this type of rather eccentric and suspicious theory that queer theorists have set about to produce.

Can you think of other examples of derogatory terms used against groups in society that have been redefined by the groups themselves? How successful do you think these attempts have been? How do you go about changing the meaning of a word or altering language? Can you think of other examples of how language has changed?

Queer theory is hybrid creature, and it can trace its beginnings to many different sources. For the sake of simplicity, we'll focus on the influence of three main movements—the gay and lesbian rights movement, postmodernism, and feminism. We've already discussed a very broad outline of feminism in the previous chapter, so here we'll look at the specific threads in feminist thought that have been most influential on queer theory. We'll also pretend, for the sake of simplicity and understanding of the origins of queer theory, that each of these movements is separate and not often characterized by considerable overlap. For example, many feminists who exerted the most influence on queer theory were those deeply immersed in postmodernism and a concern for gay, lesbian, and transgender rights; these individuals were therefore at the intersection of all three of these movements.

Origins in the Gay and Lesbian Rights Movement

Let's begin with the gay and lesbian rights movement but, first, another note on terminology. Here I will refer mostly to the gay and lesbian rights movement because, during the period we'll be referring to, this is by and large how actors within the movement conceived of themselves. Today, you'll hear references to the gay, lesbian, bisexual, and transgender movements or organizations, but at this point in the history of the movement, the focus was somewhat narrower and included mainly gay men and lesbians. Partly as a result of many of the same forces that led to the development of queer theory, movements and organizations were later reconceived to be more broadly inclusive of bisexual and transgendered people. In Chapter 5, we'll trace a history of sexuality in the developing world and examine sexuality in a cross-cultural perspective.

Right now, we'll take for granted the existence of categories of identity in Anglo-European society called gay and lesbian and focus on the specific period of the gay and lesbian rights movement that began in the United States with the Stonewall riots in New York City in 1969. Many who study gay and lesbian movements argue that this event, when gay men fought back against police who were raiding a gay bar, marks the beginning of a period of gay affirmative politics lasting into the 1970s and 1980s. During this period, gay and lesbian activists fought successfully to have homosexuality removed as a category of mental illness in the *Diagnostic and Statistical Manual,* a central document produced by the American Psychiatric Association and used by psychologists and psychiatrists to diagnose their patients. This was important to remove some of the stigma associated with homosexuality being considered a mental illness, equivalent to schizophrenia, multiple personality disorder, or anorexia. Observers of the movement argued that by the beginning of the 1980s in the United States, gays and lesbians had reached a level of social tolerance and some recognition as a valid subculture or group within the larger society (Altman, 1982).

Internal Contradictions

Like many movements that were part of the global social movement cycle of the 1960s and 1970s, the gay and lesbian rights movement began to face internal problems as well as an antigay backlash as the movement moved into the 1980s and 1990s. The backlash was brought about by the rise to power of conservative elements in the political scene in the United States—namely the election of Ronald Reagan and the brief dominance of Republicans in both houses of Congress (Seidman, 1992). This backlash was given more strength by the growing recognition of the AIDS epidemic and its association with gay communities (Seidman, 1996). Although the AIDS crisis demonstrated the strength of organizations and institutions that had been created within the gay and lesbian movement, and reinforced many alliances that had already been made, it also pushed the movement in a more radical direction. For many of those activists involved in the movement, it led to questioning the effectiveness of the politics of minority rights and inclusion (Seidman, 1996).

Connecting this to our discussion of the feminist movement in the previous chapter, many gay and lesbian activists began to doubt the ultimate usefulness of the master frame of civil rights. Early in the movement, gays and lesbians argued for rights based on their shared humanity, or on the assumption that they were fundamentally similar to heterosexuals. This is also the basic ideology that underpins liberal feminism: Women and men are fundamentally alike once societal barriers and prejudice are removed. Like many radical feminists, some gay and lesbian activists began to lose faith in this fundamental liberalist frame. In the face of the conservative backlash and the stigmas associated with increasing awareness of AIDS, many gays and lesbians began to question whether persuading the rest of society that they were just like them was the best strategy in the end.

Along with this backlash from the outside, the gay and lesbian rights movement began to face its own internal contradictions—again very similar to those faced by feminists of the second wave. These internal contradictions had to do with the necessity of organizing a social movement around some kind of core identity, and therefore assumed commonality, as opposed to the reality of many different and sometimes contradictory experiences. So, for instance, even early in the gay and lesbian rights movement, there were divisions that sometimes emerged between lesbians and gay men. The Gay Liberation Front, a social movement organization founded in New York City in the wake of the Stonewall riots, was originally composed of both gay men and lesbian women (Seidman, 1993). But tensions developed between lesbian feminists who were engaged in a critique of male domination while fighting discrimination against homosexuals and gay men who were less concerned with male domination.

These two goals sometimes put lesbian feminists at odds with gay men, who, as men, may benefit from systems of gender inequality. These tensions sometimes led to lesbian feminists forming their own, separate organizations and movements. Other tensions were similar to those encountered in the second wave of the feminist movement, which focused around racial and class differences among lesbians and gays. Women and men of color criticized the gay rights movement for devaluing their own experiences and insight and argued that the movement and its ideology reflected a white, middle-class bias (Seidman, 1996). As with feminism, questions arose about whether it was possible for one movement to represent and organize around the experiences of all lesbians and gays, given the vast differences in their experiences. The end result was a lingering question: Is there such a thing as a unified homosexual identity?

Enter Postmodernism

This question resonated not just with similar questions that had been raised in the feminist movement, but with a broad theoretical and ideological development that we will call postmodernism. Like queer theory, postmodernism spans disciplines and is manifested in many different forms by theorists, scholars, and activists. The best way to understand the complex collection of theory that is labeled postmodernism is to give a brief historical overview of the basic origins of these ideas.

The Enlightenment period in Western society fundamentally transformed the ways in which people in the global North, and eventually the global South, understood themselves, their societies, and their relationship to their environments. The Enlightenment was characterized by the emergence of science as we now know and understand it and, along with that, a profound faith in rationality and progress. Many Enlightenment thinkers, including some of the founders of sociology, believed that science would help us solve most of the problems of human societies, like poverty, war, disease, and inequality.

Great things came out of what postmodernists would refer to as the Enlightenment project. Democracy, with its trust in the ability of rational citizens to govern themselves, is a product of the Enlightenment. Scientific thinking has cured many diseases and greatly expanded the average life expectancy of people globally, although especially in the developed world. Technology evolving from science has made life easier for many people. You can see how the idea of progress is compatible with both rationality and science; postmodernists argue that during the modern period that began with the Enlightenment, the prevailing ideology saw humans engaged in a linear evolution toward a better way of living and being as humans. More simply put, things were getting better and they would continue to do so. That was part of the grand story that modernist thinking told about history.

A very basic way to understand postmodernism is as a radical rethinking of all those Enlightenment ideas and of this grand story of progress in general. Maybe we are not moving forward and getting better after all. Maybe rationality and science will not solve all of our problems; in fact, maybe they will just create new problems. Maybe grand stories of any kind that attempt to represent the truth are not to be trusted. What would lead to such a deep questioning of these modern ideas? Many point to specific historical events that led to this radical ideological turn. Although science has helped to cure diseases and give us longer lives, it has also produced nuclear and biological weapons—technologies capable of destroying all life on our planet. The same rationality that gave us democracy was also employed by Adolf Hitler and the Third Reich in their frightening application of efficiency to murdering millions of Jewish people, homosexuals, and other genetic undesirables during World War II. And although democracy as an institution seemed to work well for many people in the developed world, billions of others in the developing world continued to suffer under totalitarian regimes and were made into bystanders to the march of "progress" enjoyed by those of us in the global North. Postmodernism can be understood as a way of resolving the crisis of these contradictions—the fact that the Enlightenment project did not turn out quite as expected.

The aspect of this ideological turn that is most important to queer theory is the distrust of grand narratives, or what postmodernists sometimes call *meta-narratives* (Lyotard, 1979/1984). A **meta-narrative** is any attempt at a comprehensive and universal explanation of some phenomenon. Charles Darwin's theory of evolution is an example of a meta-narrative that attempts to explain the diversity and forms of biological life on the planet, and it has become the predominant meta-narrative in biology and, some would argue, in social theory as well. The attempts by feminist anthropologists we read about previously to develop universal explanations of male domination are meta-narratives. Science itself is a meta-narrative because it seeks to develop theories that explain the way the universe works. The story of inevitable human progress aligned with rationality and science that characterized the Enlightenment is also an example of a meta-narrative.

What's so bad about meta-narratives, you might ask? We have already begun to discover that the broader you cast your theory or explanation, the more likely you are to miss the experiences of some groups of people. Feminist theories originated by women in the developed world did not always take into account the lives of those in the developing world. Activists organized around the issues of gay men may not have taken into account or may in fact have worked against the issues of lesbian women. Trying to develop these meta-narratives inevitably leaves someone at the margin or attempts to somehow force their experiences into the grand story being told. As we mentioned earlier, meta-narratives as claims to knowledge have power implications for those who don't fit. If I define what it means to be a man in a certain way, and you don't fit that definition, you're not as likely to receive the privileges that go along with being defined as a man.

How does this connect back to some of the problems we left off with in the gay and lesbian rights movement? First, postmodernism would argue that the master frame of civil rights that underpinned liberal feminism and the early stage of the gay rights movement in the United States is another example of a meta-narrative. Remember this model made a claim to basic rights for a wide variety of groups based on their common humanity. But there are deep and problematic contradictions in this meta-narrative of liberalism and civil rights. Within this meta-narrative, basic rights are distributed based on our common humanity, but certain groups (African Americans, women, homosexuals, the disabled, etc.) are perceived as suffering as a result of their exclusion from this common humanity. Their movements to claim those rights are then based on the very "minority" identity (being Black, a woman, gay or lesbian, or disabled) that led to their exclusion in the first place. Movements organized around a civil rights frame are built on the fundamental contradiction of using the essential structure of *difference* as a basis for identity (we're all Black, we're all women, we're all homosexual, we're all disabled) to claim rights based on claims of essential *similarity* (we're all human). Within queer theory, postmodernism, and third wave feminism, theorists and activists began to wonder whether this contradiction implied that the only way for groups to gain their basic rights was for them to look like, act like, and be like the dominant group to prove that they were in fact all human (Turner, 2000). Gays and lesbians would only be given their basic human rights when they stopped acting like gays and lesbians, or when they stopped being different.

Many people, when faced with the demands for recognition made by oppressed groups (gays, lesbians, bisexuals, and transgendered, racial and ethnic minorities, women, the disabled, etc.), often wonder, why can't we all just be human? What does queer theory suggest about what is implied in this question? Does this imply that if everyone is human, everyone should be alike?

The other way in which postmodern ideas about meta-narrative intersect with both the gay and lesbian and feminist movements is in regard to identity claims themselves. Here, postmodernism asks the question as to whether anyone really can make the claims, "We're all Black," "We're all women," or "We're all gay." Here is an important point of intersection between directions in third wave feminism and those of postmodernism. In Chapter 2, we discussed third wave feminism as a reaction to the contradictions of the second wave, centered on the ways in which other identities intersected with gender. Women of color, working class women, and lesbian women felt that feminism often embodied a white, middle-class, heterosexual bias. Postcolonial theory pointed to the very different experiences of women outside of the developed world. Third wave feminism attempts to respond to these problems and questions. Feminists like Judith Butler, Eve Sedgwick, Helene Cixous, and Teresa de Lauretis attempted to answer this question: If there is no unified identity called *woman,* where do we go from here?

In very practical terms, how do you organize a social movement around a category that you suspect might not be real? As postmodernists would argue, any story about what unites a particular group of people in all times and all places under the category *woman* is a meta-narrative. The solutions various feminists proposed to these problems were various and somewhat complicated. Where we are left at the convergence of these three threads—gay and lesbian rights movement, third wave feminism, and postmodernism—is with a deep and abiding mistrust of claims to ultimate "truth," especially as they exist in the form of meta-narratives and categories of identity.

Three Key Features of Queer Theory

Queer theory, you might be thinking, had better be a fairly impressive theory to get us out of this particular theoretical and ideological corner we've backed ourselves into. But queer theory has no intention of solving any of the problems because that would involve re-establishing another meta-narrative or claim to truth. If the crises we've identified here serve to pull the metaphorical rug out from under your feet, queer theory points to the importance of realizing there was never any clear place to stand to begin with; the rug didn't really exist, anyway. This is the *queer*ness of queer theory—the tendency to view any firm theoretical footing with some suspicion—and it makes this set of theories somewhat difficult to understand compared with the kind of theories we're accustomed to. We can get a little bit more specific about the very loose group of theorists and activists working under the umbrella of queer theory, though, and identify three features of queer theory as a general orientation.

First, *queer theory as a theoretical approach is distrustful of categories*; and as a social movement, it works to do away with categories in their current form. This distrust begins with the categories specifically related to sexuality, like gay, straight, lesbian, bisexual, heterosexual, homosexual, transsexual, transvestite, intersexual, and so on. The use of *queer* as a way of self-identifying among these groups represents an "aggressive impulse of generalization" (Warner, 1993, p. xxvi) and an attempt to disrupt our conceptions of what is normal. Categories of identity, as we've discussed, are incomplete and can never successfully encompass all the diversity contained within. For example, in one of my gender classes, students read an excerpt from Kate Bornstein's (1994) book, *Gender Outlaw.* Bornstein is a male-to-female postoperative transsexual, and her book is a good example of an attempt to understand queer categories of gender and sexuality using a combination of theory and her own real-life experiences. After her sex-change surgery, Bornstein writes about having a girlfriend. Some of my students were very concerned about whether this made Bornstein a lesbian. In what queer theorists would call a very modernist vein, at least one student really wanted a definitive answer to the question. Was Bornstein, born with male chromosomes and genitalia and now in possession of surgically constructed female genitalia and female hormones, having sex and romantic relations with a biological woman who self-identified as a lesbian—was Bornstein a lesbian herself?

This student felt there was one right answer out there, but what would it be? Do lesbians have to be born as female to be lesbians? Do they have to be attracted only to other people who were born as female? Can two men who cross-dress as women and are attracted to each other be considered lesbians? What about a woman who's married to a man but has sexual fantasies about women and never acts on them? What about a woman who is heterosexual until she goes to prison and has sex with other women? What about some women in Native American cultures who live socially as men and marry other women? Native Americans don't consider them lesbians, and who they have sex with is much less important than the gender they are acting out (Whitehead, 1981). How can a category called lesbian possibly hope to take account of all these differences? Queer theory answers that they can't and that, therefore, there's not much sense in using them, at least not in the way they have been used in the past.

Queer theory doesn't just stop there, though, with the radical deconstruction of categories related to sexuality. At least in many parts of Anglo-European society, categories of sexuality assume categories of sex and gender. You can't be a lesbian or gay without the existence of some sense of gender or sex because without those categories, it makes no sense to distinguish people based on the sex or gender of those with whom they choose to have sex or to whom they are attracted. In a world where there are no sex categories (males or females), defining someone as heterosexual because they have sex with those in a different sex category makes no sense; there are no sex categories to begin with. Drawing on its feminist lineage, queer theory questions categories of gender as well. And from there, in its most transgressive forms, queer theory questions all categories because all categories have these same fundamental flaws. How do we define what it means to be a member of a particular race in the United States? The particular mode of definition based on government census categories has changed every 10 years since the government began conducting a census in 1790. What made you white in 1790 is not the same as what makes you white in the 21st century, so how can we assert there is such a thing as being white?

One solution proposed by queer theory to the problems with categories of identity is to think of these categories as always open and fluid. You might think of this *second feature of queer theory as suggesting that everyone can, in fact, be queer and that everyone already is, in fact, queer*. In her book, Kate Bornstein (1994) pointed out that gender as a system is problematic for individuals like herself who never felt they fit in but also that, "eventually the gender system lets everyone down" (p. 80). At some point everyone—straight, gay, feminine, masculine, intersexual, transgender—fails to live perfectly up to the demands placed on us by gender and is therefore hurt by this system. Heterosexual men in Anglo-European society are not supposed to show affection to other men except in appropriate ways and venues (the slap on the butt during a sporting event), and many would argue that forbidding expressions of affection among any group of people goes against our basic human tendencies and is a form of oppression. The straight man who hugs his male friend for a little too long is likely to be sanctioned in some way for not perfectly conforming to his particular category and, in this way, is let down by the gender system. The way categories of gender and sexuality are constructed affects all of us, regardless of where we fall within those categories, and this makes all of us queer in some way relative to the categories. The existence of these categories imposes on, or disciplines, all of us, and none of us conform to them perfectly.

This connects to the *third feature of queer theory: its ambition to queer many features of academic and social life that are generally considered within the bounds of normality*. Queer theory aspires to be not just a theory of sexuality, or of gender and sexuality, but to be a broad and far-reaching social theory (Seidman, 1996). This is because queer theorists believe that sexuality is an important way in which knowledge and power in society are organized and that a theory of sexuality is, therefore, a theory of society in general. This can be understood as the difference between thinking about what "gay and lesbian studies" means as compared with "queer studies." Departments and programs in gay and lesbian studies have grown in Anglo-European societies as part of the success of the gay rights movement. What would be the implied object of study of a gay and lesbian studies program? You might study the history of gay and lesbian peoples, literature about homosexuality, or literature produced by gays and lesbians. You might look at the portrayal of gays and lesbians in the media or how science has been used to explain and sometimes control homosexuality. These topics and many more might be covered in a program in gay and lesbian studies, but the main object of study would be homosexuality or, maybe, sexuality more generally.

For queer theorists, this leaves out an important part of the theoretical picture. It segments the study of homosexuality from the study of the larger society that created it and continues to re-create it. Gays and lesbians only exist in opposition to straight people. Queer theorists would argue that studying only gays and lesbians produces an incomplete picture of how sexuality works to produce identities such as straight and gay. For that reason, queer theory is just as concerned with studying heterosexuality as it is with homosexuality and with investigating how sexual practices permeate all aspects of society. A queer studies program would look at all types of literature, not just that which focuses on gays and lesbians or is written by them, arguing that sexuality is an integral part of all cultural productions. Queer studies, rather than focusing on the portrayal of gays and lesbians in the media, would also examine the portrayal of heterosexuality in the media. Science would be studied in queer studies for the ways in which it is used to create many categories of difference, rather than solely for how it applies to issues of sexuality. For queer studies, the object of study would be society itself.

An example might help to demonstrate how this works. Through an examination of literary works, Eve Sedgwick (1985) explored a concept she called *homosociality*. **Homosociality** is the continuum of cooperative relations between people of the same gender, and it ranges from actual homosexual intimacy to same-sex friendships. Sedgwick argued that there are differences in homosociality between women and men; men perceive the distance on this continuum between male friendships and actual homosexuality (sex between men) as much farther apart than do women. **Homophobia**, the fear or disavowal of homosexuality, is much more likely to arise among homosocial groups of nongay men than it is among nongay groups of women. Thus, sexuality is connected to notions of gender difference because masculinity and femininity are defined in part by these differences in homosociality. What Sedgwick's notion of homosociality does is place an understanding of sexuality firmly in the realm of what many would consider normal—heterosexual people and their nonsexual relationships with others of the same gender. The object of study is not homosexual people but same-sex relationships among straight people and how these are connected to notions of gender. This orientation turns away from a study of homosexuality or homosexual people to a more general concern with how sexuality has impacts on all of us in society. In this sense, queer theory is very ambitious in that its goal is no less than to question the truth and usefulness of categories like homosexuality and heterosexuality as a way to explain what sexuality is.

GENDER, ECONOMICS, AND DEVELOPMENT

Through examining anthropological approaches to gender, we've begun to explore gender from a cross-cultural perspective. Although anthropologists certainly are interested in the connections between different cultures and societies, and specifically between the developing world and the developed world, the next two theories we'll explore focus particular interest on gender in contemporary, global focus and on the relationships between these different parts of the world. In Chapter 1, we briefly touched on the varied vocabulary that is used to refer to a particularly important division of the world into areas of relative advantage and disadvantage. To understand gender theories related to this division, we need to understand more about the specific origins and implications of all these different terms. Beginning with the language of the "Third World," the idea of a First, Second, and Third World originated after World War II and described what were at first considered strategic political blocs of countries. So the nations of NATO (the North Atlantic Treaty Organization) and the Warsaw Pact (the treaty involving Alliance countries that ended World War II) were lumped together as the First World. The Second World, a term not heard very often anymore, referred to the areas under Communist influence, including the Soviet Union and much of Eastern Europe. The Third World evolved to describe all the countries that did not obviously fit into the first two groups, but it came also to imply countries that were less developed than the First World countries in particular.

The Development Project

Following closely on the idea of First and Third World came the division of the world into developing and developed countries. These terms are closely aligned with what some have called the **development project**. In the years before World War II, the world had for a very long time been divided into the colonized and the colonizers, and many countries of the Third World were still under colonial control after the end of World War II. Colonialism was a system of outright exploitation, and although many colonial powers argued that they brought "civilization" to the countries they occupied, the end of World War II saw an increasing acknowledgment of the injustice of the colonial system. Colonizers often assumed their presence lessened gender inequality among colonized peoples, as the gender ideologies and customs Europeans brought to the rest of the world had to be better than those that had existed simply because Europe (and perhaps the United States) was thought to be more advanced than the rest of the world.

But anthropologists and historians have increasingly demonstrated that colonization had varying impacts on gender relations among the cultures of colonized peoples, often including negative impacts. For example, among both the Igbo and Yoruba peoples in

Nigeria, women had a great deal of power relative to men in precolonial times (Denzer, 1994; Okonjo, 1976). This institutionalized power usually focused on issues of particular concern to women, which included access to markets, and this demonstrated women's important role in the economic and social life of precolonial African peoples. Much of this power was lost under colonial systems of control. Other research demonstrates the ways in which colonial land reform often resulted in women's loss of status; in placing communal land into the hands of colonial authorities, Europeans imposed their own beliefs that men should be responsible for the cultivation of land, even in cultures where this had not been the case in precolonial times (Boserup, 1970). This is just one small example of the ways in which the "civilization" that European colonizers believed they were bringing to colonized peoples was not always everything it was cracked up to be.

The years after World War II saw a shift in thinking in the First World about the proper relationship to former colonial nations. Theorists such as Talcott Parsons and Seymour Martin Lipset put forward the influential perspective of **modernization theory**, positing that the postcolonial world could become like the countries of the developed world if they followed our lead economically, politically, and socially (Lipset, 1959). The developing world could then come to enjoy the fruits of democracy, economic prosperity, improved health, and increased equality that characterized life in countries like those in Western Europe and the United States. The development project was put into action through various institutions such as the World Bank and the International Monetary Fund (IMF) to help the developing world begin the process of modernization. Harry S. Truman, then president of the United States, gave a speech that very nicely summarized the ideology of modernization and the development project in 1949:

> We must embark on a bold new program for making the benefits of our scientific advances and industrial progress available for the improvement and growth of underdeveloped areas. The old imperialism—exploitation for foreign profit—has no place in our plans. What we envision is a program of development based on the concepts of democratic fair dealing. (as quoted in Esteva, 1992)

This sounds like a pretty good deal for the developing world, and Truman's statement about the basic orientation of the developed world still holds much weight in global politics and economics. The institutions that characterized development, the World Bank and IMF, are still going strong. Although the development project was not meant to take place through physical force and occupation, some of the common justifications for the war in Iraq that emphasize bringing democracy to that country and the region in general are consistent with the goals of modernization theory. This is also still the goal of many nongovernmental organizations, or NGOs, that work in the developing world. **Nongovernmental organizations** are organizations that work on various problems that might sometimes be considered the role of government to solve—social, political, economic, or environmental problems—but that have no government affiliation and are privately funded entities. The Red Cross, Oxfam, Heifer International, and Doctors Without Borders are all NGOs with which you might be familiar that deal with issues of hunger, refugee populations, disaster relief, and providing needed medical services.

Women in Development

The development project assumed that modernization would help women and men in postcolonial nations equally, and planners as well as scholars largely ignored the unique role of women in development. Because modernization in the developed world had theoretically brought more equality between women and men, many theorists and practitioners assumed the same would be true in the developing world. But beginning in the 1970s, research increasingly began to question this neglect of women's roles, as well as the underlying model of development and modernization theory in general.

In 1970, Esther Boserup published her book *Women's Role in Economic Development,* using a wide range of empirical data about development in Africa and Asia to highlight the two simple facts that (a) a consideration of women's unique roles was important to development and (b) women were not necessarily benefitting from development in the ways many had expected. Boserup is an economist, and although as we've already discussed, many anthropologists studied gender in the developing world, Boserup brought the unique perspective of economics—a field very important to the development project. As you might have guessed from the name of the two organizations most associated with development, the World Bank and IMF, the restructuring of a society's economy is believed to be a key step in bringing about modernization, and therefore, economics is very important to an understanding of development. Although it may not have been Boserup's intention (Beneria & Sen, 1981), her analysis helped to demonstrate some important gender biases in the conceptualization of development and modernization theory.

First, Boserup (1970) pointed to the way in which both those studying development and those implementing development plans in the developing world had left women out of the economic picture. Models of development assumed, based on a European model, that men were the main economic providers for the family, primarily through their agricultural activities. In reality, Boserup found that, with some regional variations, women generally did as much agricultural work as men, if not more. As we will discuss more in Chapter 9 on gender and work, economic models often miss this kind of work done primarily by women because they focus on work that takes place in the **formal economy**, or that sector of the economy that is regulated by the government or legal systems, rather than in the **informal economy**, which is not regulated and therefore more difficult to measure and control. For example, in India, a male farmer who is paid by a multinational agribusiness firm to grow tea is in the formal economy, but his wife who spends an equal amount of time and energy growing food for her family is in the informal economy.

Many measures used by development economists to measure economic activity exclude the informal economy, disguising the huge contributions women make to the support of their families as well as to the economy in general. For example, official statistics for Egypt in 1970 indicated that only 3.6% of the agricultural workforce was female. More in-depth studies focusing on women's role in agricultural production revealed that half of all women plowed and leveled land, and three quarters participated in dairy and poultry production (Pietila & Vickers, 1994). Official statistics, including important measures of development such as gross domestic product (GDP) or gross domestic income (GDI), miss these contributions of women in the informal economy and therefore underestimate women's importance. According to one study, if women's work worldwide in the informal economy were included in official economic calculations, the world's gross domestic product would increase by almost one third (United Nations Development Programme [UNDP], 2003a; United Nations Population Fund [UNFPA], 2004).

The work of Boserup (1970) and others concerned with the role of women in development has revealed that all over the world and despite their exclusion from the economic models of development, women work more than men. Women grow three quarters of the world's food while receiving less than 10% of the agricultural assistance (UNDP, 2003a). In Africa alone, 80% of the food is grown and processed by women. Economists studying development missed this crucial fact because they considered only work that took place within the market, but if you include all the unpaid work women do, women in the developing world work on average 57 minutes more every day than men (UNDP, 2003b).

Second, in addition to highlighting the importance of women's contributions to agriculture and the economy in general, Boserup (1970) also brought attention to the ways in which the development project had sometimes hurt, rather than helped, women's status relative to that of men in the developing world. One important piece of development programs in many countries is to encourage converting more and more land toward growing **commodity crops**, or agricultural products that can be exported and sold on the world market. This process often involved cutting down native forests or replanting in fields once used in subsistence agriculture. **Subsistence agriculture** is growing crops to feed directly or provide for yourself and your family, rather than growing crops to sell to someone else in exchange for money you will use to buy food to feed yourself and your family. In addition, people in developing countries are encouraged to enter the wage economy. Thus, families move from growing their own food to feed themselves to growing food to sell to someone else, thus, entering the market system.

Boserup (1970) and others pointed out that this was not a gender-neutral process. Development encouraged men to enter the wage economy, often encouraging them to migrate away from rural areas in search of these jobs. This loss of men's contributions to the upkeep of the household was shown to increase women's work burden in many countries in the global South; with their husbands and male relatives gone doing wage work, women were left to do even more of the share of maintaining a household (Ware, 1981). In other instances, resources that were redirected toward development also made the lives of women in the developing world more difficult. In many parts of the world, women spend large parts of their day gathering the basic supplies to maintain a household, such as firewood and water. Cutting down forests to plant commodity crops often increased the distances women had to travel for their firewood, while the water demands of intensive agriculture also made access to that vital resource more difficult. Women in the developing world found more and more of their day consumed with the simple tasks of providing fuel and water for their families.

These realizations about the neglect of women's roles in development and the adverse effects of development led to a new perspective, **women in development**, sometimes abbreviated as WID. In 1975, the United Nations held the First World Conference on Women in Mexico City, and it proclaimed the next decade to be the decade of women. Government agencies as well as NGOs began to speak of integrating gender into their preexisting notions of development (Koczberski, 1998). In 1979, the United Nations passed the Convention on the Elimination of All Forms of Discrimination Against Women (CEDAW), which recognized global gender inequality and acknowledged that women's problems were unique and required attention different from that necessary to address men's problems (Walter, 2001). All these events represent the women in development approach by taking seriously the importance of gender to the larger development project as well as the ways in which development could have unintended consequences for reducing the human capability of women in the global South.

The verdict on how the women in development approach has helped women and men in the developing world is mixed. Although WID certainly put women's issues in the international spotlight, many theorists and activists critique WID because they see flaws with the underlying model of development itself, or with modernization theory. For some, the problem is that gender is being incorporated into a model—development—that is flawed from the very beginning. For others, questions are raised about whether theories that were not originally developed to deal with issues of gender can be transformed sufficiently to take gender into account. The critique of development theory itself evolved into a new theory, called *dependency theory* by economists and *world systems theory* by sociologists. The essential idea behind these theories is that the development project in many ways merely continued the legacy of exploitation from the colonial period in a different guise.

Regardless of its original intentions, development has merely continued the cycle of inequality between the global North and the global South. World systems theorists point to many examples of this dynamic, but important among them is the increasing debt trap of developing countries. The World Bank and International Monetary Fund we discussed previously as important parts of the economic piece of development are both institutions that loan money to developing countries to build their infrastructure (roads, ports, power facilities, dams, etc.). The assumption of modernization theory was that these infrastructures and technology transfers were necessary to transform the economy of developing countries into economies like those in the developed world—free market and capitalist economies. The problem for world systems theory is that the aid to the developing countries came in the form of loans, not gifts, and loans must be repaid. Many developing countries soon found themselves mired in debt to institutions such as the World Bank and IMF.

Activists, political leaders, and institutions, including the Catholic Church, have urged the World Bank and IMF to forgive the debts owed to them by developing countries. The World Bank and IMF have sometimes lowered interest rates on the loans or have worked with the governments of developing countries to restructure their loans. But these concessions come with conditions, commonly called *structural adjustment*. **Structural adjustment** has been referred to by the World Bank as shock treatment. In return for leniency on loans, these institutions demand that the governments of developing countries reduce public services, liberalize trade, emphasize export crops, eliminate subsidies, and curb inflation through raising interest rates and paying lower wages (Bell, 2004).

Previously, we discussed the effects on women's lives of moving farmers from subsistence to commodity crops, or crops for export, as part of structural adjustment. Cuts in public services means cuts in basic social services, such as medical services, education, and social welfare. Liberalizing trade and eliminating subsidies puts the developing countries at a disadvantage relative to the rest of the world economy, where nations like the United States can continue to subsidize their farmers to grow surplus crops and impose tariffs on imported goods. In all, structural adjustment has increased inequality within developing countries, greatly affecting the quality of life for individuals in the countries. World systems theorists argue that it has also increased overall inequality between the developing and the developed world. Integrating gender into this deeply flawed project from this perspective serves little useful purpose.

As theories, neither modernization theory nor world systems theory was designed to explain gender relations comprehensively, and they are therefore somewhat different than the other theories we have discussed. The women in development approach is less of a theory than a call to integrate gender into the preexisting development project. Nonetheless, these theories have had important implications for the way women and men experience gender all over the world, and especially in the developing world. For our purposes, it is important to understand that like many of the other disciplines we have discussed, theories about relationships between the developed and the developing world began by largely ignoring and bypassing gender concerns. But because gender is a crucial component of the way people structure their lives all over the world, many would argue that any theory that does not consider gender in some way is always incomplete.

ECOFEMINISM AND THE ENVIRONMENT

When you think of the ocean, do you imagine it having a gender? What about a river? A mountain? Do animals have gender beyond the particular sex of an individual animal? For example, do you tend to think of a cat as feminine and a dog as masculine? What about a bear and a deer? What about nature in general? Does nature, however you might conceive of it, have a gender? When we discussed some anthropological explanations for the universal domination of women by men (assuming we believe that to be true), one of them had to do with women's connection with the natural world, positing that women's oppression is a result of their association with nature. Ecofeminism is a theory that builds on this connection to highlight the ways in which any consideration of our relationship as humans to the environment, let alone conversations about how we should act in relation to the environment, needs to consider the role of gender.

Ecofeminism as a set of theories originated from tensions similar to those we discussed with development and world systems theory. Many women involved in the environmental movement felt gender was an important concern that was not being given a sufficient amount of attention (Birkeland, 1993). In addition, as many feminists and others turned their attention to the situation of women in the developing world, environmental issues became an important part of understanding their plight. Theoretical developments within feminist theory also contributed to the development of ecofeminism. The relationship between women and nature sometimes served as an important explanatory tool for theories of male domination, but other theories suggested a unique structure to feminine personality that made women's relationship to nature inherently different from men's. Later we will explore in depth Nancy Chodorow's (1974) theory of gender identity development, but her approach implies that women's sense of self is not dependent on a sense of separation from others in the same way men's sense of self is. Thus, patriarchy, as an expression of masculinity, reinforces a dualistic sense of nature as opposed to culture. This means nature and culture are perceived as two separate, distinct, and opposed entities or systems. Femininity is constructed to be less trapped by these dualisms, and therefore, a consideration of feminine ways of knowing and experiencing is important to understanding our relationship to nature.

All of these trends resulted in the development of a set of theories called ecofeminism or ecological feminism. You could find many competing ways to define this broad range of approaches and concerns, but in general, ecofeminism explores the links between the domination of women and our societal domination of the environment and argues that there are important connections between environmental domination and social domination of all kinds, whether it be on the grounds of gender, race, or class (Bell, 2004). Thus, ecofeminism proposes not just to be about gender and the environment but also to be about a multiple issue movement that examines how environmental domination, gender domination, class domination, and racial domination are all interconnected (Gaard, 1993). In a sense, ecofeminism is a theory that is interested in hierarchies in general, assuming that many types of hierarchies, including the one between humans and nature, are connected. Ecofeminism points to the ways in which nature is often used to reinforce the inferiority of many different groups in society. African Americans in the United States and indigenous peoples around the world have often been associated with savagery or have been considered primitive and therefore closer to nature (Bell, 2004). Historical and contemporary depictions of minority groups in the United States often portray these groups in animalistic ways (Jhally & Kilbourne, 2000; Riggs, 1987). The lower classes are also perceived as being closer to nature and therefore less able to control their animalistic emotions and aggression. And as we've already discussed, women are in many ways considered more natural and, therefore, opposed to culture, reason, and civilization.

Ecofeminism intersects in many ways with the concerns raised about women's roles in development, and like WID, it demands that attention be paid to gender issues in a theory and movement (environmentalism) that did not initially consider gender. Like queer theory, ecofeminism has ambitions to be a much broader social theory, rather than being confined only to issues of gender and the environment. Thus, ecofeminists have begun to explore the implications of their approach to development and modernization, as well as their connections with other social movements related to the environment, such as the environmental justice movement (Smith, 1997; Taylor, 1997).

Women and the Environment

In addition to these more theoretical considerations of ideologies surrounding the environment and social groups, ecofeminists also point to the concrete connections between gender and the ways in which people experience environmental impacts. The environmental justice movement has demonstrated that the negative impacts of environmental degradation are not experienced equally across all groups in society; those with the least amount of power also have little power to protect themselves from negative environmental impacts such as exposure to toxic waste, loss of natural resources, and the myriad environmental disasters brought on by human intervention, such as floods, forest and brush fires, mudslides, and landslides. In a typical city such as New Orleans, it is no coincidence that the poorest people also live in the areas of the city most prone to devastating floods. Those with wealth and power have the ability to locate themselves out of harm's way. Ecofeminists argue that this is true of women, who are generally among the less powerful groups in many societies.

What do you think of when you hear the expression, *tree hugger*? You might think of someone vaguely hippy-looking, probably white and middle or upper class, living in the United States or the developed world, and perhaps belonging to Greenpeace or the Sierra Club. Is your tree hugger a woman or a man? It's unclear exactly where the term *tree hugger* came from, but in 1974, a group of women in northern India mounted a protest known as Chipko, from the Hindi for embracing or hugging, in which they hugged trees literally and figuratively by forming circles of women around the trees to save them from being chopped down by the lumber industry. This movement was initiated by women not solely because the trees were worth saving for their intrinsic value as trees but because the trees were vitally important to the safety and livelihood of the women and their families (Curtin, 1997; Warren, 1997). The Chipko movement began in response to flooding caused by deforestation in the mountains above where the women lived because trees are important in preventing the kind of soil erosion that can cause or exacerbate flooding. Men's response to this deforestation was to seek out the wage labor jobs provided by the lumber industry.

Ecofeminists argue that women's concerns were different—not necessarily because they actually *are* inherently connected to nature. Rather, women were dependent on the trees in ways that men were not because the forest ensured a supply of water, fuel, fertilizer, and traditional medicines for the community. For these women in India, the health of the forest was linked in very concrete ways to the health of their own children and, therefore, to future generations. Men in these communities were positioned in ways that allowed them in part to escape the effects of deforestation by seeking paid work. But women's disadvantaged position led to their very different perspective on the importance of the forest. The more trees that were cut down, the further women had to walk for access to fuel (firewood) and fodder (fertilizer). Because women are often victimized by environmental degradation in ways that men are not, ecofeminism also highlights the ways in which women can become important social movement leaders in issues related to environmental concerns (Taylor,1997).

In focusing on the experiences of women and men in the developing world, ecofeminism overlaps with many of the concerns of the women in development perspective. One important difference is that ecofeminism is firmly grounded in feminist discourse in ways that WID is not. Remember that development theory did not evolve as a theory of gender and that women were added afterward to the equation. Ecofeminism locates itself firmly within the tradition of feminist theory, with a central concern for the role of gender. Ecofeminism draws on the tradition of radical feminism, which we discussed in Chapter 2, by positing that gender is built into the very basic structures of society through the ways in which we think about what is and isn't nature. Although WID approaches feel gender can be added to a preexisting theory, ecofeminists posit that considerations of gender should be central to our understanding of nature and the environment. A second important difference lies in the ecofeminist focus on the environment. Although the development project has increasingly begun to acknowledge the importance of protecting the environment in the developing world, this was not originally one of its goals. Development organizations and institutions increasingly include environmental concerns in their agenda, but unlike ecofeminism, these are relatively recent additions to a larger project.

HOW DO WE USE THEORY?

You might think of this first section of this book as a kind of theoretical boot camp. No one's barked orders at you or forced you to shave your head (yet), but in three chapters, you've been exposed to a rather intense dose of theoretical perspectives on gender. Boot camp is, among many other things, an attempt by the military as an institution to resocialize its recruits into the norms of being a soldier, including developing a sense of the importance of the chain of command and encouraging cohesiveness within military units. There's no way to accomplish that radical kind of resocialization through a book, but presenting these theories is a way to resocialize you into a new way of thinking about the role of gender in your own life.

Hopefully, reading about some of these theories has led you to question your own working explanation of gender and has allowed you to add new ways of thinking about gender that might not have occurred to you. Undoing previous ways of thinking and establishing new ones is a form of resocialization, so there's a chance that these chapters have at least begun that process. What should you now do with the wide range of theoretical perspectives with which you've been drilled? Before moving on to the remainder of the book, where we'll be revisiting how to apply these theories throughout each chapter, it's worthwhile to provide a brief survival guide to negotiating the complex world of gender theory with four useful pieces of advice for using theory.

The first piece of advice is fairly simple: Don't panic! Yes, there are a lot of theories out there about gender, more in fact than the few we've covered in these first chapters. Yes, these theories are often complicated, weird, vague, and sometimes difficult to understand fully. Yes, different theories sometimes say completely contradictory things, and that can seem frustrating and confusing. All of these facts can be overwhelming, but remember our earlier discussions about the purpose in learning theory. Learning and critically examining theories about gender are like taking your own ideas about gender for a kind of test flight. How do your own ideas about gender stand up to all the other ways in which feminists, sociologists, psychologists, anthropologists, and so on have thought about gender? Memorizing the ideas behind any theory just for the sake of being able to re-create it on a test isn't the point, or at least it isn't the point for the journey we'll be taking in this book. The point is to use these theories about gender to raise some interesting questions about what gender is and how it works. Theories are tools, and tools are instruments that help you accomplish whatever your goal might be. As with any tool, in the end you get to decide when and how to use it. So don't panic. You *can* learn how to use the tools we call gender theories.

Because theories are like tools, the second piece of advice for the survival guide to theory is to pack light and take what you need. It's okay to pick and choose which parts of theories you think make sense and which parts you think are a little bit crazy. This is a tried and true method for building your own theory, which we'll talk about as the fourth piece of advice. Learning how to think about gender involves learning which tools work best for you, and that means sifting through not just each theory as a whole but parts of each theory or approach as well. Maybe you think evolutionary psychology is onto something in looking at natural selection, but you don't agree with the particular mechanism they use to explain how evolution affects sex and gender. Fully understanding a theoretical perspective or approach makes it easier to sift out the good from the bad, what works from what doesn't. There's no rule that says you have to believe everything one theory says, so think about what aspects of different theories are most important or useful to you as you develop your own understanding of gender and how it works. Take what you need from these many different theories and feel free to leave the rest.

To not get bogged down in too much conflict is the third piece of advice on using gender theories. We discussed how sociological theories of gender are moving toward an integrative approach, trying to bring together the many different theoretical approaches within sociology to develop a more complex picture of how gender works. This process seems to be a much more productive endeavor than trying to prove once and for all that one particular perspective on gender is the undisputed best. "Give up on the idea of a world champion" might be one way to think of this piece of advice. Comparing different theoretical approaches and testing their relative strengths for explaining different gendered situations and phenomena is a good idea. But it's OK if there's no winner and if a theory that works very well at explaining the way gender works in one context (the developed world) falls somewhat flat in a different context (the developing world). If learning about gender using a sociological perspective can teach us anything, it's that the way gender works in the world is probably too complex to be reduced to one theoretical explanation. Don't get too caught up in the metaphor of gender theories as competitors in some kind of winner-take-all contest. Maybe a better metaphor is that of a theater company, where a wide variety of people with very different interests and perspectives manage to come together to create a work of art. Avoid conflict and try to instead enjoy the show.

The final piece of advice in this survival guide for gender theory is to study and learn the skills of the trade. Be adaptable and you, too, can build your own theory. The point in learning these theories is not to convert you to a particular way of thinking but to give you the skills to develop your own perspective. All of the pieces of advice build toward this final goal, which is to start you on your own path toward developing a way of making sense of the role of gender in your life. Learning how to use and apply these theories of gender is something like an apprenticeship, where you get to practice using the tools of the trade. In the end, if you've paid attention and done your fair share of work, you should have a stronger and more fully developed sense of your own theoretical perspective on gender. This isn't to say that all your questions about gender will be answered, but you should have a better sense of what questions are most important to you, as well as a sense of how to go about asking and answering those questions.

To help you work toward this final goal, we'll be working at two different levels throughout the rest of this book. First, we'll move into a topical exploration of gender, beginning with an exploration of how gender permeates our everyday lives and interactions. Although we'll be moving back and forth throughout the book, in this section on how our lives are filled with gender, we'll focus more on the individual and interactional levels. We'll look at how we learn gender, how gender and sexuality intersect, how gender matters for friendship and intimacy, and how gender impacts various issues connected to our bodies.

In the next section, we'll shift our focus to the institutional level with an examination of how gender functions in society as a whole. We'll look at the relationship between gender and institutions such as marriage, work, the media, and politics and government. Throughout these explorations of particular ways in which gender is manifested, you shouldn't forget that the theories we've covered are still out there, lurking somewhere embedded in your consciousness. At this first level, I encourage you to think throughout these chapters about how the theories we've covered work in explaining the specific examples of gender you'll be reading about. Use the details and examples from each chapter to test out the different theories you've learned and to help you work toward building your own theory. At the second level, and to help make sure you don't forget the theories we've so painstakingly explored in these first chapters, we'll be alerting you to the moments when particular theories are being used or when they are particularly applicable to the situation being discussed. These "Theory Alerts!" will demonstrate how the theories we've covered in these first three chapters can be applied to the specific topics covered in each chapter and serve as examples and models for you of how these theories can be applied to specific situations, therefore helping you identify their strengths and weaknesses. You might think of the theory alerts as a do-it-yourself demonstration of how to use theory; I'll show you how a particular theory might be applied in the hope that it will help you figure out how to eventually do it yourself.

BIG QUESTIONS

- In Chapter 2, we discussed the three different levels in sociological theories about gender—individual, interactional, and institutional—and where each of these locates gender. How do the theories we discussed in this chapter seem to line up with these three levels? Where do these different theories seem to locate gender?
- In Chapter 2, we also discussed two different types of feminism: liberal feminism and radical feminism. Which of these perspectives does each of the theories in this chapter seem to be more consistent with? To what extent do the theories in this chapter seem to be influenced by feminism?
- In this chapter, we explored gender theories in disciplines outside of sociology. How would you describe to someone who is not familiar with sociology as a discipline what seems to be unique about a sociological approach to gender? Where does a sociological approach to gender overlap with that of other disciplines? Is it useful to divide our approach to gender along disciplinary boundaries?

- Some theories in this chapter use a biosocial approach to the study of gender, assuming that there are two distinct types of bodies: male and female. Which theories would you characterize as biosocial? What are the strengths and weaknesses of these theories? What do you think is or should be the role of biology in the study of gender? How would you resolve the nature versus nurture debate as it relates to gender?
- Anthropologists as well as evolutionary psychologists assume there are important connections between the behavior of animals (and especially our primate relatives) and our own behavior. Do you agree with these connections? How much can we learn about ourselves as human beings from the animal world? What are potential limits to this approach? What does this approach presume about our place in nature? (Are humans a part of nature or outside of nature?)
- Many gender theories, including some of those we discussed in this chapter, attempt to find one unitary cause for gender inequality across all times and cultures. Do you believe it is possible to find one such explanation for gender inequality? Is this an example of a search for a meta-narrative, as described by queer theory, and what problems might that present?
- Queer theory provides a critique of meta-narratives, or big stories that attempt to explain some phenomenon. Can you have a theory that doesn't include some kind of meta-narrative? Does queer theory have its own meta-narrative, and if it doesn't, can it really be a theory?

GENDER EXERCISES

1. For each of the theories described in this chapter, identify how that perspective would answer the following questions: (a) What is gender? (b) Why does gender inequality exist? (c) What would be the best method for reducing gender inequality? Then think again (as you did at the end of Chapter 2) about what your own theory of gender would look like and how it would answer these three fundamental questions. Write a paragraph explaining how the addition of these new theories changed or did not change your own initial theory.
2. Go to some academic search engine website that includes a broad range of psychological journals. Search for the term *sex differences* and see how many articles are returned. What general trends can you identify in the areas in which psychologists look for sex differences and in the kind of sex differences that emerge? How important do these sex differences seem to be? Do the articles imply these are sex differences (and therefore rooted in biology) or gender differences (and therefore in part a result of social causes) or some mixture of the two?
3. Come up with a list of objects or phenomena that you consider part of nature, as many as you can think of. Conduct an informal survey with friends, family, and coworkers, asking them the gender of the various items on your list. For example, is a river masculine or feminine? You might also ask your respondents why they think certain objects are gendered. Does the gender assigned by your respondents to various objects seem consistent for different people? Ask them whether they think nature in general has a gender, what it is, and why? How do these

views line up with how your respondents think about the environment? How does this line up with ecofeminism as a theoretical perspective?

4. According to Kate Bornstein (1994) and queer theory, the gender system lets all of us down at some point. Free-write about times in your own life when the gender system seemed to let you down. Can you think of times when you weren't able to do something you wanted to do because of your gender? Times when you felt hurt, left out, or rejected as a result of your gender? Times when you felt you could not fully or perfectly live up to the gender expectations of your culture or group? Were there repercussions for not being able to live up to those gender expectations? Is Bornstein right in saying that we're all hurt to some extent by the gender system, and so none of us perfectly fits into categories of gender or sexuality?

5. In anthropology, gender theorists attempt to imagine a world without gender or sex, or worlds in which gender or sex are experienced very differently. This is also a project of science fiction as a genre, where directors and writers imagine worlds that are very different from our own. Think of some science fiction movies, television shows, comic books, or novels. In these imaginary worlds, is gender experienced differently from in your own culture? In what ways is gender different, and in what ways is gender the same? What about sexuality? What does science fiction say about our ability to imagine worlds in which the experience of gender is very different?

6. The United Nations Development Programme (UNDP) and the United Nations Population Fund (UNFPA) both publish annual reports on the social, economic, and environmental conditions around the world, and especially in the global South. Explore these reports at the websites listed at the end of this section. How are their findings consistent with the women in development approach and ecofeminism? Does gender seem to be an important part of many UN programs? Do the programs and goals of the UN seem to reflect the goals set by the development project and modernization theory? Do they seem influenced by world systems theory or dependency theory?

United Nations Development Programme: http://www.undp.org

United Nations Population Fund: http://www.unfpa.org/public

TERMS

Oedipus complex

castration complex

penis envy

Electra complex

superego

phallic symbols

sex difference research

meta-analysis

ethnography

Man the Hunter

Women as Gatherers

androcentric

matrilineal

patrilineal

meta-narrative

homosociality

homophobia

development project

modernization theory

nongovernmental organizations

formal economy

informal economy

commodity crops

subsistence agriculture

women in development

structural adjustment

SUGGESTED READINGS

On sex difference research

Benbow, C. P., & Stanley, J. (1983). Sex differences in mathematical reasoning: More facts. *Science, 222* (4627), 1029–1031.

Buss, D. M. (1995). Psychological sex differences. *American Psychologist, 50* (3), 164–168.

Eagly, A. H. (1995). The science and politics of comparing women and men. *American Psychologist, 50* (3), 145–158.

Epstein, C. F. (1988). *Deceptive distinctions: Sex, gender and the social order.* New York, NY: Russell Sage Foundation.

Hyde, J. S. (2005). The gender similarities hypothesis. *American Psychologist, 60* (6), 581–592.

Maccoby, E. E., & Jacklin, C. (1974). *The psychology of sex differences.* Stanford, CA: Stanford University Press.

On evolutionary psychology and sociobiology

Barash, D. P. (2002, May 24). Evolution, males and violence. *The Chronicle of Higher Education, 37,* B7–B9.

Caplan, P., Crawford, M. H., & Richardson, J. (1997). *Gender differences in human cognition.* New York, NY: Oxford University Press.

Fausto-Sterling, A. (2000). Beyond difference: Feminism and evolutionary psychology. In A. Rose (Ed.), *Alas poor Darwin: Arguments against evolutionary psychology* (pp. 209–227). New York, NY: Harmony.

Silverstein, L. B. (1996). Evolutionary psychology and the search for sex differences. *American Psychologist, 51* (2), 160–161.

Thornhill, R., & Palmer, C. T. (2000). Why men rape. *The Sciences, 40* (1), 30–36.

Udry, R. (2000). Biological limits of gender construction. *American Sociological Review, 65* (3), 443–457.

On anthropology and gender

Mascia-Lees, F. E., & Black, N. J. (2000). *Gender and anthropology.* Long Grove, IL: Waveland Press.

Mead, M. (1935). Sex and temperament in three primitive societies. New York, NY: Morrow.

Ortner, S. (1974). Is female to male as nature is to culture? In M. Z. Rosaldo & L. Lamphere (Eds.), *Woman, culture and society* (pp. 67–87). Stanford, CA: Stanford University Press.

Ortner, S. B., & Whitehead, H. (1981). *Sexual meanings: The cultural constructions of gender and sexuality.* Cambridge, MA: Cambridge University Press.

Slocum, S. (1975). Woman the gatherer: Male bias in anthropology. In R. R. Reiter (Ed.), *Toward an anthropology of women* (pp. 36–50). New York, NY: Monthly Review.

On queer theory

Bornstein, K. (1994). *Gender outlaw: On men, women and the rest of us*. New York, NY: Vintage Books.

Seidman, S. (1996). *Queer theory/sociology.* Malden, MA: Blackwell.

Warner, M. (1996). *Fear of a queer planet: Queer politics and social theory.* Minneapolis, MN: University of Minnesota Press.

On women in development

Boserup, E. (1970). *Women in economic development.* London, England: Allen & Unwin.

On ecofeminism

Gaard, G. (1993). *Ecofeminism: Women, animals, nature.* Philadelphia, PA: Temple University Press.

Warren, K. J. (1997). *Ecofeminism: Women, culture, nature.* Bloomington, IN: Indiana University Press.

WORKS CITED

Abeles, H. (2009). Are musical instrument gender associations changing? *Journal of Research in Music Education, 57*, 127–139.

Altman, D. (1982). *The homosexualization of America.* Boston, MA: Beacon Press.

Baker, D., & Jones, D. (1993). Creating gender equality: Cross-national gender stratification and mathematical performance. *Sociology of Education, 66* (2), 91–103.

Bell, M. M. (2004). *An invitation to environmental sociology.* Thousand Oaks, CA: Pine Forge Press.

Benbow, C., & Stanley, J. (1980). Sex differences in mathematical ability: Fact or artifact? *Science, 210* (4475), 1262–1264.

Benbow, C. P., & Stanley, J. (1983). Sex differences in mathematical reasoning: More facts. *Science, 222* (4627), 1029–1031.

Beneria, L., & Sen, G. (1981). Accumulation, reproduction, and "women's role in economic development": Boserup revisited. *Signs, 7* (2), 279–298.

Berman, J. C. (1999). Bad hair days in the paleolithic: Modern (re)constructions of the cave man. *American Anthropologist, 101* (2), 288–304.

Birkeland, J. (1993). Ecofeminism: Liking theory and practice. In G. Gaard (Ed.), *Ecofeminism: Women, animals, nature* (pp. 13–59). Philadelphia, PA: Temple University Press.

Bornstein, K. (1994). Gender outlaw: On men, women and the rest of us. New York, NY: Vintage Books.

Boserup, E. (1970). *Women in economic development.* London, England: Allen & Unwin.

Buchmann, C., DiPrete, T., & McDaniel, A. (2008). Gender inequalities in education. *Annual Review of Sociology, 34,* 319–337.

Caplan, P., & Caplan, J. (1997). Do sex-related cognitive differences exist, and why do people seek them out? In P. Caplan, M. H. Crawford, & J. Richardson (Eds.), *Gender differences in human cognition* (pp. 52–77). New York, NY: Oxford University Press.

Chabris, C. F., & Glickman, M. E. (2006). Sex differences in intellectual performance: Analysis of a large cohort of competitive chess players. *Psychological Science, 17* (12), 1040–1046.

Chodorow, N. (1974). Family structure and feminine personality. In M. Z. Rosaldo & L. Lamphere (Eds.), *Woman, culture, and society* (pp. 42–66). Stanford, CA: Stanford University Press.

Clawson, M. A. (1999). When women play the bass: Instrument specialization and gender interpretation in alternative rock music. *Gender & Society, 13* (2), 193–210.

Crawford, M., & Chafin, R. (1997). The meaning of difference: Cognition in social and cultural contexts. In P. Caplan, M. Crawford, J. Hyde, & J. Richardson (Eds.), *Gender differences in human cognition* (pp. 81–130). New York, NY: Oxford University Press.

Cucchiari, S. (1981). The gender revolution and the transition from the bisexual horde to patrilocal band: The origins of gender hierarchy. In S. B. Ortner & H. Whitehead (Eds.), *Sexual meanings: The cultural constructions of gender and sexuality* (pp. 31–79). Cambridge, MA: Cambridge University Press.

Curtin, D. (1997). Women's knowledge as expert knowledge: Indian women and ecodevelopment. In K. J. Warren (Ed.), *Ecofeminism: Women, culture, nature* (pp. 82–98). Bloomington, IN: Indiana University Press.

Deaux, K., & Major, B. (1990). *Theoretical perspectives on sexual differences.* New Haven, CT: Yale University Press.

Denzer, L. R. (1994). Yoruba women: A historiographical study. *International Journal of African Historical Studies, 27* (1), 1–39.

di Leonardo, M. (1991). Introduction: Gender, culture, and political economy. In M. di Leonardo (Ed.), *Gender at the crossroads of knowledge: Feminist anthropology in the postmodern era* (pp. 1–48). Berkeley: University of California Press.

Dresch, D. (1995). Chainsaw. In E. McDonnell & A. Powers (Eds.), *Rock she wrote: Women write about rock, pop, and rap* (pp. 74–75). New York, NY: Delta.

Eagly, A. H. (1987). Sex differences in social behavior: A social role interpretation. Hillsdale, NJ: Lawrence Erlbaum.

Eagly, A. H. (1995). The science and politics of comparing women and men. *American Psychologist, 50,* 145–158.

Engels, F. (1884/1972). *The origin of the family, private property, and the state.* New York, NY: International.

Epstein, C. F. (1988). *Deceptive distinctions: Sex, gender and the social order.* New York, NY: Russell Sage Foundation.

Esteva, G. (1992). Development. In W. Sachs (Ed.), *The development dictionary: A guide to knowledge as power* (pp. 6–25). London, England: Zed Books.

Feinberg, R. (1988). Margaret Mead and Samoa: Coming of age in fact and fiction. *American Anthropologist, 90,* 656–663.

Freeman, D. (1983). *Margaret Mead and Samoa: The making and unmaking of an anthropological myth.* Cambridge, MA: Harvard University Press.

Freud, S. (1933). *New introductory lectures on psychoanalysis*. New York, NY: W. W. Norton.

Gaard, G. (1993). Living interconnections with animals and nature. In G. Gaard (Ed.), *Ecofeminism: Women, animals, nature* (pp. 1–12). Philadelphia, PA: Temple University Press.

Galliano, G. (2003). *Gender: Crossing boundaries.* Belmont, CA: Thomson-Wadsworth.

Geoghegan, T. (2008, April 11). *BBC New Magazine*. Retrieved March 24, 2013, from Why Don't Girls Play Guitar?: http://news.bbc.co.uk/2/hi/uk_news/magazine/7342168.stm

Gordon, D. A. (1993). The unhappy relationship of feminism and postmodernism in anthropology. *Anthropological Quarterly, 66* (3), 109–117.

Halpern, D. (1996). Public policy implications of sex differences in cognitive abilities. *Psychology, Public Policy and Law, 2* (3–4), 564.

Hollander, J. A., & Howard, J. A. (2000). Social psychological theories on social inequalities. *Social Psychology Quarterly, 63* (4), 338–351.

Hyde, J. S. (2005). The gender similarities hypothesis. *American Psychologist, 60* (6), 581–592.

Hyde, J. S. (2007). New directions in the study of gender similarities and differences. *Current Directions in Psychological Science, 16,* 259–263.

Hyde, J., & Linn, M. (1986). *The psychology of gender: Advances through meta-analysis.* Baltimore, MD: Johns Hopkins University Press.

Jhally, S. (Director), & Kilbourne, J. (Writer/Director). (2000). *Killing us softly III* [Motion Picture]. United States: Media Education Foundation.

Koczberski, G. (1998). Women in development: A critical analysis. *Third World Quarterly, 19* (3), 395–409.

Lipset, S. M. (1959). Some social requisites of democracy: Economic development and political legitimacy. *American Political Science Review, 53,* 69–105.

Lyotard, J.-F. (1979/1984). *The postmodern condition.* Minneapolis, MN: University of Minnesota Press.

Maccoby, E. E., & Jacklin, C. (1974). *The psychology of sex differences.* Stanford, CA: Stanford University Press.

Mascia-Lees, F. E., & Black, N. J. (2000). *Gender and anthropology.* Long Grove, IL: Waveland Press.

Mead, M. (1928). *Coming of age in Samoa.* New York, NY: Morrow.

Mead, M. (1935). *Sex and temperament in three primitive societies.* New York, NY: Morrow.

National Center for Health Statistics. (2005, February). *Life expectancy hits record high. Gender gap narrows.* Atlanta, GA: Centers for Disease Control and Prevention. Retrieved February 2008 from http://www.cdc.gov/nchs/pressroom/05facts/lifeexpectancy.htm

Okonjo, K. (1976). The dual-sex political system in operation: Igbo women and community politics in midwestern Nigeria. In N. J. Hafkin & E. G. Bay (Eds.), *Women in Africa: Studies in social and economic change* (pp. 45–58). Stanford, CA: Stanford University Press.

Ortner, S. (1974). Is female to male as nature is to culture? In M. Z. Rosaldo & L. Lamphere (Eds.), *Woman, culture and society* (pp. 67–88). Stanford, CA: Stanford University Press.

Pietila, H., & Vickers, J. (1994). *Making women matter: The role of the United Nations.* London, England: Zed Books.

Riggs, M. (Director). (1987). *Ethnic notions* [Motion Picture]. United States: California News Reel.

Rosaldo, M. Z. (1974). Women, culture and society: A theoretical overview. In M. Z. Rosaldo & L. Lamphere (Eds.), *Women, culture, and society* (pp. 17–42). Stanford, CA: Stanford University Press.

Sedgwick, E. K. (1985). *Between men: English literature and male homosocial desire.* New York, NY: Columbia University Press.

Seidman, S. (1992). *Embattled eros.* New York, NY: Routledge.

Seidman, S. (1993). Identity and politics in a "postmodern" gay culture: Some historical and conceptual notes. In M. Warner (Ed.), *Fear of a queer planet: Queer politics and social theory* (pp. 105–142). Minneapolis, MN: University of Minnesota Press.

Seidman, S. (1996). Introduction. In S. Seidman (Ed.), *Queer theory/sociology* (pp. 1–29). Malden, MA: Blackwell.

Sharps, M., Welton, A., & Price, J. (1993). Gender and task in the determination of spatial cognitive performance. *Psychology of Women Quarterly, 17,* 71–83.

Shore, B. (1981). Sexuality and gender in Samoa: Conceptions and missed conceptions. In S. B. Ortner & H. Whitehead (Eds.), *Sexual meanings: The cultural construction of gender and sexuality* (pp. 192–215). Cambridge, MA: Cambridge University Press.

Slocum, S. (1975). Woman the gatherer: Male bias in anthropology. In R. R. Reiter (Ed.), *Toward an anthropology of women* (pp. 36–50). New York, NY: Monthly Review.

Smith, A. (1997). Ecofeminism through an anticolonial framework. In K. J. Warren (Ed.), *Ecofeminism: Women, culture, nature* (pp. 21–37). Bloomington, IN: Indiana University Press.

Tanner, N., & Zihlman, A. (1976). Women in evolution, part one: Innovation and selection in human origins. *Signs, 1* (3), 585–608.

Taylor, D. E. (1997). Women of color, environmental justice, and ecofeminism. In K. J. Warren (Ed.), *Ecofeminism: Women, culture, nature* (pp. 38–81). Bloomington, IN: Indiana University Press.

Turner, W. B. (2000). *A genealogy of queer theory.* Philadelphia, PA: Temple University Press.

United Nations Development Programme. (2003a). *Human development indicators.* New York, NY: Author.

United Nations Development Programme. (2003b). *Human development report.* Oxford, UK: Oxford University Press.

United Nations Population Fund. (2004). *The state of the world population.* New York, NY: Author.

Walter, L. (2001). Introduction. In L. Walter (Ed.), *Women's rights: A global view* (pp. xiii–xxvii). Westport, CT: Greenwood Press.

Ware, H. (1981). *Women, demography and development.* Canberra, Australia: Australian National University.

Warner, M. (1993). Introduction. In M. Warner (Ed.), *Fear of a queer planet: Queer politics and social theory* (pp. vi–xxxvi). Minneapolis, MN: University of Minnesota Press.

Warren, K. J. (1997). Taking empirical data seriously. In K. J. Warren (Ed.), *Ecofeminism: Women, culture, nature* (pp. 3–20). Bloomington, IN: Indiana University Press.

Washburn, S., & Lancaster, C. (1968). The evolution of hunting. In R. Lee & I. DeVore (Eds.), *Man the hunter* (pp. 293–303). Chicago, IL: Aldine Press.

Webster, P. (1975). Matriarchy: A vision of power. In R. R. Reiter (Ed.), *Toward an anthropology of women* (pp. 141–156). New York, NY: Monthly Review.

Weil, E. (2007, March 2). Teaching boys and girls separately. *New York Times Magazine.*

Whitehead, H. (1981). The bow and the burden strap: A new look at institutionalized homosexuality in native North America. In S. B. Ortner & H. Whitehead (Eds.), *Sexual meanings* (pp. 80–115). Cambridge, MA: Cambridge University Press.

Yucel, D. (2007). Cross-national perspectives on gender differences in mathematics achievement: The influence of sex-segregation in math-related occupations. Paper presented at the annual meeting of the American Sociological Association, New York.

PART II

How Are Our Lives Filled With Gender?

CHAPTER 4

How Do We Learn Gender?

Gender and Socialization

What's the very first thing you remember? How old were you, and what were you doing? Can you remember what you were wearing or who you were with? Is gender an important part of your first memory? Did it matter that you were a little boy or a little girl, or do you think that, at that point, you were aware of yourself as a boy or a girl—as a gendered human being? Can you remember the first time you thought of yourself as having a gender? Can you remember the first time someone treated you in a way that was obviously related to your gender? Do you remember a time when you didn't understand what gender was and couldn't necessarily tell the gender of the people around you? What was the gender makeup of your friends in childhood? In adolescence? Today? What kinds of games did you play on the playground, and were there gender differences in those spaces? Can you remember little boys or little girls who didn't seem to hang out with others of the same gender or didn't always act in ways appropriate to their gender? How did other kids and adults treat those children? Were you a "sissy" or a "tom-boy," or did you know other kids who were? What was the gender of the adults in your life when you were younger, and how did that affect your interactions with them? What lessons did grown-ups seem to teach you about gender? What are other ways in which you learned about gender as a child? Has the shape and form that gender takes in your life changed over the course of your life? Is being masculine different when you're 13 as compared to when you're 22? What about when you're 40, and then 65? Does gender become more or less important throughout the course of your own life? Is there ever a time when you get to stop being gendered?

These are the kinds of questions we'll explore in our examination of how we learn gender, or what sociologists call *gender socialization*. **Socialization** is a fundamental concept for sociologists in general, and it is defined as the ways in which we learn to become a member of any group, including the very large group we call humanity. The process of socialization begins the moment we are born and continues throughout our lives to the very end, as we constantly learn how to belong successfully to new groups or adjust to changes in the groups to which we already belong. It's not surprising given the importance of socialization to sociology as a whole that gender socialization is a good place to start in our examination of how gender matters in our everyday lives. In looking at gender socialization, we go back to our very beginnings, to the very moment when we were born. But we also consider all the moments since then, and throughout a person's life.

There are many different theories of exactly how gender socialization occurs, each with its own unique perspective on exactly what gender socialization is and how it happens. Nonetheless, we can formulate a general definition of **gender socialization** as the process through which individuals learn the gender norms of their society and come to develop an internal gender identity. This definition contains two other terms with which we should also become familiar, *gender norms* and *gender identity.* **Gender norms** are the sets of rules for what is appropriate masculine and feminine behavior in a given culture. In the sex role theory we discussed in Chapter 2, collections of gender norms are what make up a sex role, a set of expectations about how someone labeled a man or someone labeled a woman should behave. The way in which being feminine or masculine, a woman or a man, becomes an internalized part of the way we think about ourselves is our **gender identity**. You might think of gender identity as a way of describing how gender becomes internal—something that becomes an integral part of who we are, a part that many of us would be reluctant to abandon completely. The concept of gender identity is therefore consistent with an individual approach to gender, focusing on how gender operates from the inside (gender identity) out. Gender socialization begins in all societies from the very moment we are born, but in most societies, gender socialization presumes the ability to look at a new infant and give it a sex. In contemporary Anglo-European society, this means to put an infant into one of two categories, male or female. But before we discuss different ways of thinking about gender socialization as well as explore how this process takes place throughout our lives, let's begin with the first step of deciding who's male, who's female, and who's something else entirely.

Theory Alert!
Sex Role Theory

SORTING IT ALL OUT: GENDER SOCIALIZATION AND INTERSEXED CHILDREN

Thinking about gender socialization involves thinking about how people began to treat you as a boy or a girl from the very moment you were born. But how would people respond to a baby that is not clearly a boy or a girl? What color would parents use to decorate the baby's room, and what name would they choose? How would they talk about such a baby when gender is built into the very structure of our language (he/she, his/her)? What kind of toys would relatives and friends give to such a baby, and what would this child do when preschool teachers first instructed the children to form two lines, one for boys and one for girls? Even worse, which locker room would this child go to, and what would happen in the already anxious and insecure world of the locker room?

These may seem like hypothetical questions, but they lie at the core of an ongoing controversy about the very real cases of **intersexed** children—individuals who for a variety of reasons do not fit into the contemporary Anglo-European biological sex categories of male and female. These individuals are important to our discussions of gender socialization because they provide us with insight into a very good sociological question: How can we tell whether a baby is male or female? This is a good sociological question because at first glance, it seems like a pretty stupid question. Even a child knows the answer to that question, although you might get some interesting responses depending on the age and upbringing if you try asking some children how you can tell the difference between boys and girls. Still, many people would find it a stupid question because it seems to have a rather obvious answer. But sociology as a discipline is good at taking the stupid questions and making them a little bit more complicated than they first appear.

Genital Tubercles and Ambiguous Genitalia

So let's explore this stupid question that will take us into some interesting anatomical territory. When a baby is born, how do we tell whether it's male or female? Let's start with a case from the United States. Here, with our overall affluence and the availability of the latest medical technology, we assume that many couples can tell even before a baby is born whether it's a boy or a girl. What is it we're looking for in the grainy picture from the ultrasound in which babies often hardly resemble a human, let alone a boy or girl? The presence or absence of a penis. This is the same thing doctors are looking for when a baby is born. If the baby has a penis, clearly he's a boy. If the baby lacks a penis, clearly she's a girl (Fausto-Sterling, 2000). Case closed.

But here's another stupid question. How do you tell the difference between a penis, which we clearly think of as a part of male anatomy, and a clitoris, which is clearly something that only females have? You may think we've really gone off the deep end here, but would you be surprised to know that doctors and medical researchers have a very precise answer to that question? A baby has a penis if his genitalia are longer than 2.5 centimeters. A baby has a clitoris if her genitalia are shorter than 1.0 centimeters. Penises in males and clitorises in females develop from the same, undifferentiated organ in embryos, called a **genital tubercle**. So both organs have a common origin. What's important at birth in places like the United States is the length those organs have reached, and the existence of specific criteria for doctors tells us that the difference between those two organs is not as obvious as we might have initially assumed. And if you're paying attention, you may have noticed that there's an ambiguous space between 1.0 and 2.5 centimeters. What happens to these infants?

External genitalia are one way we believe we can tell the difference between males and females, but when infants are born with ambiguous genitalia, doctors and other medical professionals move on to other markers of biological sex. The length of an infant's clitoris/penis is not the only way in which ambiguous genitalia can occur at birth. There are cases of intersexed individuals who are born with both a penis (or enlarged clitoris, depending on your point of view) and a vagina. In all these cases of ambiguous external genitalia, doctors begin to investigate other indicators of biological sex, including the presence or absence of internal sex organs. They look for testes as indicators of maleness and ovaries and a uterus as indicators of femaleness. But this too can be a problematic way of determining biological sex. Intersexed infants can have a testis (male organ) on one side of their body and an ovary (female organ) on the other side. In other cases, the ovary and testes grow together into one organ that is indistinguishable as either an ovary or a testis and is therefore called an *ovo-testis* (Fausto-Sterling, 2000). In these cases, internal sex organs do not provide any easier answer to the question of the infant's biological sex than do external anatomy.

In the not so distant past, this may have been the scientific and technological limit of our ability to distinguish between males and females. But in a basic biology course at some point, you probably learned that there is also a genetic difference between males and females. Females are marked by a pair of XX chromosomes, while males are marked by XY chromosomes. Genetics, then, should surely be able to solve the problem of determining biological sex. But unfortunately, even at the chromosomal level, things are not so black and white. In general, females are XX and males are XY, but some individuals can be XO, which means they lack a second chromosome (usually a second X chromosome). In the case of Klinefelter syndrome, individuals have an extra X chromosome, resulting in an XXY pattern. Obviously, these genetic patterns have effects on how other measures of biological sex are expressed, so that those with XO patterns (called Turner syndrome) do not develop ovaries or the secondary sex characteristics (body changes at puberty and menstruation) associated with being female. Those with Klinefelter's syndrome are infertile and often develop breasts at puberty despite having male genitalia. Even at the level of our DNA, there is no simple answer to the question of how to tell whether a baby is male or female.

CULTURAL ARTIFACT 1: SEX CATEGORY, SPORTS, AND THE OLYMPICS

Have you ever stopped to think why almost all sports are divided by sex category? Why do we have the NBA and the WNBA, women's and men's World Cup Soccer, baseball for men and softball for women? Can you think of any sports that aren't segregated based on sex category, and then can you explain why? Neither horse racing nor race car driving are segregated by sex category. Why not? Increasing numbers of girls are choosing to wrestle in middle school, high school, and college. Why is wrestling emerging as a sport that doesn't need to be segregated by sex category? Little league baseball is often mixed sex until around puberty, when girls are funneled into softball and boys into baseball. Are there anatomical differences that make it impossible for women to throw or hit a baseball? The case of a 16-year-old girl recently drafted by a professional baseball team in Japan seems to suggest the answer is probably no ("Girl," 2008). Recently, the first woman tried out for the NFL regional draft as a kicker, and although she didn't make it, the NFL has said there is no specific rule preventing women from playing in the league (Rosenthal, 2013). In this world of strictly enforced sex segregation, how do sports officials go about ensuring that everyone is, in fact, the sex they claim to be? The Olympics began sex testing in 1968, in response to the masculine appearance of some "female" athletes, many of whom were pumped up on steroids (Saner, 2008). These tests involved detailed physical examinations by a series of doctors and were experienced as humiliating and invasive to the female athletes who had to undergo them. The career of hurdler Maria Jose

Martinez-Patino was derailed in 1986 when testing based on genetic sex revealed she had a Y chromosome (Bardin, 2012). She was subsequently banned from participating in the Olympics. In 2012, the International Olympics Committee instituted a new sex testing policy based on testosterone levels. Under this new system, women with naturally high levels of testosterone, a condition known as hyperandrogenism, would be banned (Jordan-Young & Karkazis, 2012). What about the male athletes? Is their sex category tested? The sex category of male Olympic athletes has never been universally tested or challenged. Why? The presumption is that a biological female competing among biological males would gain no advantage. A woman passing herself off as a man in Olympic competition isn't cheating in the way a man passing himself off as a woman would be. This is true despite the fact that in sports like wrestling and boxing, where competitors are sorted by weight class, there is no advantage that necessarily accrues to men. In long-distance running, women's times have been consistently catching up to men's (Lorber, 1994). Do women really need to be protected from competing with men? Can you imagine a world of sports that is not structured on segregation by sex category?

You might be thinking at this point, that's all good and fine. But how often do any of these things actually happen? How often do doctors have to measure the size of a baby's penis/clitoris, examine his/her internal sex organs, or analyze his/her DNA to determine his/her sex? There are many different ways in which individuals can be intersexed, as well as debates about exactly what makes someone intersexed, and these affect the various estimates as to the frequency of intersexuality. In addition, coming up with an exact number for frequency of intersexuality is difficult given that methods of reporting and data collection are hampered by the fact that being intersexed or having an intersexed infant is highly stigmatized and would therefore tend toward people hiding their status rather than reporting it. Nonetheless, some of the most reliable estimates put the number of infants who are born with an intersexed condition that merits some kind of surgery for genital reconstruction at 1 or 2 per 2,000 children (Preves, 2003). If you broaden the category to include not just those who require surgery at birth, but those with chromosomal, gonadal (having to do with internal sex organs), genital, or hormonal intersexed features, the prevalence in the population has been estimated as high as 2%. Other reports estimate that between 1% and 4% of the population is intersexed, and in some populations, inheritable types of intersexuality can be as common as 1 in every 300 births (Fausto-Sterling, 2000).

To compare to the prevalence of other kinds of conditions, intersexuality is more common by most estimates than albinism, or the condition of lacking any pigment in the hair or skin (Fausto-Sterling, 2000). Intersexuality occurs about as often as cystic fibrosis and Down syndrome, two conditions that are more familiar to most of us and certainly cause considerable less shame for parents and family members (Preves, 2003). If you go with an estimate as high as 4% of the population, intersexuality in all its forms would be as common as having red hair. Intersexed individuals have occurred throughout history and across many different societies. A more common term you may have heard for intersexed individuals, **hermaphrodites**, comes from the Greek name for a mythical figure formed from the fusion of a man and a woman. Even though many people may consider intersexed individuals to be abnormal, they are not *unnatural* any more than any other babies born with any other trait are perceived as unnatural; intersexuality can be an inheritable trait, and it appears with some frequency in the human population. The existence of intersexed people is only unnatural if you believe that the existence of only two biological sexes itself *is* natural.

What Can We Learn From the Stories of Intersexed People?

It is this point that makes the case of intersexed individuals important and interesting to our general discussion of gender, sex, and gender socialization. Many of those who study gender wonder why, given the existence of a rather large group of people who do not fit into the categories of male or female, we don't change the categories or acknowledge that maybe the categories don't work? If people are frequently born who are not really either male or female according to any of the biological criteria that we believe determine whether you're male or female, then are the categories of male and female really natural after all? This should sound familiar as a strong social constructionist approach to sexual dimorphism. The strong social constructionist approach posits that gender is what leads to the notion of sex. It is our belief in fundamental differences between women and men that leads us to believe there are two distinct biological categories called male and female. We stick to this notion even when the evidence of intersexed individuals contradicts that reality.

A good example of how you might argue that this works from a strong social constructionist perspective is the important criteria for penis/clitoris length we discussed previously. Why did doctors decide that 2.5 centimeters is the crucial length at which this genital organ becomes a penis? What biological imperative makes 2.5 centimeters such an important length? There are two considerations that make 2.5 an important number for doctors. First, doctors believe a penis/clitoris any shorter than 2.5 centimeters prevents little boys from peeing standing up. It doesn't interfere with their ability to rid their bodies of urine, which would be a fairly pressing medical and biological problem. Doctors feel 2.5 centimeters is an important cutoff because otherwise boys are not able to participate in the important *social* experience of peeing standing up with other little boys. The length is about avoiding that social stigma, even though there is no biological or medical reason men need to stand up to pee or reason why men should have to pee standing up. (In fact, both males and females can pee standing up, although Western toilets make this more difficult for people without penises.) The second consideration used to explain the 2.5-centimeter criteria is the ability of intersexed infants to use their penis/clitoris to have penetrative vaginal intercourse with a woman (Fausto-Sterling, 2000). This consideration exists despite the fact that some individuals, intersexed or not, have had successful sexual experiences with vaginal intercourse, including fathering a child in one instance, with penises that were shorter than the 2.5-centimeter criteria (Reilly & Woodhouse, 1989) This second criterion is also fundamentally *social* rather than based on any biological imperative. It assumes that to be a normal male, you must be heterosexual and, therefore, need to be able to have sex (and a very specific kind of sex) with a woman. It also assumes that all heterosexual males need to be able to have penetrative sex with a woman. What about males who become celibate, like priests? What about males who enjoy other forms of sexual activity, with other men or women, that do not involve vaginal penetration?

These criteria being used to determine what is supposedly a biological category reflect our deeply held social assumptions about the differences between women and men. In other words, the criteria reflect our assumptions about *gender,* and so in this specific instance, gendered ideas about what is important to male behavior informs our understanding of "biological" categories of *sex*. As the strong social constructionists would say, our gendered views of the world make us try to impose sex categories on a much more complex reality; gender creates sex. Another argument in this vein points to the ways in which doctors in contemporary Anglo-European societies focus solely on the functionality of a male penis, as opposed to other criteria from other time periods and cultures. During the late 19th century, when gender ideas were different, biological sex among the sexually ambiguous was also determined differently. The presence or absence of ovaries was the crucial litmus test for sex assignment, rather than the size of any external organ (Lorber, 1994). This was because the gender views of this time period told them that a woman is only a woman if she can procreate. In our more scientific world today, there is no consideration given to the presence of ovaries or the status of an intersexed infant's vagina and its suitability for penetrative intercourse. What might this reveal about our own assumptions about what makes males and females?

How does using penis length as the criterion for establishing sex reinforce the idea that to be male is the norm and to be female is to deviate from that norm? What does that imply about our society, and how would a society in which being female is perceived as the norm be different?

Why is this rather intimate discussion of genitalia and genetics an important starting place for a larger discussion of gender socialization? The study of intersexed individuals has often lain at the crossroads of debates about the relative importance of nature versus nurture in determining what we think of as gender. The current status quo among doctors and the medical profession regarding intersexed infants is to pick a sex and perform surgery and other medical interventions to bring the baby's gender into line with the chosen sex. So if an infant has a vagina and an oversized clitoris/penis, her clitoris will be surgically shortened and she will be raised as a female. The goal is not to preserve reproductive ability or physical sensation but to take the path that creates the maximum potential for normal-looking genitalia. Because a functional and cosmetically appropriate penis is more difficult to construct surgically, many intersexed individuals become females. In many of these cases, repeated surgeries may sometimes be necessary over the course of the individual's life, and sometimes individuals take hormones to induce appropriate secondary sex characteristics when they reach puberty. So an intersexed individual who is being raised as a male and develops breasts at puberty might be given testosterone to correct this problem. Sometimes testes, ovaries, or ovo-testes also need to be surgically removed. We'll talk about the repercussions for the development of the intersexed person in more detail later, but the process of creating a sex for an intersexed individual can be fairly involved, time-consuming, and painful. But the standard medical protocol for dealing with intersexed infants in the United States assumes that nurture (how a child is raised) can trump nature (the complexities of the sexual biology with which they may have been born).

CULTURAL ARTIFACT 2: TRANSGENDER KIDS

Eight-year-old Brandon Simms's first complete sentence was, "I like your high heels." As a toddler, Brandon would search his house for towels, doilies, and bandanas to drape over his head, which his mother now imagines was intended to give him the feeling of having long hair. In toy stores, Brandon would head straight for the Barbie aisles despite being guided by his mother toward the gender-neutral puzzles or building blocks. At two and a half, Brandon's mother finally allowed him to take one of his cousin's Barbie dolls home, and Brandon proceeded to carry it with him everywhere, even to bed. At three, Brandon's mother found him dancing naked in front of the mirror with his penis tucked between his legs, declaring, "Look, Mom, I'm a girl" (Rosin, 2008). Brandon is one of a growing number of young children diagnosed with gender identity disorder and identified as being transgender. In Anglo-European societies, the number of adults diagnosed with gender identity disorder has tripled since the 1960s. Those who treat gender identity disorder have seen the average age of their patients drop dramatically in recent years. What exactly does it mean to be transgendered or to have gender identity disorder, and how should parents deal with children like Brandon who seem determined that they are living in a body that does not correspond to their gender? For some, the increasing prevalence of gender identity disorder in young children is evidence that the brain itself is gendered; transgender children's insistence that their anatomical sex is incorrect is perceived as evidence that gender identity is influenced by some innate or immutable biological factors. Yet, no definitive research has established a biological basis for gender identity disorder or for being transgender in general, and other researchers believe gender identity disorder can be treated psychologically. For Dr. Kenneth Zucker, gender identity disorder is the result of instability or traumatic experiences in early infancy or childhood, and if caught early enough (before the age of 6), it can be treated through family therapy and intervention. Other parents are choosing to help their transgender children navigate a potentially traumatic social life by giving them drugs called puberty blockers. These drugs delay the onset of puberty with its irreversible effects on biological sex (Adam's apples or facial and body hair in boys, or the development of breasts in girls) in order to give children more time to decide on their actual gender identity. How do we explain the increasing number of children who are diagnosed with gender identity disorder and become transgender? Is this a natural phenomenon that Anglo-European societies are just beginning to recognize? What do our methods for dealing with these children reveal about our own investment in the gender system?

This debate about the relative weights of nature versus nurture is an old one, and it is big enough for its own textbook. What we do know is that there's a great deal of biological diversity out there that doesn't fit into the prescribed sex categories of Anglo-European society. We also know that there's a great deal of variety in how individual people and whole societies form the connections between sex, gender, and sexual orientation. Anne Fausto-Sterling (1993), whose critique of evolutionary psychology you read about in Chapter 3, proposed that we should have five sexes instead of two. The five sexes would be male, male hermaphrodite (merm), true hermaphrodite, female hermaphrodite (ferm), and female. Her categories suggest something less like a set of discrete categories (you're either male or female but never both) and more like a continuum (you're more or less male or female, but it's possible to possess both to varying degrees). Her article was considered by many to be a fairly radical potential solution to the situation of intersexed individuals, but it was also liberating to many intersexed individuals who read the article; for perhaps the first time, they saw themselves described as something more than an abnormality or medical curiosity. They were, instead, the touchstone for a larger conversation about the appropriateness of our existing sex categories.

The process of gender socialization begins with an infant being labeled either male or female. But although we imagine the identification of the sex of a baby as a fairly normal and unproblematic event, that is not always the case. This teaches us important lessons about the connections between gender and sex, but it also helps us with an important lesson we'll be emphasizing throughout this book. In telling a story about gender socialization, it is tempting to begin with the "typical" or "normal" story of gender socialization and then to demonstrate all the ways in which one might deviate from that typical story. But what would that typical story look like? Is the most typical story of gender socialization about a girl or a boy? That may seem like a pointless question, but remember that Sigmund Freud felt the gender development of boys was normal compared with the pathological and problematic development of girls. Would a typical story of gender socialization result in a heterosexual individual? From the very beginnings of research into what we now call homosexuality, scholars explored the idea that this behavior resulted from some kind of basic failure of gender socialization. Would the typical story of gender socialization be that of a white, middle-class child in the developed world, even though being either white, middle class, or in the developed world puts you in a numerical minority globally? (Most of the world's population is not white, middle class, or living in the developed world.)

Would the typical story of gender socialization be about an intersexed individual? You could argue for the typicalness of intersexed individuals' stories of socialization because they reveal in obvious ways what is implicit in the gender socialization stories of all of us. Our sex and gender don't always line up in the ways perfectly predicted by a typical story of gender socialization. Some of us who consider ourselves women pursue masculine careers. Some of us who consider ourselves men relate to other people in ways that are considered more feminine. The case of intersexed people helps us see gender socialization for what it is—a complex, fragile, always incomplete, and less-than-perfect process. There is no typical story of gender socialization, and so after giving an overview of some basic theories of socialization, we'll focus on stories of socialization selected not for their typicality or nontypicality but for the ways in which they help us to ask some interesting questions about this complicated, lifelong process of learning gender.

SOME THEORIES OF GENDER SOCIALIZATION

Theories of gender socialization help us to understand how the three main approaches we outlined in Chapter 2 are interconnected and overlapping. The three theoretical approaches we laid out are individual, interactional, and institutional. In discussing integrative theories of gender, we talked about how perhaps the best approach to gender is one that acknowledges the importance of all of three of these levels. Gender socialization can be understood at the individual level because it explains the ways in which gender comes to be internalized by individuals. Gender socialization is the story of how gender comes to be located inside individuals. Remember this is the emphasis of an individual approach—how gender operates from the inside out—so you could argue looking at gender socialization is a good example of an individual approach. In fact, many of the theories of gender socialization we'll examine come from the discipline of psychology, with its emphasis on individuals as a unit of analysis. But perceiving gender socialization solely as individualist leaves out an important part of the equation for how we learn gender.

Gender becomes internalized through our interactions with those around us. In sociological vocabulary, the person being socialized is the **target of socialization**. The people, groups, and institutions who are doing the socializing are the **agents of socialization**. Although the theories we'll discuss conceive of that interaction in different ways and suggest very different roles for the target of socialization, they all agree that interaction with our society is the central mechanism through which socialization takes place. So gender socialization can also be examined at an interactional level.

What about institutions? Note that in the definition of agents of socialization, we included not just other individuals but also groups and institutions. Institutions such as the family, school, and religion play important roles in the process of gender socialization. The role of the media as an institution has also become increasingly important in contemporary gender socialization. Understanding gender socialization should then employ all three of these levels at which gender operates. As you read about the following theories, think about the extent to which they examine the process of gender socialization at all three of these levels.

Social Learning Theory

Social learning theory developed in psychology from the legacy of behaviorism. You might be familiar with behaviorism as associated with B. F. Skinner and the ability to shape the behavior of rats based on a system of punishments and rewards. Behaviorism as a theoretical approach pushed psychology in a more scientific direction, or at least in a direction that many psychologists believed was more scientific than the one pointed to by Freud. This meant an emphasis on the collection of observable, empirical data (Siann, 1994). The unconscious drives Freud studied were hard to observe directly, but with behaviorism, the emphasis was on what *could* be directly observed: real human behavior—or sometimes, as with Skinner, the behavior of rats.

Behaviorism claimed that behavior in humans was learned, so behaviorists were interested in discovering exactly *how* we learn those behaviors. Their primary answer is that we learn through a process of rewards and punishments, or through a carrot-and-stick approach. When an infant smiles at her parent for the first time, she receives rewards in the form of verbal praise, attention, and affection. The behavior is rewarded, and chances are the baby will try smiling again in the very near future. When a toddler knocks his plate of food off the table, he may or may not get scolded, but at the very least, he *doesn't* get verbal praise and affection. According to social learning theory, this is basically how we learn, through the selective rewarding, withholding of rewards, or punishing of behavior. If you've been around infants, you've probably noticed that they can learn some fairly interesting behaviors from within a limited repertoire of what they're capable of at that age through this process.

Sex-typed Behaviors

It's fairly easy to see how these basic ideas apply to the specific process of gender socialization. Social learning theorists identified specific **sex-typed behaviors** (Mischel, 1970). A behavior is sex-typed when it is more expected and therefore perceived as appropriate when performed by one sex but less expected and, therefore, perceived as inappropriate when performed by the other sex. Making a list of sex-typed behaviors results in the articulation of a gender or sex role, which we've already discussed, so sex-typed behaviors are also similar to the concept of gender norms. The idea of sex-typed behaviors adds the idea, not necessarily contained in the idea of gender norms, that we very purposefully categorize behaviors as appropriate to one sex but not to the other. Gender socialization works, according to social learning theorists, by rewarding children for engaging in sex-typed behavior that is consistent with their assigned sex category. The classic example is crying; while a little girl may be soothed when she cries, a little boy may be told that boys don't cry. Crying is a sex-typed behavior, considered OK for girls and therefore not a punishable behavior. But because it is not considered an appropriate behavior for boys, the little boy may be punished or corrected for his crying behavior. Through these kinds of interactions, gender socialization occurs.

What are some other examples of sex-typed behavior? Can you remember being rewarded, not rewarded, or punished for engaging in a sex-typed behavior when you were growing up? Can you think of instances of when children you know or spend time with have engaged in sex-typed behaviors and been rewarded for those behaviors?

This original formulation of social learning theory described a somewhat conspiratorial role for the agents of socialization, people like parents, teachers, and friends. This is because social learning theory implies something of a conscious effort to reward and punish sex-typed behavior differentially. It calls to mind images of mothers and fathers developing careful plans for the behaviors that will be rewarded and punished in their sons and daughters. In reality, few parents ever sit down and make these kinds of intentional decisions. Social learning theorists subsequently added to their original formulation and said that conscious intent on the part of agents of socialization was not necessary to the process (Bandura, 1963). Latent learning can take place as a result of the way children tend to imitate those around them, regardless of whether they will be rewarded for that imitation.

This shifted the focus of social learning theory toward imitation and modeling, but it raised questions as to exactly *whom* children were imitating and modeling themselves after. Social learning theorists argue that children are more likely to model themselves on same-sex individuals by paying more attention to same-sex peers and forming a stronger bond with same-sex parents. This bond with the same-sex parent depends on a process called **identification**, where a child copies whole patterns of behavior without necessarily being trained or rewarded for doing so (Siann, 1994). This move toward modeling and imitation shifted social learning theory away from the more simplistic model of punishments and rewards as guiding behavior.

Problems emerged when these underlying assumptions of social learning theory were tested using empirical research. According to social learning theory, children should pay more attention to peers of the same sex and identify more with their same-sex parent. But research suggests that neither of these is necessarily the case. Children in experiments do not always pick a same-sex playmate to model nor do they always identify with a parent of the same sex (Maccoby & Jacklin, 1974; Williams, 1987). Other critiques of social learning theory pointed out the very passive role of targets of socialization in this formulation. Children are perceived as largely passive recipients of their culture's ideas about gender and sex-typing, and there is little room for considering children as playing an active role in this process.

Cognitive-Development Theory

The next theory emerges from this specific critique of social learning theory and reflects another general shift in psychology as a discipline. Behaviorism faded in importance within psychology during the middle part of the 20th century (Siann, 1994). The specific area in psychology dealing with socialization came under the influence of the Swiss theorist Jean Piaget. Piaget (1954) directly contradicted social learning theory by emphasizing children's active role in their own socialization rather than considering them to be passive recipients of socialization. In addition, Piaget brought to the psychological study of socialization an emphasis on the stages of children's cognitive development. This was a somewhat radical approach to child development because it implied that children were not "little adults," but they were fundamentally different in the way they think, feel, and understand the world around them.

Lawrence Kohlberg (1966) used Piaget's models of child development to create a new psychological theory of gender socialization, commonly called **cognitive-development theory**. This theory seeks to explain the ways in which children acquire a sense of a gender identity and the ability to gender-type themselves and others. We've already discussed gender identity as the internalized sense of yourself as male or female. Gender-typing is another term for sex-typing, for identifying behaviors that are perceived as appropriate for one sex or gender but not the other. Children acquire gender identity and learn to gender-type as they progress through a series of discrete, fixed developmental stages. The emphasis throughout is on how children actively develop an understanding of gender and, then, based on that understanding, how they actively socialize themselves, rather than serving as passive objects of socialization.

Stages of Gender Socialization

The first stage happens between the ages of two and a half and three when children acquire a gender identity. Children of this age should be able to identify their own gender as well as identify the gender of others around them. If you've been around small children, you may have noticed that we're certainly not born with the ability to distinguish gender—it is something that has to be learned. Hence, the two-year-old who points to people on TV or, sometimes more embarrassingly (although it's interesting to think about why it's embarrassing when a child cannot correctly identify someone's gender), a friend or relative and asks, "Is he a boy or a girl?" By the age of five, children have acquired gender stability, according to cognitive-development theory. With **gender stability**, children know that their gender is permanent, and that it is the gender they will be for the rest of their lives.

This seems fairly obvious to us but, interestingly, not to children. The son of one of my students is pretty typical in having expressed the desire at one point to grow up and become a mommy, implying that he would be able to change his gender at some point in the future. Once a little boy has achieved gender stability, he understands that by and large he cannot become a mommy. It is not until the age of seven, according to cognitive-development theory, that children reach the final phase of gender understanding: gender constancy. With gender constancy, children develop the complicated understanding that even a male wearing a dress, a wig, or makeup is still fundamentally a male. **Gender constancy** brings an understanding that even changing the outward physical appearance of a person does not change his or her underlying sex category. Up until this stage, children's understanding of gender is still limited and based on very concrete rules (e.g., girls have long hair, and boys have beards) (Siann, 1994).

Although the process of gender socialization for cognitive developmentalists begins when children develop a gender identity, at the age of two and a half to three years, actual gender-typing does not begin until children achieve gender constancy at age seven. At that point, children actively begin to select from their environment the behaviors that they consider consistent with their gender identity. The basic idea is that once a little girl begins to perceive herself and others as gendered, she will be self-motivated to engage in feminine behaviors and to model herself on the other people she identifies as women in her environment. This is driven in part by children's need for cognitive consistency; if children know what their gender is, then what they do and think should line up with that gender (Bem, 1983). Children work to achieve **gender congruency** and, in the process, achieve gender socialization. Children do not become fully sex-typed until they have achieved the final stage of gender constancy at around age seven. Cognitive development theory does not completely dismiss the importance of the external environment or of society itself. Society obviously provides the material from which children pick and choose to achieve gender-congruency. But it does locate much more of the power in the process of socialization with the targets (children) rather than with the agents of socialization.

Both social learning theory and cognitive development theory largely predate the broad influence of second wave feminism on academic disciplines like psychology. Both theories are general theories of socialization that can be applied to the specific question of gender socialization. Given that these theories developed before feminists' entrance into psychology, it's probably not surprising that one of the critiques of cognitive-development theory is its male bias. Kohlberg (1966) focused his theory mostly on the case of young boys, and you can see how the theory can work better for little boys than little girls.

From this perspective, children are self-motivated to gender themselves, without the need for much external pressure. It makes sense in a male-dominated world that boys would be motivated to adopt a masculine gender identity. Being masculine brings with it power and privilege. But if children are savvy enough to work on developing gender congruency, they're probably also smart enough to work out that being feminine is not quite as valued in our society as being masculine. Cognitive-development theory had difficulties explaining this differential dynamic between little girls and little boys. Other critiques based in empirical research point out that children begin to demonstrate preferences for objects and activities based on gender by the age of three (Unger & Crawford, 1992). This contradicts the predictions of cognitive-development theory that children will not begin to engage in gender-typing until gender constancy is achieved at age seven. If you've been around children younger than age seven, you might have noticed that as a group they're fairly invested in gender. Generally, cognitive-development theory seems to place the process of gender development fairly late in childhood.

Gender Schema Theory

The next two theories emerge from the specific context of an increasing influence of feminism in psychology and of feminist psychologists bringing their own perspective to the topic of gender socialization. The first is **gender schema theory**, which builds on the frameworks of both cognitive-development and social learning theory to formulate an explanation that is specific to gender socialization, rather than to socialization as a more general process. This theory was developed by Sandra Bem, and one of her critiques of cognitive-development theory was that it provided no explanation for why children socialized themselves based on sex as a category in particular. Bem (1983, 1993) questioned why sex became the important organizing principle around which children built their identities, rather than other readily available categories such as race, religion, or even eye color.

Bem called this the "why sex?" question; why is it that sex becomes such an important difference in the lives of very young children. Because the theory of cognitive development does not provide an answer or address how its theory might address these other categories of difference as well, Bem (1983) argued there's a presumption that sex differences are "naturally and inevitably" (p. 602) more important to children than other differences. This is connected to a more general problem with cognitive-development theory; the theory errs too far on the side of privileging individual behavior and decisions. As Bem (1983) noted, in cognitive-development theory, children seem in large part to socialize themselves in the "absence of any external pressure to behave in a sex-stereotyped manner" (p. 601). Gender schema theory seeks to correct this swing away from the external pressures important to social learning theory by finding a balance between the two.

What Is a Schema?

How does gender schema theory find this balance? As the name of the theory suggests, it begins with the concept of schemas. A **schema** is a cognitive structure and network of associations that helps to organize an individual's perception of the world (Bem, 1983). According to gender schema theory, schemas help us organize incoming information and perceptions from the outside world; they serve a kind of sorting and organizing function. More importantly, they shape the way we look for and experience information as well as what kind of information we integrate into our way of thinking. In this sense, schemas are more than just a system of organization, like a file folder that exists inside our head. Schemas also include the complicated ways in which the files are interconnected, as the file for mother is probably also connected to female, woman, feminine, nurturer, caretaker, and so on. The existence of this schema then shapes the very way in which we perceive the world around us; if our schema about mothers is made up of all these characteristics, we might pay more attention to the mothers we see who line up with all these other qualities, while becoming less likely to perceive the mothers who do not.

Schemas are ways of organizing information, but they also become, as Bem (1993) pointed out, lenses that shape the way in which we see the world. A **gender schema**, then, is a cognitive structure that enables us to sort characteristics and behaviors into masculine and feminine categories and then creates various other associations with those categories. Like schemas more generally, gender schemas also eventually come to shape the ways in which we perceive the world around us, through the lenses of gender. Gender schema theory poses that rather than rose-colored glasses, we all live with gendered-colored glasses that lead us to see the world in some very specifically gendered ways.

Socialization occurs as children assimilate their self-concept, the way they think about themselves, to their gender schema (Bem, 1983). Children learn the content of their particular society's gender schema, or the network of associations around the characteristics of masculine and feminine. They also know that they fall into one or other of those categories based on their own sex. When they begin to think of themselves as masculine or feminine, that particular gender schema is also associated with their sense of identity. They learn that when they are picking behaviors and ways of thinking to assimilate into their own sense of selves, they should limit themselves to the particular subset of behaviors and attitudes appropriate to their own gender. As with cognitive-development theory, children are motivated to socialize themselves—but now through the mechanism of the power of gender schema.

So far, gender schema theory doesn't sound that much different from cognitive-development theory in its emphasis on socialization from the inside out. Remember that one critique of cognitive-development theory was that it swings too far away from attention to the role of culture and society. One of Bem's (1983) critiques of cognitive-development theory was that it lacked any explanation for why sex in particular becomes the most important category of organization. Gender schema theory addresses these concerns, and it is here that the theory shifts to processes beyond the individual to the level of society.

Characteristics of Gender Schema

Gender in particular becomes an important organizing category because it is considered by almost all cultures to be functionally important to society. Gender schemas exist because cultures are structured in such a way as to convince us that society cannot function without the existence of sex and gender categories. Because of the importance placed on gender by most cultures, a very broad set of associations between the categories masculine and feminine and many other attributes, behaviors, and categories comes to exist. In other words, gender pervades the way we think about the world and crosscuts many other categories. An example Bem (1983) provided was that people are perfectly and consistently capable in experiments of sorting seemingly gender-neutral terms and objects into masculine and feminine categories. In experiments, people will spontaneously sort *tender* and *nightingale* as feminine and *assertive* and *eagle* as masculine, despite the fact that these terms have no clearly gendered content. Gender schemas are particularly important, then, because culture creates and enforces that importance.

Does it make sense to think of objects and feelings (like *tender, assertive, eagle,* and *nightingale*) as gendered (masculine or feminine)? In some languages, even words like *cat* and *dog* have a gender. Is this true in a more subtle way for English, which no longer has gendered forms for many nouns?

In later work, Bem (1993) outlined some of the specific content of gender schemas as they existed in Anglo-European societies, including a history of their cultural evolution. She identified *androcentrism* and *gender polarization* as two important lenses that shape the way we see and understand gender in many parts of the developed world. **Androcentrism** (which we also discussed in Chapter 3) is the belief that masculinity and what men do in our culture is superior to femininity and what women do. Femininity and all it entails are considered deviations from the universal standard of masculinity. Bem traced a long history of androcentrism in Anglo-European culture, some examples of which are already familiar to us. Freud's theory of gender development is a good example, where having a penis was perceived as the norm and women's development was perceived as inferior and abnormal as a result of women's lack of a penis. Androcentrism is also a useful concept for explaining the many ways in which it is sometimes more acceptable for women to engage in masculine behavior than it is for men to engage in feminine behavior. In the United States, most men will get a lot more flack for wearing a skirt or makeup than a woman will receive for wearing men's pants or a man's hat. We will see in the next section how psychoanalytic theory, another theory of gender socialization, has a different explanation for the same types of behaviors.

Can you think of other examples of androcentrism in society, or ways in which it's more OK for women to act masculine or do masculine things than it is for men to act feminine or do feminine things? Can you think of examples that don't fit this pattern, or of examples of times when it's considered OK for men to act feminine but not OK for women to act masculine?

Gender polarization, the second important part of how we perceive gender in Anglo-European society, describes the way in which behaviors and attitudes that are viewed as appropriate for men are viewed as inappropriate for women and vice versa. Bem (1993) argued that gender polarization operates in two ways. First, it creates two mutually exclusive scripts for being female and male. This means that the script that is appropriate for males is only ever appropriate for males, and no script can ever be appropriate for both males and females. Second, gender polarization problematizes any person who deviates from these mutually exclusive scripts as unnatural, immoral, abnormal, or pathological, depending on the particular system of thought being used. Gender polarization is an important way in which the strong link among sex (as biology), gender, and sexuality is maintained. To be female is to be heterosexual and to be attracted to males, and so lesbian women would be an example of people who are considered to be unnatural, immoral, abnormal, or pathological as a result of gender polarization. In this later work on the lenses of androcentrism and gender polarization, Bem focused even more attention on the question of **enculturation**, or on how culture comes to reside inside individuals. This shift in gender schema theory brings us back to a balance between the importance of external agents of socialization and active targets shaping their own process of learning gender through the mechanism of gender schemas.

Can you think of other types of individuals or groups of people who would be problematic according to the system of gender polarization?

Psychoanalytic Theory

The final theory of gender socialization we will explore also draws on psychology as a discipline but on a very different kind of psychology. We discussed Nancy Chodorow in Chapter 3 as an example of a feminist theorist attempting to provide an explanation for women's universal subordination that is based on a social, rather than on a biological, explanation. Chodorow (1978) laid out her answer to this question in her book *The Reproduction of Mothering,* and you should be able to guess from the title where she locates her explanation for women's universal subordination. Like gender schema theory, psychoanalytic theory is an explanation specific to the process of gender socialization, rather than beginning as an exploration of the process of socialization more generally. Rather than drawing on cognitive or behaviorist theory, Chodorow began with Freud's legacy of psychoanalysis as important to explaining the key causal factor in women's subordinate position: their status as mothers.

Psychoanalytic theory begins with the importance of women's status as mothers and uses principles from Freud and others in the psychoanalytic tradition to explain the ways in which gender becomes deeply embedded in the psychic structure of our personalities. This is important to distinguishing psychoanalytic theory from other theories of gender socialization in which gender is a behavioral acquisition, something children pick up in the process of socialization. For psychoanalysts, gender is something that becomes deeply embedded in our personality structures very early in our development in ways that other theories of gender socialization do not adequately describe.

To understand exactly how this happens, it is necessary to understand a few concepts essential to psychoanalytic theory. The first is identification, here used somewhat differently than it was in the context of social learning theory. **Psychoanalytic identification** is the way in which a child modifies her own sense of self to incorporate some ability, attribute, or power she see in others (usually a parent) around them. When a child is developing a sense of right or wrong, he does not just internalize a kind of miniature version of the parent who tells him what is right or wrong. Rather, in identification, that ability to distinguish between right and wrong becomes a part of the child's own sense of self; it becomes a sense of inner regulation for the child (Chodorow, 1978, p. 43). The other important concept is that of **ego boundaries**, another term borrowed from Freud, which describes the sense of personal psychological division between ourselves and the world around us (Chodorow, 1978, p. 68). Ego boundaries are what help us figure out where the stuff called "me" stops and everything else begins. This may seem pretty self-evident, but from a psychoanalytic perspective, it's not as simple as it might seem. We're not born with ego boundaries; they are something we learn and develop in early childhood.

In her formulation of psychoanalytic theory, Chodorow (1978) maintained many of the basic ideas of Freudian theory, including the Oedipal complex. But the end result of these processes is different; rather than focusing on the attainment of gender identity and heterosexuality, psychoanalytic theory is interested in the "relational potential" produced in people of different genders (p. 166). Identification occurs for both boys and girls with their mothers initially, as a result of the complete dependence of the infant on the mother as primary caregiver. In this very early phase of our development, occurring beyond our ability to remember consciously as adults, we have no sense of ego boundaries between ourselves and our mothers. According to psychoanalytic theory, infants at this stage do not experience themselves as separate from their mothers. Eventually, though, infants come to see that although they are completely dependent on their mother for their survival, the reverse is not true; even the most dedicated of mothers has other concerns beyond her infant, and psychoanalytic theory focuses on how boys and girls resolve the tensions caused by this realization. In other words, if Mom has concerns that are not consistent with my own, Mom must actually be separate from me. The process is qualitatively different for boys and girls, which gives us radically different personality structures for men and women as adults.

The Gender of Ego Boundaries

Not to state the obvious, but female infants are of the same gender as their mother. Because of this similarity, they can experience a sense of connection with their mothers for longer than male infants. This is because, conveniently enough, the gender identity they need to learn is available to them much more readily than it is to boys; girls can develop a sense of gender identity through their direct personal relationship with their mothers. In addition, Chodorow (1978) argued that because mothers themselves have already internalized a sense of gender identity, they experience their infant daughters as more similar to them than their infant sons. On some unconscious level, mothers then push their sons away in ways that they do not push their daughters. As a result of these early psychological dynamics, girls emerge with a personality structure characterized by empathy and with less of an ability to differentiate themselves from others. Feminine personality structure has less developed ego boundaries.

Male infants have the task before them of acquiring a masculine gender identity, despite the fact that their primary identification is with their mother, who represents feminine gender identity. How do boys acquire this masculine identity in the absence of an initial masculine identification? According to Chodorow, this is a problematic dynamic for masculine development as boys learn masculinity in the absence of an ongoing relationship with a male figure. In addition, to become masculine, boys must sever their sense of connection to and identification with their mothers. As a result of these underlying dynamics, the masculine personality structure emerges with a much more well-developed sense of their separation from others. Men have stronger ego boundaries than do women. Masculinity is learned by boys in part as a rejection of what is feminine, including their identification with their mother. In the absence of this kind of strong relationship with other men, masculinity is learned by boys through the use of cultural stereotypes, rather than through the kind of direct observation that girls experience with their mothers. This results in two important features of masculine gender identity: It is less stable than feminine gender identity, and it contains, as a basic element, a devaluation of all things feminine.

There are two important features to highlight about psychoanalytic theory. First, it explains not just how gender socialization occurs but also how the same process of gender socialization re-creates itself across generations. Girls who emerge from this developmental process more empathetic and with less of a sense of ego boundaries are predisposed to seek out the kind of nurturing involved in mothering, therefore, reproducing the same personality structure in their children. These processes of gender development don't just produce generic gender differences, they also produce a new generation of women whose personalities lead them to want to mother, and therefore to reproduce again in their own sons and daughters the same inevitable process. In Chapter 8, we will discuss the persistence of women's roles as primary caregivers to children, even in families who consciously attempt a more equitable division of labor. Chodorow's (1978) theory helps explain this persistence because the desire to mother is a fundamental part of feminine personalities. The second feature to note in psychoanalytic theory is that it also helps to explain the subordination of women through the development of masculine personality. Masculinity has a devaluation of women and therefore of the feminine built into its very structure. This neatly explains why women seem to be universally subordinate to men. Considering women to be inferior is an essential part of what it means to be masculine, according to identification theory.

Psychoanalytic theory has had widespread influence and has inspired many studies to explore these dimensions of masculine and feminine personality (Belenky, Clincy, Goldberger, & Tarule, 1997; Gilligan, 1982; Williams, 1991). Carol Gilligan (1982) used identification theory to argue for a uniquely feminine approach to issues of justice and morality. You might remember from our earlier discussion of Freud that he predicted women would have a much less developed sense of justice and morality. Gilligan used identification theory to argue that women's morality is structured by the fact that they experience less of a sense of separation between themselves and others in their environment. So although traditional ideas of justice assume that right and wrong must be determined by an objective devaluation of empathy and compassion, a more feminine sense of justice is deeply entwined with the idea of being able to take the position of others. Masculine ideas of justice are blind and assume that one can only determine what is just by ignoring the particulars of a person's situation. Feminine justice assumes that the unique set of particulars must be considered. Other studies have used psychoanalytic theory to explain the experiences of men in predominately female occupations, the attraction of young boys to sports, and gender differences in how women and men learn (Belenky et al., 1997; Messner, 1990; Williams, 1991).

Psychoanalytic theory is a good example of the unique perspective feminist theory can bring to preexisting and gender-biased modes of thinking. Chodorow (1978) took Freudian theory's emphasis on women's problematic development and flipped it on its head, arguing that in some ways, women's psychological development is less fraught with difficulties than that of men. Both feminine and masculine personalities have their difficulties, but psychoanalytic theory reverses the tendency of Freudian theory to normalize masculinity while problematizing femininity. But by drawing on Freud and psychoanalysis as a model, psychoanalytic theory is subject to some of the same critiques. In psychoanalytic theory, most of the important events of gender socialization happen at a very early age, resulting in a relatively fixed gender identity by the time we are about two to three years of age.

In addition, although psychoanalytic theory emphasizes the importance of social factors, namely the structure of the family, in its emphasis on unconscious processes that occur so early in our development, it can be considered an essentialist theory in its implications. That is, psychoanalytic theory implies that because gender differences become deeply embedded in the structure of our personalities, they are part of our essential natures and difficult to change. Other critics point out the difficulty of verifying the assumptions and predictions of psychoanalytic theory using empirical research. How do you prove that an infant experiences no sense of separation between himself or herself and his or her mother? How do you demonstrate the inherent instability of masculinity? Although some of the studies discussed attempt to demonstrate this in adults, proving the initial dynamics described in psychoanalytic theory is a difficult task.

How are gender schema theory and psychoanalytic theory, which are theories influenced by feminism, different from social learning theory and cognitive-development theory, which were not developed with the influence of feminism? What questions do the latter two theories ask that seem consistent with what you've learned about feminism so far?

THE EARLY YEARS: PRIMARY SOCIALIZATION INTO GENDER

Now that we have a familiarity with some of the more important theories of how gender socialization *should* happen according to these theories, we can begin to explore accounts of how gender socialization actually *does* happen in a variety of different settings. We'll start at the beginning, with what sociologists call *primary socialization*. **Primary socialization** is simply the initial process of learning the ways of a society or group that occurs in infancy and childhood and is transmitted through the primary groups to which we belong. **Primary groups** are characterized by intimate, enduring, unspecialized relationships among small groups who generally spend a great deal of time together (Cooley, 1909). This definition should sound like what most people understand a family to be, although other groups, such as childhood friends, can also make up a primary group. One way to think about a primary group is if you drew a ring of concentric circles around yourself that represented increasing levels of intimacy and importance to you in your life, the innermost circle would be your primary group. All the theories we have discussed deal with primary socialization, and this includes a focus on what happens very early in our lives in the specific environment of our families.

Primary Socialization

Primary socialization begins before a baby is even born because the baby is born into a society with certain assumptions about gender that existed long before he or she came along. Important to newborns' socialization is the relative value placed on either gender in society. In many societies around the world, more value is placed on male infants than on female infants (DeLoache & Gottlieb, 2000). In many cultures, the family name can only be carried on by sons because daughters take on their husband's name when they marry. Sons may be responsible for taking care of elderly parents or for rituals important to family and religious life. For example, in the Hindu religious tradition, important rituals surrounding the death of a parent should generally be carried out by a son, including proper preparation of the body for cremation. The reasons for this preference for boys vary, but it has important impacts on the gender socialization of boys and girls. The United Nations has estimated that female infants and children suffer higher rates of abuse and neglect in countries such as China, India, Papua New Guinea, and the Maldives because of the preference for male children (United Nations, 2000). Historically, in these countries, female infants survived at much lower rates than male infants because parents fed the girls less, sometimes left them exposed to the elements, and generally neglected their female children. In a moment, we'll explore some more subtle examples of gender socialization, but if you can imagine being a girl in one of these societies, what would be the end product of this type of gender socialization? Assuming you did survive as a girl, what kind of gender identity might you end up with?

Does your culture show a preference for male infants? As technology develops that allows parents to influence the sex of their child, will parents demonstrate a preference for male or female children?

Boys are not considered more desirable in every society across the globe. The Mukogodo people of Kenya place more value on female infants because of the importance of bride wealth that a girl child can bring to the family (Cronk, 1993). The hunting activities of male

Mukogodo are also stigmatized in comparison with the farming activities of surrounding tribes. Because of the advantages female children bring to the family relative to male children, the Mukogodo women breastfeed their daughters longer and take them to receive medical care more often. There are also historical variations in the extent to which Americans valued girls and boys. In the rural 1800s, boys were considered more valuable than girls because of their assumed ability to contribute to farm work. But in urban areas of the same time period, girls were considered more valuable because they could work in factory jobs and contribute to the family's income (Coltrane, 1998). In cultures like these two, the experience of gender socialization for girls would be very different from the experience of girls in places like China and India.

This is the first point for us to realize about gender socialization. The way gender socialization takes place and the particular form of gender that children learn is deeply related to the type of culture in which they find themselves. The relative value placed on boys and girls is just one piece of that picture. As psychoanalytic theory indicates, the structure of family life is also important to the form of gender socialization. Do children grow up in what has come to be perceived as the typical model in the United States: the nuclear family consisting of mom, dad, and children? Or are they raised in a multigenerational family in urban India, with three generations of a family often living together in one apartment? Within that family structure, the way child care and other duties are distributed is also important to the process of gender socialization. Who is primarily responsible for taking care of children in a society? How much importance is placed on the task of caring for children relative to other activities in the society? What other kinds of duties does the person primarily responsible for child care have, and how compatible are those other duties with the task of caring for a child? If we begin just with this basic set of questions that have to do with socialization that happens within family structures, it should already be obvious that the social context matters. Let's begin with the specific case of gender socialization in one setting to explore some of these questions about the social and cultural context of gender socialization.

The One-Child Policy and Gender in China

The family is an important source of primary socialization. The role of families is especially emphasized in social learning theory, which has been described as a top-down explanation of gender socialization. Literally, the ones on the top (the taller people) are primarily responsible for imparting gender to those on the bottom, the shorter humans we call *children*. From this top-down perspective, parents and other adults in a child's life matter a great deal for gender socialization. Adults take an active role in shaping the way children think about and internalize gender. Taking this top-down approach, the case of China and its one-child policy presents an interesting example of how parents and family structure can be important to the process of gender socialization.

As we discussed, Chinese culture has a long history of preference for male children over female children. Within a traditionally patrilineal and patrilocal system, daughters marry into their husbands' family and are responsible for the support of their in-laws in old age, rather than for their own parents (Tsui & Rich, 2002). Sons therefore become important to parents for the support they will provide to them in their old age, as well as for their ability to farm in rural areas. This meant that for much of Chinese history, as in many parts of the world, women were not educated at the same rates as men. Because of these parental priorities, part of the gender socialization experiences of boys in China would have been that education was for them but not for their sisters. In this top-down model of socialization, the priorities of parents are instilled in the motivations and expectations of their boy and girl children.

Much of this changed during Communist rule in China, which began in 1949. Improving the status of women was among one of the goals of the Communist government in China, and this included denouncing the more traditional ideas about gender that had existed up until this point. These goals were enforced through a gender-neutral, state-assigned employment system that helped to encourage more Chinese women to enter the workforce as well as to diminish the pay gap between women and men in China. This meant the government literally decided who was hired for various jobs rather than putting that power in the hands of the employers themselves. In addition, education was nearly free at the primary and secondary level, removing the cost barrier to parents educating their daughters. Prior to Communist rule, the gap between men's and women's levels of education was as high as 40% (83% of women compared with 40% of men were illiterate in 1949). In 1990, among urban Chinese born under Communist rule, the average length of schooling for men was 10.4 years compared with 9.4 years for women, a considerable reduction in this inequality (Tsui & Rich, 2002). Ideologically, the Communist party instituted slogans such as "Women hold up half the sky" to encourage the cultural values of gender equality.

The social system of Communism in China is a fascinating example of what happens when a government seeks to very intentionally and radically alter preexisting values and patterns in a culture. Communist leaders in China wanted to reduce gender and class inequality, but the nation also faced the pressing problem of overpopulation. In Chapter 7, we'll discuss the gendered implications of government attempts to limit population growth, but in China, the main strategy for population control became the **one-child policy**. Under this government-mandated system, which began in 1979, couples were limited to having only one child per family. This might seem rather extreme from your own cultural perspective, and especially so if you consider control over reproduction and the structure of families to be a basic human right. In addition, 20 years of Communist rule encouraging the ideals of gender equality have not been enough to erase completely centuries of male domination in China. Because many couples still value having sons over daughters, the one-child policy has led to sex-selective abortion and the abandonment of female infants (Tsui & Rich, 2002). Urban couples limited to having only one child find out the sex of the baby before it's born and abort female children, while in rural areas, couples simply abandon female infants after they are born.

Gender Implications of the One-child Policy

In their research, Ming Tsui and Lynne Rich (2002) pointed out that in addition to these negative consequences, the one-child policy has also had an unintended positive consequence for girls born to one-child families in urban areas in China. They focus specifically on urban families in China, rather than focusing on the very different experiences of those in rural areas and smaller cities where educational and employment opportunities, as well as the extent to which the one-child policy is followed, are different. Their research on only-girl and only-boy families in Wuhan, the most populous city in central China, demonstrates that a variety of cultural, economic, and governmental factors come together to reduce significantly gender differences in the educational aspirations of boys and girls. Culturally, Confucianism as a strong philosophical tradition in China places a great deal of importance on education; teachers are highly revered in Chinese society in a way that they are not in Anglo-European societies like the United States.

At the same time, economically many parents look to their children as a kind of future "welfare agency" for them as they age. Although older persons were often well provided for under Communism with pensions and old-age payments, more recent changes to the economy have made parents more nervous about how they will be provided for in their old age. Free-market economic reforms have undermined pensions, while inflation and widespread layoffs have led to increasing distrust in the government's ability to provide for people in their old age. This means parents increasingly turn to their children as crucial to providing for their welfare when they can no longer work. These pressures, combined with the reality of the one-child policy, mean that urban families can no longer afford to encourage only their sons to pursue an education. This research in Wuhan demonstrates that an unintended consequence of the one-child policy is that it helped to promote education for girls in one-child families as a result of these pressures as well as of the absence of a brother or other siblings to compete with for support and encouragement.

Their analysis of survey results from 1,021 eighth-grade boys and girls in Wuhan reveals no significant gender differences in a wide range of variables that deal with actual academic achievement as well as with the amount of resources invested in the education of boys and girls. There were no significant gender differences in mathematical scores between the boys and the girls, despite what we've already learned about the tendency in many cultures for boys to be more encouraged to pursue and excel in math than girls. This demonstrates further support for the idea that this particular sex difference in mathematical ability is probably more cultural than biological. Girls and boys also did not differ in their educational aspirations for themselves. In addition, the expectations parents had for their children's education did not show any significant differences by the gender of the child for these one-child families, and parents of boys were no more likely to invest in their child's education than were the parents of girls. Tsui and Rich (2002) demonstrated that among this particular group of urban, one-child families, gender differences in educational expectations and performance seem to have been greatly reduced, if not to have disappeared altogether.

Thinking about your own family, how do you think the structure impacted your own gender socialization? Were there gender differences in how your parents treated you and siblings of a different gender? Were there more women or men in your family (including parents and other members of your household), and how did that affect your own gender socialization?

What does the case of the one-child policy in China teach us about gender socialization in general? First, the case of the one-child policy in China demonstrates how important the wider structure of society is to what we might be tempted to think of as a fairly intimate and personal process—learning how to become a gendered person. In this case, we see the overlap of many different factors in dictating the gender socialization experiences of boys and girls. There is first the backdrop of Chinese culture in general, with its historical preference for boys, as well as the Confucian view on the importance of education. But culture is never static, and Communist rule in China very intentionally sought to change aspects of this gender ideology. Gender socialization must take into account the dynamic nature of changing values in a given society. So in looking at the socialization of children in China, or in any context, it is important to avoid the temptation to talk about "Chinese culture" in general; in this study, Tsui and Rich (2002) examined Chinese culture in urban cities, among one-child families, in the beginning of the 21st century, and all those specifics are important. In addition to cultural factors that impact socialization, there are also economic and governmental policies that reach all the way into the intimate realm of parents and children. The combination of an economic decision in the move toward a free-market economy along with the governmental one-child policy can both affect the dynamics of gender socialization in Chinese families. Thus, institutions can have an important impact at the interactional and the individual level on the process of gender socialization. Programs and policies, like the one-child policy, can have unintended gender consequences.

Theory Alert!
Interactional and Individual Level

This case also provides a specific situation through which to examine some of the theories we laid out at the beginning of the chapter. Applying social learning theory to this case helps to demonstrate its strengths and weaknesses as an approach. From a social learning perspective, the attitudes of parents toward girl and boy children matters very much in this top-down perspective. If parents believe the educational success of their daughters is just as important as the educational success of their sons, this will be reflected in the ways in which parents reward various behaviors and in how they think of various behaviors as sex-typed. One variable Tsui and Rich (2002) explored was whether children talked to their parents about school. You can imagine, through the lens of social learning theory, how parents who are equally invested in the educational success of their daughters and sons would respond to their child discussing school; they would respond enthusiastically to both girls and boys, serving as a reward for their daughter's or son's focus on their education. But if parents are more invested in the education of their sons, they might be indifferent to their daughter when she talks to them about school, sending her the message that education should not necessarily be considered important for her. In social learning theory, parents become the conduits through which the predominant views of society flow into their children. On the other hand, remember that social learning theory did not have much to say about how certain behaviors come to be sex-typed, or about how the larger social structure is involved in the process of socialization. The socialization taking place in these Chinese families is impossible to understand without that larger context, and so this is one potential weakness of social learning theory to understanding gender socialization.

Doctors Teaching Gender: Intersex Socialization

The one-child policy in China demonstrates how cultural aspects of a particular society, such as a preference for male children, can have an effect on gender socialization. One particular aspect of cultural beliefs about gender in the Anglo-European world is our belief in sexual dimorphism, as discussed in Chapter 1. Intersexed children do not clearly fit into the established order of two sexes and two genders, and so their experience of gender socialization is influenced by one of Western society's most fundamental cultural beliefs—the existence of two sexes. What does gender socialization look like in the case of intersexed individuals who are made to fit into our preexisting sex and gender categories?

For many intersexed children, in addition to the important agents of socialization of parents, school, church, and the media, the medical community is added as an important element in how they learn about gender. In her interviews with 37 intersexed individuals, Sharon E. Preves (2003) discovered that the experience of repeated surgeries and medical examinations had important effects on the socialization of people who did not clearly fit into existing sex categories. The gender socialization of intersexed individuals is supposed to proceed "normally" based on whatever sex is chosen for the children by the doctors. Leaving aside the question for now as to whether there is such a thing as normal gender socialization, parents and family are instructed to raise the child as a normal boy or girl and to hide the truth of her or his medical condition from the child. But for most of the individuals in Preves's sample, there was nothing normal about the repeated medical procedures to which these children were subjected; 95% of her sample underwent some kind of surgical intervention. These experiences taught intersexed children that there was something wrong with them, although they were never certain exactly what it was. Many came to see themselves as freakish or monstrous. What are especially interesting in Preves's study are the cases of the few individuals who did not undergo any surgical intervention at a young age. Surgery is prescribed by doctors to avoid the social stigma that they assume will result from having ambiguous genitalia as a child. But one of the few individuals who was born with ambiguous genitalia and did *not* undergo surgery described other children's reaction to seeing her/his naked body with the following description:

> It was at some point in my youth when I was playing doctor with other kids, or playing take off your clothes and show and tell, and realizing that I was different from anybody else there. And I also remember it wasn't a big deal at all. Everybody was like, "Wow! That's cool. Hey, you look like this, I look like this. Oh, yeah, cool, fine, whatever." And that wasn't really a big deal at all. (Preves, 2003, pp. 64–65)

In at least this case, the experience of having ambiguous genitalia was not at all stigmatizing to this individual as a child. According to Preves's (2003) study, the individuals who had experienced repeated surgeries and medical examinations often compared their experiences to sexual abuse, feeling that as children they were turned into a "freak show" or "dog and pony show" (p. 66) for doctors and other medical professionals. One individual who had been raised as a female was told by a doctor that her ovaries (which were actually undeveloped testes) had to be removed because they were precancerous and that, if they were not removed, she would develop cancer very soon. She spent much of her childhood worrying about dying from cancer. Another individual was told that if she stopped taking her medicine (hormone pills to prevent the development of secondary sex characteristics), she would die. In general, Preves's research demonstrates that the psychological damage done to these individuals' sense of self might far outweigh any of the possible negative experiences of living with the bodies into which they are born, without the intervention of surgery.

Doctors and Gender Socialization

The extent to which doctors are involved as agents of gender socialization in the lives of intersexed children is demonstrated by two experiences Preves (2003) identified among individuals in her study. First, doctors' examinations of intersexed children involved repeatedly putting the genitals or naked bodies of these children on display. This included photographs that were taken for journal articles, books, or other medical publications. One intersexed individual tells a story about being hospitalized for one of her surgeries; she began counting the number of people who were paraded through her room to examine her and her genitals but stopped when she had reached more than 100. In addition, as part of the physical examinations, doctors sometimes stimulated genitals (through physical contact) in order to test for responsiveness. These experiences had long-term effects on how these individuals thought about themselves, their bodies, and their sexuality.

The second way in which doctors served as agents of socialization was in the way they often counseled intersexed individuals about their intimate relationships. As we already mentioned in this chapter, one measure of success for sex reassignment for intersexed individuals is that they achieve normal gender identity, and that often includes the achievement of heterosexuality. This is because an important component of feminine gender identity for women is to be sexually, emotionally, and psychologically attracted to men. Doctors are therefore somewhat invested in the intimate relationships of intersexed patients. In Preves's (2003) interviews, individuals told how doctors coached them with stories to provide to intimate partners to explain their conditions in ways that sought to conceal the truth of their intersexed status. For example, one doctor provided his patient with an alternative medical condition to use as an explanation for her short vagina, a condition that was not a result of intersexuality. Another doctor advised his patient, when she started dating, simply to bring her boyfriend in to talk to him rather than trying to explain her condition or be honest with him about her intersexed status.

Coming Out as Intersexed

Preves (2003) compared the experiences of her research subjects to that of many within the homosexual community and to the important experience of coming out. Like those within the contemporary gay and lesbian community, many intersexual individuals are increasingly coming out of the closet about their identities. This is a result of the increased visibility of intersexuality through groups like the Intersexed Society of North America. Many of the individuals in Preves's interviews experienced a sense of relief from the anxiety caused by their repeated encounters with the medical community and the air of secrecy about their situation only when they finally found out what was really "wrong" with them. At least for the individuals in Preves's study, finding out about the true nature of their intersexuality was not, as doctors assumed, a traumatic event.

When many of the individuals in Preves's (2003) study discovered their identity as intersexed, they faced questions about whether or how to share this information with others in their lives. Some individuals described their experiences in terms of trying to pass as someone of one gender or another because they know their true biological sex does not line up with either gender in expected ways. One intersexed individual described this experience of passing in the following way: "I was trying really hard to act like I thought men acted, so I watched my voice very carefully and I watched my mannerisms, and I would observe how men were" (as quoted in Preves, 2003, p. 83). Some developed stories to explain certain features of their intersexed status, like scars from past surgeries or their inability to have children. Preves argued that, in various forms, these individuals searched for some external validation of their gender status, and they pursued a variety of strategies toward this end. One intersexed individual described her wedding day as important to this overall task of feeling fully and normally gendered as a woman:

> When I got married was when I really felt so feminine with my bride's dress, my wedding dress, and a veil. I felt so pretty. I was so proud. That was a turning point and that was about the most significant time in my life that was special to me, where I felt like totally female, totally womanly. (Preves, 2003, p. 85)

The same individual goes on to describe how her husband now helps her to feel feminine through his love and sexual attraction, in addition to making her feel physically small compared to his large size.

Why are the experiences of these individuals important to our general understanding of gender socialization? Can we say that if there is such a thing as normal or typical gender socialization, the experience of intersexed individuals has nothing in common with that norm? Assuming that you yourself are not intersexed, was there anything about the experiences of this group of people that sounded familiar to your own experience of gender socialization? Or does the experience of this group of people seem so foreign to your own experiences with gender and socialization as to be completely unrecognizable? The intersexed are important to the question of nature versus nurture. In theory, their stories could show us what socialization might be like in the absence of sex. But as we have discussed, this is not what happens in the Anglo-European world with intersexuality. Biological sex is medically imposed on these individuals. What we learn from Preves's (2003) study are the damaging effects of this particular framework for dealing with intersexed individuals.

But take a step back and think about what intersexed individuals learn from their interaction with doctors and the medical establishment. The lesson they are being taught, even though they may not become aware of it until later, is that to be born without an obvious sex is both a dark and terrifying secret and a fascinating medical and biological phenomenon. The experiences of these individuals teach us that, at least in Anglo-European society, intersexuality is both fascinating and terrifying. When doctors encourage a culture of secrecy and lies, lying to parents of the intersexed and to the intersexed children themselves, and eventually encouraging the intersexed themselves to lie, they are patrolling society's entrenched beliefs in the existence of sex and gender. What is so dangerous about knowledge of intersexuality? The experiences of these children teach us that part of the context within which socialization occurs in many cultures is a rigid sex/gender system that needs to be protected from any individuals, like the intersexed, who disturb its smooth operation. This is why the truth of intersexuality is so frightening; it calls into question too many of the basic assumptions of this particular society.

Once intersexed individuals find out about their status as intersexed, they may feel like they are attempting to pass as a true gender. Here's how one intersexual individual in Preves's (2003) study described this feeling: "I think you are constantly monitoring yourself. Is this the way women act? Is this what women think? Is this how women are? It's maybe not something that you do consciously, but it's a constant kind of trying to fit in" (as quoted in Preves, 2003, p. 82). This description might sound familiar thinking back to our discussion of the doing gender perspective. Remember, according to this point of view, all of us are passing as a gender through making our actions accountable to others as gendered.

Theory Alert!
Doing Gender

You might argue from this perspective that the experiences of intersexed individuals and the ways in which they experience gender socialization are different only because their unique position makes them more aware of gender as a performance, as an attempt to pass as a woman or a man. A few intersexed individuals Preves interviewed came to describe their experience of gender as being like the performance of drag, or performing a series of roles that don't reflect any underlying reality. As one intersexed individual described, "And in this society, all I really do is drag. I do the execu-dyke drag, I do girl drag, I do boy drag. This is girl drag, what I'm wearing today" (as quoted in Preves, 2003, p. 86). The doing gender perspective would argue that what we have in common with intersexed children is that socialization merely teaches us this ability to perform. Although intersexed individuals may be more aware of this performative nature of gender, all of us are to some extent doing drag.

Can you think of moments you have experienced that are similar to those described by some of these intersexed individuals, when you questioned whether your actions or thoughts were consistent with your sex category and gender? Have you ever thought to yourself, "Is this what women do?" or "Is this what men do?" and been uncertain about whether your own behavior fit within expectations about your gender?

Research on adolescents who are not intersexed reveals similar dynamics, especially for girls as they go through puberty. Girls experiencing their first menstrual period, developing breast and pubic hair, all wonder about whether their experiences and their bodies are, in fact, normal (Martin, 1996). Although intersexed individuals are an extreme case of these concerns, their stories teach us that all of us are plagued by concerns about our normality. We worry about our normality in terms of our biological sex, our performance of gender, the direction of our desires and attractions (toward the same sex, opposite sex, or some combination of both), and a thousand other gender-related aspects of our lives. Queer theorists might argue that the categories of sex, gender, and sexual orientation make all of us at one time or another feel like a freak or a monster. This is not to say all of us suffer in quite the same way or to the same degree as intersexed children. But queer theorists would point out that gender makes freaks of us all, and so we have much more in common with intersexed people than you might at first expect. Queer theory as a perspective suggests

Theory Alert!
Queer Theory

that the experiences of the intersexed merely help them become more aware of these connections that all of us have experienced at one time or another.

Gender Camp: The Importance of Peer Groups

When I ask students in my classes to write personal accounts of their own gender socialization, they usually focus on stories about how their parents or other adults encouraged or discouraged them to engage in various activities. There's usually at least one story about how a sibling of the other gender was treated differently—a brother allowed to stay out later than a sister or a sister being scolded less harshly than a brother. There's almost always some discussion of toys, Barbie versus Power Rangers, and memories of rooms being painted pink or blue. Sometimes there's a particularly traumatic story about a time when someone misidentified their gender and thought they were a boy when they were really a girl or vice versa. There's usually not a lot of focus on the next aspect of socialization we'll discuss: the importance of peer groups to gender socialization. Maybe this is an indication of the intuitive appeal of social learning theory, with its emphasis on the family and adults in general in the process of socialization. But if you remember from Chapter 2, social network theory draws our attention to the important, although small, initial differences in the play groups of girls and boys. Boys with their larger play groups come to share a whole set of information that is completely different from what girls are talking about in their own smaller groups. These differences set the basis for the gradual accumulation of gender differences over the course of our lives.

Theory Alert! Social Network Theory

Theory Alert! Cognitive Development

The shift in psychology and sociology from a social learning model to a cognitive model led many researchers to begin to study the peer groups of children in childhood and how interaction among children is important to the process of gender socialization. This follows from the shift to cognitive-development and later gender schema theory because the focus shifts from what parents and other adults do to children's active role in their own socialization. Cognitive-development and gender schema theories are more of a bottom-up approach to socialization, emphasizing the active role of children themselves in their socialization. If this is the focus, it makes sense to study children in the types of settings where they get to exercise more control over their environments. Where do children have more independence and control? In play groups made up of their own peers and characterized by less adult supervision.

Barrie Thorne's (1993) ethnography of girls and boys in their early school years (kindergarten through fifth grade) was one of the first studies in this area, and she found that gender is important to the organization of play among young children at school. On the playgrounds of the two working-class American schools that Thorne observed, girls and boys were often involved in gender-segregated activities; girls played games that took up less space, like jumping rope, doing tricks on the monkey bars, or four-square, while boys ranged over wider areas of space and tended to play competitive sports like soccer, basketball, or football. Thorne also brought attention to the moments when play and interaction were not gender segregated; boys and girls interacted across genders in several different ways, sometimes reaffirming gender boundaries (through games of pollution, like "cooties") and enacting heterosexuality (using taunts like "Jimmy likes Betty" or "Betty likes Jimmy"). But Thorne also found certain situations in which gender as an organizing category becomes relatively less important. One example was when children were engaged in a particularly absorbing project that required cooperation, like a group art project or organizing a radio show. Thorne's research overall drew attention to the ways in which children engaged in behavior that often reinforced traditional gender ideas but sometimes challenged or ignored those ideas as well. In other words, children are active shapers of gender socialization rather than passive recipients, and they do not simply absorb gender as it is channeled through their parents and other adults.

Since Thorne's (2003) ethnography, a variety of research in different settings has been conducted on the importance of peer group interaction to gender socialization. In one example of such a study, C. Shawn McGuffey and B. Lindsay Rich (1999) examined how children negotiated gender within the context of a diverse day camp for children between the ages of 5 and 12. These researchers used the concept of hegemonic masculinity to explore how girls and boys patrol notions of what it means to be masculine and feminine, focusing on behaviors that occur in what they call the **gender transgression zone**. The gender transgression zone is not literally one physical space but any activities or behaviors that have the potential to be perceived as violating gender norms in some way. The gender transgression zone is the social space between gender-typed behaviors or between what's considered appropriate for boys and appropriate for girls. McGuffey and Rich were interested in exactly what happens when boys or girls at this particular day camp entered this zone.

McGuffey and Rich (1999) found that even at that relatively young age, gender inequality and the power of hegemonic masculinity were reflected in the micro-politics of the setting. **Hegemonic masculinity** is a concept that comes from R. W. Connell's (1995/2005) exploration of how our dominant ideas about what it means to be a man influence the behaviors of actual men in any given society. According to Connell's theory, there is no one male role, as might be assumed in sex-role theory, but a variety of masculinities that interact with each other in hierarchical and contested ways. Men enact different versions of this masculinity depending on where they are located in social hierarchies of power. Hegemonic masculinity is the type of gender practice that, in any given space and time, exists at the top of those hierarchies. It is important in Connell's formulation to keep in mind that there is not one dominant way of being masculine but only masculinities—of which hegemonic masculinity is but one example. In addition, the particular version of masculinity that is hegemonic changes across times, cultures, and subcultures. The masculinity that is considered hegemonic in the summer day camp that forms McGuffey and Rich's (1999) research setting is not necessarily hegemonic for adults, for American men 50 years ago, for Mexican men, or even for the boys at another summer day camp down the road. But once a particular version of masculinity becomes hegemonic, it can be used to patrol the behaviors of men, or boys, within that particular setting.

Boys and Girls, Together and Apart

The social world of both girls and boys at this day camp was stratified within each gender. Within the boy's group, being high status was associated with athletic ability, as well as with general attitudes of detachment, competitiveness, and attention-drawing behavior. These qualities then made up the content of hegemonic masculinity in this particular setting. Girls lacked the existence of one standard hierarchy that applied across all girls; each small group of girls might have one or two high-status girls, but there was more flexibility about what determined status among the girls. Between the two genders, the status of boys seemed generally to outrank that of girls, and this was demonstrated in the power boys had to patrol both other boys *and* girls in the gender transgression zone.

Boys entered the gender transgression zone when they did not conform to the established requirements of masculinity as defined by this particular group of boys. For example, one boy, Phillip, was marginalized by the other boys because of his small stature, lack of coordination, feminine appearance (shoulder-length hair), and preference for stereotypically feminine activities like jump rope. Phillip was treated as a social pariah by the other boys; called a faggot, fag, or gay; and excluded from all boy activities and boy social circles. Phillip got along fine with girls, but the researchers tell a story in which one boy refused to be paired with Phillip in a game even under the threat of being forced to sit out the whole summer camp; the other boy declared, "I don't care if I have to sit out the whole summer 'cause I'm not going to let that faggot touch me" (McGuffey & Rich, 1999, p. 619). This rejection of Phillip is a way in which boys patrolled the gender transgression zone, demonstrating what happens when you do not follow the norms dictated by hegemonic masculinity as it is defined for this group. Boys also used Phillip to warn other boys if their behaviors became too dangerously outside the bounds of normalcy, using the threat of associating them with Phillip. None of the boys wanted to act like Phillip, and the previous story shows how strongly they felt about touching Phillip, let alone interacting with him.

Boys also patrolled girls in the gender transgression zone. This was important to the preservation of hegemonic masculinity because if too many girls could make claims to masculinity, masculinity would begin to lose its meaning as something exclusively within the male domain. Girls who entered the boys' sphere were either marginalized or masculinized by being adopted into their particular boy culture. Within this particular day camp, several African American girls were actually better athletes than some of the boys—recall that athletic ability was crucial to boys' status in this setting. The athletic ability of these girls clearly represented a kind of gender transgression, especially as these girls still also participated in many feminine behaviors. When boys were asked why they did not associate with these African American girls, their explanations centered on these girls' "weirdness" and differentness. Hegemonic masculinity was maintained by describing the girls who actually fit some of its requirements (like athletic ability) as weird or abnormal; marginalizing girls in this way preserved the idea that athletic ability was really a masculine trait in the end, despite the contradiction that some girls also seemed to possess this trait.

At least one girl in the research setting was included in the boys' activities and treated to some degree as an honorary boy. Patricia was very athletic and could outperform the boys in the important game of basketball. One day when the boys created an obstacle course that they claimed would test their manhood, Patricia was able to complete the manhood tasks better and faster than many of the boys. The boys proclaimed Patricia a man, and she was by and large accepted into the boys group. Patricia also had the added benefit of demonstrating an appropriately masculine level of detachment when she was interacting with the boys. When asked why Patricia was accepted among the boys, one boy answered, "Well, Patricia is not really a girl. Technically she is, but not really. I mean, come on, she acts like a boy most of the time. She even passed the 'manhood' test, remember" (McGuffey & Rich, 1999, p. 620)?

The researchers argued that girls like Patricia who ventured into the gender transgression zone by demonstrating masculine behaviors could join the boys' group but only by giving up their femininity. The boys allowed Patricia to join them not as a girl but only because she was not *really* a girl. To allow Patricia to join while retaining her femininity would have jeopardized the idea of hegemonic masculinity by giving girls access to it. Girls like Patricia became degendered, and this maintained how masculinity in this day camp dictated not just how the boys behaved but how the girls defined femininity as well. Boys in this setting maintained hegemonic masculinity both by controlling the behaviors of other boys and by dictating the behaviors of girls.

Theory Alert! Social Network Theory

What about girls and how they patrolled each other, as well as boys who entered the gender transgression zone? This research demonstrated that girls had less power to do this kind of patrolling than the boys because of their smaller cliques. This should sound familiar from our discussion of social network theory in Chapter 2. Remember that girls generally tend to have smaller networks as children. A high-status girl in one of these small cliques could certainly patrol femininity within her own small group, but she had less power to control femininity among *all* the girls at the day camp in the way boys did. When a girl entered the gender transgression zone by intruding on boys' turf, other girls were more likely to band together as a large group to support her than they were to punish or discourage her behavior. For example, girls rallied around Corisa, who was able to beat Travis at Connect Four, after he had been bragging about how easy it was for him to beat girls in general. The other girls at this day camp were so proud of Corisa that one girl introduced her to her mother that day in the following way: "Mommy, this is Corisa. She beats boys in Connect Four" (McGuffey & Rich, 1999, p. 622). When a group of girls was asked about the behavior of Patricia, the honorary man, they merely noted that she liked different stuff, but she was still "nice." Girls transgressing femininity did not seem to be the same kind of violation for girls as gender transgression was for boys. But girls also did not perceive boys who transgressed gender behavior in the same way as other boys. Girls were generally accepting of transgressive boys such as Phillip, asking only that they adhere to feminine norms of niceness, such as sharing candy with friends.

When you think back to your own experiences at school or camp, are they similar to those described by Thorne (2003) or McGuffey and Rich (1999)? How are they different? How might you explain those differences?

Peer Groups and Gender Socialization

There are two interesting and interconnected questions raised by McGuffey and Rich's (1999) research. The first is, "What are these children learning and teaching themselves about the structure of gender in general?" With hegemonic masculinity, remember that no one really completely conforms to the ideal of hegemonic masculinity, but that it's a kind of yard stick against which men are forced to measure themselves. McGuffey and Rich argued that this was a large part of what this particular group of boys was learning through their interactions at the day camp. Through the example of boys like Phillip and through the way girls were largely excluded from their play groups, boys learned the content of hegemonic masculinity. Equally as important, they learned the consequences for violating those norms of hegemonic masculinity. They also learned that masculinity carries power—the power to regulate both the behavior of other boys as well as the behavior of girls.

One particular example of this power came when a high-status boy was able to use his power to change the content of the gender transgression zone. Adam, the highest ranked boy in the group, learned how to do a hand-clapping routine one afternoon while he was waiting for his parents to pick him up from the camp. These routines were clearly feminine activities because only girls engaged in them, and so learning hand-clapping routines was a clear example of gender transgression. Adam was able to transgress because of his high status among the boys. Although boys were initially puzzled when they saw Adam engaging in this feminine behavior, they eventually followed his lead, and soon girls and boys were interacting in what had previously been defined as a girls-only activity. This incident demonstrated that boys in this particular setting had the power not just to patrol the borders of gender but also to alter and shift those boundaries. We can argue, then, that an important lesson these children learned was about gender inequality—that boys have access to power to control social interaction in ways that girls do not.

The second question raised by McGuffey and Rich's (1999) research has to do with how the case of these children serves as an example of the importance of peer groups to gender socialization. Note that there is no mention of parents, teachers, camp counselors, or other adults in this account of gender socialization at day camp. This is not to say that adults were not present on a daily basis in the day camp, including the researchers themselves asking children questions about their interactions. But although gender is clearly important to the interactions at day camp, what is absent is any sense in which children are being rewarded or not rewarded by adults for various gendered behaviors. Boys seemed to be motivated to act in ways consistent with the particular form of hegemonic masculinity as a result of their real fears of what would happen to them if they did not conform to this norm. Where does the boys' clear sense of the boundaries of what is appropriate and inappropriate for them as masculine come from?

Theory Alert! Gender Schema Theory

Bem (1983, 1993) might argue the patrolling done by boys at this day camp is clear evidence of gender schema they have come to internalize. These boys have learned to see the world through the specific gender lenses provided by their society. Gender schema theory would point out the ways in which the content of gender socialization in this day camp conformed to the characteristics of gender polarization and androcentrism. The boys had a clear sense that there were some behaviors (hand-clapping) that were appropriate for girls but not for boys. The way in which they responded to boys who exhibited any "feminine" behaviors, like Phillip, demonstrated androcentrism; Phillip lowered his status by engaging in feminine behaviors because feminine behaviors were clearly perceived as inferior to masculine ones. This was not the case for girls because they seemed to have no problem with boys engaging in feminine behavior. From the perspective of gender schema theory, we can observe clear examples of children socializing themselves through the mechanism of gender schema, or the gendered ways in which they see the world.

If you were paying careful attention, though, you might have noticed that the behavior of the children in the day camp did not always perfectly conform to the content of gender schema as described by Bem (1983, 1993) and others. Although hand-clapping was at first considered clearly OK for girls but not for boys, the story of Adam shows how this group of children transformed that norm. In this particular day camp, a high-status boy was able to undo the gender polarization of that particular activity. In addition, Adam's adoption of hand-clapping somewhat altered the structure of hegemonic masculinity, even though the researchers also found that boys eventually masculinized hand-clapping by changing lyrics in not-so-subtle ways (from the girls original, "All the birdies on J-Bird Street like to hear the robin go tweet, tweet, tweet," to the boys adaptation, "All the birdies on J-Bird Street like to hear robin say eat my meat" (McGuffey & Rich, 1999, p. 624). What does this tell us about the process of gender socialization? It is not a simple process of transmission or recreation of predetermined, and therefore unchanging, gender norms and ideology. The exact gender polarization of activities can change in somewhat spontaneous ways, largely dictated by the culture of children's peer groups. The content of gender children learn is not static and set in stone but is in a process of change and adaptation, much of which is the result of children's own creativity. In other words, from a bottom-up approach, children have the ability to alter the exact content of gender socialization in interesting ways.

This is true of socialization as considered from the top-down approach as well. In social learning theory, parents and other adults have the power to alter the content of gender they teach to children. In McGuffey and Rich's (1999) research, Adam was motivated to learn hand-clapping in part because he saw the male researcher engaging in hand-clapping activities with the girls. Adam's high status was also important, but the role of the adult researcher shows how the behavior of adults can also lead to an alteration of gender norms. This research helps us to see that gender socialization is probably the result of complex interactions among adults, family structure, and the control children themselves exhibit over the ways in which they learn gender. Although it is useful to understand the unique perspective of each of the four theories we have discussed, it is also useful to observe the ways in which they can work together to explain the process of gender socialization.

LEARNING GENDER NEVER ENDS: SECONDARY SOCIALIZATION

Up until this point, we have dealt mainly with primary socialization—the ways in which we initially learn gender from those close primary groups such as family and childhood friends. For many of us, the first years of our lives very much revolved around these primary groups. And throughout our lives, although our primary groups may have changed (as we formed our own families in a multitude of different ways), these collections of people remain important sources of gender socialization. However, for many of us, the process of growing up involved increasing contact with secondary groups and therefore moved us into the realm of secondary socialization. **Secondary groups** are generally larger, more temporary, more impersonal, and more specialized than primary groups. They tend to be more specialized in that they focus on one or two primary goals, rather than on the unspecialized set of goals (if they can even be called that) that characterize a family as a group. On the other hand, a business organization as a secondary group has a definite purpose: to make money. A college class is another good example of a secondary group.

Every time we join a group, we need to learn the norms and rules of that particular collection of individuals or of that institution. So **secondary socialization** is the learning process that takes place each time we join one of these new secondary groups. In our discussion of work in Chapter 9, we will explore the ways in which women and men become socialized into various occupations and the important implications this has for inequality. Unless you're a real hermit, hiding out in the woods somewhere and intentionally avoiding people, you continue to join new secondary, as well as primary, groups throughout your life. These groups may have their own ideas about gender and may subtly or not so subtly affect the way in which you think about gender. As a result, the process of gender socialization is by no means fixed at a certain age; it changes greatly throughout the course of a person's life.

Learning to Be American: Socialization Through Immigration

We've already begun to see that the range of agents of socialization that help us learn gender are varied and extensive. There are many possible sources for our knowledge about how to be gendered, including parents, doctors, and other children. The experiences of immigrants and their children adapting to the culture of a new country serves as a good example of how socialization extends beyond childhood and how institutions as large as nations themselves can serve as agents of gender socialization. It may seem strange at first to think about nations as a source of knowledge for how to be gendered. But if you have traveled to other countries or even seen foreign movies or TV shows, you might have noticed some obvious differences in how gender works in other places. In France, men kiss each other without casting any doubts on their sexuality and women grow their armpit hair while still being seen as feminine. In Scotland and India, men wear skirts, while in some nations in the Middle East, Asia, and Africa, Muslim women wear veils. These are some of the obvious cultural differences that demonstrate how your experience of gender might vary depending on the country in which you were born. When people move from one country to another, both these obvious differences and more subtle ones become important in negotiating how to adapt to life in a new country.

Gender and Hybrid Identities

Lubna Chaudhry (2006) documented how four young Pakistani Muslim women in the United States created hybrid identities as they learned how to become at least partial members of the very large secondary group of people called Americans. Having grown up in Pakistan, these young women were socialized in the particular gender norms of Pakistani and Islamic culture. In the United States, they were faced with a new set of gender norms that often existed in opposition to those enforced by their parents and other Pakistani Muslim immigrants. Chaudhry demonstrated through the lives of these four women how they were subtly shaped by their encounters with American culture and their decisions about how to do gender in their new country. Nida came from a background of family wealth in Pakistan and the United States, where her father owned a prosperous business.

When she transferred to American schools from Pakistan, Nida's teachers encouraged her to pursue an advanced degree, reflecting an American gender norm regarding women and education. However, the gender norms of her particular family's clan background dictated that women marry as early as possible after the onset of puberty.

Nida reconciled these conflicting values by marrying early and having a baby at 18 but also continuing to pursue her educational goals, with the support of her husband and her own family. In this sense, Nida chose to adapt to American values, which emphasized the importance of education and upward mobility largely regardless of gender. Her overextended schedule in college as student, wife, and mother was not enjoyable, but she saw it as reflecting part of what it meant to be an American woman. But by marrying and having a child at such a young age by American standards, Nida also chose to uphold the gender norms of her particular Pakistani culture. In this sense, Nida carefully negotiated the boundaries between American and Pakistani identity.

In her study of Filipina women in the United States, Yen Le Espiritu (2005) identified a similar dynamic. From the perspective of immigrants from the Philippines in the United States, family is more important to Filipinos than it is to many white Americans. One Filipino immigrant described it as follows:

> Our [Filipino] culture is different. We are more close-knit. We tend to help one another. Americans, ya know, they are all right, but they don't help each other that much. As a matter of fact, if the parents are old, they take them to a convalescent home and let them rot there. We would never do that in our culture. We would nurse them; we would help them until the end. (p. 234)

According to Espiritu (2005), within this framework, women are considered especially important to upholding the importance of family in Filipino culture. The responsibility of Filipino men lies largely in marrying a Filipina woman. But Filipina women who try to pursue education, jobs, and upward mobility for themselves sometimes feel inhibited by the gender norms placed on them by Filipino culture. One Filipina wife and mother explained these conflicting gender norms through her experience returning to school to pursue a doctoral degree in nursing:

> The Filipinos, we are very collective, very connected. Going through the doctoral program, sometimes I think it is better just to forget about my relatives and just concentrate on school. All that connectedness, it steals parts of myself because all of my energies are devoted to my family. And that is the reason why I think Americans are successful. The majority of American people they can do what they want. They don't feel guilty because they only have a few people to relate to. For us Filipinos, it's like roots under the tree. You have all these connections. The Americans are more like the trunk. I am still trying to go up to the trunk of the tree but it is too hard. I want to be more independent, more like the Americans. (pp. 235–236)

Like Nida, this Filipina woman struggled with trying to achieve the goals set out by American culture for women—independence and success—while still retaining her ties to her Filipino family and community. The gender norms prescribed by Filipino culture made achieving the gender norms set out by American culture more difficult.

The cases of Pakistani Muslim and Filipina women demonstrate how secondary socialization can have important impacts on the way in which we experience gender. Nida's immigration to the United States exposed her to new secondary groups in her schools, including her teachers and female classmates. Learning American gender norms and values transformed the way she thought about what it meant to be a woman. The Filipina women were similarly exposed to the more individualistic values of American women and found themselves feeling torn between family obligations and their own desire for independence. Immigrating to a country means joining a whole host of new secondary groups while, in the case of these women, maintaining primary groups in the form of families.

Their experiences demonstrated the complexity of socialization, in that often we receive messages about how to be gendered from many different sources, and those messages often contradict each other. Not all of us may have had the particular experience of immigrating to a new country, but all of us have joined new groups at some point in our lives and have found ourselves negotiating the new set of norms, some of them related to gender. As suggested by both cognitive-development theory and Bem's (1983, 1993) gender schema theory, women like Nida have some ability to socialize themselves by picking and choosing which of the gender norms of their new culture they will follow and which gender norms from their country of origin they will retain. The case of Pakistani Muslim and Filipina immigrants demonstrates that socialization is a process that we actively shape throughout our lifetimes as we join new groups.

Can you think of an experience when you joined a new group and had to adapt to a new set of norms? Were there specific norms related to gender that were different in your new group? How did you negotiate these new gender norms?

WHAT HAPPENS TO GENDER AS WE AGE?

We have already begun to explore the ways in which gender is never really a free-floating concept that exists without being attached to a multitude of other identities and cultural and historical contexts. There is no normal experience of what it means to be gendered but only many multiple and often contradictory ways of experiencing this particular social category. To explore how this is true as it relates to gender socialization, think about an older person you know, perhaps a grandfather, grandmother, or other older relative. Do you think of this person as gendered? Is gender an important part of the way in which you perceive this person? Do you think gender is an important part of his or her life? Is gender more or less important in her or his life than it is in your own? If you were to make a list of some of the general characteristics of what makes someone feminine or masculine, how many would this older person correctly fit? These questions helps us begin to think about the ways in which gender changes as we age and how what it means to be a woman or a man, masculine or feminine, is very different at age 10, age 30, age 50, and age 70.

Looking at how key experiences and social roles influence individuals' lives is the focus of a **life-course perspective**. Each life course consists of a biography that transforms through exposure to new social settings, and this unfolds in an interactive trajectory. So, having been married—one life-course event—will affect how an individual experiences retirement or old age. Research that combines a life-course perspective with an intersectional approach to the study of gender ideology reveals some results that highlight the importance of taking into account how other identities interact with gender.

Important life events like marriage, parenthood, and work have different effects on gender ideology depending on racial identity (Vespa, 2009). Getting married is associated with a more egalitarian gender ideology among African Americans but less egalitarian ideology among whites, and this relationship is strongest among white men. Having more children leads to less egalitarian gender ideologies among Black men and white women. But the presence of young children actually increases Black women's gender egalitarian ideology. Finally, with increases in full-time work, men become less egalitarian, while women become more egalitarian (Vespa, 2009).

This research demonstrates how gender socialization is an ongoing process that continues well beyond childhood and adolescence. Our particular gender ideologies are not set in stone when we graduate from high school or college. Life experiences continue to change the way we think about gender in our own lives. As intersectional theory suggests, these changes are dependent on how gender interacts with other categories like our racial and ethnic identities.

Gender and Aging: Does Gender Fade Away?

For now, regardless of the length or particular content of this adult period in a person's life, there remains the question as to what happens when these roles fade or disappear. And what happens when new roles emerge to take their place? As many parents will tell you, you never really stop being a father or a mother. Many demographers argue that this is especially true in many parts of the global North where the period during which children are financially dependent on their parents has grown increasingly longer. For some parents in the United States with 30-year-old children still living at home, that you never stop being a mother or father seems especially and perhaps painfully true. But even in places like India, where some mothers send their daughters away to live with their husband's family and may have little contact with them afterward, occupying the social role of mother does not end there. What *does* change for many people is the nature of the role of father or mother, partly through generally consuming much less time, effort, and energy as children move away or become more self-sufficient. The roles of father and mother change, while the roles of wife and husband can come to an end with the death of a spouse. This is especially true for married women, who are much more likely to end up as widows than their married male counterparts as a result of a combination of factors, including women's longer average life expectancy around the world and the tendency for women to marry older men.

Aging can also bring an end to the social role of worker in its many forms, although there are variations in how this social role intersects with age, as well. The concept familiar to many in the developed world of retirement is a fairly recent and isolated phenomenon. The idea that paid workers should be able to stop working and be provided for economically in their old age is an invention of the 20th century, and for most of those 100 years, it dealt primarily with the work experiences of men. With increasing longevity and quality of life at advanced ages, the retirement age is being pushed farther and farther back in many parts of the global North, but growing old often means a shift in the set of social roles that come along with work, whatever form that work may have taken. The idea of retirement is still a relatively foreign concept in many other parts of the world: In the mountains of Peru, there are expectations that individuals do hard physical labor well into their eighties. In some other cultures, older people are provided for through institutions beyond the government or the economy; in some Navajo tribes, they are allowed to pick a grandchild to move in with and that grandchild then takes care of their daily chores (Cruikshank, 2003). Whatever particular form it may take, for many but not all people across the world, aging also signals a shift in the important role of worker.

Exactly what do these shifts in many of the most central social roles we occupy in adulthood look like as men and women age? In the past, and perhaps for many still today, there exists a belief that old age brings with it a process of androgenization, or a kind of gender crossover. To be androgynous is to possess both feminine and masculine characteristics, so the process of **androgenization** means adopting some of the qualities of the opposite gender. The basic idea, which worked its way into popular culture during the 1980s and 1990s, is that while women become bolder and more aggressive as they age, while men become kinder and gentler (Gutman, 1987). The actual empirical truth of these assumptions is difficult to verify because during the same time period when much of the research took place, important changes were occurring in the gender roles of women and men in general.

Examining attitudes among older men and women involves sorting out differences that are a result of their particular position in the life cycle as opposed to a result of generational or cohort differences. In other words, which characteristics are specifically a result of being older and which characteristics are a result of the unique historical experiences of particular generations of people (for example, the unique experiences of the "Greatest Generation" in the United States and Europe who survived the Depression and World War II). Subsequent research seems to indicate that the important factor is not age in and of itself but important events that come with aging, such as having fewer parenting responsibilities or achieving high levels of occupational success (Carlson & Videka-Sherman, 1990). In addition, other identities besides gender can be crucial to understanding the operation of gender in later life; a cross-national study found that middle-class women did experience an increase in perceived power as they aged but that lower class women did not (Friedman & Pines, 1992; Friedman & Todd, 1998). These middle-class women become more "androgynous" because the extra status of age and affluence made it less necessary for them to seem meek or less powerful (qualities often associated with femininity) in public. Because of the many variations in the experiences of aging globally, it is difficult to make many generalizations about the exact impact on gender of growing older.

Playing Old Maid: The Gender of Widowhood

As a small child, you might have played a card game called Old Maid. You might have even come across a special deck of cards designed specifically for the game of Old Maid. These special decks usually had at least one card with at least one visual depiction of what an Old Maid looked like. The basic form of this game is played in many places around the world, but in the United States, it goes under this specific name, and the images on these cards tell us a great deal about our expectations for women who reach a certain age in life without having been married. The Old Maid is generally depicted surrounded by cats and, perhaps, knitting. She looks kind and benevolent enough, but it's important to notice that, in general, she has no male equivalent. The phrase itself is a good example of a **language asymmetry**, or a way in which the structure and vocabulary of a language reflects and helps to re-create the social inequalities of the culture in which it exists. What do you call an older man who has never been married? There's *bachelor,* but that term is not age specific in the way old maid is; a bachelor could range in age from 18 to 80, while an old maid, depending on your own sense of what makes someone old, is generally older. In addition, *bachelor* doesn't have a particularly negative connotation attached to it. Most men probably wouldn't mind being called a bachelor, but how do you think many women would respond to being called an old maid?

The existence of the old maid, both as a type of playing card and as a cultural ideal, reveals for us the specific importance of marriage for women. Marriage is a central institution in many cultures. Marriage certainly involves both men and women, but the old maid is a clue to how this institution in places like the United States might be perceived as more important for women than for men. The old maid represents a stigmatization of women who do not marry that has no particular equivalent for men. Understanding the importance of marriage for women helps give us a context for understanding a common status for women in later life: widowhood. The old maid who has either chosen or by some set of circumstances outside of her own control remained unmarried is different from the older woman who has lost her married status through the death of her husband. The old maid is somewhat unthinkable in many cultures, where marriage is much more compulsory for women than it is in places like the United States, but looking at the role of the widow in these places tells us a story about the importance of marriage, as well as about how gender changes over the course of our lives.

When you think of the status of widowhood, you may imagine a feminine image similar to that of the old maid, even though men can also be widowed. Widowers (men who have lost their spouse) are generally rarer than widows (women who have lost their husband), but virtually all widows around the world lose status when they lose their husbands. This loss of status takes various forms. In many parts of India, widows literally become *inauspicious,* people whose presence is undesirable because of a sense that their bad luck in losing their husbands might be contagious to others (Lopata, 1996). In South Korea, widows may move in with their adult sons, but they find themselves with reduced status in the household relative to that of their daughters-in-law. In places like the United States and Israel, widows suffer from a loss of economic status. What do all these different types of status loss tell us about women and marriage? In many different ways, marriage is a valued status for women and a social role that comes with some degree of power, even if that power has varying levels of limits placed on it. Widows are forced to exist outside the relative safety of this institution, and they are viewed differently because of that new location.

This loss of status is an intensification of what happens in general to women as they age in many societies. Inequalities that persist between women and men over the whole course of their lives become intensified in old age, resulting in what some gerontologists (people who study aging) call **cumulative disadvantage** (Cruikshank, 2003). The woman in the United States in the past who may not have worked outside of the home for much of her life, or who worked in a job that paid significantly less than her male counterparts, is inevitably at a disadvantage relative to men when she reaches old age. It is only recently that a whole cohort of women who have worked full time for most of their lives in jobs that allow for a retirement or pension will begin to move into old age. Although many women have worked full time for paid work outside of the home over the history of the United States, these jobs generally did not provide for or allow women the ability to save for retirement, which as we discussed is a relatively recent social phenomenon. This baby boom generation of working women who will have the economic ability to retire is unprecedented, and researchers are curious to see what their experiences of old age will be like. A person's status in old age becomes a mirror reflection of what happened over the course of their lives, and all status positions they occupied are represented. These inequalities are reflected in the statistics on poverty and aging. In the United States, one third of all older women are likely to be poor or near poor. Among African American women and Latinas, the number increases to 58% and 47% of older women, respectively. Poverty is intensified among widowed women because being a widow greatly increases the likelihood of being in poverty; half of all poor widows were not poor before the death of their husbands (Cruikshank, 2003).

Certainly, women in old age and widowhood experience loss of status. This demonstrates for us the centrality of institutions such as marriage to the status of women, as well as the ways in which inequalities are intensified in old age. The social role of wife, although it is changing rapidly in many parts of the world, is still a highly gendered one. If widowhood represents an exit from this highly gendered role, does it not also bring with it a life that is less subject to gender norms and restrictions? Do older, widowed women have the ability to step outside of some of these gendered expectations?

The answers to these questions are complicated and contradictory. For some widows in India, typical markers of feminine appearance are not allowed for widowed women. Widows cannot wear flowers, bangles (bracelets), nose rings, or apply *kumkum* (the red dot on the forehead of married Hindu women). In some areas, the widow's head is shaved (Lopata, 1996). These appearance norms have relaxed in many areas of India, especially in more urban areas, but they do represent a kind of defeminization of widows. The purpose of these rituals is to make sure that no men are attracted to widows, as remarriage of widows is strongly discouraged in Indian society. The removal of these marks of femininity is considered to be more of a punishment than any kind of reward, and they exist against the historical (and sometimes still contemporary) backdrop of *suttee.* **Suttee** is a Hindu practice of ritual self-immolation (setting oneself on fire), which has been illegal in India since 1829, although cases have occurred as recently as 1987. Widows are literally thought to be to blame for the death of their husbands because of wrongdoing either in this life or in previous lives, and they are therefore called on to fulfill their duty by throwing themselves onto their husband's funeral pyre. So although widowhood in India does diminish femininity to some extent, it is not at all perceived as a positive status for women.

How to Be an Old Maid: Roleless Roles

This is true for many women in India because of the immense importance of marriage to Indian society. In the contemporary United States, marriage is still important but certainly less significant than for many women in India. Although most Americans still *will* marry at some point in their lives, the number of single people in the United States is also on the rise. Within this context, widowhood in the United States is less culturally defined than it is in places like India. There are few set expectations and rules about exactly what widows should do or how they should behave, and this has led some researchers to describe being a widow as a **roleless role** (O'Bryant, 1994). One way to demonstrate this is to think about the role requirements involved in *wife* and *mother;* you could probably pretty easily come up with a list of what makes a good wife or mother. But how would you answer the question, what makes a good widow? You'd probably be hard-pressed to generate that list, and many of the characteristics would probably have to do with adapting emotionally and psychologically to being without a husband. Adjusting to widowhood, for many women in the United States, is linked to the process of grieving. There are also changes to self-concept as well as social networks for widowed women in the United States.

When being part of a marriage has become an important part of a person's identity, how do you adjust to the loss of that identity? This is perhaps more difficult for widows, who did not choose to exit that identity as divorced women did. Especially for older widows, the possible range of identity choices may be limited. Some older widows may still be mothers, but if all their children have left the home, it will be difficult for this to become a central identity. Older widows with satisfying jobs outside of the home may be able to rely on their work identity, but for current older generations, this is less likely to be the case. This lack of readily available identities to choose from can lead to a kind of disengagement among widowed women.

In addition to the ways in which they think about themselves, widowed women often find themselves having to negotiate the world without the benefit of the interdependent relationship they may have had with their husband. In many marriages, and especially more traditional marriages of past generations, important duties related to managing a household were split along gendered lines. Men often took care of duties such as car maintenance and finances, and some widows are challenged by the need to learn about these more "masculine" concerns at a later stage of life (Lopata, 1996). Dealing with mostly male repairmen, as well as negotiating legal and financial details, is a way in which widowhood forces women to move outside of their previous, comfortable sphere of typical femininity. Although this can be initially stressful, as women progress in their widowhood, they begin to perceive these and other changes more positively. In this sense, the increased self-sufficiency of widows provides an opportunity to move beyond strict gender roles and can be perceived as a positive situation, even if it is somewhat forced on widows by their circumstances.

Widowhood and old age represent a kind of exiting from one of the main stages for what doing gender would call the performance of gender during the course of our lives. The period of partnering, parenting, and making a living places specific demands on our performances of gender. Widowhood and old age might be perceived as a chance to escape those demands, but a closer examination reveals an important dimension of the way those demands are experienced by real people in their everyday lives. Although being a father or a husband or worker may carry with it many responsibilities, it also carries with it many rewards. Responsibilities and rewards alike become important parts of how we think about ourselves, important parts of our identity. We become quite comfortable living in our gender costumes, you might say, and, therefore, taking them off in old age is not easy or even necessarily desirable. Even if old age has the potential to "free" us from gender expectations, this may not be a freedom that many of us want.

Certainly in the extreme case of Indian widows, being "freed" from the gendered expectations of marriage is not a positive turn of events. In the United States, widowed women also struggle with their loss of identity when they lose their husbands. What seems to be true is that, although the particular content of the gender costume we wear throughout our lives can change, we become attached to the costume itself and are in no hurry to take it off altogether. For many people in the Anglo-European world, we literally wear these gendered costumes into the grave because women and men are usually dressed up in their very best clothes for burial, a dress for women and a suit for men. Although gender is the source of striking inequalities between men and women at the end of their lives, the loss of gendered roles can also be painful and a cause for adjustment. This reveals one of the core contradictions in gender as a category and social system. Although gender as a category creates inequality and often serves as a limit placed on us as individuals, it can also become an identity to which we feel a real sense of attachment. You might go so far as to say that many of us like being gendered.

SUMMING UP

Perhaps after what we have explored in this chapter, our attachment to our gender identities should come as no surprise. Our gendered lives begin even before we leave the womb because we enter into societies that have already made decisions about how our sex and gender will be determined. In addition, the content of the cultures into which we are born will partially dictate exactly what gender lessons we learn. Within our families, parents, siblings, and other relatives surround us with subtle (and sometimes not-so-subtle) messages about what it means to be a gendered person. Even those whose biological makeup does not fit within established sex categories—the intersexed—are still indoctrinated within a gender system.

Gender is all around us, and that it quickly becomes internalized by infants and children should therefore come as no surprise. We can understand how gender identities become important to our identities through the socialization theories that emphasize how children are self-motivated to acquire gender. Cognitive-development theory and gender schema theory both emphasize the active role of children in accomplishing their own socialization. If we are motivated early on as children to become gendered beings, then it makes sense to think that we would also be sad about losing parts of our gendered identity later in our lives. Exploring gender socialization helps us to think about the nature of gender as a social phenomenon. If gender is a source of inequality, then certainly part of our mission should be to change gender as a social system, or even to do away with it altogether. But how are we to accomplish such tasks if people are attached to their gender identities in real and important ways? This is just one of the questions we will continue to explore in the next chapters.

BIG QUESTIONS

- In this chapter, we talked about the difficulty of defining a normal or typical story of gender socialization. How does this fit in with some of the theories and concerns regarding gender you've read about in previous chapters? For example, how does it line up with some of the critiques of feminism by women of color, lesbian women, and women from the global South? Why is an understanding of the specific social context important to understanding the process of gender socialization?
- Theories of gender socialization are examples of individual-level perspectives on gender because they emphasize how gender becomes internalized. What are potential problems with perceiving gender as largely something that operates from the inside out, or as something that internally motivates people?
- Assume you wanted to change your current gender system or do away with gender as a system altogether. Looking at each of the theories of gender socialization covered in this chapter, how would each theory imply you go about doing that? In other words, what course of action does each theory imply if you want to change the way we experience gender? Is this harder or easier to do from the perspective of one of these theories compared with another?
- How would you define what successful socialization is? What criteria would you use to measure whether an adult has been successfully gender socialized? What would gender socialization that was not successful look like? How would these answers change depending on the time period or culture?
- One of the questions raised by the case of intersexed children is what it would mean to raise a child who does not fit into existing sex or gender categories. How could you raise a child without a sex or gender? What problems would this present? How would other people treat this child? How might it affect the way the child thinks about itself and the outside world?
- Some of the theories we discussed in this chapter were developed specifically to explain gender socialization, while others were developed to explain the process of socialization more generally. Are there important differences in these two types of theories? Are there advantages and disadvantages to each approach? Could the gender-specific theories of socialization be used to explain how we learn about other categories, such as race, social class, or sexuality?
- In Chapter 1, we discussed biosocial and strong social constructionist approaches to the relationship between sex and gender. How would you sort these theories into one of these two approaches? Which theories seem to be based on the assumption that there are two different types of people, male and female, and which are based on the assumption that sex itself is socially constructed?

GENDER EXERCISES

1. Try to remember a specific incident or event from your own childhood or adolescence in which you learned what it meant to be masculine or feminine, a boy or a girl. Write a story or an account of that incident. Which theory seems to explain best the process of gender socialization in your own story? Would this particular theory explain all the stories of gender socialization from your own history or just this one in particular?
2. Interview several people whom you believe are likely to have experiences of gender socialization that are very different from your own. This could be someone of a different gender, different race, different social class, different sexual orientation, different age, different culture, and so on. Ask them to tell you stories about how they learned what it meant to be a boy or a girl. How do their experiences compare with your own? How do their stories demonstrate the ways in which the specific context in which gender socialization takes place is important?
3. According to gender schema theory, objects, feelings, and descriptions can all be perceived as gendered. Come up with a list of objects, feelings, and descriptions, and ask several people to give all the items on your list a gender. Do people have trouble assigning a gender to the items on your list, or does it seem like an easy thing for them to do? Do consistencies emerge in the gender assigned to the items on your list, or are their variations? What does your experiment suggest about the strength of gender schema theory?
4. In this chapter, we discussed the sometimes traumatic socialization experiences of intersexed individuals. Many organizations have been founded to defend the rights and represent the perspectives of intersexed individuals, including the Intersexed Society of North America (ISNA). Go to their website (http://www.isna.org/) and explore some of the materials and information there. How would you characterize their perspective on what it means to be intersexed? What suggestions does the organization make about the best way for dealing with intersexuality? What are some of the issues ISNA raises related to being intersexed? Does their perspective seem to assume that everyone should be made to fit into one of two sex categories (male or female) or that there might be other possibilities beyond male and female?
5. Find some examples of popular media targeted toward children or adolescents. This could include TV commercials, TV shows, cartoons, movies, food (like McDonald's Happy Meals), toys, video games, radio stations, and websites. What kind of messages do these media seem to be sending to children or adolescents about gender? Do they seem to depict both women and men equally? Do people of different genders seem to be engaged in different kinds of activities? Do you think the messages being sent in these media can be considered to challenge the gender status quo or reinforce it?
6. If you have access to some children (your own, relatives, or the children of friends), spend some time observing them, either playing with them yourself or observing children playing with each other. Or observe some children playing at a local park. Is there a gender to the play you observe? Can you see examples of the gender transgression zone as described in this chapter, or of children patrolling this zone? Do there seem to be differences in gendered behavior among children of different ages? How does that line up with the predictions of some of the theories we discussed in this chapter?

TERMS

socialization

gender socialization

gender norms

gender identity

intersexed

genital tubercle

hermaphrodites

target of socialization

agents of socialization

social learning theory

sex-typed behaviors

identification

cognitive-development theory

gender stability

gender constancy

gender congruency

gender schema theory

schema

gender schema

androcentrism

gender polarization

enculturation

psychoanalytic theory

psychoanalytic identification

ego boundaries

primary socialization

primary groups

one-child policy

gender transgression zone

hegemonic masculinity

secondary groups

secondary socialization

life-course perspective

androgenization

language asymmetry

cumulative disadvantage

suttee

roleless role

SUGGESTED READINGS

On social learning theory

Bandura, A. (1963). *Social learning and personality development.* New York, NY: Holt, Rinehart & Winston.

On cognitive-development and gender schema theory

Bem, S. L. (1983). Gender schema theory and its implications for child development: Raising gender-aschematic children in a gender-schematic society. *Signs, 8* (4), 598–616.

Bem, S. L. (1993). *The lenses of gender: Transforming the debate on sexual equality.* New Haven, CT: Yale University Press.

Kohlberg, L. (1966). A cognitive-developmental analysis of children's sex-role concepts and attitudes. In E. Maccoby (Ed.), *The development of sex differences* (pp. 82–123). Stanford, CA: Stanford University Press.

On identification theory

Belenky, M., Clincy, B., Goldberger, N., & Tarule, J. (1997). *Women's ways of knowing: The development of self, voice and mind.* New York, NY: Basic Books.

Chodorow, N. (1978). *The reproduction of mothering: Psychoanalysis and the sociology of gender.* Berkeley: University of California Press.

Gilligan, C. (1982). *In a different voice: Psychological theory and women's development.* Cambridge, MA: Harvard University Press.

Williams, C. L. (1991). *Gender differences at work: Women and men in non-traditional occupations.* Berkeley: University of California Press.

On intersexuality

Colapinto, J. (1997, December 11). The true story of John/Joan. *Rolling Stone,* pp. 54–72.

Preves, S. E. (2003). *Intersex and identity.* New Brunswick, NJ: Rutgers University Press.

Reilly, J., & Woodhouse, C. (1989). Small penis and the male sexual role. *Journal of Urology, 142* (2), 569–571.

On socialization across cultures

DeLoache, J., & Gottlieb, A. (2000). *A world of babies: Imagined childcare guides for seven societies.* New York, NY: Cambridge University Press.

Tsui, M., & Rich, L. (2002). The only child and educational opportunity for girls in urban China. *Gender & Society, 16* (1), 74–92.

On socialization in secondary groups

McGuffey, C. S., & Rich, B. L. (1999). Playing in the gender transgression zone. *Gender & Society, 13* (5), 608–627.

Messner, M. (1990). Boyhood, organized sports and the construction of masculinities. *Journal of Contemporary Ethnography, 18,* 416–444.

Thorne, B. (1993). *Gender play: Girls and boys in school.* New Brunswick, NJ: Rutgers University Press.

WORKS CITED

Bandura, A. (1963). *Social learning and personality development.* New York, NY: Holt, Rinehart & Winston.

Bardin, J. (2012, July 30). *LA Times.* Retrieved April 16, 2013, from Olympic games and the tricky science of telling men from women: http://articles.latimes.com/2012/jul/30/science/la-sci-olympics-gender-20120730

Belenky, M., Clincy, B., Goldberger, N., & Tarule, J. (1997). *Women's ways of knowing: The development of self, voice and mind.* New York, NY: Basic Books.

Bem, S. L. (1983). Gender schema theory and its implications for child development: Raising gender-aschematic children in a gender-schematic society. *Signs, 8* (4), 598–616.

Bem, S. L. (1993). *The lenses of gender: Transforming the debate on sexual equality.* New Haven, CT: Yale University Press.

Carlson, B., & Videka-Sherman, L. (1990). An empirical test of androgyny in the middle years: Evidence from a national survey. *Sex Roles, 25* (5–6), 305–324.

Chaudhry, L. (2006). We are graceful swans who can also be crows: Hybrid identities of Pakistani Muslim women. In J. O'Brien (Ed.), *The production of reality: Essays and readings on social interaction* (pp. 465–475). Thousand Oaks, CA: Pine Forge Press.

Chodorow, N. (1978). *The reproduction of mothering: Psychoanalysis and the sociology of gender.* Berkeley: University of California Press.

Coltrane, S. (1998). Theorizing masculinities in contemporary social science. In D. Anselmi & A. Law (Eds.), *Questions of gender: Perspectives and paradoxes* (pp. 76–88). New York, NY: McGraw-Hill.

Connell, R. (1995/2005). *Masculinities.* Berkeley: University of California Press.

Cooley, C. H. (1909). *Social organization.* New York, CA: Schoken Books.

Cronk, L. (1993). Parental favoritism toward daughters. *American Scientist, 81* (3), 272–279.

Cruikshank, M. (2003). *Learning to be old: Gender, culture and aging.* Lanham, MD: Rowman & Littlefield.

DeLoache, J., & Gottlieb, A. (2000). *A world of babies: Imagined childcare guides for seven societies.* New York, NY: Cambridge University Press.

Espiritu, Y. L. (2005). Americans have a different attitude: Family, sexuality, and gender in Filipina American lives. In M. B. Zinn, P. Hondagneu-Sotelo, & M. A. Messner (Eds.), *Gender through the prism of differences* (pp. 233–241). New York, NY: Oxford University Press.

Fausto-Sterling, A. (1993,). The five sexes: Why male and female are not enough. *The Sciences, 33* (2), 20–24.

Fausto-Sterling, A. (2000). *Sexing the body: Gender politics and the construction of sexuality.* New York, NY: Basic Books.

Friedman, A., & Pines, A. (1992). Increase in Arab women's perceived power in the second half of life. *Sex Roles, 26,* 1–9.

Friedman, A., & Todd, J. (1998). *The effect of modernization on women's power: Kenyan women tell a story.* Tel Aviv, Israel: Tel Aviv University Press.

Gilligan, C. (1982). *In a different voice: Psychological theory and women's development.* Cambridge, MA: Harvard University Press.

Girl, 16, who throws knuckleball, drafted by Japanese pro team. (2008, November 18). *Associated Press.* Retrieved May 21, 2009, from http://sports.espn.go.com/mlb/news/story?id = 3709884

Gutman, H. (1987). *Reclaimed powers: Toward a new psychology of men and women in later life.* New York, NY: Basic Books.

Jordan-Young, R., & Karkazis, K. (2012, June 17). *New York Times.* Retrieved April 16, 2013, from You Say You're A Woman? That Should Be Enough: http://www.nytimes.com/2012/06/18/sports/olympics/olympic-sex-verification-you-say-youre-a-woman-that-should-be-enough.html?_r = 0

Kohlberg, L. (1966). A cognitive-developmental analysis of children's sex-role concepts and attitudes. In E. Maccoby (Ed.), *The development of sex differences* (pp. 82–123). Stanford, CA: Stanford University Press.

Lopata, H. Z. (1996). *Current widowhood: Myths and realities.* Thousand Oaks, CA: Sage.

Lorber, J. (1994). *Paradoxes of gender.* New Haven, CT: Yale University Press.

Maccoby, E., & Jacklin, C. (1974). *The psychology of sex differences.* Stanford, CA: Stanford University Press.

Martin, K. A. (1996). *Puberty, sexuality and the self: Girls and boys at adolescence.* New York, NY: Routledge.

McGuffey, C. S., & Rich, B. L. (1999). Playing in the gender transgression zone. *Gender & Society, 13* (5), 608–627.

Messner, M. (1990). Boyhood, organized sports and the construction of masculinities. *Journal of Contemporary Ethnography, 18,* 416–444.

Mischel, W. (1970). Sex-typing and socialization. In P. H. Mussen (Ed.), *Carmichael's manual of child psychology* (pp. 3–72). New York, NY: Wiley.

O'Bryant, S. (1994). Widowhood in later life: An opportunity to become androgynous. In M. R. Stevenson (Ed.), *Gender roles through the lifespan: A multidisciplinary perspective* (pp. 283–299). Muncie, IN: Ball State University Press.

Piaget, J. (1954). *The construction of reality in the child.* New York, NY: Basic Books.

Preves, S. E. (2003). *Intersex and identity.* New Brunswick, NJ: Rutgers University Press.

Reilly, J., & Woodhouse, C. (1989). Small penis and the male sexual role. *Journal of Urology, 142* (2), 569–571.

Rosenthal, G. (2013, February 19). *NFL.com.* Retrieved April 16, 2013, from Female will compete at regional combine for the first time: http://www.nfl.com/news/story/0ap1000000140423/article/female-will-compete-at-regional-combine-for-first-time

Rosin, H. (2008, November). A boy's life. *The Atlantic.* Retrieved May 11, 2009, from http://www.theatlantic.com

Saner, E. (2008, July 30). The gender trap. *The Guardian.* Retrieved from http://www.guardian.co.uk

Siann, G. (1994). *Gender, sex and sexuality: Contemporary psychological perspectives.* London, England: Taylor & Francis.

Thorne, B. (1993). *Gender play: Girls and boys in school.* New Brunswick, NJ: Rutgers University Press.

Tsui, M., & Rich, L. (2002). The only child and educational opportunity for girls in urban China. *Gender & Society, 16,* 74–92.

Unger, R., & Crawford, M. (1992). *Women and gender: A feminist psychology.* London, England: McGraw-Hill.

United Nations. (2000). *The worlds' women, 2000.* New York, NY: Author.

Vespa, J. (2009). Gender ideology construction: A life course and intersectional approach. *Gender & Society, 23,* 363–387.

Williams, C. L. (1991). *Gender differences at work: Women and men in non-traditional occupations.* Berkeley: University of California Press.

Williams, J. (1987). *The psychology of women: Behavior in a biosocial context.* New York, NY: W.W. Norton.

CHAPTER 5

How Does Gender Matter for Whom We Want and Desire?

The Gender of Sexuality

When was the first time you knew about something out there called *sex?* How did you figure out what sex was, and was that initial picture right? Can you answer the question for yourself now? What is sex? What's the difference between *having* sex, *doing* sex, and being *sexual?* How are all of those things connected to your gender? What kind of people were involved in your initial picture of sex? Women, men, young people, old people, attractive people, white people? Who are appropriate partners for sexual relationships, and how are they different from appropriate partners for friendship or romantic relationships? How are sex, love, and attraction all related to each other? What makes someone homosexual, heterosexual, or bisexual? Does having sex with someone of the opposite sex make you heterosexual? Does having sex with someone of the same sex make you homosexual? Is being heterosexual something you do (have sex with people of the opposite sex) or something you are (a basis for identity)? Can you be straight at one point and then gay at another within the course of one lifetime? What are all the different possible ways in which we could create categories around sexuality, and would they all have to do with sex category or gender? How does sexuality overlap with other institutions, like marriage and the family? Are there differences in the ways you experience sex based on your gender? How would you draw a line around the parts of who you are that have to do with your sexuality and the parts that don't? Or is it even possible to isolate sexuality in that way? And how would those parts of you connected to sexuality overlap with those parts of you that are connected to gender? What would any of these things—sex, sexuality, feeling sexual—mean in a world without gender? Can you answer any of the questions we've posed without referring in some way to gender? What are the ways in which gender and sexuality are related?

As we'll explore in this chapter, it's difficult to have much of a conversation about gender without also talking about sexuality, even if we may not be quite sure about what we mean by sexuality. Sexuality is something that's discussed throughout this book. In this chapter, we'll take a more intentional and in-depth look at the topic of sexuality and how it relates to gender. We'll also begin to look for the answers to some of the questions we've posed. Sexuality, as you may have begun to realize in the previous list of questions, can include a broad range of behaviors, feelings, and identities, and almost all of them have some kind of gendered dimension in Anglo-European societies. As we will explore in this chapter, the way we feel sexually and the ways in which we are expected to act sexually are determined in important ways by our gender. In addition, the way we understand categories of sexuality in Anglo-European societies are also dependent on the existence of categories of sex and gender; a society without sex categories or gender would be a society without homosexuals, heterosexuals, or bisexuals. A world without gender or sex categories would be a world in which the way we thought about sexuality would look very different from our current modes of thinking, and this begins to suggest the deep interconnections between these two concepts.

Before we dive into this complicated discussion, it might be useful to establish some working terminology. In the English language, the word *sex* is an especially confusing one. It can refer to the biological category to which you are purported to belong based on your anatomy, chromosomes, or hormones (male or female). It can also refer to any kind of act that is deemed to be *sexual* in nature, although usually when people talk about having sex, they are referring to heterosexual intercourse (a penis penetrating a vagina). As we discussed in Chapter 1, in this book when we're referring to the first definition of sex (male and female), we generally use the term "sex category." **Sex** then becomes any act that is defined as sexual—which has the potential to become much broader than just intercourse. We'll use **sexual identity** to refer to the particular category into which people place themselves based on the current, Anglo-European division of the world into heterosexuals, homosexuals, and bisexuals. We use this term in part because it avoids the implications of more common terms like "sexual orientation" or "sexual preference," the first of which often presumes something one was born with and the second, a choice one makes about whom to have sex with. As we'll see in this chapter, for sociologists the question is less about why some people are gay or straight and more about why those questions seem so important to us in the first place—which often involves looking at where the categories gay and straight came from to begin with. As with other identities such as race and gender, the important question for sociologists is not why someone is African American or feminine, but where those particular sets of options came from and what their implications are for our social lives.

Come up with your own definition of sex and sexuality. Does your definition emphasize feelings, identity, emotions, behaviors, or some combination of all of the above? Is gender or sex category a part of your definition?

Sexual identity, then, avoids a lot of complicating issues we'll discuss in determining who is homosexual or heterosexual by defining sexuality based on how an individual identifies him- or herself. This leaves us with the last term to define: *sexual desire.* This term is perhaps the most difficult. What exactly does it mean to feel sexual desire? Is it the specific desire to be engaging in a sexual act, or is it more generally the way one feels in her or his body? Does sexual desire have to be aimed at some particular person or thing, what Sigmund Freud would call the love object? Is sexual desire best measured biologically (through pulse rate or blood flow to genitalia) or by asking someone whether he or she feels aroused? These are just a few of the questions that make it so difficult to define sexual desire, and we haven't even touched on how the categories of sex category, gender, and sexual identity complicate these questions. Is the sexual desire experienced by males or females, men or women, or homosexuals or heterosexuals somehow different? For our purposes, we'll define **sexual desire** as a combination of objective physical responses and subjective psychological or emotional responses to some internal or external stimulus. Sexual desire certainly has a physical component, but as we will discuss in more depth, it can also be deeply influenced by our culture, which can instruct us in when, where, and with whom it is appropriate to feel sexual desire.

DOES SEXUALITY HAVE A GENDER?

In his essay on the relationship between gender, sex, and sexuality, John Stoltenberg (2006) begins by imagining a world without sex categories or gender. In this world, the inhabitants emphasize the way in which every person is unique—unique in their genitalia, their particular mixture of hormones, their DNA, and their reproductive capability. When a baby is born in this imaginary world, there is no announcement—"It's a boy" or "It's a girl"—which emphasizes membership in a particular category of people who are presumed to be alike. Rather, they celebrate any newborn with an especially rare mixture of chromosomes or genitalia or hormones as yet another indication of how unique every individual is. Although in this world individuals are born with a wide range of different types of genital tissue, they emphasize the common origins of all these different organs in a prenatal nub of embryonic tissue called a genital tubercle. Thinking in this vein, these individuals are likely to focus on the fact that what someone else felt in their genital tubercle is probably very similar to what they feel in their own genital tubercle because their common origins mean that all these different types of organs are wired in fairly similar ways (p. 254).

In Stoltenberg's (2006) imaginary world, then, there are no sex categories. There are no males with penises or females with vaginas or intersexed people who don't fit into either category. There are just highly unique people who are different each from the other in a broad variety of ways. What does sex itself look like in this world without sex categories? The individuals in this world *have* sex without having *a* sex category. They have "rolling and rollicking and robust sex, and sweaty and slippery and sticky sex, and trembling and quaking and tumultuous sex, and tender and tingling and transcendent sex" (p. 254). They have sex that involves their genital tubercles but also sex that involves many other body parts. What is important in Stoltenberg's imaginary world is that in their erotic lives, these people are not acting out their status in any particular category when they have sex. They are not required to act in feminine or masculine ways while they are having sex; they are free to simply be themselves as genuine individuals. This is how Stoltenberg imagines sex would be for people who were not males or females, masculine or feminine.

The trick in Stoltenberg's (2006) description is that everything he describes about the biology and anatomy of his inhabitants is true for us in *this* world; the difference is in how we make sense of those biological realities. As we read in Chapter 4, we all begin this world with genital tubercles in the womb, and the same basic material makes up the genitals everyone ends up with. There's a great deal of variation in chromosomal sex (XX, XY, but also XXY and XO) and in the relative amount of different types of hormones in our bodies. Everyone on this planet is unable to reproduce for some period of their lives (for everyone before puberty; and for those of us lucky to live long enough, we will become infertile again as we age), and so the wide variations in who can and cannot reproduce are not an important basis for distinction. Onto the same biological reality, those of us in Anglo-European societies impose two distinct and discrete categories, and we make a whole host of assumptions about what it means to be in those categories, including what it means to be in those categories sexually. Stoltenberg attempts to explore what sex would be like in such a world and, in the process, implies what sex *is* like in our world with sex categories and gender.

Does the idea of sex in a world without sex category or gender sound appealing to you? How would sexual desire work in such a world? What kinds of things would we find attractive?

Namely, Stoltenberg and other gender scholars argue that having sex is one of the ways in which we create the idea that there is such a thing as sex categories and gender. We create the idea of gender in part through our expectations about what it means to be sexually as a woman or a man. This might be a strange way to think about sex at first, but Stoltenberg (2006) argued that because sex categories and gender are not actually real, we have to act in ways that make them feel like something real, and having sex is one way in which we accomplish that. This should sound familiar as a good example of a doing gender perspective. In this theoretical perspective, gender is omnirelevant, which means we're performing gender all the time and even in the most intimate of settings. Having sex itself becomes a way in which to create an accountable performance of gender.

Theory Alert!
Doing Gender

Heteronormativity and Compulsive Heterosexuality

Stoltenberg's (2006) point is one that's important to our understanding of the connections among sex category, gender, and sexuality. To be masculine and feminine in Anglo-European societies is essentially to be heterosexual. Children who are bullied for being "gay" are often being teased not for any expression of same-sex sexual desire. Rather, they are singled out for their gender nonconformity. Boys are called "sissy" and "fag" because they read books, are nice, don't like looking at pornography, like certain kinds of music, or walk with a certain "swish." Does this list of activities have anything to do with whom a boy wants to have sex with, or is it actually about his gender performance?

In Anglo-European societies, this clear link between gender and heterosexuality is taken for granted. This is part of what we mean when we talk about heteronormativity. **Heteronormativity** is the way in which heterosexuality is viewed as the normal, natural way of being. If you are heterosexual, you may have never spent much time thinking about the ways in which your society is structured based on an assumed heterosexual norm. The heterosexual questionnaire is a list of questions that highlights heteronormativity by taking questions often directed at homosexual people and turning them on their head (Rochlin, 1972). Take a look at the first nine questions from the questionnaire below. The very first question asks, "What do you think caused your heterosexuality?" This is a question that gay and lesbian people get asked frequently, but it is not one many straight people have ever considered. This question shows how homosexuality is something that needs to be explained while heterosexuality simply is.

There are several other assumptions about homosexuality that the questionnaire powerfully reveals. Question #5 showcases how we often believe that gays and lesbians would be straight if they only gave heterosexual sex a try. In Question #7, we see the fear some people have that homosexuals would like to change the sexual orientation of others. Question #8 raises an interesting question—do straight people flaunt their heterosexuality? What would it mean to do so?

1. What do you think caused your heterosexuality?
2. When and how did you first decide you were a heterosexual?
3. Is it possible your heterosexuality is just a phase you may grow out of?
4. Could it be that your heterosexuality stems from a neurotic fear of others of the same sex?
5. If you've never slept with a person of the same sex, how can you be sure you wouldn't prefer that?
6. To whom have you disclosed your heterosexual tendencies? How did they react?
7. Why do heterosexuals feel compelled to seduce others into their lifestyle?
8. Why do you insist on flaunting your heterosexuality? Can't you just be what you are and keep it quiet?
9. Would you want your children to be heterosexual, knowing the problems they'd face?

One potential answer to the first question in the questionnaire, what causes heterosexuality, is the concept of compulsive heterosexuality. **Compulsive heterosexuality** describes the way in which heterosexuality becomes institutionalized into the practices of daily life and therefore enforced as a way of regulating our behaviors and distributing power and privilege (Pascoe, 2007). The concept comes from an essay by Adrienne Rich (1986/1993), who used compulsive heterosexuality to explain the ways in which heterosexuality as an institution serves to ensure male physical, emotional, and economic access to women. In her essay, Rich identified the important role played by sexuality in reinforcing gender inequality.

As we have already begun to explore, sexuality is deeply gendered, and therefore, heterosexuality as an institution does not serve the exact same purpose when enforced on men as it does when enforced on women. The important premise of compulsory heterosexuality is that sexuality is more than just individual dispositions or behaviors; rather, it is an institution. As an institution, heterosexuality has the power to dictate norms (to be masculine is to be heterosexual) and to distribute power and privilege (those who are heterosexual have powers and privileges that are not available to those who are not). Compulsive heterosexuality for women may be about men maintaining their power over women, but for men, compulsive heterosexuality is more about men maintaining access to their power *as* men. Being called a fag is dangerous because it is not just about a boy's sexual identity or sexual desire, but it is about his *masculinity*. If heterosexuality is a social institution, then it has a history that we can explore.

A Brief History of Heterosexuality

It may seem strange to talk about a history of heterosexuality. Surely heterosexuality is as old as human society and therefore requires no explanation. This is, in fact, the approach many early social scientists took in their study of sexuality from a social perspective. The study of sex, or **sexology**, traditionally focused on heterosexuality but not in any way that suggested heterosexuality was not the natural state of being (Nagel, 2003). Early sexology was divided into two main tasks: documenting the practices of heterosexuals, not in order to demonstrate how heterosexuality is socially constructed but rather to establish the boundaries of "normal" and "abnormal" sexuality; and studying "deviant" sexualities, including homosexuality (Nagel, 2003). In this formulation, homosexuality is the phenomenon that needs to be explained, while heterosexuality is the taken-for-granted norm.

There are many problems with this particular approach to sexuality, including that it falls into the same trap we discussed in Chapter 2 in relation to the sociological approach to gender in sociology before the influence of feminism. Time and again sociologists and other social scientists have realized that for any given social category (gender, race, class, sexual identity, etc.), they must "study up" as well as down. **Studying up** in sociological language refers to the need to study those at the top of any particular power structure (Messner, 2005). Sociologists have tended to study down first, focusing on racial minorities, the working class, the poor, women, and homosexuals instead of white people, the middle class, the upper class, the wealthy, men, and heterosexuals. Most recently in sociology, studying up focuses on the study of whiteness, masculinity, and heterosexuality. This is one important reason to begin with a history of heterosexuality: It reminds us that there is a reason some categories are perceived as normal, natural, and therefore, unnecessary to explain.

The Invention of Heterosexuality

Beginning with a history of heterosexuality is also important because there was no such thing as heterosexuality in Anglo-European society until the 19th century. Neither was there anything called homosexuality. You may find this hard to believe until you understand an important distinction many historians of sexuality make between the words *heterosexual* and *homosexual* and all they have come to represent as compared with the actual sexual behaviors in which people engage. People have always engaged in opposite-sex sexual behavior, but no one until about 1886 would have labeled this behavior an example of heterosexuality or the people engaged in such behavior as heterosexuals. Similarly, increasing evidence suggests that same-sex sexual behavior is common across historical time periods and different cultures, but no one would have labeled a person engaging in same-sex sexual behavior a homosexual until about 120 years ago. *Heterosexuality* comes into the English language at the same time as its counterpart, *homosexuality,* because the two terms make no sense without each other. The two terms were made popular by a sexologist, Richard Krafft-Ebing, who wrote about homosexuality as a personality disorder in his book, *Psychopathia Sexualis,* published in 1886.

In the early Victorian period in the United States that immediately preceded the introduction of the terms *heterosexuality* and *homosexual* by Kraft-Ebbing in 1886, married women and men were expected to exhibit the idealized True Love in marriage (Katz, 2009). True Love, as it was represented in popular literary and religious texts, was certainly meant to exist between a man and a woman, and so we might be tempted to call it heterosexuality. But True Love was a state characterized by purity and freedom from sensuality. True Love did not include kisses, let alone actual sexual intercourse, and it was confined only to the proper institution of marriage. Can we call this particular perspective heterosexuality when it is largely not about sexual behavior? Within the Victorian discourse on sexuality, the only purpose of sex was procreation between a married couple; proper sexual desire, then, was desire that led to procreation, and everything else was improper. Certainly this would have included same-sex sexual behavior, but it also included married couples having sex that was not exclusively directed toward procreation. The idea that having sex on a regular basis for non-procreative purposes among married couples was part of a healthy relationship was completely foreign to the Victorian ideology of True Love. The Victorians shared some current views on the general lustiness of men, but they did not think the sexual desire of either men or women was exclusively or naturally directed toward the other sex. A man who had sex with another man was merely failing to contain his roving lust, and it was this lack of control, rather than the sex category or gender of his partner, that made this behavior objectionable. From this perspective, same-sex sexual behavior among men and soliciting prostitutes were perceived as similar offenses; they both resulted from a man's inability to control his lust. The particular lines drawn by the early Victorians in the United States hardly seem similar to the way heterosexuality is defined in contemporary Anglo-European societies.

CULTURAL ARTIFACT 1: WHAT DOES DISNEY TEACH US ABOUT HETEROSEXUALITY?

Heteronormativity is something that is taught in Anglo-European society and absorbed at a young age by children. Where do they learn that heterosexuality is normal and what exactly heterosexuality is all about? Parents, teachers, and peers send subtle and not-so-subtle messages about sexuality.

(Continued)

(Continued)

The media is also an important source of information about what it means to be sexual. In their content analysis of children's G-rated films, Karin A. Martin and Emily Kazyak (Marin & Kazyak, 2009) find that the depiction of heterosexuality on the big screen is complicated along gender and racial lines. Only 2 of the 20 films in their sample did not have significant hetero-romantic references. In these hetero-romantic narratives, heterosexuality is depicted as both magical and natural. Mrs. Pots—a character in Disney's *Beauty and the Beast*—tells Belle that if a "spark" is there, all that needs to be done is "let nature take its course." The magical nature of hetero-romantic love is depicted in cartoons by literal sparks, leaves, or fireworks that occur while the characters stare into each others eyes or kiss for the first time. Same-sex friendships for girls in these movies are rare unless the friends are mother figures, while lead male characters are often paired with comic buddies. Even more interesting is the depiction of the heterosexual gaze in these films. Women in animated features are drawn with cleavage, bare stomachs, and bare legs. Women of color are more likely to be depicted as grown women with large breasts and ample hips, while white women are imagined as delicate girls. There are many scenes where women are "almost caught" naked and men are depicted ogling women's bodies, often the bodies of women of color. In *The Hunchback of Notre Dame,* the dark-skinned gypsy, Esmerelda, performs a dance in front of the cathedral that resembles a strip tease. There were fewer examples in their sample of white women characters being depicted as "sexy." Overall, these films send the message that sexiness is something women possess and use for getting men's attention. The researchers reveal that although we may think G-rated movies are free of sexual content, there are actually very explicit messages being sent about the naturalness of heterosexuality and the particular sexual roles of women and men. Can you think of other messages about heterosexuality in children's movies?

This distinction between the actual term *heterosexuality* and sexual behavior between two people of the opposite sex (keeping in mind that from a strong social constructionist perspective, the notion of opposite sex itself is problematic) may seem a bit, well, nitpicky. They may not have given things the same names, but how important is the name we give to something after all? Would heterosexuality by any other name be the same, to paraphrase Shakespeare? It is important when we take a look at the ways that existed before the late 19th century for describing this range of sexual behaviors. Sexual behavior between two men of the same sex before 1886 would have been called *buggery* or *sodomy,* and one of the first laws in Anglo-European history to address same-sex behavior was passed by Henry VIII in 1533; it condemned all acts of buggery as being "against nature" (Weeks, 1996). The law was not designed to address same-sex behavior specifically, though; acts of buggery were defined as any acts that were "against nature," regardless of whether they were between man and woman, man and beast, or man and man.

This means that a man and a woman engaging in a sexual behavior that was considered against nature (usually anal sex) was defined as buggery. If people engaging in a sexual act that we would now call heterosexual were perceived as committing buggery, then buggery really does not mean the same thing as homosexuality at all. As with ancient Greek society, these buggery laws reflect a system for categorizing sexuality that is not based primarily on sex category or gender. What do same-sex sexual behaviors, bestiality, and anal sex all have in common that would warrant them being placed in the same category? They are all forms of non-procreative sex, meaning that they cannot lead to the birth of a child. For much of Anglo-European history, the division between procreative and non-procreative sex was one of the most important divisions, and it therefore makes sense that categories of importance (buggery versus legal and acceptable types of sex) reflected these divisions.

One Sex or Two?

What does this tell us about the sex category and gender system as well as the distribution of power in these historical societies? Many historians of gender and sexuality suggest that the fact that categories of sexuality were not based on sex category or gender reflect a different view of these phenomena. In fact, for much of Anglo-European society, the predominant view of women and men was a one-sex model, rather than our current two-sex model. The **one-sex model** of sex categories comes from the ancient Greeks and views women, not as a completely different type of creature, but as an inferior version of men. The one-sex model is rooted in a view of society as a gradient hierarchy; at the top of this hierarchy for the Greeks were the gods and then men; then below men were women, slaves, and other "deviants" such as dwarves. From this perspective, women are not fundamentally different from men, but they are merely inferior versions of men, just as men are inferior versions of the gods. If women are just inferior versions of men, rather than completely different types of beings, then making distinctions based on the sex category of whom you have sex with makes little sense.

In the current **two-sex model**, women and men are believed to be two completely different types of people, and sex categories are viewed as discrete (you're either male or female, and there's nothing in between). The one-sex model helps to understand the way in which distinctions based on sexual behavior were made for much of Anglo-European history, as well as how power was distributed. In the one-sex model, women are still considered to be at the bottom of the hierarchy, although the hierarchy itself may be differently shaped. For this reason, categories of sexuality focused almost exclusively on the behavior of men. This is not to say that same-sex sexual behavior didn't happen between women but that women's sexual behavior with other women was considered largely unimportant. What was most important in this model of understanding sexuality was not the sex category of the person you were having sex with, but where that person was on the hierarchy and whether you took the appropriately dominant role in sexual activity.

Sexuality in Cross-Cultural Perspective

If we've already established that the idea of heterosexuality and homosexuality are relatively recent historical inventions of the Anglo-European world, then it probably follows that these concepts are unique to the Anglo-European world as well. In the areas of gay and lesbian studies, feminist scholarship, and queer theory, research investigates how other cultures conceive of sexuality and the particular categories they create to classify types of sexual behavior. Although initially these projects were often described as searching for examples of homosexuality or homosexual behavior in other cultures, more recent work focuses on same-sex sexual behavior instead. This acknowledges that given different cultures with their own unique systems for organizing sex categories, gender, and sexuality, the terms "homosexuality" and "heterosexuality" very well may not make sense. These terms presume a network of relationships and ways of understanding sex category, gender, and sexuality that, as we will discuss, are simply not true for many other cultures. Given the amazing variety in the types of cultures found around the world, it shouldn't be surprising also to find a great deal of diversity in the particular forms of sexuality that exist.

Cross-cultural research on same-sex sexual behavior has found that social status in the form of age is often important in dictating appropriate and inappropriate forms of sexual behavior. A practice common among an entire grouping of cultures ranging from Papua New Guinea to some Australian Aborigine tribes to the Solomon Islands is a *boy-inseminating ritual* (Herdt, 1997). Among the Sambia of New Guinea, this ritual is an important part of boys' transition from childhood to becoming an adult. The Sambia believe that semen is a crucial substance for fertility and reproduction. Not only is semen what obviously results in the conception of a baby, but semen is also believed to have the magical power to transform itself into mother's milk. The Sambians also believe that although women are born with the important *feminine* reproductive fluids—breast milk and menstrual blood—men are not born with semen. For the Sambians, the male body does not naturally create semen, and therefore, semen must be provided by an outside source. The boy-inseminating ritual serves this purpose, and it involves oral sex between younger and older boys (between the ages of 10 and 15).

This same-sex sexual behavior is highly structured by strict ritual rules. Boys begin engaging in the behavior after undergoing an initial purification ritual to have their "feminine" traces removed. They are then ready to be initiated into masculinity by performing oral sex on older boys and, thus, becoming semen recipients. When they enter puberty, they switch from being semen recipients to becoming semen donors as the next group of younger boys performs oral sex on them. Young Sambian men engage in this behavior until they are married, and at that point, same-sex sexual behavior ends for most Sambian men. These rituals, which the Sambian believe are crucial to the very reproduction of themselves as a people, are kept secret from women, who may have little idea of the content of these behaviors. Undergoing these rituals is what allows Sambian men to become husbands and eventually fathers. Engaging in same-sex sexual behavior for Sambian boys is what allows them to become truly masculine.

Another interesting system for categorizing sexual behavior comes from many North American Native cultures. The **berdache** role, with some variations, exists across a wide range of Native American communities and cultures. The word itself, *berdache,* is from the French for male prostitute and reflects how early French explorers and settlers made sense of this particular aspect of the Native American communities they encountered. Within various Native American tribal groups, the exact term for the berdache varies, and the term more commonly involves men than it does women. A person becomes a berdache in different ways depending on the norms of the particular community, but the status usually followed from some expressed preference for cross-sex behavior. This could be expressed voluntarily on the part of a young boy or girl who showed more interest in learning the traditional activities of the other gender. Berdaches were also sometimes identified through tests of infants or small children in which some characteristically male and female implements were set in front of the child. If the child chose the particular implement of a different gender, they became a berdache and were raised as the opposite gender. A male berdache would dress as a woman, perform all the tribal duties associated with women, and in some cases have sexual relations with or marry a man. A male berdache for all purposes became a social woman within the community, even though he was anatomically still male. Although some early anthropologists assumed that berdache must actually be intersexed individuals, this was not the case.

The status of berdache within Native American communities varied. Among the Navajo, Cheyenne, and Mojave, berdache were believed to have exceptional abilities as matchmakers, love magicians, or curers of venereal diseases (Whitehead, 1981). The berdache had special ritual functions among the Crow, Papago, and Cheyenne. Male berdaches were also sometimes stigmatized as being cowards or as less than a full male. Their role within the societies, regardless of its particular content, was a well-established, taken-for-granted part of Native American culture. Berdaches were not perceived as "freaks" or as existing outside the norms of Native American societies but simply as filling one particular social niche in the fabric of their larger society. The most important part of their identity was their particular gender, and against this identity, their sexual behaviors were considered unimportant.

The cases of berdaches and boy-inseminating rituals demonstrate the many unique forms of sexuality across human history and cultures, including many distinct ways of thinking about same-sex sexual behavior. A more extensive survey would direct us to the *mahu* among native Hawaiian cultures in the United States (Matzner, 2001); the *bichas, viados,* or *travestis* of Brazil (Nanda, 2000); the Kathoey of Thailand (Nanda, 2000); the sworn virgins of the Balkans (Nanda, 2000); and the *hijra* of India (Nanda, 1998). In ancient Japan, samurai warriors took younger male lovers called *nenja* (Herdt, 1997). Among the Basotho people of Africa, women engaged in "mummy-baby" relationships, which involved a public ritual celebrating a bond between two women that might include emotional exclusivity, kissing, body rubbing, and genital contact (Gay, 1989). The only commonality among all these categories is some sexual behavior between individuals of the same sex. Beyond that, some of these relationships exist alongside heterosexual marriage, as in the mummy-baby relationships in Africa. Others, like the *hijra* of India, are isolated in communities of others like them and therefore do not participate in societal institutions such as marriage. In some instances, same-sex sexual behavior is considered perfectly consistent with one's gender role, as in the case of the samurai and the Sambia. Engaging in these same-sex relationships for men reinforced rather than threatened their masculinity. In other cases, as in some instances of berdache, gender identity did come into question.

Mixing It Up: Sex Category, Gender, and Sexuality

The important lesson to take away from all these different cases is that there is nothing given about the particular relationship between sex category, gender, and sexuality that is presumed in current Anglo-European society. Although many of the explorers, missionaries, and anthropologists who initially provided the accounts of these sexual behaviors and institutions tried to subsume them into their existing paradigm, they simply did not fit. Are the Sambian, berdache, *mahu, hijra,* or samurai all groups of "homosexuals"? To answer yes would be to engage in a classic act of ethnocentrism. **Ethnocentrism** is seeing one's own culture as better, more correct, or right relative to another culture. Clearly, in the dominant Anglo-European view of sexuality, many of these same-sex sexual behaviors seem to be homosexual. But labeling them as such is ethnocentric because it views the Anglo-European way of understanding sexual behavior as *the* correct way. If we simply label these behaviors as homosexual, we ignore the ways in which these behaviors are firmly grounded in a cultural context very different from that of Anglo-European cultures, including different ways of conceiving of sex categories and gender.

For example, anthropologists studying Sambian culture initially labeled these rituals as examples of homosexual behavior. But to see the ways in which that specific label of homosexual might not apply to the Sambian, we must untangle exactly what is implied in Anglo-European cultures' definition of homosexuality. In Anglo-European societies, gender and sexuality are closely linked, such that a woman who doesn't act feminine might be assumed to be a lesbian. This is in part the legacy of another 19th-century term and idea—*sexual inversion.* As defined by early sexologists in the late 19th and early 20th centuries, **sexual inversion** attributed homosexuality to an inborn inversion of gender traits. Male homosexuals were essentially women trapped in male bodies, and this explained their sexual desire for other men. This also explained their supposedly feminine characteristics, behaviors, and tendencies.

Although few people today would conceive of homosexuality in these exact terms, the idea that homosexual men are less masculine and homosexual women less feminine is still very much a part of Anglo-European definitions of homosexuality. How does this work for the Sambian? Sambian boys become masculine through engaging in same-sex sexual behavior; performing and receiving oral sex from other boys does not threaten their masculinity, but it is actually essential to the creation of that masculinity. The same is true for Japanese samurai, for whom the taking of *nenja,* younger male lovers, enhanced their status as a samurai and therefore their masculinity. To call the people in either of these groups homosexual is to ignore the very complex relationships between these behaviors and their own, unique gender system—relationships that are very different from those presumed in the meaning of homosexuality in Anglo-European cultures.

Ethnocentrism poses the same problems when we turn to the case of the berdache among Native American groups. Early anthropologists who examined this phenomenon concluded that the berdache role could be understood as an institutionalized outlet for homosexual behavior among Native American groups. The role of the berdache provided a way for individuals with same-sex desire to satisfy those desires by taking on the particular institutionalized role of berdache. One problem with this explanation of berdache is that same-sex sexual desire or behavior was not what determined whether you became a berdache. Attraction to individuals of the same sex is never mentioned in the anthropological literature as an indication of berdache status. Cross-gender behavior, exhibiting some desire or predisposition to engage in the activities normally associated with the other gender, *was* considered the main indicator of berdache status.

In the Anglo-European conception of sexuality, these two behaviors would be considered two sides of the same coin; if a little girl indicated that she liked doing masculine activities better than feminine activities, we would assume that tells us that she must also have some sexual desire for other women. But evidence suggests that Native Americans saw gender and sexuality as largely separate phenomena. The fact that a berdache preferred the activities of the opposite gender told them nothing about the form and content of their sexual desire. That this was true is evidenced in the different ways many Native American groups dealt with same-sex sexual behavior, which did occur from time to time outside of the context of the berdache role. The ways in which various tribes and communities viewed same-sex sexual behavior varied, sanctioning it sometimes as evil or in other places merely as foolish (Whitehead, 1981). But same-sex sexual behavior was not considered a sign of an enduring disposition among those who engaged in it and was considered unrelated to their particular gender identity. Same-sex sexual behavior was a phenomenon quite distinct from the role of the berdache.

Although from our perspective the berdache were clearly engaging in same-sex sexual behavior, anthropologists argue that the cross-gender status of berdache caused Native Americans to interpret their sexual behavior very differently. By choosing to be or being designated as berdache, these individuals existed outside of the normal rules enforcing opposite-sex sexual behavior as the norm. The berdache had a special status in regard to his or her sexual behavior, and whether the berdache had sex with someone of the same sex or opposite sex was unimportant. In addition, the nonberdache person who was intimately involved with the berdache was not reclassified (thought of as berdache) or perceived of as deserving the stigma of having engaged in same-sex sexual behavior. In other words, a man who married and presumably had sex with a male berdache (a biological male living as a social woman) would not have been perceived as engaging in same-sex sexual behavior, while a male nonberdache having sex with another male nonberdache *would* be perceived as engaging in same-sex sexual behavior. Anthropological research reveals no instances of a berdache having sex with another berdache.

What does all this reveal about the way Native Americans constructed their own system of sex category, gender, and sexuality? Although Native Americans certainly perceived links among sex category, gender, and sexual behavior, the linkage among these social phenomena was not quite as strong as it is for many in Anglo-European societies. For berdaches, it was most important that their sexual behavior was consistent with their gender. We might say that in the case of two nonberdache men engaging in same-sex sexual behavior, the problem is not so much that two biological males (sex category) are having sex with each other as it is that two social men (gender) are having sex with each other. If one of the males were a berdache, then the same-sex sexual behavior would be perceived as a completely different category of behavior. To wrap our minds around this, we'd have to imagine in Anglo-European society that sex between a biological male and a biological male who lives as a woman would not be perceived as homosexual. Gender would trump sex category and be the important criteria for establishing what is and isn't appropriate sexual behavior.

Together, all these cases return us to the potential complexity of the relationship between gender and sexuality. The sexual categories that have developed in Anglo-European society and have begun to spread around the world through the process of globalization reflect a specific way of thinking about the relationship between sex category, gender, and sexuality. But this is just one among many diverse ways of understanding these relationships and of forming categories around the complex constellation of sexuality. Gender has not always formed the most important basis for understanding categories of sexuality, as we saw in the case of classical antiquity. In some contemporary societies, gender can still be less important in the process of classifying sexual behaviors.

The berdache role has largely disappeared from contemporary Native American cultures, but in some Latin American cultures, categories of sexuality are still based on dominance more than on gender. Among Costa Rican male prostitutes, the attribution of homosexuality is made on the basis of being active versus passive; a male prostitute who is penetrated is homosexual, while a male prostitute who always does the penetrating is heterosexual, regardless of the sex of whom he is penetrating (Nagel, 2003, p. 205). In Nicaragua, the word *cochon* has been translated as synonymous with homosexual, but it really refers to a man who is the passive recipient (the penetrated) in anal sex, and not to the man who is the active participant. Putting aside our tendency to view other cultures through an ethnocentric lens, we see the many different kinds of relationships between gender and sexuality.

MEASURING SEXUALITY: WHAT IS SEXUAL ORIENTATION?

It's all very good and fine to look at how other cultures and other time periods understand the relationships between sex category, gender, and sexuality. But what about today? How do you know in contemporary, Anglo-European culture who is gay and who is straight?

The modern American history of the scientific research of sexuality begins in Indiana with Alfred Kinsey. Kinsey, a biologist by training who studied gall wasps before taking up an interest in human sexuality, undertook research in the 1940s to find out what people do, sexually speaking (Kinsey, Pomeroy, Martin, & Gebhard, 1948, 1953). After surveying a large sample of American women and men, Kinsey published two volumes on human sexuality, the first focusing on sexual behavior in the human male and the second on the sexuality of human females. The picture of sexuality that Kinsey revealed was disturbing to many because it did not fit into a neat picture of gay, straight, or bisexual. Because Kinsey was interested in the balance of homosexual and heterosexual behaviors within individuals, he asked them about their activities and thoughts, rather than about their underlying identity.

Kinsey's volume on female sexuality generated much more controversy than his volume on male sexuality. Why do you think this might have been? Do you think in today's society it is still more controversial to discuss women's sexuality as opposed to men's sexuality?

The result of Kinsey's research was the Kinsey scale, a new way of measuring sexual behavior. Focusing on thoughts and behaviors, Kinsey proposed that human sexuality was best described as not discrete categories of homosexuality and heterosexuality but as a continuum. As you can see from the scale, at one end, a person is exclusively heterosexual and at the other, exclusively homosexual. Kinsey found that most of his subjects fit somewhere in between, having had some same-sex and opposite-sex behavior or desire.

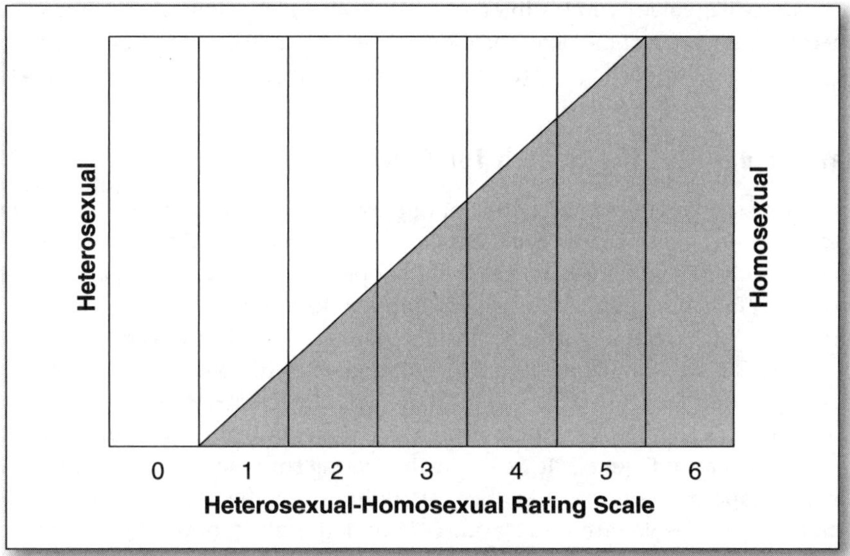

Source: Reprinted by permission of The Kinsey Institute for Research in Sex, Gender and Reproduction, Inc.

Kinsey's research, especially his volume on female sexuality, was controversial, and there was little in the way of scientific research on sexuality until the 1990s, when sociologist E. O. Laumann took another look at human sexual behavior (Laumann, Gagnon, Michael, & Michaels, 1994). Consistent with Kinsey, Laumann found very little consistency between how people identified their sexual orientation, their sexual behaviors, and their sexual desires. So, some of the people in his survey had same-sex desires and had engaged in same-sex behaviors, but they did not identify themselves as gay. Others had same-sex encounters but reported neither a **homosexual identity** (self-identifying as gay or lesbian) nor having same-sex desire. This research, like that of Kinsey 40 years earlier, points to the fact that sexuality is a multidimensional phenomenon and therefore one that is difficult to measure.

Herein lies the problem in scientific studies that search for a "cause" of homosexuality, whether that cause be genetic, hormonal, psychological, social, or some combination of all of the above (Fausto-Sterling, 2012). The first problem is figuring out who in your study will be considered "homosexual" and who will be considered "heterosexual." If you want to include "bisexual" as a category, or "asexual," you introduce even more complications. Sociologists like Laumann have the benefit of being able to ask live subjects about the history of their behaviors, desires, and identities; surveys allow them to access large samples of the population. But if you're collecting tissue, doing brain scans, or performing autopsies, it's much more difficult and much more expensive to cast so wide a net. Biological research on homosexuality recruits subjects at AIDS clinics, gay rights parades, or gay bars. You can begin to imagine how the two different approaches would yield very different kinds of samples and very different kinds of results.

The best science of today provides us with no definitive answers to many of the questions we have about sexual desire, gender, and sex category. Many would agree, though, that the answers will involve complex interactions between physiological and reproductive development, as well as psychological, sociological, and cultural development. As is true with many aspects of gender, sorting out the role of each of these variables is a formidable task.

Chocolate or Vanilla: The Search for Answers

Understanding the sexual behavior of human animals—as is true for all types of human behavior—is a complicated business with a great deal at stake. As complicated as the scientific study of sexuality is, what would happen if a scientist "proved" that sexual orientation is, in fact, biological. What would happen if scientists identified what has sometimes been referred to as the "gay gene"? What would the implications of such a discovery be?

Perhaps if we knew that sexual orientation were biologically based, the prejudice and discrimination faced by gays and lesbians would decrease. After all, if whom we desire is hardwired into our DNA or predetermined in the womb, then it is something we have little control over as individuals. It would be like hating someone for the color of his or her eyes or the shape of his or her earlobes, or perhaps, for the color of the skin, which is also something that is determined by our DNA and, therefore, beyond our control.

Yet, the list of horrors that have been visited on people based largely on skin color is a long and frightening one. For most of the 20th century, Americans believed that whites and Blacks in the United States were biologically distinct races; you were born white or Black in the same way you might be born gay or straight. As history tells us, this didn't prevent people and governments from discriminating against people based on the color of their skin. In fact, that people of color were perceived as biologically different from white people was often one of the primary reasons used to justify discrimination. History tells us that believing that a category is biological does not at all preclude the possibility of being discriminated against on the basis of that category.

The belief that race was a biological category is part of what gave rise to the eugenics movement in the United States. This widely popular philosophy based on evolution and Mendelian genetics held that the purest races should be encouraged to reproduce, although only with others from pure racial backgrounds. Racial mixing would pollute racial purity, and so it should be avoided at all costs. Those of suspect racial backgrounds, which in the United States included anyone who was not white, should be discouraged from reproducing. In many places in the United States, this took the form of forced sterilization programs. Eugenics was not a hidden movement that took place behind closed doors, but it was a way of thinking that reached to the very upper levels of society.

If we came to believe that sexual orientation were a category based solely in biology, would there be attempts to eradicate homosexuality from the population? Would parents be faced with deciding whether to abort their homosexual fetus? As social theorist Michel Foucault famously demonstrated, the claim to knowledge about a subject is also a claim to power. That beginning in the 19th century sexuality came to be known as a matter best understood through the lens of science is not without important consequences. The need of scientists such as Kinsey to describe and categorize sexuality is a means to control our sexual behaviors. Foucault would want us to ask ourselves, why do we want an answer to the question—why are some people gay and some people straight? Why is this such an important question? Why are we not equally as interested in whether someone likes chocolate or vanilla ice cream?

Well, of course, you might say, no one cares about whether you like chocolate or vanilla ice cream. Like sexuality, our ice cream preferences are about desire and pleasure. If someone asked you why you like vanilla or chocolate ice cream, you might be hard-pressed to give a definite answer. You might even say that sometimes you like chocolate and sometimes you like vanilla. Ice cream preference, you might explain, is complicated.

Why do we spend money researching who someone wants to have sex with but not the kind of ice cream they want to eat? Because we have placed a great deal of social significance on whom you want to have sex with but not as much significance on your ice cream preferences. We have constructed important social categories around the matter of sexual preference but not (as far as I know) around ice cream preference. No one would probably expect scientists to be able to identify a gene for vanilla or chocolate ice cream preference. And yet many believe that we may have already discovered a gene for sexual preference. What does our interest in these questions reveal about the assumptions of the societies in which we live?

Men as Sexual Subjects

In the particular Anglo-European view of sexuality, the links between gender and sexuality are important for how women and men experience sexuality. Men are perceived as powerful **sexual subjects**, meaning they have a sense of power and agency in their own bodies that allows them to *act in* their bodies rather than being *acted on* (Martin, 1996). In contemporary Anglo-European society, it is assumed that men will have strong sexual desires, whether they are soldiers in a foreign country or boys in middle school and high school. One example of men as sexual subjects is from Joane Nagel's (2003) book about the connections between sexuality, ethnicity, and race. She describes the *military-sexual complex* as part of the ways in which sex itself is militarized. Providing sex for soldiers is an essential part of how armies motivate men to fight, while rape committed during war is a way to create solidarity among troops through creating mutual guilt. What assumptions about masculine sexuality underlie phenomena like the military-sexual complex?

One assumption is that men have a need for sex so great that they can be motivated to do things like kill and die partly in exchange for ready access to sex. Certainly, the provision of women and prostitutes is not the only or the main reason soldiers fight and die, but the militarization of sex suggests a belief that the sexual desire of men is a pressing need that must be met. In encouraging and sometimes sanctioning the network of sex work institutions that surround military bases, governments are treating the provision of sex as just one more basic need of their armies that they are obligated to provide. Armies, historically and still primarily composed of men, quite simply *need* sex. That U.N. peacekeepers would solicit young, female prostitutes is considered just something that men do, and it is therefore excusable under the expression, "boys will be boys." Within this particular cultural framework, the assumption is that the sexual desire of men is, at the very least, greater than that of women and, at the most extreme, powerful, natural, and beyond control. Men are considered sexual beings, although in a way very different from that of women.

If armies, or other large groups, were composed primarily of women, would governments and military officials feel compelled to provide sex workers to meet the needs of their female soldiers? Why or why not?

As we will discuss later, women are also considered sexual beings, but they are more likely to be sexualized and made into sexual objects. Being a sexual subject seems like a good thing because it becomes a source of power and agency for men. Rape as a weapon of war, although rarely discussed, is part of the way in which sexuality is used to gain or reinforce power. In this most extreme version of sexual subjectivity, men act on the bodies of women or other men and these bodies become like the physical territory over which battles are fought. But the last lesson about masculinity and sexuality to be learned from these stories is that the power of sexual subjectivity and sexual desire also comes with a price. The presumption of men's strong sexual desires puts sometimes unrealistic expectations on men, and the connection between sex and power may deprive men of the ability to have different kinds of intimate connections. As explored by Susan Bordo (1990), if a penis is really like a power tool, then it should also be expected to perform like one; it's relatively easy to turn on and should function perfectly until someone turns it off. If it's "broken," it should easily be fixed. To admit that maybe your own sexual desire doesn't work like that is to be perceived as abnormal and defective, or even as literally powerless. If so much of masculinity is defined in current Anglo-European societies through sexual desire and sexual performance, this is bound to create some pressure on the actual men who live in those societies.

In addition, to describe masculine sexuality using machine metaphors implies a disconnect between sexuality and other aspects of what it means to be human. A machine does not require warmth, trust, affection, respect, love, or connection to function. And a machine exists only to serve a function or accomplish a task. What does it mean to think about having sex in these ways? Are we to believe that the act of sex for men is really so devoid of any connections to other emotions? Is sex for men really merely a performance, perhaps to be judged good or bad, but existing outside of any larger context or relationship? In his essay on sexuality, Stoltenberg (2006) argued that men are deprived of the opportunity to experience sex as something more than just a performance of masculinity because of their need to have *a* sex through having sex. Sexual subjectivity may be a source of power for men, but many would argue that it can also create pressures and place limits on masculine sexuality.

Can you think of slang terms for female genitalia? Are there as many slang terms for women's anatomy? What do these terms for female anatomy imply? What kind of metaphors do they involve, and what do those say about feminine sexuality?

Women as Sexual Objects

An exploration of stories about women and sexuality reveals a very different set of lessons than those about masculinity and sexuality. One of the legacies of the military-sexual complex outlined by Nagel (2003) in her book is a booming global business in sex tourism. The infrastructure developed to support military bases in many countries such as Thailand has been used to build an economy in sex tourism, including guidebooks that instruct tourists on how to negotiate for massages and buy bar dancers for the night. Some governments have even taken to advertising sex tourism, treating the availability of sex as a kind of natural resource to be developed. Unlike the military-sexual complex that involves mostly men, sex tourism appeals to women from the global North as sex tourists, as well. Nagel is interested in the ethnosexual nature of these encounters because sex tourism often involves white women from developed countries encountering men who are racially or ethnically Other in developing countries. Sex tourism for both women and men often satisfies sexual fantasies that are distinctly racialized, reflecting deeply held views about the sexuality of darker skinned peoples. In Jamaica, Rasta men are particularly popular for female sex tourists because they are believed to be "more passionate, more emotional, more natural and sexually tempting" (Nagel, 2003, p. 207), presumably as compared with their counterparts in the global North.

This interest in exotic, racialized Others is shared by male and female sex tourists. What is distinct about women as sexual tourists is their interest in romance as well as in sex, prompting Nagel and others to refer to women's sex tourism as *romance tourism*. **Romance tourism** describes the ways in which many women as sex tourists are "looking more to be swept away by men than to assert their strong control over their paid male counterparts" (Nagel, 2003, p. 207). Women are interested in more than a quick sexual encounter on their vacations to exotic locales, and research demonstrates that these women are much more likely to establish relationships with the men they "date" on these vacations. Women as romance tourists often keep in contact with the men they meet on vacation after they have returned home, sometimes send money, and are likely to return year after year to spend their vacation with their offshore "boyfriends." In addition, note in the previous description what makes Jamaican men appealing to romance tourists. They are perceived as more sexual, which is similar to how many male sex tourists view exotic women. But they are also considered to be "more passionate, more emotional, more natural." These suggest that romance tourists perceive these men as better at *romance* and as more sexual than their male counterparts at home.

In the same high school where C. J. Pascoe (2007) found high-school boys bragging about their sexual exploits and using the term *fag* to highlight any unmasculine behavior, she found distinctly unfeminine and lesbian young women being elected homecoming queen, an institution deeply characterized by traditional gender norms and heteronormativity. The institution of the homecoming court is heteronormative because it assumes a king and a queen, a heterosexual pairing who are often required to dance together after they are crowned. The queen is usually wearing a formal dress with all the current feminine accessories (makeup, a stylish hairdo, jewelry, high heels, etc.), while the king wears a suit or a tuxedo. Jessie, the young woman who was elected homecoming queen in the high school Pascoe studied, typically wore boy's clothes and was out in her high school as a lesbian. Another lesbian who played for the basketball team was also a popular student at the high school, along with a group of girl basketball players who wore baggy, hip-hop-style clothing and rarely dated boys.

Boys at the high school responded to the sexuality of these girls with statements like, "Hey, that's cool," a far cry from their reactions to behavior among other boys that did not conform to the specific gender norms of the high school. Girls who identified themselves as straight had no problem with being touched by Jessie, the homecoming queen, or with jokes made by Jessie and the basketball girls about being a "pimp" or having the ability to turn straight girls gay (Pascoe, 2007). Although both of these girls did behave in ways that violated the norms of femininity at the high school, they did not seem to be punished for this behavior, and their sexuality seemed to be accepted by both boys and girls; in fact, they were two of the most popular girls in the high school. On the other hand, several girls in the high school who were out as lesbians and affiliated with the GSA (Gay/Straight Alliance club) were not popular and found their expressions of sexual identity suppressed and discouraged by the school administration. Pascoe's research seems to indicate that the acceptance of lesbian sexuality in this high school was contingent on the adoption, and the reaffirmation, of masculine identity.

Sex, Romance, and Ideal Love

In her own study of the experiences of boys and girls at puberty and with their first sexual encounter, Karin A. Martin (1996) found that many of the girls she interviewed saw sex as something that happened to them, rather than as something they actively desired and pursued. Many of the girls in Martin's study experienced pressure to have sex with their boyfriends to maintain the relationship. Martin described adolescent girls as experiencing ideal love with their first boyfriends. **Ideal love** is "submission to and adoration of an idealized other whom one would like to be like and from whom one wants confirmation and recognition" (Martin, 1996, p. 61). As you might imagine, being the object of ideal love gives these adolescent boys a great deal of control. Many adolescent boys in Martin's study looked forward to having sex for the first time and saw this as an important accomplishment, especially among their male peers. These dynamics paired adolescent boys who were eager to have sex with adolescent girls who were often quite reluctant and anxious about having sex, but who idolized their boyfriends because of ideal love. When adolescent girls anticipated what their first sexual experiences would be like, the expectations of the majority centered on potential pain and fear. In these relationships, boys who were eager and excited about having sex pressured their girlfriends, who were reluctant if not scared about having sex.

When Martin asked her female teenage respondents why it was hard for some girls to say no when they didn't really want to have sex, they gave answers like, "Cause the boys break up with you or something like that. Or they say, 'You don't love me'" (Martin, 1996, p. 73). Ideal love for their boyfriends leads many adolescent girls to have sex out of fear of losing their boyfriends, and given these motives, it's not surprising that the experience of first sex is often not pleasant for these girls. The girls in Martin's study told stories about not just the physical discomfort of their first sexual experiences but also about the psychological and emotional confusion they experienced afterward.

In part because of these dynamics, when the adolescent girls in Martin's (1996) study described their first sexual encounter, they often described the event as something that happened to them rather than as something that they made an active decision to do. Several girls in Martin's study used the exact same phrase when they described how they decided to have sex—"it just happened." For example, here's Elaine explaining how she decided to have sex with her boyfriend:

> It just happened really. I mean, I didn't want to 'cause I couldn't ever picture myself having sex, but umm, all my friends did, and umm, so it just happened and he was my first so . . . I thought it was right cause we were going out for two years before we did. (Martin, 1996, p. 72)

The Double Standard

In these descriptions, girls turned themselves into sexual objects rather than into sexual subjects. Sex is something that happened *to* them, rather than something they made an active decision to engage in because of sexual desire or the anticipation of pleasure. To view oneself as a **sexual object** is just that—to view oneself as the passive recipient of sexual behavior and sexual desire, or to be the one who is sexually acted on and sexually desired, rather than the one doing the sexual acting and sexual desiring. This is in line with what Martin (1996) found about adolescent girls' attitudes toward their bodies in general. In their descriptions of their experiences of puberty and their general attitudes toward their bodies, girls felt their physical bodies were something separate from them and often outside of their own control. Rather than a source of pleasure, their bodies were often sources of pain, shame, embarrassment, and disappointment. These experiences combined to make girls much less likely than boys to *act in* their bodies and more likely to view their bodies as things that are *acted on*.

What assumptions does this set of stories reveal about feminine sexuality? What gendered assumptions do we make about women's sexuality? The first assumption is that sexual desire for women is not as easily divorced from other aspects of what it means to be human as it is for men. Romance tourists go to Jamaica not just because they consider the men there to be more sexual, but because they are also considered to be more romantic and emotional. When these female sex tourists seek ethnosexual encounters with those in other countries, they tend to establish relationships that are, at least on the surface, about more than sex. As one of my American students relayed, she had been told growing up that women have sex to get love, while men give love to have sex. This truism implies a lot about the existing cultural assumptions about sex and love but also about how each gender prioritizes the two. For women, it is love, not sex, that is the important thing to gain, while for men the opposite is true. One of the cultural beliefs about feminine sexuality, then, is that it is more diffuse and more connected with other emotions like trust, intimacy, warmth, love, and affection.

Is feminine sexuality, like masculine sexuality, also assumed to be heterosexual? Are lesbians themselves and lesbian behavior considered unfeminine? As with masculinity, there is certainly a connection that is perceived between sexual identity and gender identity for women. Some people might describe lesbians as women who really just want to be men. But the stigma associated with lesbianism in Anglo-European societies is different than that attached to homosexuality among men. One group of lesbian girls who were out in the high school studied by C. J. Pascoe (2007) seemed to suffer few negative consequences as a result of their sexuality, as long as they also adopted masculine behaviors. Same-sex activity between two women was considered by high-school boys as hot, in large part because of the portrayal of same-sex sexual activity between women in much heterosexual pornography. But straight girls at the high school also seemed unthreatened by one particular group of lesbians and non-gender-conforming women. The lesbian women at the high school who did seem threatening were those who made intentional efforts to challenge norms of both gender and sexuality. The GSA girls were not popular, and they found their efforts to be recognized in the high school blocked by the administration. This might suggest that for women, lesbian identity and behavior are considered acceptable as long as they do not threaten the existing gender order as well.

Perhaps this can be explained in part through the third assumption we can gather from these stories. Sex, in the end, is something that belongs to men. Lesbian sex isn't particularly threatening because it's not real sex, and it's not real sex because no men are involved. If sex is something that is done to someone, and only men can be the doers, then what two women do together by its very nature isn't sex. In other words, perhaps women are always perceived as sexual objects, even when engaged in same-sex behavior with other women. This tendency to view women as sexual objects rather than as sexual subjects is revealed in the stories told by adolescent girls in Martin's (1996) study of first sexual encounters.

Their stories reveal a common assumption about feminine sexuality: that women have less sexual desire than do men. Although some of the girls in Martin's study did begin to experience pleasure in sex after their first experience with intercourse, none of them pursued their first sexual experience as something that was important because it would bring pleasure to them. Sex was something that would please their boyfriends, but it was largely not something that was perceived as enjoyable for the girls themselves. Martin argues that these initial experiences with sex can have important effects on the way women view sexuality throughout their lives. In one large-scale study of sexual behavior conducted in the United States in the mid-1990s, 1 in 10 women reported lack of interest in sex, inability to achieve orgasm, finding sex not pleasurable, having difficulty lubricating, experiencing pain during intercourse, and anxiety about performance during the past year (Laumann et al., 1994). In a less scientific survey by Ann Landers, 72% of women readers said they preferred "hugs and cuddles" as adequate forms of sexual expression (Martin, 1996). This evidence suggests that beginning in puberty and perhaps throughout their lives, women are perceived as sexual beings, but sexual in the sense that they are the objects of sexual desire, rather than the people doing most of the desiring.

How do you define what it means to *have sex?* Is this sexual intercourse? Oral sex? Any sexual behavior that results in orgasm? What are the gendered implications of different ways of defining what it means to have sex?

One of the important consequences of women as sexual objects rather than as sexual subjects is what we call the double standard. The **double standard** is a cultural belief in Anglo-European society that the exact same sexual behaviors or feelings are OK for one gender (men) but not for the other (women). This usually means that it's considered OK for men to think more about sex, have more sex, have more casual sex, and have more non-marital sex than it is for women to do so. You can see how the double standard connects back to the idea of women as sexual objects and men as sexual subjects. The double standard both reflects this belief and helps to maintain it by making for fewer social sanctions for men than there are for women when they engage in sexual behavior. Thus, when adolescent girls have sex for the first time, the double standard tells them that they have done something wrong in a way that is not true for their male counterparts. Adolescent boys having sex is perfectly in line with gender expectations, but many girls must struggle with figuring out whether they have violated some important norm of what it means to be a girl, or at the very least a "good" girl. Martin noted that the double standard has been modified somewhat for these girls in that girls are not considered sluts merely for the act of having sex. Rather, a slut is a girl who has sex with someone she doesn't love or who has sex when she is too young (generally perceived as having sex at the age of 12 or 13) (Martin, 1996, p. 86). Nonetheless, the double standard is an important aspect of gendered sexuality in that it discourages women from expressing and acting on their sexual desires while expecting men to do so.

PLAYING THE PART? SEXUAL SCRIPTS

These assumptions we've laid out about masculine and feminine sexuality are, in the end, just that—assumptions. Maybe as you were reading them, you thought some of them seemed kind of true but others were way off base compared with your own particular experiences. Our list certainly doesn't cover the extensive list we could develop for all the gendered assumptions we might make about sexuality, and some of them, as we'll discuss, would directly contradict the ones we've just laid out. This is especially true as we begin to move across cultures and time periods; you might think of the particular list of assumptions we outlined as hegemonic for this particular time period—meaning they help define the ideal combinations of gender and sexuality in this specific time and place. But calling them hegemonic means that even in this very precise context, not everyone conforms to these particular ideals. Sexuality is a complex ideal, and just because some lesbian high-school women in Pascoe's (2007) study seemed to escape the kind of stigmatization directed at unmasculine high-school boys doesn't mean this is true for every setting or even every high school. Having set up some of these assumptions, let's see how they hold up when we look at sexuality across cultures and historical periods.

One way in which to answer these questions about the gendered nature of sexuality is to look at sexual scripts. Sexual scripts answer this important question: What exactly are we supposed to do, sexually speaking? **Sexual scripts** are the learned guidelines for sexual expression that provide individuals with a sense of appropriate sexual behaviors and sexual desires for their particular culture (Galliano, 2003). Sexual scripts can tell us both whom we should have sexual desire for or engage in sexual behavior with as well as what exactly we should do when we have sex. Sexual scripts are a concept firmly based in a social constructionist understanding of sexuality. The question is, when we get into the bedroom or whatever particular location is considered the appropriate venue for sexual activity (which is another part of sexual scripts), what do we actually do? If you believe that our sexuality is mostly the result of biology, then this isn't much of a question. Our sex drive or genes or hormones or anatomy tells us what to do, and there's no need for sexual scripts. The wide range of sexual behaviors that people do engage in suggests that there is at the very least some social dimension telling us what exactly to do sexually. Where do these sexual scripts come from if they are not hard-wired into our bodies? As our definition suggested, sexual scripts are derived from values and beliefs in the larger culture. Sexual scripts draw on beliefs about sexuality, but these intersect with beliefs about sex category, gender, and other hierarchical categories like age, class, and race.

Grinding and Sexual Scripts on the College Dance Floor

In Anglo-European society, sexual scripts suggest that men should be the initiators of sexual behavior. In her study of grinding behaviors on the dance floors of college campuses, Shelly Ronen (2010) found that there were unspoken roles dictating this particular style of highly sexualized dancing. Grinding, also known as "freaking," "freak-dancing," "dirty dancing," "bumping," and "booty-dancing," is a kind of dancing associated with club culture and hip-hop, which dominates the college party scene. In her study, Ronen found that grinding was always heterosexual and most commonly involved a woman with her back against a man's torso rubbing her buttocks into his groin. Face-to-face grinding was less common, but it involved women and men pushing their groins and chests into each other. At these parties, women would initially dance suggestively together in groups with other women. Men never danced with other groups or men or by themselves, as to engage in such behavior would potentially label them as homosexual.

In almost all the instances of grinding observed, men intiated the behavior. This usually happened nonverbally, with a man coming up behind a woman and beginning to grind against her. Women could refuse to accept this intiation of grinding behavior, but they almost always did so in ways that were communal and pleasant. For example, women might grab another female friend on the dance floor and begin dancing with her to disengage from a male partner. They would sometimes dance "limply," refusing to cooperate in the grinding behavior until the man gave up and walked away. Some would wait for the current song to end. In no cases did the researchers observe women confronting men directly with a rejection of their grinding advances.

In the few instances when women did ask a man do dance, the interaction usually ended quickly. The correct script for grinding, then, was for women to dance suggestively on the dance floor with other women until a man approached her to grind, most commonly sneaking up on her literally from behind. As the researchers point out, this behavior very closely mirrors our own sexual scripts for gender and sexuality. Men are the active initiators of sexual activity. Women have the power to accept or reject their approaches but not the power to initiate them. When they do reject sexual advances, they must do so in a way that minimizes the potential embarrassment and discomfort for the men. In this context, men remain the sexual subjects and women the sexual objects.

Sexuality From an Islamic Perspective

As cultural ideas about the connections between sex category, gender, and sexuality vary, sexual scripts will vary as well. Fatima Mernissi (2002) examined Muslim sexual scripts about male activity and female passivity. In Muslim traditions, sexual instincts in and of themselves are neither good nor bad. They exist and are understood without the kind of division between flesh and spirit made in Christianity, where the flesh is weak and sinful and the spirit is closer to God and therefore perfection. Sexual desire in the Muslim tradition, when properly controlled by religious teachings, is good in that it can help to serve God's purpose in three important ways. Sex leads to procreation, thereby ensuring the continuance of the human race. Sex also serves as a "foretaste of the delights secured for men in Paradise" (p. 297), thereby providing incentive for men to follow the religious dictates of Islam. Finally, sex helps to serve God's purpose in allowing for intellectual effort. According to some Islamic teachings discussed by Mernissi, one of the most precious gifts given to humans by God is reason. For men to use this reason fully, they must reduce the tensions created by sexual desire both within and without their bodies and be able to avoid the distractions of indulging in earthly pleasures (Mernissi, 2002, p. 305). Releasing these tensions through sexual activity allows men to avoid distractions and engage their reason. In these three ways, sex from the Muslim perspective is considered a part of God's plan, rather than as the work of the devil, as it is often considered in Christian traditions.

This particular sexual script as explored by Mernissi (2002) says sex is a good thing. How does this sexual script vary for women and men? In Muslim theology, women's sexuality is not generally perceived as passive. Ali, the husband of the Prophet Muhammad's favorite daughter, Fatima, and the founder of Shiite Islam, said of sexuality and gender, "Almighty God created sexual desire in ten parts; then he gave nine parts to women and one to men" (as quoted in Brooks, 1995). In other words, this creation story tells us that women got most of the cosmic share of desire to be doled out, nine tenths of it to men's measly one tenth. Women experience greater levels of sexual desire than men, and the writings of the Islamic scholar, Ghazali, studied by Mernissi (2002) confirm this view. In his instructions for husbands, Ghazali emphasized the important duty of a husband to satisfy his wife or wives' sometimes overwhelming sexual desire. It is a husband's duty to meet his wife's sexual needs to preserve her virtue because a woman who is not sexually satisfied by her husband will seek sexual gratification elsewhere. Ghazali recommended for the husband with four wives that he rotate between them so that each wife has sex once every four nights, although this schedule might need to be adjusted should the particular needs of his wives dictate.

If this isn't exhausting enough, Ghazali and the Prophet Muhammad both instruct men never to rush toward the act of intercourse. Both Ghazali and Muhammad emphasized the importance of foreplay to marital relations, as in the following passage where Ghazali quotes Muhammad:

> The Prophet said, "No one among you should throw himself on his wife like beasts do. There should be, prior to coitus, a messenger between you and her." People asked him, "What sort of messenger?" The Prophet answered, "Kisses and words." (as quoted in Mernissi, 2002, p. 303)

Muhammad was once quoted in a discussion of what constitutes cruelty as giving intercourse without foreplay as an example (Brooks, 1995). In these passages, female sexuality is clearly perceived as active. Who wants sex more according to this particular formulation of Islamic thought? Women. Who gets to say what happens during sex? Islamic religious authorities seem to be instructing men to do what women want to satisfy them sexually.

In fact, the idea that women's sexual desire greatly exceeds that of men is not at all unique to Islamic thought. In Greek mythology, the blind prophet Teiresias, who lived part of his life as a man and part as a woman, was asked to settle a dispute between the king of the gods, Zeus, and his wife, Hera, about who enjoyed the greater pleasure in sex or the better orgasm, women or men. Teiresias answered, "If the parts of love-pleasure be counted as ten, Thrice three go to women, one only to men" (Blackledge, 2009, p. 281). The number is eerily similar to that given by Fatima's husband in Islamic tradition, but the idea that women gain more pleasure from sex than men is also echoed in Hindu tradition. In the Indian epic, the *Mahabharata,* the powerful king Bhangasvana, like Teiresias, is turned into a woman by an angry god, Indra. Years later, Bhangasvana is forgiven by Indra and asked whether he would like to become a man again or remain as a woman. Bhangasvana answers, "The woman has in union with man always greater joy, that is why . . . I choose to be a woman. I feel greater pleasure in love as a woman, that is the truth, best among the gods" (Blackledge, 2009, p. 281). Even the Old Testament in the Judeo-Christian tradition from the 3rd century B.C. recognizes the power of female sexual desire with the following lines: "There are three things that are never sated . . . Hell, the mouth of the vulva, and the earth" (Blackledge, 2009, p. 281).

In this broader cultural and historical perspective, the sexual script in current Anglo-European society that women are passive sexually is something of an anomaly. Sexual scripts in other cultural traditions, as well as cultural traditions (like the Bible) within Anglo-European society, view women's sexuality as much more active than men's. Why are our current ideas so different? Some feminists suggest that the idea of passive, feminine sexuality is an attempt to control women's power in society. In her study of adolescent girls, Martin (1996) argues that girls' first sexual experiences and their negative feelings about puberty explain the drop in self-esteem attributed to girls at around this age. Feeling like a sexual object and as if one's body is something beyond one's control has an effect on how girls' think about themselves in general, not just sexually. Controlling a person's sexuality is therefore a means to control their own sense of agency and efficacy in the world, and perhaps this explains the existence of this particular sexual script.

In general, this brief exploration of sexual scripts provides strong support for the role of social life in influencing our sexuality. How might the experiences of the girls in Martin's

(1996) study have been different if they had lived in a culture that saw women as sexually active rather than as passive, as sexual subjects rather than as sexual objects? As with gender itself, the exact ways in which we think about the sexuality of women and men depend a great deal on the cultural context in which we find ourselves. This is true of the ways in which gender affects how we think about sexuality in general and also for the ways in which our categories of sexuality depend on the existence of sex categories and gender.

Stabane and Sexuality in South Africa

Can we ever escape the particular sexual scripts of sexual subjects and sexual objects? What happens to these dynamics in sexual interactions that do not involve biological males and biological females? Is that different from what might happen in sexual pairings that involve two people who are both masculine or both feminine? When two masculine individuals engage in sexual behavior, who is the sexual subject and who is the sexual object?

To answer these questions, let's consider the particular alignment of sex category, gender, and sexuality among Black South Africans. South Africa's constitution was the first in the world to offer protection to its citizens on the basis of sexual orientation, gender, and sex. But this legal protection of gays and lesbians is complicated by the particular cultural understandings of sex category and sexuality. *Stabane* is a Zulu vernacular word used to describe an intersexed individual (Swarr, 2009). Someone who is *stabane* is assumed to have both a penis and a vagina. However, in contemporary South Africa, the concept of *stabane* becomes conflated with homosexuality so that gays and lesbians are assumed to be intersexed.

Gays and lesbians in South African townships who express same-sex desire are often forcibly examined to verify their status as stabane. As one woman reports of her own experiences, "They assumed that because I had proposed to the girl I must be a stabane, a hermaphrodite, with both male and female genitals. I was taken to a hut where a woman forced me to undress and examined me. When they discovered I was 'normal' the chief ordered that I be lashed." (Swarr, 2009, p. 531). Similarly to the concept of sexual inversion, the concept of stabane reflects that same-sex sexual desire must be located in some internal sex category abnormality.

These beliefs about the nature of homosexuality affect that particular sexual scripts followed by gays and lesbians in South Africa. In Soweto, having sex with someone of the same sex does not necessarily change how an individual might label her own sexual orientation or how others might label it. So straight women who have sex with a butch lesbian may not be considered lesbians themselves. When butch lesbians have sex with straight women, the stabane concept necessitates that they find a way to penetrate their female sexual partners because they are expected to have a penis. As one butch lesbian in Soweto describes, "And one day she'll start a silly, confusing conversation, 'Do you have a penis as well?' A woman can be shy like that. Thirdly, she'll maybe want you to just penetrate her, not to play around. You know this foreplay thing. And then if you don't do that as quick as she wants, she expects it, and then you are a failure in bed." (Swarr, 2009, p. 535). In a similar vein, gay men might hide their penis or use smell and other means to convince their straight male sexual partners that they have a vagina.

In both these instances, the particular sexual scripts associated with heterosexuality are replicated in the context of same-sex sexual interactions. Like men, butch lesbians are expected to take control of sexual interactions and penetrate their partners with little attention to foreplay. Gay men are penetrated by their straight male lovers. The stabane concept preserves the idea that even when men are having sex with men and women are having sex with women these are still heterosexual interactions.

SEX AND SOCIETY

The idea that sexuality is an object to be studied scientifically, whether biologically or socially, is a relatively new idea in Anglo-European history—which is not to say that sex categories, gender, sexual desire, and sex haven't always been around. But in exploring the ways in which sexuality is gendered in current Anglo-European society, it helps to have some perspective on the differing ways in which other cultures have conceived of sexuality. This exploration reveals an important aspect of sexuality that may be harder for us to understand when we're firmly planted in our own cultural context. The way in which any particular society structures sexuality reveals something important about the working of that society in general. Notice, this is a deeply social approach to sexuality. From this point of view, sexuality is more than just an internal, psychological, or biological aspect of our selves; it is also a part of the structure of society, built into its institutions, statuses, and social roles. Far from being something that is isolated from other aspects of our social lives, like family, education, or work, sexuality is something that is inevitably linked to other institutions within society. Most of us today probably believe that sexuality exists apart from these other areas of life, and if we think about the relationships at all, we are probably inclined to argue that our sexuality affects those other aspects of social life but not the reverse. That is, if we have a crappy sex life, we may believe that can cause us to be unhappy in our work or family relationships. But we generally wouldn't believe the opposite, that our work or politics are reflected in the way we behave sexually. But when we take a broader perspective, by examining other cultures, we can begin to realize the ways in which our sex lives tell us a great deal about who we are as a culture in general.

Sex and Social Status

Social status and power are important components of sexual behavior in many cultures. One example of how this social status and power flows along gendered lines is the premium often placed on female virginity prior to marriage in ways not generally true for male virginity. In other words, it is more important for women than for men to be virgins when they marry, and in fact, many people think of a young girl (and not a young boy) when they think of the "typical" virgin. In some Polynesian societies, the daughter of the chief is called a *taupo* or *sacred maid,* and her virginity is important not just to her family but also to the tribal community as a whole. The sacred maid receives special honors and privileges, but these are in return for a strict enforcement of her continuing virginal status, and if she loses that virginity, it is a matter dealt with by the tribal council as a whole. In some instances, the consummation of her marriage may be public, but even if it is not, proof of her virginity in the form of a blood-stained mat is displayed to the whole community. There is no equivalent male role in these societies, and although sex before marriage is fairly common among the general population of young women and men in these societies, a young man having sex with a young woman is perceived as having stolen from the woman's father and brothers (Ortner, 1981).

These specific cases demonstrate a status inequality in sexual relationships between those of the opposite sex, which is common across many cultures. The sexuality of women is controlled and dictated in ways that the sexuality of men is not. In the instance of the sacred maid, the sexual behavior of these young women is literally the property and concern of the entire community. In perceiving other young women's loss of virginity as a theft from her father or brothers, her sexual decisions are still conceived of as something belonging to the male members of her family. Norms surrounding what is and is not appropriate in these cultures tell a story about gender inequality. Men, by and large, own and control their own sexuality, while the sexuality of women is owned by their families or, in the case of the sacred maid, the entire community.

Looking at ancient Greek society serves as another example of the connections between social structure on the whole and the structure of sexuality specifically. First, a little background on what Greek society was like is necessary. Society in classical Athens was divided on a strict hierarchy, with male citizens at the top and women, children, foreigners, and slaves all lacking full rights and therefore existing beneath male citizens in the hierarchy (Halperin, 1989). Sexual relations in Athens were structured very strictly in line with this hierarchy, so that the Athenians did not think of sex the way many of us do, as a joint and mutual enterprise. Rather, sex for the Athenians was understood as an action performed always by a social superior on a social inferior. Sex was a deeply asymmetrical act that divided the world into penetrator and penetrated, and these categories were in no way interchangeable. Those deemed socially superior (male citizens) were the penetrators, and those deemed socially inferior (women, children, foreigners, and slaves) were the penetrated. These were strict rules to be followed in terms of acceptable behavior, and so what people did sexually did not reflect any personal preferences or some unique aspect of their identity, as might be believed in current Anglo-European society. Sexual acts rather reinforced the hierarchical system in which the Greeks lived, and although real Athenians no doubt felt real sexual desire, their desire was inevitably shaped by the dominant principles of Athenian political life. Sex in this sense was neither a part of one's personality nor something that existed in a sphere separate from the important stuff of Athenian social and political life.

To what extent do our contemporary concepts of sexuality reinforce existing hierarchies in our society? Do we have norms about what is proper and improper sexual behavior that seem to reflect who is considered superior and inferior in our society?

The central hierarchy that was enacted and enforced in sexual scripts for the ancient Greeks was not focused primarily on sex category or gender. A man having sex with another man was only considered "wrong" if the man doing the penetrating was below the man being penetrated in the social hierarchy. In contemporary Anglo-European society, sex category or gender is central to the ways in which we understand what is and is not "appropriate" sexual desire and sexual behavior. Sexuality is gendered in that to be masculine is also to be heterosexual, but being heterosexual to begin with makes no sense without the existence of sex categories or gender. Here it's useful to ask another seemingly stupid sociological question: What exactly do we mean when we say that someone is heterosexual? How does someone know that they're heterosexual, and why are some people heterosexual and some people not? Perhaps the simplest definition we could lay out at the outset is that a heterosexual is someone who experiences sexual desire for someone of a different sex category than their own. This answer works out pretty well if you assume a biosocial approach to the relationship between sex category and gender. If a type of person who is biologically male has sexual desire for the type of person who is biologically female, then that person is heterosexual. Under this definition, sexual desire is primarily about anatomy, hormones, and genetics. Heterosexuals are people with penises who sexually desire people with vaginas or people with vaginas who desire people with penises. It all seems very neat and tidy, if you leave aside the messy fact that most of us (who are not walking around naked or with our chromosomal sex tattooed across our heads) don't have definitive evidence of whether the person we desire *does,* in fact, have a penis or vagina until further along in the relationship. To what exactly, then, are we attracted? Is it sex category (that the person has a penis or a vagina) or gender (that the person looks and acts in feminine or masculine ways)?

Theory Alert! Biosocial Perspective

There's some confusion under the assumptions of the biosocial approach, but sexuality gets even messier if you assume, like strong social constructionists, that sexual dimorphism is a claim rather than a fact. If there are not really two types of people, males and females, then what does it mean to be heterosexual? If sex categories are essentially imposed on a more complicated underlying biological reality, then why would it make sense to build any other categories (heterosexual, homosexual, bisexual) based on what is, in fact, a more complicated reality? Where would an intersexed individual fit into the categories of heterosexual or homosexual, given that he or she doesn't necessarily fit into the categories of female and male to begin with? What are transsexual individuals? Is a person who lives as a woman and has sexual desire for other women but has a penis and XY chromosomes a lesbian or a straight man? If there are many different types of people, not just males and females, then how can we possibly describe their sexuality with such limited boxes as heterosexual, homosexual, or even bisexual?

CULTURAL ARTIFACT 2: COULD A GAY SINGER WIN *AMERICAN IDOL?*

"Gay bars across the world are watching Idol Tuesday and Wednesday nights" (Hernandez, 2007). In an article in *The Advocate,* Greg Hernandez argues that *American Idol* is not only one of the most popular shows on American television but also one of the gayest. It's complete with divas; an ongoing flirtation between judge Simon Cowell and host Ryan Seacrest; and contestants—like Clay Aiken—who set off the gaydar of many viewers. But is there an unspoken rule on *American Idol*—and perhaps other reality shows as well—that contestants keep their sexuality under wraps while they're competing? Finalist R. J. Helton, who came in fifth in Season One's competition, came out only after his time on the show, saying, "I did tell some of the assistant producers because I felt like it was eating me alive. But I was advised to just keep it to myself. The reason they gave me was that it wouldn't be a good idea for my career." Although *American Idol* producers claim they put no pressure on contestants to conceal their sexuality, others wonder whether the show isn't concerned about preserving its wholesome image. Would having openly gay contestants create problems in appealing to the widest demographic possible? Others argue that reality television, starting with MTV's *Real World* series, has exposed millions of Americans to gay and lesbian characters, thereby increasing acceptance. Recently, a gay man competed on CMT's (Country Music Television) *Redneck Island,* a niche version of *Survivor,* which included hunting, bowling, and fishing (Blake, 2012). What's more, Adam Freeman, an openly gay father of five, came in second place on the show. If a gay man can do so well on *Redneck Island,* surely *American Idol* can't be far behind. Would a gay *American Idol* winner be hurt by her or his sexual orientation?

In the particular formulation of current Anglo-European society, the important categories for describing sexuality are deeply dependent on the existence of sex categories or gender. This has not always been the case in Anglo-European history, as we discussed with ancient Greece, and it is certainly not always the case in other cultural traditions. Michel Foucault, a theorist whose views on sexuality have been influential in the areas of feminism, cultural studies, postmodernism, and queer theory, argued in his own history of sexuality that we must pay special attention to the ways in which discourses of sexuality are also discourses of power and control. A **discourse** for Foucault is "the means by which institutions wield their power through a process of definition and exclusion" (as quoted in Storey, 1988, p. 96). Discourses on sexuality, then, define what is and is not acceptable in relation to sexual behavior, as well as what is and is not defined as sexual behavior. In ancient Greek society, discourses on sexuality reinforced the existing hierarchy in society and therefore were part of a larger mechanism used by one particular group in society to maintain control. Therefore, an exploration of sexuality in different historical time periods and across different cultures will tell us a great deal about that particular society's means of categorizing sex and gender, as well as reveal a great deal about who is in power and how that power is used in a given society. We'll begin with an exploration of sexuality in Anglo-European history.

Boston Marriages and Sexuality

In a similar vein are the *Boston marriages* or *romantic friendships* of 18th-and 19th-century middle- and upper-class white women in Anglo-European societies (Rupp, 2004). **Boston marriages** describe passionate attachments between women in this time period, which often lasted through marriages and, as relationships, were generally not looked on with disapproval. In the early 1870s, Molly Hallock Foote wrote to her friend Helena,

"I wanted so to put my arms around my girl of all the girls in the world and tell her . . . I love her as wives do love their husbands, as *friends* who have taken each other for life" (Rupp, 2004, p. 303). The expression of this kind of passionate love, a love comparable to that which a wife would feel for her husband, was also common during this time period. Rose Elizabeth Cleveland, the sister of U.S. President Grover Cleveland, shared a long and passionate correspondence with her friend, Evangeline Simpson Whipple, describing how she was "heavy with emotion" for her friend, trembling at the thought of being in her arms. Although Whipple married briefly (to a 74-year-old Episcopalian bishop), the two women eventually settled together in Italy after the death of Whipple's husband (Roy, 2001). It's clear from many of these letters that women in these romantic friendships would kiss, caress, and sleep with each other (in the same bed), but it is impossible to tell whether they actually engaged in same-sex sexual behavior. Were these women lesbians? How do we make sense of these behaviors?

Both of these examples reveal a historical world in which passionate friendships between people of the same sex were the norm. As we will explore in Chapters 6 and 8, for much of its history, and still in many places today, marriage had very little to do with passionate love, let alone sexual desire. In Anglo-European history, the highest form of love was for many years believed to be that which existed between two men. This made sense in a very gender-segregated world where men spent most of their lives in the company of other men and women in the company of other women. How could such very different experiences form the basis for a strong and passionate attachment in marriage? That we look at passages from these historical letters and make presumptions about the sexuality of the authors tells us a great deal about how different our concept of homosocial relationships is today than it was in the past. The term **homosocial** refers merely to social relationships between those of the same sex. The relationships between men in a fraternity, mothers and daughters, or all male soldiers in a platoon are all homosocial relationships. Many historians argue that in Anglo-European history, with the invention of heterosexuality and homosexuality as well as the rise of the companionate marriage (marriage based on compatibility rather than on financial or political criteria), a shift developed in the way people thought about homosocial relationships. Much of the passion that had once been channeled into friendships with those of the same sex was diverted into the institution of marriage; the companionate marriage assumes that your husband or wife is also your close, if not best, friend. At the same time, the idea of homosexuality, that sexual behavior marked a kind of person rather than a set of behaviors, changed the ways in which people viewed homosocial relationships. To express passionate love for someone of the same sex became associated with sexual behaviors under the new system of sexuality.

This historical tale again points us to the important links between gender and sexuality. Sexuality as an institution is affected by the ways in which we draw the lines between those of different genders and how we understand what are and are not appropriate types of relationships. In some time periods, passionate, romantic love between women is viewed as just one particular manifestation of homosocial relationships. Today, we still expect that women's friendships are closer than those of men, as we will discuss in the next chapter, but the letter from Molly Hallock Foote would still probably be perceived as excessive. The degree to which we connect romantic love and attachment to our notions of sexuality is also influenced by how we understand gender in a particular time and place.

Bisexuality: Somewhere in Between

Although "heterosexuality" and "homosexuality" as terms date back to the late 19th century, "bisexuality" is an even more recent concept in social life. In one of its early uses by an American doctor, *heterosexual* was used to describe someone with "inclinations to both sexes," which is how we currently understand what it means to be bisexual and demonstrates how initially the meaning of terms like "heterosexual" and "homosexual" were unclear (Katz, 2009). Freud believed that all of us as humans are potentially bisexual, capable, in Freud's language, of having same-sex and opposite-sex love objects. It is the civilizing force of society that channels this original flexibility into an object choice that is determined by sex. Dividing the world into heterosexuals and homosexuals for Freud is like a fall from grace from our original, and more flexible, underlying nature. But Freud's view of bisexuality is one among many competing ways of thinking about the complex formulation of sexual desire. In contemporary Anglo-European society, bisexuality is increasingly recognized as a category of sexual identity alongside heterosexuality and homosexuality. Bisexuality does not come without its own continuing stigmas, though. Bisexuals can receive negative reactions from both homosexuals and heterosexuals. In her study of the way in which lesbian feminists conceive of bisexual women, Amber Ault (1996) identified four techniques of neutralization through which some lesbian feminists express their hostility toward bisexual women. Through suppression, lesbians insist that there is no such thing as bisexuality. Bisexuals are either confused lesbians or heterosexuals who are experimenting. This is connected to the second technique: incorporation. Bisexual women are lesbians who are not yet aware of their lesbian identity or are "on a bi now gay later plan" (Ault, 1996, p. 314). Marginalization acknowledges that bisexual women exist, but it considers them to be unimportant to lesbian politics and lesbian communities; bisexuals are on the fence and therefore not to be counted. Deligitimation, the final technique of neutralization, casts aspersions on the characters of bisexual women as promiscuous, disloyal bed-hoppers. The animosity felt by some lesbian women toward bisexuals leads them to describe bi women as "sexually promiscuous, personally deceived, immature, in denial, perverted and unable to form stable familial bonds" (Ault, 1996, p. 314). These descriptions are all labels that heterosexual culture also directs at lesbians, and Ault points to the ways in which bisexuals are stigmatized by a group (lesbians) that are themselves already stigmatized.

The bisexual women in Ault's (1996) study point out that on the other hand, they are often subject to the exact same discrimination and hostility as their lesbian counterparts from the straight world. A bisexual woman walking down the street with her female lover is harassed as a lesbian, and her harassers are unlikely to stop and ask her whether she's bisexual before they do so. Courts and legal institutions also often lump bisexuals together with gays and lesbians. Colorado's Amendment 2 prohibited any laws in the state that would have given protection to homosexuals, lesbians, or those with bisexual orientations. In other words, this law ensured that these groups would not be protected from discrimination based on sexual orientation as individuals are ensured protection based on their gender, race, nationality, and religion. In the eyes of these lawmakers, bisexuals are not considered distinct from gays and lesbians. Like homosexuality, bisexuality is placed clearly outside the boundaries of what makes "normal" sexuality and treated by our heteronormative culture as similar to homosexuality.

The tension between those within gay and lesbian communities and bisexuals is something that has faded somewhat recently, partly because of the influence of queer theory and queer activism. Increasingly, activist organizations on college campuses and in other places use the LGBT label, which includes lesbians, gays, bisexuals, and transgender individuals. This demonstrates an inclusiveness that was not as evident in Ault's (1996) research with lesbians. Why does the category of bisexual have the potential to evoke such hostility on the parts of both gay and straight people? What essential notion of sexuality does the existence of people who call themselves bisexual call into question? Some bisexuals suggest an answer to this question when they create their own categories for labeling sexuality that put bisexuality at the center and all other types of sexual orientation at the margins. In Ault's (1996) study, some bisexual women referred to straights as well as gays and lesbians as monosexuals. In this new dichotomy, bisexuals, who do not limit themselves to only one sex in their choice of romantic partners, are juxtaposed with monosexuals, who have sex or romantic relationships with only one sex. Because this particular system originates with bisexuals, there is a disparaging quality to the new label of monosexual, and both straight people as well as gays and lesbians are considered to be the Other to the category of bisexual. These new categories can be viewed as an attempt by bisexuals as a community to marginalize those other groups, heterosexuals and homosexuals, who have already marginalized them.

SEXUALITY AND ITS CONSEQUENCES

This consideration of the position of bisexuals as an identity and a community points us to an especially important point about categories of sexuality. We have thoroughly explored many different ways of forming categories of sexuality. As sociologists, we can understand that categories of sexuality like heterosexuality, homosexuality, and bisexuality are all socially constructed, but this doesn't mean the categories don't have an important impact on our lives. The attempt by bisexual individuals to redefine the world along the lines of bisexuals and monosexuals matters only if they can convince the rest of society to believe in this particular system of categorization. In the meantime, bisexuals are often still considered to be fence sitters, and their status relative to the existing system of sexuality has consequences for how they are viewed and treated in society. Understanding that there is nothing given about the particular system of heterosexual, homosexual, and bisexual in contemporary Anglo-European societies does not negate the important consequences these categories have for people living in those societies.

When Did You Know You Were Straight? Heterosexism and Heteronormativity

We can begin thinking about the consequences of these categories with the question posed in this subsection's title: When did you know you were straight? It's a hypothetical question, as we aren't actually assuming that everyone reading this chapter right now is, in fact, straight. It might sound like a strange question, or like one of the words is wrong. Shouldn't the question be, "When did you know you were gay?" This is a question that many gay or lesbian people might get asked, sometimes with some frequency. What happens when you change the question in this way? How does someone come to know that they're heterosexual?

If you're like many of the straight students in my classes, this question is something of a stumper. If you do consider yourself heterosexual, you probably haven't ever had to think much about the answer to that question. The privilege of not being called on to explain your sexual desires and behavior is a part of heteronormativity, as we discussed earlier. A **privilege** is an (not necessarily) earned right that is attached to a social status. You can think of it as the goodies or bonus that comes with occupying a certain kind of status in society. As a professor, you get to tell your students what to do, and that's a privilege. Privileges are also attached to statuses like race, social class, gender, and sexual identity. **Hetero-privilege** is the set of unearned rights that are given to heterosexuals in many societies. One privilege received by heterosexuals in Anglo-European society is to never (or rarely) have to explain their sexuality to other people; it's a privilege not to have to explain to someone why you have sexual desire for someone of the opposite sex. Institutions, norms, and values in heteronormative societies reinforce the perceived normality of heterosexuality. The obvious institution in many places in the United States that currently reinforces heteronormativity is marriage because in most states marriage between two people of the same sex is illegal. But in a heteronormative society, all institutions privilege heterosexuals and assume heterosexuality as the norm. Take a minute to think about some of the ways in which heterosexuality may be built into the structure of your particular college or university. Are men and women allowed to live together in one dorm room? At a few colleges and universities, the answer may be yes, but generally even in coed (mixed sex) dorms, women live with women in individual dorm rooms. Why do many colleges and universities prevent men and women from living together in the same dorm room, and what assumptions are they making about the sexuality of their students? This practice is heteronormative in that it assumes that to prevent romantic or sexually involved couples from living together on college campuses, women and men must be prevented from living in the same room but *not* men with men or women with women.

Can you think of other examples of heteronormativity on your college or university campus, or other ways in which practices and values reinforce the idea that heterosexuality is normal and right?

Other researchers have documented the ways in which middle schools and high schools as educational institutions control sexuality and enforce heterosexuality on their students (Eder, Evans, & Parker, 1995; Miceli, 2009; Pascoe, 2007). Heterosexuality is part of the **hegemonic curriculum** in these schools, in that the practices in these institutions both legitimize the dominant culture and marginalize or reject other cultures and forms of knowledge (Miceli, 2009, p. 344). This is to say that schools teach the centrality of heterosexuality, not just or primarily through the formal curriculum (textbooks, exams, lectures, assignments, etc.) but through the informal practices of the institutions. Perhaps the simplest example of this heteronormativity is the institution of prom. Melinda S. Miceli (2009) argued that prom represents the culmination of the *hetero-romantic* norm that dominates high schools and middle schools. These **hetero-romantic norms** prescribe specific behavioral norms for males and females that are important to proving their masculinity or femininity, as well as gaining acceptance in the peer culture of high school (Miceli, 2009, p. 345). Prom, exemplifying the complex relationship between gender and sexuality, reinforces heterosexuality and prevailing gender norms in one neat little package. Prom is heteronormative most obviously in that, until recently, and still for many high schools, a prom date consists of a young man and a young woman; although the two most popular individuals in any given high school may be two young men or two young women, the prom court always consists of a king and a queen rather than of a queen and queen. Prom is an opportunity for high-school women and men to demonstrate fully their gender identities, as young women focus on their appearance, their dress, and a date who will treat them well. For young men, prom is about securing an attractive date and, often, ending the evening with sex. A whole genre of popular films is dedicated to enshrining these norms surrounding prom as the culminating event in a young person's life up to that point. These are all good examples of the ways in which our schools are gendered organizations.

Theory Alert! Gendered Organizations

Prom is just one small example of the hegemonic curriculum in schools enforcing heterosexuality. Beginning in elementary school and continuing through high school, students are constantly divided into groups of boys and girls for various activities. Academic awards are often distributed to one young man and one young woman. Heterosexual expressions of public affection are often tolerated by fellow students and school officials in ways that homosexual expressions of affection would not be. Sex education in schools generally assumes heterosexuality as the norm and does not explore the possibility of other sexual ways of being, let alone addressing the many issues gay and lesbian teenagers may face. Any attempts to include models of sexuality that differ from the heterosexual norm in sex education are generally met with fierce resistance by parents and other community members. On top of all these institutional practices, heterosexuality is enforced in schools by ongoing harassment of gay and lesbian students. In one survey, 84% of LGBT students reported that they had been verbally harassed, while 82.9% added that teachers and administrators rarely, if ever, intervened when they heard homophobic comments (Miceli, 2009, p. 348). Sixty-four percent of LGBT students reported that they felt unsafe at their school because of their sexual orientation.

Nationalism and Heteronormativity

Negotiating what it means to live outside the established norms of a heteronormative society becomes even more complicated when issues of race, ethnicity, and nationalism enter into the picture. In our globalized world, the Anglo-European categories of heterosexuality and homosexuality are increasingly becoming the dominant way of thinking about sexuality. It's no surprise, then, that many people in the global South associate homosexual behavior and homosexuality with the West and often with Western imperialism. Although same-sex sexual behavior may have existed in these societies, the particular formulation of homosexuality and heterosexuality *is* a new concept to many cultures. In areas of the world where nationalism is strong, homosexuality poses a particular problem to the project of nation building. Nationalism is "a genre of claims, understandings, and grounds for recognizing, promoting, and legitimizing peoplehood, identity, and sovereignty" (LiPuma, 1997, p. 36). Nationalism is similar to ethnicity in that it involves a group of people who make a claim to some common sense of peoplehood, but unlike ethnicity, nationalism includes some political claim—to land or sovereignty. Nationalism in most of its manifestations is both gendered and sexualized. Nationalism is inherently masculine, as nationalist scripts are generally written primarily for men, about men, and by men (Nagel, 2003, p. 159). The nation-state is a masculine institution, with its strictly hierarchical structure and male dominance of decision-making positions. In addition, the culture of nationalism emphasizes and resonates with masculine themes such as honor, patriotism, cowardice, bravery, and duty (Nagel, 2003, p. 160). It is men's job to form, define, and defend the nation, and because of this masculine formulation, both feminists and homosexuals constitute a threat to the cause of nationalism.

Women's role in nationalism is largely reproductive and symbolic. Women's bodies can come to represent symbolically a nation, and therefore, the sexual purity of women becomes important to nationalist causes. A woman who has sex with someone outside the national group is considered a traitor, defiling the honor of the nation as a whole, in a way that a man violating those boundaries is not. This partly explains the use of rape as a weapon of war and ethnic/national conflict; women's bodies become symbolic of the actual territory of the nation, and so to violate (through rape) a woman's body is to gain a kind of military victory. Women's reproductive role is to increase the population of the nation itself. In nationalist settings like Fascist Italy and the ethnic dictatorship of the former Yugoslavia, governments curtailed women's control of their reproduction in hopes of increasing their national population. Feminist claims for the interests of women are perceived as competing with the agenda of the nation as a whole, and such women face challenges to their loyalty to the nationalist cause and their sexuality; they may be labeled as lesbians or as in league with colonialist oppressors and acting under the influence of Western feminism (Nagel, 2003, p. 161). Obviously, to the extent that feminists also argue for women's control of their own reproduction, they also come into conflict with the nationalist agenda.

Homosexuality is considered problematic to nationalist goals for a variety of reasons. First, nationalism is an inherently backward-looking vision of the world, often advocating a return to some real or imagined, precolonial past, before outside oppressors polluted the "true" culture of the people. This backward-looking orientation is at odds with the generally forward-looking orientation of both feminism and gay rights movements, for whom this idealized past is probably not so ideal (Nagel, 2003, p. 163). Homosexuality among men also threatens the masculinity of nationalist movements, sometimes leading to "homosexual panics" as exemplified in the Nazi persecution of homosexuals during World War II (who were branded with a pink triangle much like the gold star for Jews in Germany), or the Cold War practice of expunging supposed homosexuals from the U.S. and British government. Senator Joseph McCarthy of the United States is famous for his relentless search for Communists during the 1950s, but McCarthy was also interested in homosexuals in the government, presumably because these individuals posed a blackmail threat (Nagel, 2003, pp. 163–164). Homosexuality is also a threat to nationalism because, like feminism, it is often considered to be a product of the West and colonialism and, therefore, as incompatible with the return to a precolonial utopia.

All these complex interconnections between nationalism, ethnicity, gender, and sexuality are lived out in the experiences of many gays and lesbians living in or emigrating from parts of the global South. In her study of Arab American femininities, Nadine Naher (2009) used oral histories of Arab American women living in the United States to examine these intersections of nationality, ethnicity, religion, gender, and sexuality. In the oral history of Lulu, a Syrian-American Muslim woman living in San Francisco, she described how, among Muslims in her parents' native Syria, cultural beliefs dictated that women should be virgins when they marry, and although Arab men may have sex before they marry, they are having sex with "Christian women" in Syria, and not with Arab women. Lulu identified that in her family narrative, "whores" are either Christian women or American women. When Lulu came out as a lesbian to her Syrian family, her sexuality was framed within a perspective that views homosexuality as American. Lulu's uncles told her, "This (lesbianism) doesn't happen in our culture. You've been brainwashed by Americans. You've taken too many feminist classes, you joined NOW (National Organization for Women), you hate men, you have a backlash against men." Lulu said, "It was like . . . 'This is what American society has done to our daughter'" (Naher, 2009, p. 189). In this particular blending of nationalism, gender, and sexuality, lesbianism is something that is perceived as nonexistent in Arab cultures, and Lulu's new identity must therefore be a product of her enculturation into American society and her exposure to feminism.

Theory Alert! Intersectionality

In these types of encounters, gay men and lesbians receive messages that they cannot be both Arab and homosexual, which is a good example of how gender overlaps with categories like ethnicity and nationality. Lulu's family tells her that lesbianism is something that doesn't exist in Arab culture, and it's not surprising that Lulu herself felt that she could not be both Arab and gay. As she explains in her oral history, "When that was the reaction, I totally disassociated myself from Arabs. I felt I couldn't be gay and Arab. I felt that either I have to go home and be straight or totally out and pass as white" (Naher, 2009, p. 189). Lulu eventually found a network of other Arab lesbians and came to realize that she did not have to give up either parts of her identity, being lesbian and being Arab. But her family, who once told Lulu she was "the only gay Arab in the world," explained the existence of other lesbian Arab women by explaining that these other women must be American or Christian (Naher, 2009, pp. 189–190). In this nationalist perspective, both lesbianism and feminism are associated with American nationality, and not with Syrian or Arab nationality, as well as with Christianity, rather than with Islamic faith.

Lulu's story, and the stories of other people who face a complex constellation of race, religion, ethnicity, nationality, gender, and sexuality, demonstrate the ways in which sexuality itself can be used to reinforce and draw other boundaries. Americans as a nation are different from Syrians and Arabs in part because of their sexual practices. In Naher's (2009) larger study, she revealed a dichotomy used by Arab families to police their daughters' behavior in which "good Arab girls" were opposed to "bad American(ized) girls," or "Arab virgin" opposed to "Amercan(ized) whore" (Naher, 2009, p. 187). In addition, homosexuality is American or Christian and, therefore, not part of authentic Arab culture or true Islamic tradition.

Using sexuality to draw these kinds of boundaries is hardly unique to Arab cultures, and it is a central, but largely unexamined, facet of American history. Early explorers and settlers used sexuality to draw boundaries between themselves and other groups like Native Americans and Africans. Native American women were described as beast-like and eager to copulate, defile, and prostitute themselves with early explorers. African women were similarly described as lecherous and beastly, while some peoples were described as having men among them who lived as women and married men (Nagel, 2003, p. 93). Some accounts by male European explorers focused on the unusually large size of African men's genitalia and, not surprisingly, concluded this was yet another indication of their bestiality and moral inferiority. The supposedly more primitive sexual practices and lax morality of Native Americans and Africans became part of the ideology that justified their virtual destruction as a people and enslavement, respectively. During the period of American enslavement of African Americans, beliefs about the sexuality of slaves helped to justify the systematic rape of Black women by their white male owners.

These examples point to the ways in which sexuality, like gender, can be an important tool of power. With hindsight, we know that many of the differences "observed" by European explorers and settlers were differences they imposed because of their own particular viewpoint. Explorers and settlers wanted, to some extent, to believe that these groups were different from and inferior to them, and it's therefore no surprise that this is what they found. Their beliefs about the sexuality of Africans and Native Americans were socially constructed, but those social constructions had devastating effects. Under the system of slavery, a Black woman could not legally be raped by anyone, her master or another slave; as legal property, she had no rights to dictate whom she did and did not have sex with, and historical court records as well as slave narratives demonstrate that under today's definitions, the rape of female slaves was a common practice.

SEX AND POWER

Perhaps this is something of a depressing conclusion to come to in an examination of sexuality. Although it has certainly not always been the case, contemporary perspectives on sexuality in Anglo-European societies tend to emphasize sexuality as a positive aspect of our lives. Countless magazines, books, television shows, and websites exist to instruct us in how to have good and pleasurable sex, and these cultural sources increasingly include a wide range of sexualities. Many gender scholars argue that one legacy of the feminist movement, intended or not, has been an increased sense of sexual liberation for women in many parts of the global North. The double standard, although it has not disappeared, has certainly changed its particular form. Although many people would still view the sex-loving, sexually prolific, generally commitment-phobic character Samantha on *Sex and the City* as a slut, many other people would not. Shows like *Sex and the City* that depict women as sexual subjects rather than as sexual objects, as well as the influence of feminism, may have made sexuality something that is positive for women *and* men. These are good signs, but there is also no denying that sexuality has been and continues to be a means for controlling the lives and behaviors of people. Acts that many would like to consider the most intimate and private are not beyond the reach of societal influence. Sociology is important in the bedroom, as well. Power and sexuality are inevitably linked.

Red = Top; Black = Bottom

Theory Alert! Queer Theory

We began this chapter with one vision of what sexuality might look like in a world without gender. John Stoltenberg's (2006) vision of having sex without having *a* sex is also a world where the solidity of the links between sexuality and gender becomes a little more fluid. A queer theory approach gives us another vision of what a sexuality might look like that was unhooked from sex category and gender. Kate Bornstein (1994) is a male-to-female transsexual who, in her book *Gender Outlaw,* suggested what desire might look like uncoupled from these other categories. Instead of basing our sexual choices on the particular genitalia someone does or does not possess, which is assumed under the categories of homosexual and heterosexual, what if we chose based on the particular type of sexual activity people preferred? Bornstein proposed a set of colored bracelets that could be worn to indicate the type of sexual activity preferred by individuals as well as their preferred position of dominance or passivity. An orange bracelet would indicate "anything goes," while a light blue bracelet would indicate a preference for oral sex. Surely the kind of things one likes to do sexually is an important component of sexual compatibility, and it is perhaps as, if not more, important than someone's sex category or gender. Someone's sex category or gender doesn't necessarily tell us anything about what they do and don't like to do sexually.

Bornstein's (1994) proposed system is consistent with queer theory in its attempt to expose the problems with many of our existing categories. Knowing someone is genitally female or a woman does not tell you what particular kind of sexual act she might like to engage in or whether she likes to engage in sexual acts at all. We use sex category and gender to organize our sexual desire and attraction because we believe these categories do tell us something true and important about another person that will make them compatible with us in some way. Gay men believe there is some quality in other men that makes them more compatible than they are with women. This is placing a great deal of faith in these categories, as well as assuming that there is something that is consistently true enough about all the people we call *men* that makes them all potential romantic or sexual partners. Queer theory reminds us to question whether this is true. Is there anything that is consistently true about all the people in the world who are labeled men? This is queer theory questioning gender categories. Queer theory also proposes treating categories as open and fluid, and Bornstein's bracelet system is a good example of this. Bracelets can easily be put on and taken off, depending on the particular type of mood you're in. This means that power in sexuality does not necessarily disappear. If you wear a bracelet that indicates you like to dominate in sexual activities, you're likely to have the power in that encounter. But it *would* mean that individuals could choose whether they wanted to be the one with or without the power in any given situation, rather than being assumed to be passive or submissive based on something that's largely out of their control (their sex category or gender).

Under this system, power doesn't disappear altogether, but it becomes more free-floating, a matter of individual choice rather than institutional prerogative. Unlike gender or sex category, which we're generally stuck with and about which we have little choice, in Bornstein's system, your particular position of power in sexual encounters could change from day to day. And power (who's submissive and who's passive) would be just one small way of describing the full complexity of our sexual desires. Sexuality again becomes about acts (what you like to do) rather than types of people (males/females, men/women, heterosexuals/homosexuals, etc.).

Both Bornstein (1994) and Stoltenberg's (2006) ideas about different systems of sexuality may seem fairly strange, although we've observed that different ways of linking sex category, gender, and sexuality have existed historically and still exist in other cultures today. Outside of utopian visions, do we have any examples of what sex would be like without sex category and gender? Gay male and lesbian relationships certainly do not exist outside the world of gender; two men or two women in a relationship are still gendered beings, even if their relationship is not a heterosexual one. We should not be so naïve as to think that gender's influence on sexuality somehow disappears in the sexual lives of gay men and lesbians, given its pervasiveness in all societies. But what might sex look like where the dynamics of masculine activity and feminine passivity that are taken for granted in heterosexuality aren't as readily available? We discussed the gendered nature of sexual scripts earlier in this chapter. Do the same sexual scripts apply to homosexual and heterosexual relationships? Does the double standard work only if a woman is having sex with a man, or can "too much" same-sex sexual behavior among women qualify as "slutty"? Is it more okay for gay men to have a lot of casual sex than it is for lesbian women? How do these sexual scripts matter among gays and lesbians?

As we discussed in our examination of the history of the category of homosexuality, common notions of what it means to be homosexual that have come down to us from early ideas of inversion presume that gay men and lesbians will not conform to gender expectations. Remember, inversion assumes that a gay man is a female soul in a male body, so that a gay man is expected to act in feminine ways. Presumably, that would include following feminine, rather than masculine, sexual scripts. Do gay men worry about a double standard then? Do lesbian women have a more active and therefore masculine sexuality? Not really. Research on the actual sexual practices of everyone—heterosexual or homosexual—are hard to come by for the simple fact that it is difficult to gather a large, random survey about people's sexual practices. This is, not surprisingly, because most people would not feel comfortable being asked questions about their sex lives by a complete stranger or social scientist. Some of the most comprehensive data come from the National Health and Social Life Survey suggests that there are some differences between the sexual lives of gay men and lesbians. Men who have had some same-gender partners since the age of 18 (one way of measuring sexual identity in the survey, given the confusions we already talked about in this chapter) had 44 partners on average since the age of 18, while women who have had some same-gender partners since the age of 18 had on average 20 partners since the age of 18 (Laumann et al., 1994). This suggests that, like their heterosexual counterparts, gay men do have more sexual partners on average than lesbians. However, the frequency of sex does not vary significantly between gay men and lesbians; gay men have sex about 5.7 times per month on average, while lesbians have sex about 6.1 times per month.

Moving outside the bounds of heterosexuality, then, does not necessarily seem to include a movement outside of the bounds of gender expectations. Are homosexuality and heterosexuality really that different in their sexual scripts? There is one area in which sexual behavior differs between heterosexuals and homosexuals. Homosexual relationships tend to be more egalitarian, and this extends into sexual relationships. A relationship that is more **egalitarian** is one that is embedded in a belief in equality among all people. In heterosexual relationships, men are far more likely to initiate sex. But in relationships among lesbians and gay men, there is no division of labor in terms of who initiates sex; each partner is equally likely to initiate sex. In addition, gay men and lesbians generally have more variation in their sexual activities than do heterosexual couples. This is especially true in regard to nonpenetrative sex. Lesbian and gay couples, in other words, have different sexual scripts in terms of defining exactly what sex is, and these definitions are broader than for heterosexual couples. Perhaps because of these broader definitions, some research suggests that gay couples have longer lovemaking sessions than do heterosexual couples (Laumann et al., 1994; Masters, Johnson, & Kolodny, 1978).

In this sense, the gender of partners does seem to have at least a partial effect on sexual behaviors. Although we have laid out many examples to choose from in this chapter, it is still impossible for us to say for sure what sexuality would look like in a world without gender and sex categories. What would our sexual scripts be? Would our sex lives be better for it? Would sexuality continue to be a way to maintain power and draw boundaries between different groups? These questions are difficult to answer. What we can say is that although sexuality is an aspect of our lives that we might like to think of as the truest expression of ourselves, it is still deeply influenced by our social world and, therefore, our ideas about gender. Whether sexuality is the driving force creating our ideas about gender or gender is what gives us our concepts of sexuality, it is impossible to deny the ways in which these two concepts are deeply and inextricably intertwined.

BIG QUESTIONS

- In this chapter, we explored what it means to say that sexuality is socially constructed, even though social scientists have only recently begun to explore these ideas. Why do you think the social construction of sexuality is something that has just recently begun to be explored? Why might it be more difficult to think of sexuality as socially constructed?
- How is the construction of feminine sexuality as passive and masculine sexuality as active important to larger issues of gender inequality? Some gender scholars argue that sexuality is *the* key to understanding gender inequality. How might you make this argument, that the ability to control someone's sexuality is an important aspect of power?
- How are both women and men harmed by gendered beliefs about sexuality? What are the advantages and disadvantages of being perceived as a sexual object or as a sexual subject?
- In this chapter, we discussed contradictory ideologies about women's sexuality. Although women are often perceived as sexually passive, other traditions suggest that women have

more sexual desire than men. How might you explain these contradictions? What are the gendered implications of these two different perspectives on feminine sexuality?

- We suggested in this chapter that homosexuality is not at all the same as same-sex sexual behavior and that the names we use to identity social behaviors and values matter. Do you agree that terms like "homosexuality" and "heterosexuality" should only apply to certain cultural contexts and that to attempt to describe behaviors outside of that context using these terms confuses the different meanings of these behaviors in different cultural contexts?
- Given what you now know about the social construction of sexuality, what might be problems with research that attempts to explain why some people are homosexual and some people are heterosexual from a biosocial perspective? For example, if you were doing such a study, how would you define the criteria for what makes someone heterosexual and what makes someone homosexual?

GENDER EXERCISES

1. In this chapter, we discussed the double standard regarding masculine and feminine sexuality. The double standard has a linguistic component in the form of language asymmetry, as we discussed in Chapter 4. Test this out by generating a list for yourself of all the terms (slang and otherwise) used to describe promiscuity or immoral sexual behavior for women, and then make a similar list for men. Ask other people to generate similar lists, and then compare. Is there an asymmetry in the number of words available to describe promiscuous or immoral sexual behavior in women as compared with men? How does this relate to the double standard?
2. How are our ideas about appropriate sexual behavior for women and appropriate sexual behavior for men different? Conduct your own informal surveys or interviews on this topic by devising a list of questions on how people would view certain behaviors in women and men. For example, you might ask people what an acceptable number of sexual partners is for women as compared with men—or how they would view a man cheating on his partner as compared with a woman. Interview or survey people from different backgrounds to compare how they think about these issues.
3. In this chapter, we attempted to describe a homosexual role, or the basic expectations that go along with being homosexual in current Anglo-European society, as well as the difficulty in establishing what it means to be homosexual or heterosexual. Ask a group of people what they believe makes someone heterosexual or homosexual. Is it sexual behavior, sexual desire, or sexual identity? Be specific. Is someone who has one opposite-sex sexual experience automatically heterosexual? How many opposite-sex sexual experiences are required to make someone "officially" heterosexual?
4. Pick an institution or group to which you belong. This could be a formal group, like a club, or an informal group, like your friends or family. Write an essay reflecting on ways in which this group or institution is heteronormative. You might think about how someone who is not

heterosexual would experience belonging to this group or institution. Try to uncover the taken-for-granted ways in which this institution or group assumes that its members are heterosexual or that heterosexuality is the norm (this exercise will probably be easier if you are not heterosexual and have had these experiences yourself). How could this group or institution change in ways that made it no longer heteronormative?

5. Gender and sexuality are connected in that many homosexuals are assumed to have a different gender identity than heterosexuals. Gay men are assumed to be less masculine and lesbian women less feminine. Pick some films, television shows, magazines, or books that depict gay and lesbian people. Conduct a content analysis of these cultural artifacts, identifying how often gay and lesbian characters are depicted as acting in ways that are gender-consistent as opposed to gender-inconsistent. What does this content analysis tell you about our views on gender and sexuality?
6. There are increasing numbers of media outlets that cater specifically to gay and lesbian communities. Magazines include *The Advocate, Out, Curve,* and *Gay Life.* On television, LOGO is a channel that caters to gay and lesbian communities. Conduct a content analysis of some of these cultural artifacts, examining the ways in which gender is depicted in these magazines. For example, you might compare the articles about women and men in *The Advocate* with articles in "straight" men's and women's magazines such as *GQ* and *Cosmo.* Are there differences in gender depictions in these gay and straight media outlets?
7. Interview someone from a cultural background different from your own about his or her culture's perspectives on sexuality, as well as his or her own views. You might ask questions about the person's views on the differences between men and women's sexuality, how he or she thinks about what is appropriate and inappropriate sexually, and how he or she thinks about homosexuality and heterosexuality.

TERMS

sex
sexual identity
sexual desire
heteronormativity
compulsive heterosexuality
sexology
studying up
one-sex model
two-sex model
berdache
ethnocentrism
sexual inversion
homosexual identity
romance tourism
ideal love
sexual object
double standard
sexual scripts
discourse
Boston marriages
homosocial
privilege

hetero-privilege
hetero-romantic norms
hegemonic curriculum
egalitarian

SUGGESTED READINGS

Sexuality in general

Ferber, A. L., Holcomb, K., & Wentling, T. (2008). *Sex, gender and sexuality.* New York, NY: Oxford University Press.

Laumann, E. O., Gagnon, J. H., Michael, R. T., & Michaels, S. (1994). *The social organization of sexuality: Sexual practices in the United States.* Chicago, IL: University of Chicago Press.

Martin, K. A. (1996). *Puberty, sexuality and the self: Girls and boys at adolescence.* New York, NY: Routledge.

Sociological theory and sexuality

Messner, M. A. (2005). Becoming 100% straight. In M. B. Zinn, P. Hondagneu-Sotelo, & M. A. Messner (Eds.), *Gender through the prism of difference* (pp. 227–232). New York, NY: Oxford University Press.

Miceli, M. S. (2009). Schools and the social control of sexuality. In T. E. Ore (Ed.), *The social construction of difference and inequality: Race, class, gender and sexuality* (pp. 344–353). New York, NY: McGraw-Hill.

Pascoe, C. (2007). *Dude, You're a fag: Masculinity and sexuality in high school.* Berkeley: University of California Press.

Seidman, S. (1996). *Queer theory/sociology.* Malden, MA: Blackwell.

Stoltenberg, J. (2006). How men have (a) sex. In E. Disch (Ed.), *Reconstructing gender: A multicultural approach* (pp. 264–274). Boston, MA: McGraw-Hill.

Cross-cultural and historical perspectives on sexuality

Gay, J. (1989). "Mummies and babies" and friends and lovers. *Journal of Homosexuality, 11* (3/4), 97–116.

Herdt, G. (1997). *Same sex, different cultures: Gays and lesbians across cultures.* Boulder, CO: Westview Press.

Mernisi, F. (2002). The Muslim concept of active female sexuality. In C. Williams & A. Stein (Eds.), *Sexuality and gender* (pp. 296–307). Malden, MA: Blackwell.

Nagel, J. (2003). *Race, ethnicity, and sexuality: Intimate intersections, forbidden frontiers.* New York, NY: Oxford University Press.

Nanda, S. (1998). *Neither man nor woman: The hijra of India.* New York, NY: Wadsworth.

Ortner, S. B. (1981). Gender and sexuality in hierarchical societies: The case of Polynesia and some comparative implications. In S. B. Ortner & H. Whitehead (Eds.), *Sexual meanings: The cultural construction of gender and sexuality* (pp. 359–409). Cambridge, England: Cambridge University Press.

Rich, A. (1993/1986). Compulsory heterosexuality and lesbian existence. In L. Richardson & V. Taylor (Eds.), *Feminist frontiers III* (pp. 158–179). New York, NY: McGraw-Hill.

Rupp, L. J. (2004). Finding the lesbians in lesbian history: Reflections on female same-sex sexuality in the western world. In L. Richardson, V. Taylor, & N. Whittier (Eds.), *Feminist frontiers* (pp. 302–312). New York, NY: McGraw-Hill.

Whitehead, H. (1981). The bow and the burden strap: A new look at institutionalized homosexuality in native North America. In S. B. Ortner & H. Whitehead (Eds.), *Sexual meanings* (pp. 80–115). Cambridge, England: Cambridge University Press.

WORKS CITED

Ault, A. (1996). The dilemma of identity: Bi women's negotiations. In S. Seidman (Ed.), *Queer theory/sociology* (pp. 311–330). Malden, MA: Blackwell.

Blackledge, C. (2009). The function of the orgasm. In A. L. Ferber, K. Holcomb, & T. Wentling (Eds.), *Sex, gender and sexuality* (pp. 272–284). New York, NY: Oxford University Press.

Blake, M. (2012, November 30). *The Los Angeles Times.* Retrieved April 29, 2013, from For Gays, Reality TV Can Be a Path to Acceptance: http://articles.latimes.com/2012/nov/03/entertainment/la-et-st-reality-tv-gays-acceptance-20121103

Bordo, S. (1999). *The male body: A new look at men in public and in private.* New York, NY: Farrar, Straus and Giroux.

Bornstein, K. (1994). *Gender outlaw: On men, women and the rest of us.* New York, NY: Vintage Books.

Brooks, G. (1995). *Nine parts of desire: The hidden world of Islamic women.* New York, NY: Anchor Books.

Eder, D., Evans, C. C., & Parker, S. (1995). *School talk: Gender and adolescent culture.* Piscataway, NJ: Rutgers University Press.

Fausto-Sterling, A. (2012). *Sex/Gender: Biology in a Social World.* New York, NY: Routledge.

Galliano, G. (2003). *Gender: Crossing boundaries.* Belmont, CA: Thomson-Wadsworth.

Gay, J. (1989). "Mummies and babies" and friends and lovers. *Journal of Homosexuality, 11* (3/4), 97–116.

Halperin, D. M. (1989). Is there a history of sexuality? *History and Theory, 28* (3), 257–274.

Hernandez, G. (2007, April 24). American Idol's big gay closet. *Advocate,* pp. 34–39.

Herdt, G. (1997). *Same sex, different cultures: Gays and lesbians across cultures.* Boulder, CO: Westview Press.

Katz, J. N. (2009). The invention of heterosexuality. In T. E. Ore (Ed.), *The social construction of difference and inequality: Race, class, gender and sexuality* (pp. 150–162). New York, NY: McGraw-Hill.

Kinsey, A., Pomeroy, W., Martin, C., & Gebhard, P. (1948). *Sexual behavior in the human male.* Philadelphia, PA: W.B. Saunders.

Kinsey, A., Pomeroy, W., Martin, C., & Gebhard, P. (1953). *Sexual behavior in the human female.* Philadelphia, PA: W.B. Saunders.

Laumann, E. O., Gagnon, J. H., Michael, R. T., & Michaels, S. (1994). *The social organization of sexuality: Sexual practices in the United States.* Chicago, IL: University of Chicago Press.

LiPuma, E. (1997). The formation of nation-states and national cultures in Oceania. In R. J. Foster (Ed.), *Nation making: Emergent identities in postcolonial Melanesia* (pp. 33–68). Ann Arbor: University of Michigan Press.

Marin, K. A., & Kazyak, E. (2009). Hetero-romantic love and heterosexiness in children's G-rated films. *Gender & Society, 23,* 315–336.

Martin, K. A. (1996). *Puberty, sexuality and the self: Girls and boys at adolescence.* New York, NY: Routledge.

Masters, W., Johnson, V., & Kolodny, R. (1978). *Human sexuality.* New York, NY: Harper and Row.

Matzner, A. (2001). *'O Au No Keia: Voices from Hawai'i's Mahu and transgender communities.* Philadelphia, PA: Xlibris Corporation.

Mernissi, F. (2002). The Muslim concept of active female sexuality. In C. Williams & A. Stein (Eds.), *Sexuality and gender* (pp. 296–307). Malden, MA: Blackwell.

Messner, M. A. (2005). Becoming 100% straight. In M. B. Zinn, P. Hondagneu-Sotelo & M. A. Messner (Eds.), *Gender through the prism of difference* (pp. 227–232). New York, NY: Oxford University Press.

Miceli, M. S. (2009). Schools and the social control of sexuality. In T. E. Ore (Ed.), *The social construction of difference and inequality: Race, class, gender and sexuality* (pp. 344–353). New York, NY: McGraw-Hill.

Nagel, J. (2003). *Race, ethnicity, and sexuality: Intimate intersections, forbidden frontiers.* New York, NY: Oxford University Press.

Naher, N. (2009). Arab American femininities: Beyond Arab virgin/American(ized) whore. In A. L. Ferber, K. Holcomb, & T. Wentling (Eds.), *Sex, gender and sexuality* (pp. 186–192). New York, NY: Oxford University Press.

Nanda, S. (1998). *Neither man nor woman: The hijra of India.* New York, NY: Wadsworth.

Nanda, S. (2000). *Gender diversity: Crosscultural variations.* Prospect Heights, IL: Waveland Press.

Ortner, S. B. (1981). Gender and sexuality in hierarchical societies: The case of Polynesia and some comparative implications. In S. B. Ortner & H. Whitehead (Eds.), *Sexual meanings: The cultural construction of gender and sexuality* (pp. 359–409). Cambridge, England: Cambridge University Press.

Pascoe, C. J. (2007). *Dude, You're a fag: Masculinity and sexuality in high school.* Berkeley: University of California Press.

Rich, A. (1986/1993). Compulsory heterosexuality and lesbian existence. In L. Richardson & V. Taylor (Eds.), *Feminist frontiers III* (pp. 158–179). New York, NY: McGraw-Hill.

Rochlin, M. (1972). The heterosexual questionaire. Retrieved from http://www.utexas.edu/diversity/ddce/gsc/downloads/resources/Heterosexual_Questionnaire.pdf

Ronen, S. (2010). Grinding on the dance floor: Gendered scripts and sexualized dancing at college parties. *Gender & Society, 24* (3), 355–377.

Roy, W. G. (2001). *Making societies.* Thousand Oaks, CA: Pine Forge Press.

Rupp, L. J. (2004). Finding the lesbians in lesbian history: Reflections on female same-sex sexuality in the western world. In L. Richardson, V. Taylor, & N. Whittier (Eds.), *Feminist frontiers* (pp. 302–312). New York, NY: McGraw-Hill.

Stoltenberg, J. (2006). How men have (a) sex. In E. Disch (Ed.), *Reconstructing gender: A multicultural approach* (pp. 253–262). Boston, MA: McGraw-Hill.

Storey, J. (1988). *An introduction to cultural theory and popular culture.* Athens: University of Georgia Press.

Swarr, A. L. (2009). "Stabane," intersexuality and same-sex relationships in South Africa. *Feminist Studies, 35* (3), 524–548.

Weeks, J. (1996). The construction of homosexuality. In S. Seidman (Ed.), *Queer theory/sociology* (pp. 41–63). Malden, MA: Blackwell.

Whitehead, H. (1981). The bow and the burden strap: A new look at institutionalized homosexuality in native North America. In S. B. Ortner & H. Whitehead (Eds.), *Sexual meanings* (pp. 80–115). Cambridge, England: Cambridge University Press.

CHAPTER 6

How Does Gender Impact the People You Spend Your Time With?

The Gender of Friendship and Dating

Who was the very first friend you ever had? How did you know they were your friend? What does it mean to call someone a friend? How would that first friendship and other friendships you've had be different if you lived in a different time or a different place? Was your very first friend of the same gender or of a different gender? Did that matter to the friendship? What about your friends now? Are they mostly of the same gender, different gender, or some mixture in between? In your experience, do women or men make better friends? How do you know the difference between being what we call "just friends" with someone as opposed to being "more than friends"? What does it imply to be more than friends, and how do we draw the boundaries between friendships and other kinds of relationships in our lives? Assuming you've been on a date with someone, was that person already your friend? What's the difference between dating someone and being friends with them? How do you even know what a date is for sure, and what's the point of dating? Does dating have to lead to marriage, and do you have to date before you get married? What do you do if marriage isn't even in the cards for you because you're either vehemently opposed to it or legally prevented from getting married? Where did this whole idea of dating come from in the first place? What happens in places where people date, and how does that affect the way men and women interact with each other? And one last whopper of a question: How does love fit into this whole picture?

LOVE, INSIDE AND OUTSIDE THE FAMILY

These are some of the questions we'll be exploring in this chapter on friendship and romantic love. You could say this chapter is about all the kinds of love that exist between human beings that are mainly outside of what we call the nuclear family in the social sciences. The **nuclear family**, as we'll discuss more in Chapter 8, refers to a historically and culturally specific way of organizing family life that is reputed to have evolved in Anglo-European societies during the middle of the 20th century. The nuclear family, as it is conceived, consists of only two generations, parents and children. Everyone else—grandparents, aunts, uncles, cousins, grandchildren, and so on—become extended family. In this chapter, we'll make a separation between what we might call nuclear and nonnuclear love, although as you'll see throughout this chapter, that division is not exactly cut and dry. In some places in Africa, especially close friends become twins—demonstrating the closeness and importance of this bond. This makes sense historically in some Western traditions as well, such as when St. Augustine called his friend "thou half of my soul" (Brain, 1976).

Do friends who are so close that they're "like brothers" or literally twins fit in the chapter on friendship or the chapter on family? What about many societies, including our own, where networks of friendship and kin often overlap (although perhaps less than in the past or in other cultures). **Kin** refers to the people with whom you're related either by actual blood (e.g., your biological mother or biological sister) or through symbolic ties (an adopted mother or father, stepmothers, stepfathers, step-siblings, etc.). In New Guinea, men are said to marry their wives not for any love or special attraction to them but to have their brothers as best friends—thus demonstrating the interconnections between kin and friends (Brain, 1976). As is true throughout our exploration of gender, the line in the sand between love in the nuclear family and love outside its bounds is not so easy to draw. For the purposes of this chapter, we'll be leaning in the direction of relationships that exist outside of the family.

Is there any intersection between the people who qualify as your kin and the people you call friends? Would you consider brothers, sisters, cousins, aunts, or uncles friends? What makes someone not just kin but also a friend?

Does it make sense to begin a discussion of friendship with a discussion of love? Or should that be reserved for the chapter on family? What is the difference between the love you feel for a friend and the love you feel for a member of your family? Is one more important than the other? These questions are related to a particular problem sociologists and other researchers encounter in the study and examination of friendship. What exactly is it? Many of us think of love as a heavyweight emotion, largely reserved for romantic love and the intimate bonds of family. But other societies and cultures have held up the love for friends as the very highest form of love there is, a love that is superior for being more altruistic and voluntary than other types of love (Allan, 1990). C. S. Lewis called it one of the four primary types of love, and many Anglo-European traditions celebrate great friends such as Achilles and Patroclus in Greek myth, Jonathan and David in the Bible, or any of the countless buddy movies beginning with Paul Newman and Robert Redford in *Butch Cassidy and the Sundance Kid* and extending to Jackie Chan and Chris Tucker in the *Rush Hour* franchise of movies.

Although this may be changing in many modern societies, by and large, we don't get to choose our families, and they often expect a lot from us; philosophers like Lewis have argued that these two criteria (choice and obligation) are part of what makes friendship unique and perhaps better than other types of relationships. In places like the United States, we pretty much *do* get to choose our romantic partners, but entering into a relationship such as marriage is something that is more or less voluntary depending on who you are and where you live. There are also a great many expectations within romantic relationships that may not be true of friendship. But the idea behind sentiments such as those of C. S. Lewis and St. Augustine is that we choose our friends with a freedom that doesn't exist in our other relationships. Once those choices are made, friendship, so the argument goes, is mostly about the simple enjoyment of another person's company, no strings attached. In contemporary Anglo-European society, friendship might not be held up as the highest form of love as it was in the past, but we can probably agree in principle that friendship is both voluntary and largely altruistic. If you're friends with someone primarily because of what you can get out of the relationship, then that relationship really ceases to be appropriately labeled a friendship in this way of thinking (Allan, 1990).

So we can agree that friendship is voluntary and altruistic, right? And therefore this is part of how we describe the particular form of love that exists between friends. Maybe. As we will explore throughout this chapter, whether your friendships are voluntarily chosen or not may be subject to specific limitations. A sociological approach to the matter of friendship, which seems like a matter of personal choice and preference, helps us to think about the ways in which friendship, like many other behaviors in our society, is highly influenced by social structure. Gender as an important aspect of the social structure in all societies therefore becomes an important aspect of understanding how much we actually *do* get to choose the people with whom we're friends.

In addition, to say that we expect nothing at all in return for friendship is probably a bit of an exaggeration. As we'll discover in friendships across gender lines, balance is an important part of how many of us define friendship (Allan, 1990). Friendship in this way of thinking is what exists between two equals, and whatever the particular system of exchange is in that friendship, a certain balance should be maintained. You may have had a friendship in which you shared a great deal of intimate information with the other person but felt the other person never really shared much intimate information with you. Or maybe you had a friend who constantly asked you for a ride or help on a homework assignment and never seemed willing to help you out when you were in need of similar kinds of assistance. Both of these are forms of maintaining balance in a friendship, and for many of us, when it's off, friendships can be in trouble.

THE TROUBLE WITH FRIENDSHIP

For many years, sociology as a discipline didn't have a lot to say about friendship and, therefore, also not much to say about the relationship between gender and friendship. You might call friendship something of a blind spot for many sociologists, and the general neglect probably reflects many of the prevailing attitudes about friendship in the culture of many developed nations. Within sociology, there are several possible explanations for this neglect. First, other topics seem a lot more important than friendship. Social stratification, urbanization, crime, and deviance all seem like pressing social matters for sociologists to consider. But friendship seems like something that mostly takes care of itself, and it can't easily be defined as a social problem that needs to be solved. Friendship has also often been understood as a matter of personal choice and preference, and this is a second reason it has been understudied in sociology. When you actually formulate the sentence, to say that friendship exists outside of social considerations sounds a little crazy, but viewing friendship as something we idiosyncratically choose based on the vagaries of our personalities and experiences makes a sociological approach difficult.

Of course, friendship is not solely a matter of personal preference and personality. This is obvious in anthropological accounts of other cultures where friends are sometimes selected at birth or where certain relationships are expected to develop into friendships. But friends are not merely a matter of personal preference and personality in places like the United States, either. As we will explore, both the people we can be friends with as well as the type of friendships we have are constrained by social structures, including our gender, our class background, our sexual identity, and our racial or ethnic background.

Make a list of your 10 closest friends. What is the gender makeup of this group? How many are of the same gender as you, and how many are of a different gender? How many of them are about the same age? How many have the same sexual identity? Same race? Same nationality? Same religion?

The third reason sociologists may have stayed away from friendship is something we've already begun to discover—it's hard to figure out exactly what we're talking about when we say someone is our friend. It's difficult to study something when you're not really sure what it is, and this is an especially big problem methodologically speaking. Take the example of Lillian Rubin's (1985) in-depth study of friendship in her book, *Just Friends.* Rubin points out that the English language gives us only one word—friend—to cover relationships that can be as different from each other as night and day. Other researchers have noted that the term *friend,* unlike *aunt, coworker,* or *neighbor,* is more a description of a way of relating than it is a categorical label. It's clear that your aunt is a sister to one of your parents, that a coworker is someone with whom you share an employer, and that a neighbor is probably someone who lives relatively close to you. A friend is marked not by any of these well-defined structural relationships but, rather, by the particular manner in which you relate to a friend, characteristics that we will discuss later (Allan, 1990).

This makes it difficult to develop some objective way of measuring exactly what makes a friend, and it creates the problem for researchers of being uncertain as to whether all their research subjects are defining friendship in the same way as each other, let alone as the researcher herself. One respondent in Rubin's study swore that he had 15 to 20 intimate friends, and he stuck to this number with repeated probing and prodding from Rubin as the interviewer. This same respondent identified as his best friend someone he had not seen in 10 years, whom he had spoken with maybe a couple of times over the phone in those 10 years, and although he felt certain he would help him if he showed up unexpectedly on his doorstop, he was no longer sure where that doorstop would be; the friend had moved, perhaps a couple years back, and the respondent was no longer sure where he lived (Rubin, 1985, pp. 5–6). For cases like these, Rubin sometimes followed up on this "friendship" by contacting the person identified by her initial respondent as a friend. Among the 132 people she contacted after they were identified as friends by her respondents, 64% made no mention of the person who had mentioned them in their interview as a friend. When prompted by Rubin, they would remember the person but convey that the person was not an important relationship for them. This aspect of Rubin's study demonstrates the methodological difficulties of studying friendship when for many of her subjects, the friendships seemed to exist largely for one of the "friends" and not the other.

The last explanation for why sociologists have not paid much attention to friendship is rooted in the particular culture in which sociology exists. There are some variations across the global North, but in many countries like the United States, friendship doesn't seem to be viewed as a very important relationship. As many of those who have explored the subject of friendship in their research point out, we have no institutions surrounding friendship. In the United States, there are no ceremonies that celebrate friendship. The closest we come is children who become blood brothers or blood sisters. We do have the colloquialism, "Blood is thicker than water," to remind us of the importance of family ties over friendship. This is demonstrated for many of us in our most important ceremonies. At weddings and funerals, a clear line is drawn between family and everyone else.

Despite the fact that relationships with friends may be much more intimate than the relationships within families, in wedding ceremonies, there is an expectation that family be given priority in filling the important ceremonial roles—bridesmaid, groomsmen, ushers, and so on. As we will see in our cross-cultural explorations, this lack of institutional support for friendship is not true in many other cultures, and it perhaps reflects a kind of neglect of friendship in developed countries like the United States. From this perspective, our lack of institutions or ceremonies surrounding friendship is a reflection of their lesser importance in society because societies institutionalize those behaviors or beliefs that are most valued. Some scholars attribute this to the rise of the companionate marriage in developed countries, as we discussed in Chapter 5, or to the decline in importance of other types of relationships in general. Marriage and the nuclear family in places like the United States are now expected to meet a wealth of emotional, material, and psychological needs that in other cultures are much more widely distributed across a network of relationships, kin and nonkin. Perhaps as the nuclear family expands, the importance of friendship shrinks, and so although there are many sociology classes on marriage and families, there are very few on the sociology of friendship.

Do your own experiences with friendship and family conform to this description? How would you rank the relative importance or intimacy of relationships with your friends as compared with relationships with your family?

Defining Friendship

Despite these factors contributing to a general lack of attention to friendship, there is a growing body of research exploring this type of relationship. Not surprisingly given the difficulties we've just discussed, one of the first things researchers in this area attempt to do is define exactly what people think of as friendship. For now, we'll focus specifically on these definitions as they evolve where many sociologists do their research, in the global North. One common way of defining friendship is as a way of relating usually characterized as voluntary and altruistic, as we discussed earlier. Friendship involves sentiment and sociability (Allan, 1990; Brain, 1976); we have some emotional attachment to our friends, and we like spending time with them merely for the sake of enjoying their company. This concept of **sociability** comes from the social theorist Georg Simmel and is associated with friendship because it conveys the concept of being social purely for the sake of enjoyment, with no ulterior motive or specific content prescribed. While the point of dating someone may be to figure out if you want to spend your entire life with them, and the point of a mother spending time with her children is to socialize them, the point of friends spending time together is just that—to spend time together as friends. When Rubin (1985) asked her respondents the abstract question, "What is a friend?" this idealized definition included trust, honesty, respect, commitment, safety, support, generosity, loyalty, mutuality, constancy, understanding, and acceptance (Rubin, 1985, p. 7).

A common definition of **friendship** in sociological literature is "a voluntary, informal, and personal relationship" (Allan, 1990, p. 17), although there are problems with this simple definition. We'll say more about how voluntary friendships are, but there are also questions about how informal and personal friendships are. Many friendships may form in formal settings, such as the workplace, and some friendships exist only within the context of those settings. In addition, Rubin's (1985) research on the reciprocity of people identified as friends casts some doubt on the personal nature of these relationships. Friendship is also more likely to be a relationship between status equals, two people who are on roughly the same social level. This helps with the idea of maintaining balance and reciprocity in a friendship relationship, and it is an issue we'll return to in our discussion that follows. Obviously, no two friends are perfectly equal in all things, but there's a general sense that equality is an important ideal to the voluntary coming together of two individuals as friends. What should we do with all these contradictions in regard to defining friendship? One strategy we'll undertake is to move away from generalities and into the nitty-gritty of specific groups of people and their friends. So we'll talk about women's friendships, men's friendships, children's friendships, and so on. But even within those contexts, it's good to keep in mind that not everyone may be thinking of the exactly same thing when they talk about what friendship is and that, as with many sociological concepts, defining exactly what it is we're studying is never as simple as it first seems. As we'll discuss, this problem of defining friendship has come to have some important gender implications as well.

MY FRIEND JANE VERSUS MY FRIEND JOE: WHO'S BETTER AT BEING FRIENDS?

As an entrance into this discussion of friendship in specific contexts, let's start with an interesting question on gender and friendship. Who's better at friendship, women or men?

You can conduct your own informal survey to test this out, but chances are that in today's society, many people would give women the blue ribbon in the friendship competition. The characteristics of friendship that are emphasized in Anglo-European societies are intimacy, trust, caring, and nurturing (O'Connor, 1992). Who would most people identify as better at intimacy, trust, caring, and nurturing—women or men? Research backs up this conclusion by documenting the many ways in which men have difficulty making these kinds of friendships as well as maintaining them once they're established (Goldberg, 1976; Inman, 1986; Kilmartin, 1994; Pleck, 1976; Tognoli, 1980). Some of these researchers go so far as to claim that most men have never had a close male friendship that was not plagued with guilt and fear (Lewis, 1978), while others merely describe male adults friendships as weak and lacking (Pleck, 1976). Either way, the picture that evolves is one of men as fairly incompetent at making, being, and keeping friends.

Friendship in Historical Perspective

This is the general picture you'd get in the 21st century United States, but we're beginning to learn about the importance of historical context to ways of thinking about gender. Who wins the friendship contest if we turn the clock backward? In many time periods, the hierarchy reverses and male friendships are viewed as the model, while women are largely incapable of this higher form of association. You'll notice that all of the famous friends through history listed earlier in the chapter were male. The important characteristics linked to friendship in many historical writings are bravery, loyalty, duty, and heroism; Achilles is a good friend to Patroclus in Homer's *Iliad* because when his friend dies, he flies into a fit of grief, finally leaves the tent in which he's been waiting out the Trojan War, and cuts off the head of Patroclus's murderer. These are very different from some of the more contemporary ideas about friendship, but into the 19th century, "manly love" (Nardi, 1992, p. 41) was an important part of the definition of masculinity. Male factory workers exchanged letters in which they expressed intimacy and tenderness toward their male friends that would probably make many men today uncomfortable (Galliano, 2003).

Other references to the veneration of male friendship in Western history abound. In a Biblical story, Jonathan "fell in love" with David, and his loyalty to his friend trumped both his political interests and his loyalty to family in the form of his father, King Saul, whose wishes he thwarted to help his friend (Brain, 1976). Benjamin Disraeli, prime minister of England during the 19th century, said of the friendships English boys develop during their school years, "All the loves of after life can never bring its rapture, or its wretchedness; no bliss so absorbing, no pangs of jealousy or despair so crushing and so keen" (Brain, 1976, p. 42). Women, on the other hand, were considered to be unable to commit to other women and incapable of forming close bonds. Women's friendships were perceived as likely to be easily abandoned in favor of male companionship and plagued by jealousy and competition over men (Johnson, 1996; Rose, 1995).

This view, which lingers as a cultural assumption into the 20th century, completely contradicts the research describing men as virtually incapable of participating in or sustaining friendship. The quote also begins to suggest the flip side of this argument: that many feminine qualities were historically understood to make women lousy at being friends. Prior to this shift in models of friendship, women were thought to be unable to form the bonds necessary for friendship and were considered less capable of commitment.

Women, so the story went, were likely to abandon female friends for the superior option of male companionship (Galliano, 2003). As the quote suggests, women's friendships were also liable to fall victim to petty squabbles or jealousies. These qualities made women inferior at friendship, as well as the fact that if friendship involves being able to cut off the head of your friend's murderer, that's historically been a little bit more difficult, although not at all impossible, for women to accomplish.

CULTURAL ARTIFACT 1: FEMININE FRIENDSHIP IN THE "REAL" WORLD

How much does the world of "reality" television tell us about, well, reality? Exactly how realistic are the behaviors of people on these shows, or like other popular culture outlets, does reality television trade in gender stereotypes? What do reality shows such as *The Real Housewives of Orange County* or *The Real World* tell us about women's friendships? Shows like *The Real Housewives* generally focus on a group of wealthy women whom we are made to presume are friends but who proceed to argue, backstab, and betray each other, usually culminating in a series of dramatic confrontations. In MTV's *The Real World*, women are depicted as superficial and overly dramatic, more consumed with competing over men in the house than in forming lasting relationships with each other. Psychologist Laura S. Brown (2005) worked as a consultant for one reality television show, and she was struck by the contrast between the complicated and generally likable people she interacted with in person and the edited portrayals that she watched when the television show eventually aired. She believed that the producers, directors, and editors of many reality television shows edit a more complicated gender reality to fit certain gender stereotypes. For example, strong women on reality television shows are often depicted as manipulative "bitches." In couple competitions like *The Amazing Race*, women in heterosexual couples are often made to appear domineering, insensitive, and pushy, rather than simply more powerful than their male partners. When women try to outwit or outplay their competitors on shows like *Survivor*, they are labeled in online discussion boards as "two-faced," "manipulative," or "mean," while their male counterparts are perceived as merely engaging in a particular strategy for winning (Brown, 2005, p. 78). Reality shows like *Survivor* and *Big Brother* seem to encourage contestants to act in deceptive and manipulative ways. Are there gender differences in the behaviors of female and male contestants on these shows? Do you think editors purposefully emphasize gender differences in their choices of which scenes to include? Do audiences respond differently to certain behaviors based on whether the person is a man or a woman? Do these shows reflect the reality of women's friendships or an exaggerated version edited to create higher ratings?

How do we account for this rather radical shift in the commonly perceived relationship between gender and friendship? First, as some researchers have pointed out, men's friendships, as with men's lives in general, were more public, and therefore generally viewed as more important than the relationships in the private sphere of women (Johnson, 1996). This is part of the general doctrine of separate spheres that we will discuss more in Chapter 9 but that views the public sphere of politics and the economy as inherently masculine. It is only recently, and with the emergence of feminism, that what happens in women's lives between women has become important to research or to society in general. Whether women actually were lousy at friendships is as subject to debate as the more recent conclusion that men are lousy at friendships, although the truth of women's historical status is more difficult to uncover; these women and their friends are no longer alive to tell their tales firsthand, and as we will see, talking in depth to people about their friendships can be an important way of uncovering the "truth" about friendships. Women's friendships were less visible, but this view of women as inferior friends can also be understood as evidence of a long history of patriarchy in Western society. Women were considered to be inadequate to the task of forming friendships because friendships were important, and women were perceived as generally inferior at anything that society deemed important. This theory might also help us explain our current views of friendship. Perhaps women are now perceived as better at friendship precisely because friendship in general is viewed as less important than it was in the past.

Gender Differences in Friendship

Luckily, we don't live in the past, and we have access to a growing body of research on friendships among both women and men. For now, we'll limit ourselves specifically to friendships within genders, before moving into the world of cross-gender friendships. You shouldn't be surprised by now to learn that there are many contradictions and points of disagreement in research about men's and women's friendships. Even if the concept under study here, friendship, were fairly straightforward and easily defined, there are always different ways of approaching its relationship to gender. We'll start with the extensive body of literature on the gender differences that exist in friendship.

Theory Alert!
Sex Differences

In Chapter 3, we discussed sex differences research, and here is another example of a body of literature that uses various methods to demonstrate how men and women orient differently toward friendship in important and meaningful ways. As in our earlier discussion of sex difference research, it is important to note that there are important similarities in women's and men's styles of friendships, as well as variations in friendship styles among different groups of women and different groups of men. In studies of the way women and men define what makes friendship, crucial similarities emerge. Both men and women generally want the same set of abstract qualities in their friends—intimacy, acceptance, trust, and help.

Face-to-face and Side-by-side

The differences between men and women in relation to friendship are sometimes described by identifying women's friendships as "face-to-face," while men's friendships are "side-by-side" (Nardi, 1992). Women tend to look for an intimate confidante, with whom they can share their feelings, while men tend to seek someone with whom to share activities and interests. The **face-to-face** of women's friendships is then meant to describe the degree of conversation and emotional sharing that takes place in women's friendships. Men are **side-by-side** in that they are participating in activities together but not necessarily engaging in the kind of conversational intimacy of women. Interestingly, Deborah Tannen's (2001) video research of men and women talking to each other across different age groups demonstrates that face-to-face and side-by-side are also true in the spatial arrangement of men and women. In her study, she placed girls and boys across a range of ages (from young girls and boys to adolescents) in a room with a friend of the same gender and instructed them just to talk. Girls in the study consistently moved their chairs so that they could face each other in order to have a conversation, looking at each other face to face. Most of the boys in her study given the same instructions moved their chairs not so they faced each other, but into a side-by-side arrangement, and therefore, they did not look at each other as they talked (Tannen, 2001).

Some of the things women and men are looking for in a friend are different, and it therefore follows that the friendships they have would turn out to be different. There is research on differences between self-disclosure in friendship that indicates that men are less likely to disclose intimate information about themselves to their friends than are women. It also seems that women are more uncomfortable when their friendships lack intimate self-disclosure than are men (Sherrod, 1989). This research suggests that men are not as good at verbal intimacy, but other researchers have asked whether, perhaps, men's intimacy involves nonverbal cues and expressions. Perhaps men hug each other more or can read their friend's emotions through their facial expressions without need for verbal self-disclosure. Evidence on this aspect of friendship also indicates that women are more comfortable with physical affection with their friends and feel they are better able to understand their friends through nonverbal communication (Sherrod, 1989). This is a fairly short list of the differences between women's and men's styles of friendship that researchers have reported.

Theory Alert! Sex Differences

If you remember from Chapter 3 and our discussion of sex difference research in general, although much work has been done documenting the many differences between women and men in countless behaviors and values, sex differences research spends less time attempting to explain these differences. In the area of friendship, this is less the case, and there are several different explanations that attempt to account for the kinds of differences we observe between women and men. Many of these will be familiar to you from earlier chapters on theory, beginning with the importance of socialization to men's and women's orientation toward friendship. Socialization is the primary explanation used by Lillian Rubin (1985) in her comprehensive study of the differences between men's and women's friendships. In earlier chapters, we discussed several different explanations for exactly how gender socialization happens, and Rubin draws mostly on psychoanalytic theory.

Psychoanalytic theory is adapted from psychoanalysis and builds on Sigmund Freud's ideas about the importance of early family dynamics in children's psychological development. Rubin (1985), drawing on psychoanalytic theory, notes the importance of mothering to both boys and girls, meaning that for both boys and girls, their first and very closest relationship is with a woman. The problem in psychoanalytic theory is this: How do both boys and girls then learn their appropriate gender identity? Girls are able to develop a sense of femininity directly through their relationship with their mother, who is of the same gender. They never have to sever this close connection with their mothers to become feminine, and in their adult friendships, this makes connection with other women much easier. In Freudian language, this is because women have relatively weak ego boundaries.

Boys, who are also deeply attached to their mothers at an early age, must learn masculinity. Psychoanalytic theory argues that boys learn masculinity by denying their connection to their mothers and so learn masculinity not through direct experience, but by identifying with absent or less present fathers. Breaking their initial connection with their mothers means that boys have much more developed ego boundaries than their female counterparts. Later in life, these ego boundaries make connecting intimately with others more difficult, and this explains the emphasis on doing things together rather than talking, as well as the lack of self-disclosure and reduced affection in men's friendships. Among adult women and men, these early childhood patterns make connecting much easier for women than for men.

Theory Alert! Hegemonic Masculinity

This is one explanation that relies on socialization, especially as it occurs very early in a child's development, because psychoanalytic theory posits that the essential dynamics of gender identity are established at a fairly early age. Another explanation that also relies on socialization emphasizes the specific content of masculinity as a gender identity. This particular view is in line with hegemonic masculinity, which was discussed in Chapter 4. Hegemonic masculinity sets up a certain bar that men must meet to pass a kind of cultural masculinity test. Hegemonic masculinity changes over time and is more properly called hegemonic masculinities because the particular form of masculinity men are called on to conform to varies a great deal from place to place and time to time. The basic idea is that men feel some pressure to conform to hegemonic masculinities, and there are repercussions for straying too far from this dominant form of masculinity. In contemporary societies where men are demonstrated to have less intimate friendships, hegemonic masculinity dictates that men be competitive and rational (Cohen, 2001). These and other qualities of hegemonic masculinity make intimate disclosure in friendships more difficult.

If competition is an important aspect of demonstrating masculinity, then revealing your vulnerabilities and weaknesses to another man will probably not be an easy thing to do. In addition, in contemporary societies like the United States, hegemonic masculinity has a strong heterosexual component. Real men are not gay, and moreover, hegemonic masculinity suggests, regardless of whether all men follow it, that real men not engage in any behavior that could in the least bit suggest any affection for other men. Close friendships and intimacy with other men introduce just that possibility—questions about a man's sexuality and suggestions of homosexuality (Nardi, 1992). Rubin (1985) found in her study of friendship that men commonly associated male friendship with homosexuality. Although this has not always been the case (recall that in the 19th century manly love was an important part of the definition of what it meant to be masculine), in our society open and intimate expressions of emotion are often perceived as something gay men do and, therefore, are outside the bounds for men trying to conform to hegemonic masculinity. Boys learn the content of this gender identity at an early age, as we saw in Chapter 4, through the patrolling of masculinity in the gender transgression zone among boys and girls at day camp. This inevitably carries over into the way men relate with each other as friends into adulthood.

Friendship and Social Structure

Theory Alert! Social Network Theory

Many theories of socialization explain how gender operates as coming from the inside out. Gender becomes an internalized part of our identities, of who we are and how we think about ourselves. This motivates us to act in gendered ways. Other explanations of gender differences in friendships take an approach that more greatly emphasizes the importance of social structure. This approach most closely resembles the social network theory of gender that we discussed in Chapter 2. Men and women develop different types of friendships not because of any internalized sense of gender identity but because of the social structure of their lives, including, but not limited to, gender as a social structure. As we will discuss, other social structures also impact the types of friendships men and women have as well as the content of those friendships. This explanation emphasizes the real constraints that are placed on friendships. Part of the reason sociologists ignored friendship for so long was the sense that our friends are largely a result of personal choice and, therefore, rooted in our idiosyncratic personalities.

But there are certain general structural requirements to friendship that have little to do with our own personal friendship preferences. You have to meet someone before you can become friends with them, and that generally requires being in the same physical space. This may be changing some with communication technologies such as Facebook and MySpace, which allow all of us potentially to have many "friends" we have never met. But most of us probably still consider some time period spent in the same physical space with our friends important, even if we may not have been physically with some friends for many years. Friends now living in different places are maintained at a distance but at some point shared a physical space. Even if we go down the road of online networking, there are certain structural constraints on friendship. If all your friends come from Facebook, you have still already limited yourself to the set of people in the world who have access to the Internet and so have automatically excluded a large percentage of the world's population from your potential friendship pool.

Geography generally provides one structural constraint on friendship, and at its most simple level, this means you're less likely to become friends with someone on the other side of the planet. But geography can matter even if you're in the same city or town because most places around the world demonstrate differing levels of **social segregation**, which is simply the separation of some realm of social life into different groups on the basis of some category. **Residential segregation** describes this kind of separation in terms of where people live, and in the United States, communities are especially segregated by race and class. In other cities around the world, segregation based on religion or ethnicity may be more prominent. For many Americans—Latino, Black, Asian American, or white—although they may live within yards or feet of people of different racial backgrounds, socially constructed spatial boundaries can result in different groups rarely coming in contact with each other. In some areas of especially high poverty concentrations and racial segregation, inner city residents may have never traveled outside the boundaries of their neighborhood. The chances that these individuals will become friends with someone of a different race or class background are fairly low.

Class background can constrain our friendship choices through geography and in other ways connected to our access to material resources. Have you ever had a friend who had significantly less or more money than you? Did it put a strain on the relationship? Research demonstrates that status inequality tends to put a strain on relationships in general, and especially friendships (Harris, 1997). This is partly because friendships often contain a norm of equality, as we discussed briefly earlier. Friends can maintain the kind of balance necessary to a friendship if they are closer to equal, and inequalities tend to cause tensions.

A concrete way to think about this in terms of class is money and access to other resources. If a wealthy woman's idea of a fun day is to go shopping in New York City and buy a $14,000 pair of Manolo Blahnik alligator-skin boots, she might find it difficult to be friends with another woman whose yearly income is about the same as the price of her boots. Friends generally spend time doing things they enjoy together, and things we enjoy doing, whether we like it or not, are often shaped by what our incomes make it possible for us to do. Class can have an impact on friendship through the experience of different types of work, as well. The worker on the late shift in a hospital or factory might find it difficult to be friends with the nine-to-five office worker. These different demands and expectations tied to different types of jobs influence the types of friendship choices we make (Allan, 1990).

There is one type of work in many parts of the global North that is almost always a job involving long hours of isolation from other adults and, therefore, provides little opportunity to make friends. That job is the housewife, or the home worker. As we will discuss in Chapter 8, the responsibilities involved in keeping a household running and raising children is a type of work, even though it is work that often goes unpaid. The paid domestic worker may have some company, but the unpaid mothers and wives work largely alone except for the company of small children. This points us to the very important role of gender in shaping the constraints placed on our ability to make and keep friends. Women who do not do paid work outside of the home can face the constraints of having reduced contact with people in that outside world (Allan, 1990). This can be offset by the contact with other parents that comes when children begin attending school, playing organized sports, and participating in other extracurricular activities, as much research demonstrates children's activities as an important sense of community connection for parents (Fischer, 1982).

Fewer and fewer women in much of the developed world are working just within the home as housewives, so how do their experiences inside the workplace affect their friendships? Women may come into contact with more people in their work lives and, therefore, have more opportunity to form friendships. But friendships also require a time commitment, and all the research on working women in married relationships confirms that they shoulder a larger burden of the domestic and child-care work. Some sociologists, as we will discuss in Chapter 8, call this the third shift to describe the extra shift experienced by working women at home cooking dinner, doing housework, and taking care of children. Structurally, this puts another constraint on women's ability to make friends.

These differences help to explain the constraints on the number of friendships women and men have and on whom they might be friends with. They help explain the different content of men and women's friendships. Men have larger social networks because their experiences in the workplace bring them into contact with a broader range of people, although these larger networks are true of boys in childhood, as well. Women have smaller networks on average, and there tend to be more kin in those networks. For these reasons, men's networks seem more helpful to them in the world of work, in finding jobs and obtaining promotions.

Does looking at gender as a result of social networks and social position help to explain the differences in the content of men's and women's friendships we discussed? Men do things together, while women engage in conversational exchanges of intimate information. If we think about the kind of demands made of men in the workplace in many industrialized societies, these include a kind of self-sufficiency, ambition, and independence (Allan, 1990). A successful man in the business world cannot become too dependent on anyone—they could become a potential rival—and so a friendship style that does not emphasize emotional intimacy makes sense given this structural position. If a woman's success has historically been tied to her role within the household, an emphasis on affection and intimacy can be perceived as carrying over from her structural role as emotional caretaker of her family. A person defined primarily as a wife and a mother has less need to be competitive, independent, or ambitious. Social network explanations of gender differences in friendship therefore also seem able to account for the differences in what goes on in friendships between women and between men.

Theory Alert! Social Network Theory

Gender Similarities in Friendship: Are Women and Men Really All That Different?

Examining how social structure is important in explaining both the different content and types of friendships between women and men as groups points us to larger questions about what these differences in general mean. Research into sex or gender differences often relies on fairly small statistical differences between any behavior or predisposition in the two genders, describing what are mostly overlapping ways of acting and thinking. On average, men may be less likely to disclose intimate information in their friendships with other men, but that does not mean that a great number of men do *not* share their innermost feelings with their friends. Remember, one critique of research into sex or gender differences is that when we set out to look for something, like a difference, we set up a predisposition to find it and make something of a big deal out of it, even if it's a relatively small difference. Other researchers believe we should be looking for differences within each gender, which often leads us in the direction of a search for gender similarity. Are the friendship styles of all or even most women more like other women's friendship styles, or do some women have much more in common with men's ways of being and having friends?

Theory Alert! Social Network Theory

Looking at social structure points us in this direction because the social network theory of gender makes the claim that gender is nothing more than a residual effect of the particular shape and content of the network in which we are embedded. If you want to change someone's gender, in the sense of the way in which they see the world and the way they choose to act in it, you can simply change their position within social networks and what we call gender will disappear. Gender differences in friendships are the result of men and women's different networks, and evidence for this perspective exists in studies that compare women and men who are positioned more similarly to each other in terms of their social networks and their positions in the social structure.

For example, some research seems to demonstrate that women's friendships are significantly different from men's only for those women who do not work in the paid labor force or whose family takes priority over their paid work life (Walker, 1994). When middle-class women who are oriented toward their careers are compared with their male counterparts, both groups are more likely to report an aversion to being intimate with their same-gender friends. These findings demonstrate the importance of looking for differences within each gender (middle-class women are just as different from working-class women as they are from men) and suggest the importance of an individual's social-structural position in determining how gendered she or he acts and thinks. This is consistent with a belief that gender is largely about the particular position in the social structure that you happen to occupy.

There is further evidence to cast some doubt on the power and importance of gender differences in terms of friendship. When researchers examine longer lasting friendships, the differences between women and men begin to disappear (Wright, 1998). This is also true as women and men age because friendships among elderly men and women also become more similar in many ways (Antonucci & Akiyama, 1987; Roberto & Scott, 1989). Single men tend to be more involved with their male friends than are married men, and this is especially true for married men with children (Cohen, 1992; Shulman, 1975); married men also report being less close to their friends and that they disclose less to them than their single counterparts (Fischer & Phillips, 1982; Tschann, 1988). These differences carry through the life cycle because older widowed men spend more time with their friends than do older married men (Powers & Bultena, 1976). In addition, across all groups of men and women, regardless of their particular, unique position in the social structure, many core values of friendship are shared across gender boundaries. Both women and men value trust and authenticity in their friendships, and their descriptions of ideal friendships both focus on the importance of communication, intimacy, trust, and interpersonal sensitivity (Galliano, 2003).

Are Our Friendships Really All That Different?

One interesting study of women's and men's friendships demonstrates how all the "differences" researchers have documented along the lines of gender may have to do with the particular method used to study friendships. One methodological problem with studying friendship is that many studies rely on self-reports by subjects. This means researchers ask respondents about both friendship in general and their own experiences with the people they call friends. You'll remember from earlier in the chapter that this especially causes problems because people define friendship in many different ways. In fact, some scholars of friendship point to the inadequacy of the English language in general for describing the complex ways we have of relating with each other, throwing all of these relationships together under the blanket term of *friendship,* although that one word disguises a great deal of underlying variety. Karen Walker's (1994) research suggests another methodological problem when it comes to the study of gender and friendship, although she also suggests ways to overcome this difficulty.

Walker (1994) conducted in-depth interviews with 52 working-class and middle-class men and women in the United States using a referral system to include friendship pairings in her study. This means that all but three of her respondents had at least one other friend who was also interviewed as part of her study. One problem Walker sought to address in her study was the tendency to approach friendship by either asking respondents global questions about friendship or asking them to focus on *best* friendships. Global questions ask respondents to either describe friendships in the abstract ("How do you define friendship?" or "Do you think men's friendships differ from women's friendships?") or answer questions about their own experiences with friendship in general, as if all their friendships were similar. A global question of the second type would ask a respondent something like, "Are you likely to tell your friends when you are feeling sad or upset?" This question is difficult to answer because although you may be likely to tell one particular friend when you're feeling sad or upset, you may never tell other people whom you still label friends. Another possible approach, asking respondents *only* about their best friendships, gives us a limited view of the wide spectrum of friendship types and limits our ability to discuss friendship as a general concept.

In her study, Walker (1994) asked both global questions about friendship and a series of detailed questions about specific friendship relationships mentioned by the respondent. So in addition to asking her respondents whether they believed there were differences in men's and women's friendships, she also asked them in great detail about their particular friendship with Susan, Juan, or Shanti. Global questions, Walker argued, tell us a great deal about our cultural expectations of friendship, even when our actual experiences do not conform to those expectations. So when Walker asked her respondents about the differences between friendships among men and women, about 65% of them agreed with the dominant cultural ideology about gender and friendships, expressing a belief in the idea that women are more intimate and talk more in their friendships than men, who are more likely to engage in activities absent much sharing of feelings as compared to women. Walker found many of her respondents' answers assumed some fairly fundamental differences between women and men that were being reflected and manifested in what they believed to be different about women's and men's friendships.

These findings on their own confirm the idea that there are important differences between the ways men and women go about being and having friends. Walker's (1994) study got more interesting when she reported what her respondents said about their *actual* experiences in *specific* friendships. For example, one man in her study, when asked about what friendship is in general, gave a fairly stereotypical list of shared masculine activities. In masculine friendships, according to this respondent, men share activities such as "fishing, baseball, you know, maybe playing softball in the field, huntin'" (Walker, 1994, p. 249). When Walker asked the same male respondent about the specific details of his friendships, she found he never went hunting or fishing with his male friends. His working-class lifestyle with a wife and two children made it too difficult to find the time or financial resources to hunt or fish. When his male friends came over, they drank and watched TV together. The friends he socialized with most often were those he worked with as a janitor at the local high school, and with these friends, he discussed retirement, their wives, their children, and their shared interest in sports.

When Walker asked this same man what he had talked about with his closest friend at work that day, he revealed a conversation about their wives' preferences for courtship and physical intimacy. The respondent's wife liked spontaneity rather than planning as a prelude to sex, while his friend's wife liked going out to eat and shopping on nights to set the scene for sex. If you read a description of two friends discussing their partners' preferences for intimacy in intimate detail, you'd probably assume the two friends were both women. This one specific instance among many documented by Walker demonstrates that this man's experiences of friendship not only doesn't conform to the general trends identified in research but also doesn't conform to his own stated perception of what friendships among men are like.

In Walker's (1994) research, a full 75% of men reported engaging in nongendered behaviors with their male friends, including sharing intimacies about their spouses and other family members as well as sharing their feelings in general. Three young lawyers in her study all talked about the importance in men's friendships of doing sports together, including going to football games together and talking about sports on the phone. When Walker asked them about what they did in specific friendships, she found sports had little importance in their friendships beyond a kind of symbolic importance. The friends did get together for perhaps one or two sports events every year, and two of the men had run together in races on two occasions. Their most frequent interactions, though, were during the holidays and usually involved parties, going out to dinner, or getting together with their wives. Although in their bimonthly or weekly telephone conversations they did sometimes discuss sports, they also discussed work and family, including one man's impending decision on what to do about a transatlantic romantic relationship. Sports, Walker argued, were important symbolically to these three friends, even though sports were not a particularly important part of their common, daily pattern of friendship interaction.

Another man reported going shopping with his male friend like "two old ladies" comparing prices to get the better bargain. One middle-class man, Gene, could not discuss moisturizer with his straight male friends because it was "faggotty," but on a visit with his gay male friend in another city, he went shopping to buy moisturizer and vegetable pate, activities he thought of as stereotypically gay and therefore not appropriate for heterosexual men like himself (Walker, 1994). Gene remained friends with his gay male friend after he came out to him in part to conform to his ideals about friendship as exemplifying loyalty; maintaining his friendship with his gay friend was a way of publicly demonstrating the importance of loyalty. In addition, the ability of Gene's gay friend to engage in typically masculine joking behavior made maintaining this friendship, one that many heterosexual men might find threatening, easier to do.

What Walker (1994) argued all these examples have in common is a complex interplay between what society tells people they should do based on their assigned gender identity and their actual experiences as individuals in real friendships. In the case of Gene, for example, we've already discussed the potential danger of homosexuality that lurks in the background of heterosexual male friendship in Anglo-European societies like the United States. Gene's friendship with a gay man was even more dangerous in light of this societal expectation that men be very careful about the amount of affection they show to other men. But Gene negotiated this friendship through drawing on norms of loyalty in friendship and by maintaining joking as a norm of masculine friendship in his relationship with his friend. Although other men Walker interviewed engaged in intimate sharing about their private lives with their friends, they still supported a general belief in the gendered nature of friendships on the whole. Walker's study is an excellent example of the ways in which our ideologies sometimes contradict our actual experiences.

Walker (1994) found these contradictions to be just as true among the women in the United States she interviewed. When asked global questions about what women's friendships are like, women reported the ability to tell friends anything and rarely emphasized the importance of shared activities among female friends. Yet as with the men, 65% of the women Walker interviewed reported engaging in some behavior with their friends that did not conform to the dominant gender ideology. Examples of these nonconforming behaviors included going to sporting events together, which women never mentioned unless Walker asked specifically whether the women went to sporting events or played sports together. Although 25% of her female respondents worked out together, took aerobics classes together, or belonged to a local sports team, they defined the actual activities they were engaging in as unimportant. These sports activities were defined by the women as ways to spend time together and talk, and therefore, the specific content of these activities were considered unimportant. Another 15% of the women Walker interviewed reported friendship interactions that took place while friends were engaged in some other activity. Many belonged to clubs of various sorts, and thus, their activities together as friends centered on their activities in these clubs. These shared activities, as with many men's friendships, served as the basis of their friendship.

In addition to finding that women often seem likely to have friendships centered on activities and doing things together like men, Walker (1994) also found that some women followed the stereotypical masculine model of friendship in their unwillingness to share intimate information. About 25% of the women Walker interviewed reported that they thought of certain information as private and would not share that information with their friends. One professional woman interviewed by Walker—Ilana—described friendship as involving openness and caring. But when asked detailed questions about her specific friendships, she admitted to finding it difficult to talk to her female friends about things, including her romantic relationships with men. Although Ilana seemed to feel that her reluctance to disclose private details of her life to her female friends made her unusual for her gender, she also noted that some men she knew seemed willing to disclose too much to each other. Thus, she constructed masculinity as entailing gossiping behavior and defined herself in opposition to that quality.

Two friends who were both female academics also reported that they did not often share information about their romantic relationships with friends; both of these friends felt that they did not want to burden people with their worries or problems. Some of the women who confessed to not being able to open up to their female friends also admitted to feeling like this meant there was something wrong with them. That they did not have friends to whom they could "say anything" made them bad friends, or incapable of having the correct kind of friendships. These anxieties about being inadequate at friendship were not something any of the men in the study reported, suggesting that it is correct to say that being open in a friendship is a norm set up more specifically for women in friendship than for men. On the other hand, none of the women in Walker's interview reported worrying about the activities they engaged in with their friends being perceived as "faggotty," either, suggesting that the fear of being perceived as homosexual in their friendship relationships is not as pressing of a concern for women.

The Intersectionality of Friendship and Gender

How did Walker (1994) explain these findings, which seem to contradict very radically much of the research we've already discussed on gender differences in the friendships of men and women? She argued that social class needs to be taken into account when examining gender differences in friendships. We have already discussed some ways in which social class can have important impacts on the number of friends, who those friends are, and what you do with those friends. Walker suggested that the friendships of professional men and professional women who are oriented toward their jobs tend to be similar. These groups of people tend to be geographically mobile, like a doctor moving from medical school to residency to her first job, and this makes friendships more difficult to maintain. In addition, once these middle-class people find themselves in professions such as medicine, making new friends to replace the ones they physically left behind becomes increasingly difficult. One female doctor interviewed by Walker talked about both the lack of time for making friends and the difficulty of forming professional friendships in the highly competitive world of medicine. Finally, Walker pointed out that women like this doctor assume that friends should be peer-equals; they assume that their friends will be of roughly the same status. As professional and middle-class individuals move up the social ladder in terms of their income, education, and prestige, they find there are fewer and fewer status equals from which to choose from in their network of acquaintances.

These factors impact the friendships of professional men and women in similar ways. They help to explain the reluctance to disclose intimate information among many professional women first because, like professional men, these women work in competitive careers that make establishing intimacy (and therefore vulnerability) with a potential rival difficult. Intimacy is also more difficult for professional women because an intimate relationship is assumed to require some time to develop. The professional women Walker interviewed required time to develop intimacy in a friendship, and this was time that was consumed by their work and therefore not available to spend on friends. This lack of time was even more of a problem for professional women who were married with children, as well as for professional men with wives and children. When professional women did fit more with the stereotypes about women's friendships, they were generally women who stayed at home with small children and therefore had friends drawn from their neighborhoods rather than from their workplaces. Walker suggested that these women had more opportunities to see their friends alone rather than in the more public space of the work world, and these made for more opportunities for intimate discussion. This finding points to the importance of not just social status as working class or middle class but also to one's position either inside or outside of the world of paid employment.

The influence of social class on the lives of working-class women and men also seemed to be more important than gender differences. Working-class respondents tended to have fewer friends than their middle-class counterparts but to have had those fewer friends for longer periods of time. Remember that research shows that many of the differences between men's and women's friendships disappear with the length of the friendship, and so it makes sense that both female and male working-class individuals reported talking to their friends about intimate matters. This makes intuitive sense in that the longer you've been friends with someone, the more likely you probably are to tell him or her intimate things about yourself. Working-class women and men saw their friends more frequently and more informally in each others' homes or neighborhoods, and they therefore knew much more about each others' lives than did their middle-class counterparts. Interestingly, Walker (1994) noted that the lives of working-class women and men contained many more of what most of us would consider problems; her working-class respondents talked about more financial difficulties, substance abuse problems, family problems, and health issues than middle-class respondents. But Walker argued that in part because of the ability working-class individuals had to discuss their problems with their more intimate, long-time friends, their lives were more stable in many respects than middle-class individuals in her study.

Walker's (1994) research demonstrated some of the potentially misleading traps in an approach that begins with an assumption of gender differences in friendships. If her research asked respondents only to describe their own sense of gender and friendship, their responses would confirm pretty exactly the dominant ideology about men and women in friendship. The reality of how these men and women behaved in their actual friendships was more complicated. Walker emphasized the ways in which her research revealed the ongoing social construction of gender in day-to-day life and was, therefore, drawing on the doing gender approach we discussed in Chapter 2. A doing gender approach draws on the tradition of ethnomethodology to excavate the ways in which gender is a performance we stage constantly in our everyday interactions. As with the social network theory of gender, gender is not something that is internal to individuals, but it is something that is located in the realm of interaction. A "successful" performance of gender does not have to conform to some long and complicated cultural rulebook but only to be perceived as accountable by everyone involved in the interaction.

Theory Alert!
Doing Gender

Walker argued that her research was an especially good example of this aspect of doing gender; we can violate the dominant gender ideology in our friendships without really compromising either our own sense of ourselves as gendered or others' sense of us as masculine or feminine. In the cases we discussed previously, both men and women violated the norms for friendship that were perceived as appropriate to their gender, and that they themselves claimed to perceive as appropriate to their gender. And yet none of these respondents seemed to believe these violations meant that they were not really a man or a woman. That Ilana didn't disclose private information to her female friends did not cause her to doubt her femininity. Instead, she adjusted her own understanding of masculinity (as gossipy) and defined herself in opposition to that version of masculinity. In one interesting case among three working-class women in Walker's study, one woman in the friendship triad was reluctant to disclose information and was urged by the other two friends to overcome this tendency, which they viewed as a failing on her part. Walker pointed out that the friends doing the urging to be more disclosing both re-created ideas about femininity, pressuring their reluctant friend to conform while allowing for their friend's deviation without questioning her underlying femininity. In this dynamic, one woman's choice to deviate from the feminine norm for friendship was a clear violation of gender norms but did not affect her ability to be accountable as a woman. In fact, her violation served as an opportunity for her other female friends to reaffirm this norm as important by urging her to conform. Gender can be accomplished interactionally even when people are not following the rules prescribed by our dominant gender ideology.

Walker (1994) also suggested that it is important to pay attention to the particular cultural and historical context within which people are doing gender. As we noted at the beginning of this chapter, in response to her global questions about feminine and masculine friendship, the individuals she interviewed were reflecting the more recent view that considers women to be generally superior at friendship. Besides the few individuals who suggested that women's friendships are more tenuous, most of her respondents conveyed no sense of women as incapable of the qualities of loyalty necessary to friendship.

Walker suggested that the explanation for this shift from masculine superiority to feminine superiority in friendship can be traced directly to the second wave feminist movement in the global North. Feminist research on feminine styles of friendship responded to the previous view of feminine inferiority in friendship with an account that reversed the story. Walker argued that these new views on gender and friendship filtered down into our culture and began to influence the way men and women think about gender in their everyday lives. They interpret their behavior through the lenses of these ideologies, and they orient their behavior toward these beliefs—even if their experiences don't always perfectly conform. This exploration raises interesting questions about the gendered nature of friendships in societies outside of the developed world, away from areas where feminism as a specific ideology emerged and has had its strongest influence.

FRIENDSHIP IN GLOBAL PERSPECTIVE

If friendship is a difficult concept to define when we confine ourselves to the developed world and the more limited context of the United States, imagine how confusing friendship becomes when we examine these relationships across a wide variety of cultures. The English language has essentially one word for friendship, which disguises the great variety of relationships that we fit under that single term. In other languages, there are many different words for relationships that in many ways resemble what is called friendship in Anglo-European societies. But just as we need to be cautious in doing research while assuming that our research subjects are thinking of friendship in the exact same way we are, we need to at least acknowledge at the outset that the extent to which some of the relationships we will examine fit under the broad label of friendship is also an assumption we're making. Having acknowledged this caveat at the outset, we can move on to some ways in which friendships around the world are similar and different from the basic idea of friendship in the Anglo-European societies we've outlined.

Choosing Your Friends

In global perspective, friendships are similar to those in places like the United States in that they are generally assumed to involve a sense of sentiment and loyalty. In most cultures, the relationships that most resemble friendship are composed of individuals who like each other and have a sense of obligation to each other's welfare; this is consistent with the ways in which we've already described friendship. We did not mention, however, the voluntary nature of friendships that is an important belief in Anglo-European society; we like to believe that we choose our friends based purely on sentiment, similar interests, personality, and so on. We've already demonstrated that this is not completely true in the context of Anglo-European societies, where social structure limits the possibilities of whom we can befriend. But that friends are voluntarily chosen is an important ideal regarding friendship in Anglo-European societies, and it is often less true in other cultures. In Cameroon, among the Bangwa people, children are given both a friend and a wife upon their birth. Among the Sarakstani, a group of herding peoples who live in mainland Greece, men develop friendships only with their first cousins. In New Guinea, friends are often selected from the men your sisters marry, and so your friend is likely also to be your brother-in-law (Brain, 1976). If these friendship relationships are not completely voluntary, but in some cases assigned at birth, how can they involve genuine sentiment and loyalty? Surely affection of this sort is tied to our ability to handpick our friends?

From the perspective of many of us in Anglo-European cultures, the idea of having someone else pick your friend for you seems to contradict the most basic notions of what a friend is. On the other hand, we are expected to feel sentiment and loyalty to a whole group of people whom we by and large also do not choose to be socially attached to: our family. The notion that sentiments like love, for both friends and romantic partners, can evolve only when the person to be loved is freely chosen is not a notion that is universally shared. As we will discuss in the context of dating, our standard model is that love evolves as part of a careful process of sorting people out, whether friends or lovers, to find the *right* person to love. But love and affection can also develop among people who are *put* together rather than *choosing* to be together, especially when these choices tend to group together people who already have a great deal in common. In this instance, we must be careful to avoid ethnocentrism, or the tendency to view the values and norms of one's own culture as normal and correct. Friendship and love that is not chosen, from an Anglo-European perspective, seems inferior or maybe just weird but, at any rate, seems in many ways not the right way to go about things. But for people in other cultures, the idea of freely choosing your friends or romantic partners seems equally strange.

The voluntary nature of friendship differs across cultures, as do the institutions and ceremonies that surround friendship. In the United States, we have few, if any, rituals that celebrate friendship. The closest thing we have is the relationship of godparents or, among Latinos, *compadrazgo,* which translates as co-parents. Among Latinos in the United States and in many Latin American cultures, the *compadrazgo* relationship cements a preexisting friendship between the godparents and biological parents of the child. At the child's baptism, the friendship between the parents and godparents is ritualized through the church and becomes institutionalized as *compadrazgo;* the friends become **fictive kin**, or symbolic family members. Godparents are common among non-Latino Catholics as well, although these relationships tend to place more emphasis on the godparents' responsibilities to the child, rather than on their relationships with the child's parents.

These rituals evolved during the Middle Ages when the Catholic Church sought to incorporate what had already existed as a pagan blood ritual. Priests performed a ceremony to sacralize (make sacred) a friendship, and the ritual included the literal mixing of blood. The blood ritual in the church is long gone, but the tradition of godparents and *compadrazgo* persists as one of the few rituals celebrating friendship in places like the United States. Blood ceremonies are common in many cultures as a way to cement friendships ritually, and their importance stays with us through phrases like *blood brothers.* A ritual exchange of blood symbolically makes friends into kin and gives a magical property to the friendship; if you share blood with a friend, then any potential betrayal of that friend could poison your very veins, according to the beliefs of some cultures.

As in Anglo-European societies until relatively recently, friendships within genders are generally more common than cross-gender friendships in a global context, and this is especially true of relationships that are institutionalized in some way. Many early explorers and missionaries traveling outside of the Western world misidentified many of these same-gender relationships as involving sexual contact as well. This is especially true in cultures that institutionalized same-gender friendships through various kinds of marriage ceremonies. Among the Nzema of southern Ghana in Africa, men will sometimes marry other men. In two ethnographic accounts of these institutions, two men meet and one finds the other attractive, often meaning physically attractive. In one account, an older man was especially impressed by the beauty of a younger man's hands. In another, one man married another to "have a beautiful person near him" (Brain, 1976).

Although the physical attraction seems important to these marriages, there seems to be no evidence of sexual behavior; however, you'll remember from Chapter 5 that it's not always easy to say what exactly is and isn't sexual behavior. In these marriages, the two men live with each other and often sleep in the same bed with each other. Often, the man serving as the husband in these relationships already has other wives who are women (the Nzema society is polygamous, meaning a man marries more than one woman), and sometimes the man who was originally the wife goes on to marry a woman for his own wife. The husband and wife in these situations are determined by who pays the bridewealth, and this demonstrates how these marriages are perceived as similar to marriages between women and men. In a marriage between a man and a woman among the Nzema, the man pays the bride's family a price in the form of money or goods—the **bridewealth**. The same rules of exchange are followed for marriage between two men, and being able to afford the bridewealth for an attractive young man can be a sign of status for the husband. Similar woman–woman marriages occur among the Bangwa, where a woman becomes the husband to another woman by paying bridewealth to her family. In the case of woman–woman marriage, the husband has the additional advantage of having legal rights to the children of her legally sanctioned wife. How exactly does the wife in a woman–woman marriage produce children? She may come with her own children from a previous marriage, who legally become the husband's upon marriage. But woman–woman marriages also largely do not involve sex between the two women, and they allow for the wives to have sex freely with men. The woman who takes these wives gains legal rights to children as well as an additional source of help and potential income in her household.

Can we call these relationships between two people who are recognized by their particular culture under the law and religious systems as married as also examples of friendships? Can two people who are married to each other be friends? Certainly in Anglo-European societies, the answer is yes. Although we focused our own discussion of friendship primarily on nonkin relations, many people in the United States would call their spouse or romantic partner their best friend and would expect that friendship can lead to romantic interest and eventually to marriage. Especially in the men–men marriages among the Nzema, there seems to be genuine affection between the two men, and they often describe the feelings they have for each other as love. It might seem strange to us for two male friends to sleep in the same bed together, but this strangeness is a fairly recent phenomenon in Anglo-European society. Abraham Lincoln sometimes slept in the same bed with his male friend, and during the 19th and 20th centuries, married female friends sometimes displaced their husbands to sleep in the same bed when they visited each other. Although the women–women marriages are somewhat more instrumental in their orientation, the two women generally get along and have a relatively close relationship.

Thinking about these relationships serves as a good way to begin considering an important question we raised at the beginning of the chapter: Where do we draw the line between friendship and other types of relationships that involve love and affection? Marriage is an institution that we in the United States associate with romantic love (another relatively recent phenomenon), but does marriage have to involve romantic love? In our language, reflecting its clear privileging of romantic love over other forms, when do *just friends* become something *more than friends?* The cases of man–man and woman–woman marriages demonstrate that although we believe the lines between friend and lover are clear-cut and easy to draw, making these distinctions may not always be so simple. We can further explore these boundaries by looking at cross-gender friendship and at how same-gender friendship is negotiated among gays and lesbians.

DRAWING A FINE LINE: SEX, ROMANCE, AND FRIENDSHIP

In discussing the relationships between best friends, Lillian Rubin (1985) wrote the following:

> More than others, best friends are drawn together in much the same way as lovers—by something ineffable, something to which, most people say, it is almost impossible to give words. . . . People often talk as if something happened to them in the same way they "happened" to fall in love and marry. (p. 179)

Rubin (1985) goes on to describe several examples of the stories best friends tell about their first meetings or the beginnings of their relationships, and they have the same magnified moment quality as stories of romantic love. Other scholars have speculated that there is an erotic component to most close friendships, but that this erotic aspect is disturbing and therefore repressed by most individuals (Seiden & Bart, 1975). In heteronormative societies, to think of an erotic component to friendship may be strange when considering same-gender friendships, but what happens when we consider the context of cross-gender friendships? Is this erotic component of friendship even more intensified in relationships between individuals of different genders? To answer yes to this question would be to confirm a heteronormative perspective. If the sexual possibilities of friendship are perceived as somehow more important to consider in the context of cross-gender friendships, then we are assuming a world in which most or all people are sexually attracted to someone of a different gender. We know that this is not the case, and we are already aware of the erotic potential that exists in same-gender friendship, as acknowledged in the man–man marriage among the Bangwa, which depends on some physical attraction between the two male friends.

Research suggests that figuring out the boundary lines between friend, partner, and lover may not be significant problems within gay communities to the extent that they are for cross-gender friendships among heterosexual individuals. Friendships within gay male communities may have their tensions, but anecdotal accounts by gay men, especially in cities with large gay communities, suggest that friendship is an important social glue that helps to provide a source of positive identity and support for a group of people who are marginalized by the rest of society. Thus, some gays and lesbians suggest that because of this marginalization, friendship within gay communities becomes even more important than it is in the straight world.

Theory Alert! Intersectionality

Peter Nardi (1999) argued that the ways in which gay men mix sex and friendship demonstrates how they both challenge and conform to hegemonic masculinity. As we discussed earlier, in male, straight friendships, the fear of homosexuality can be a looming concern. Michel Foucault, the influential philosopher and historian we discussed in Chapter 5, said in an interview that "the disappearance of friendship as a social institution, and the declaration of homosexuality as a social/political/medical problem, are the same process" (as quoted in Owens, 1987). Foucault was pointing out the way in which the rise of the institution called homosexuality made friendship, most specifically friendship between men, more and more difficult. For straight men, any feelings of closeness, affection, or passion for another man are always in danger of being interpreted as homosexual. This was not the case among the gay men Nardi (1999) studied, who were largely freed from these fears and were therefore able to have close, intimate, and sometimes passionate friendships with other men. One example of this is the amount of physical touching that takes place between gay male friends, as 78% reported that they "frequently or usually" expressed affection through touching their best male friend (Nardi, 1999, p. 83). These percentages were higher than the rates of touching for gay men whose friends were women, rather than other gay men, and are probably considerably higher than the rate of physical affection we would observe between straight men in the United States. Gay men subvert hegemonic masculinity, then, in the way their friendships with other men can exist freed from the constraints placed on many heterosexual men.

Theory Alert! Hegemonic Masculinity

On the other hand, gay men in their friendships do not seem completely free from the influence of hegemonic masculinity. Nardi (1999) investigated the relationship between sex and friendship among gay men and found that although many friendships did involve some sexual intimacy, this period of sexual involvement usually came at the beginning of the relationship. Although many of the men had had sex with their best or close male friends, for most, the sex ended fairly early in their relationship as it transitioned into friendship. Most of the gay men Nardi interviewed expressed fears about mixing friendship and sex as difficult and dangerous for their relationships. This belief is much like the belief within cross-gender friendships that women and men cannot really be friends; sexual attraction gets in the way. Among the gay men in Nardi's study, gay men can be friends, but having sex with each other will make the friendship more difficult. Nardi argued that these beliefs represent a separation of sexuality from intimacy. It is OK to be intimate with a friend, but it is not OK to mix that intimacy with sexuality.

This is an important component of the gender order and hegemonic masculinity. Teenage boys are taught that sex will lead to intimacy, while teenage girls come to believe that intimacy must be established before any sexual involvement occurs. In this formulation, hegemonic masculinity teaches men that sex and intimacy are incompatible, and Nardi believed this is why gay men believe sex and friendship should be avoided. In this way, gay men's friendships reproduce hegemonic masculinity in its prescriptions about the way sex and intimacy should be mixed. Although many people may believe that gays and lesbians in their everyday lives are able to or, in fact aspiring to, escape the power of gender norms to influence behavior, gender continues to be an important force shaping their friendships, just as it is for heterosexual men and women.

THE RULES OF ATTRACTION

When we move from the world of friendship to the complex of other types of attraction, the picture gets even messier. Friendship is a concept that we can at least very roughly translate into a global perspective. We may have to make necessary adjustments for friendships that are less voluntary and that sometimes blur the lines between friendship and other types of intimacy, such as the man–man marriages among the Bangwa. But when we begin to discuss other types of intimacy as they exist in Anglo-European societies, a global perspective becomes much more difficult. This potential for messiness can be pretty easily demonstrated if we ask what seems like a fairly basic question: What makes people attractive to each other? Surely this is an important and fairly simple component of relationships that lead to things like dating and marriage. People need to be attracted to each other before they decide to date or get married, right? Well, not really, as we'll see. But let's stick with the basic idea of attraction. What makes people look physically attractive to other members of their society?

The answer to that question all depends on where and when you are. As we'll discuss in more detail in the next chapter on gender and bodies, thinness is generally valued as an attractive quality for women in the United States, and perhaps increasingly for men as well. But if you've been to an art museum and have seen paintings of women from previous time periods, you may have noticed that very few of those women could do very well on *America's Next Top Model* or *Project Runway.* Nor would thinness get you very far in places like Brazil, where until the growing influence of Western beauty values, the ideal woman was shaped like a guitar, rather than a fence pole. Among the Padaung of Burma, attractiveness in women is tied to the length of their necks because women use a series of rings added throughout their lives to elongate their necks artificially. In some places (increasingly including some segments of the U.S. population), piercings and perforations of the body are considered the height of attractiveness. Among the long list of attractive features across the globe are stretched lips and earlobes, filed teeth, and flattened skulls (Turner, 2003).

Physical attraction in global perspective poses its own set of problems, but when you include emotional elements of attraction, the discussion becomes even more convoluted. This is a result of the problematic notion of romantic love and exactly how it fits into the structure of a larger society. In the United States, we generally believe in a thing called romantic love. You could say we've built up something of a cult following around the phenomenon. We have at least one holiday solely devoted to the ideal of romantic love in Valentine's Day (and two if you count the more recent addition of Sweetheart's Day). Add to that countless books, movies, television shows, Internet sites, songs, plays, and magazines, and an outsider might wonder whether people in the United States think about much of anything *but* romantic love.

In the United States, romantic love has been conceived of as primarily heterosexual and as something that, at least in the happily ever after version, eventually leads to marriage. A more global definition of **romantic love** is an intense attraction in which the love object is idealized, has an erotic context, and the expectation of lasting for some time into the future (Jankowiak & Fischer, 1992, p. 150). Anthropologists debate whether this particular ideal, which in the West we find so central to our own lives and happiness, is even something that can be said to exist outside of Anglo-European cultures (Smith, 2001). The debate about whether romantic love is universal is ongoing, but even if we assume for the moment that it is, there are still variations in how romantic love lines up with other societal institutions across the world.

In many parts of the global North, the assumption is that romantic love comes before marriage, and being in love is like the green light to proceed to marriage, at least for those who are heterosexual and invested in getting married. This ordering is different in other cultures, where romantic love is something that is assumed to come after marriage—if it's something that's appropriate at all to feel in a marriage. Marriage is an economic or political arrangement; love and intimacy are to be found in friends or other relatives but not with the person you will marry or to whom you're already married. This makes discussing romantic love a rather complicated matter, and so we'll start with a brief history of Anglo-European traditions before discussing the increasing influence of this idea in other parts of the world.

CULTURAL ARTIFACT 2: THE GENDER OF VALENTINE'S DAY

Does Valentine's Day have a gender? Is it a feminine holiday? The exact origins of Valentine's Day are murky, but like many holidays, it was probably pagan in origin and then later incorporated by the Christian church. The Romans celebrated Valentine's Day by placing the names of all the young women in an urn, from which all the city's bachelors would draw. For that year, the man would be paired with the young woman whose name he drew, and these couplings were often expected to lead

(Continued)

(Continued)

to marriage. The Catholic Church, not surprisingly, did away with this particular ritual, and the tradition of exchanging tokens among friends and lovers emerged later in Great Britain in the mid-18th century. Today, lovers on Valentine's Day expect chocolates, flowers, and romantic dinners. But is there a gender to who generally does the giving and who does the receiving on Valentine's Day? In heterosexual couples, do women buy men gifts of chocolate and flowers or take them out for romantic dinners? Is the burden of Valentine's Day unfairly placed on men so that Valentine's Day is essentially a holiday for women? On the Internet and talk radio, some men have suggested this is the case, arguing that men need their own holiday—Steak and Blow Job Day. As stated on one website dedicated to this new holiday for men, "Simple, effective, and self explanatory, this holiday has been created so you ladies finally have a day to show your man how much you care for him" (Birdsey, n.d.). What does Steak and Blow Job Day say about masculinity, femininity, and heterosexuality?

Friendship and Values

To understand how a culture or society approaches love and intimacy, it's important to understand how these phenomena fit into the value system as a whole. Countries like the United States and other developed nations are, by and large, individualistic societies. In an **individualistic** society, the collective needs are generally considered to be less important than the needs of the individual (Dion & Dion, 1993). This is evident in the United States politically in our strong emphasis on individual rights; we often protect the rights of individuals even when the exercise of those rights may not benefit specific groups or even society as a whole. You can understand how an individualistic approach applies to the world of love and intimacy if you ponder how you would feel about allowing your parents, or some other family members, to select all of your potential romantic partners without any of your own input or choice in the matter. To most people in the United States, this proposal would violate a fundamental sense that, outside of the family we are born with, we have the right to choose whom we love. The particular collective need our family might have to see us date someone who is economically or socially desirable is less important to us than our own individual needs to be happy in our relationship. We value our needs as individuals as more important and, what's more, as distinctly different from and perhaps at odds with the needs of our families or other groups. This is different from the collectivistic societies we will discuss later.

In cultures like the United States, we expect that we will be able to choose our intimate partners, and we expect that that selection process will at least in part involve the notion of romantic love, even though that has not always been the case historically. It is only as recently as the 18th century in Europe that romantic love became strongly tied to marriage. This shift came about as marriage became less about political and economic arrangements, and it is conceived of as a bond of personal affection between two individuals. By the 21st century in the United States, the idea that romantic love is an important part of our lives, funneling individuals toward fulfillment, happiness, and (if you're heterosexual) marriage, is somewhat inescapable. The question remains: How do we get there? How does one go about finding and identifying "the one"? Or in a culture where single people for the first time outnumber those who are married, is romantic love still considered a step toward marriage, or is it now considered an end in itself? Once a culture begins to make the assumption that the person you choose to love romantically is, in fact, a matter of personal choice, what kind of system do we use for helping us to make that choice, and how is gender involved?

Courtship to Dating: A Brief History

For many, although these norms may be in the process of change, the clear answer to the question of how one finds true love is through dating. On some college campuses, dating may already be becoming a thing of the past thanks to the emergence of a new hook-up culture, which will we discuss later. If dating is, in fact, on its way out, it will have been just another transformation in a long list of the ways in which rituals of dating and courtship have changed rapidly since about the 18th century. In her history of courtship and dating in the United States, Beth Bailey (1988) described the specific gender and social class implications of transformations in our romantic rituals during the 20th century. Looking at how courting institutions have changed over time and the gender implications of those changes is a good example of an institutional approach. During the 19th century in America, there was no such thing as dating as we generally think of it today. The word *date,* used in the context we're using it here as a potentially romantic encounter, probably got its first use around 1896 (Bailey, 1988).

Theory Alert! Institutional Approach

Before then, courtship in 19th-century America involved men coming to call at the homes of their potential love interests, a form of courtship appropriately named **calling**. The men were called suitors, as you'll know if you've seen *Gone With the Wind* or other movies that depict this often highly formalized process of courtship. Men came to women, and couples rarely *went* anywhere; instead, they spent the evening chaperoned by their parents, usually on the front porch or in the family parlor. In addition, a man could not come calling on a woman unless she had invited him to do so and he had received the approval of her mother. For a man to *ask* permission to call on his love interest was highly inappropriate, and advice columns of the time period cautioned young men against taking this step. In this way, the process of courtship was initiated by women rather than by men. Courting took place mostly in the private realm of the girl's home, and this was a woman's world. A family's prestige and social standing came from the father's position in society, but during the late 19th century, the separation of the spheres was well under way, and so the household was considered to be a woman's domain. The system of calling was a system that was almost entirely under the control of women.

The particular courtship system of calling on a young woman had its widest prominence among the middle and upper classes in the United States. Bailey (1988) argued that the system evolved in part out of class motives, as a way for the middle and upper classes to maintain some control over their daughters' marriages in an age when telling who was who was becoming increasingly difficult. With industrialization, the strict class hierarchies that had existed in many places in the United States were beginning to fade, and people were beginning to move around more, both socially and geographically. This made discovering whether a young man was the right kind of person to bring into the family increasingly difficult to do. Calling allowed mothers, as the social gatekeepers of the family, to patrol the appropriateness of suitors through the ever useful convention of being "not at home." According to the detailed rules of courting, a young man who came calling could be told, usually by the maid who answered the door, that the lady in question was not at home. A young man deemed less than desirable by a mother or daughter would be told that the young lady was not at home regardless of whether this happened to be true. In this way, women were conveniently able to screen potential suitors and subtly to send the message that his hopes of courting the young woman were not going to be fulfilled.

Although those in the lower classes might have aspired to the system of calling and were certainly influenced by it, it was more difficult if not sometimes completely impossible for them to participate in the same way as the middle and upper classes. Calling required space—a parlor, front porch, or some room where the courting couple could ideally sit together, chaperoned by the parents or with chaperones close by in the next room. Working-class families, especially those crowded together in urban neighborhoods, had no such space for entertaining in their homes. As we will read in Chapter 9, during the early period of industrialization in the United States, many of the factory workers were single women, usually recent immigrants or farmer's daughters who were sent to work in the factories until they married while their brothers worked the farm. These factory girls usually lived in one room in a boarding house or apartment and certainly had no space for a gentleman caller. Although one magazine from the time period tells of one group of factory girls' creative attempt to create such a space by pooling their resources to rent a room to share for hosting callers, this option was out of the reach of many working-class families. Many women of the lower class, especially those of the new waves of immigrant families, simply desired to escape out of the house and into the freedom of the public world, even if the possibility of calling *was* available to them.

Theory Alert! Intersectionality

Dating, on the other hand, emerged first from the lower classes; it then trickled up into the upper classes and, finally, down into the middle classes. The transition happened fairly quickly, but it certainly would not have been an easy one for many middle- and upper-class women. The doctrine of separate spheres at this time enforced a strict division between the world of the home and the public world outside of those doors. Working-class girls worked in factories, but the rest of the public realm was perceived as deeply inappropriate for women. Dance halls, restaurants, and movie houses were not places where respectable women of the middle and upper classes went. When upper-class women did go out, they went to private parties with carefully selected guest lists or to the movie houses for private parties from which the general public was excluded. Dating moved the act of courtship into the public world, a world that defined as masculine at the time and, therefore, not the proper place for women to be. Privileged young women of the upper classes eventually came to realize the advantages offered to them by dating and being able to experience more fully the increasing commercial amusements of urban life. Working-class women had been dating in dance halls and restaurants for a while, and eventually these trends affected the behaviors of the middle class as well.

Dating moved courtship out of women's realm inside the home and into the public realm of men. Power in courtship shifted from women to men in another way with the emergence of dating. Dating added a whole new economic twist to the question of courtship. In the past, the economic standing of a potential suitor was obviously important, and it was one factor that was probably often used to weed out potential husbands. But the act of courting itself didn't require any output of money; when you went and visited a woman in her home, you didn't have to pay for or buy anything. With dating, this situation changed. Suddenly, courting a woman required quite a significant outlay of cash, and during this transition period (and perhaps still today), men often deeply resented the financial burden that came with a woman's company. Bailey (1988) pointed out that one usage of the word "date" during this early time period was to describe a visit to a prostitute. Bailey argued, and some letters in popular magazines of the time from men agree, that dating became a system of exchange similar to that of prostitution in that men were essentially paying for the time they spent with a young woman.

The young men of this time period were not particularly happy about this new system of exchange. In a letter to *American Magazine,* one man tabulated that he had spent more than $5,000 on dates that year and had come to the conclusion that female companionship, as pleasant as it might be, was not worth that price. What men received in exchange for this outlay of money, Bailey (1988) argued, was new power in the realm of courtship. It may have cost them more, but the outlay of money could carry certain expectations of reciprocity from the young woman in the form of "favors," including sexual favors. Power shifted toward men under dating also in that it became the responsibility of men to pursue women through asking them out on dates. Remember, under the rules of calling, a man could visit a young woman only upon receiving her invitation. As dating evolved, it became quite clear that women were, under no circumstances, to ask a man out on a date. This, too, was tied to the economics of dating. Although there was some confusion about this reversal in the first 20 or so years of the transition from calling to dating, advice columnists strictly instructed women never to suggest a date to a man; men bore the financial responsibilities of the date, and so suggesting a date was inappropriate for a woman because you were asking a man to spend money on you (Bailey, 1988, p. 21). Under the system of dating as it evolved, and still to a large extent today, women were left to sit around and wait to be asked on a date.

Nonetheless, from roughly the beginning of the 20th century on, dating was the predominant mode of courtship in the United States. The term has become fairly blurred in its meaning over time, in part because of the ways in which exactly what it means to date has continually shifted. In the early part of the 20th century, dating among teenagers was primarily about volume rather than about length or quality of an individual relationship. Girls and boys dated many different people in what sociologist Willard Waller (1937) called the **rating-dating-mating complex**. Dating was based on a system of ranking in which both boys and girls aspired to date someone whom they rated as a bit higher than themselves in terms of dating desirability, thus, using dating to raise their own status. The competition was with peers of the same gender because girls increased their status among other girls by being able to date higher status boys and by being perceived as good dates.

A problem Waller identified was that the definition of a "good" date, toward which both girls and boys were aspiring, was very different depending on your gender. For girls, a good date preserved her reputation by not engaging in any sexual activity, while for boys being good meant gaining sexual experience and sophistication. In this model of dating, the goal is to improve your status and create a reputation for yourself as a good date, and this is accomplished by dating a lot of people. The rating-dating-mating complex may sound a bit cutthroat to you, or perhaps not depending on your own experiences with dating. Where's the love and passion in this system of using dates to increase your status? Willard argued that affection and love didn't have much to do with this particular system of dating, nor was it necessarily about finding a compatible individual to marry. Dating in this time period was primarily a way to make yourself more popular, and in this sense, dating had less to do with romantic love.

Historians of courtship argue that this changed during the period around World War II when dating became less about a continual parade of boys or girls and morphed into *going steady*. Bailey (1988) provided evidence of this transition in a magazine story from the time where an older uncle asked his niece about her experiences at prom. His question, "Dance with a lot of boys?" reflected the old dating-rating-mating system of dating, where a girl's goal at a dance would have been to be continually dancing with many different partners (p. 32). The worst imaginable fate for a young woman under the old system of dating would have been getting stuck dancing with one boy. This meant you were not desirable enough for another boy to cut in on your partner, and as you could not leave the dance floor yourself, both the girl and the boy were stuck in the undesirable position of having to dance only with each other. By the time of the magazine story about the uncle and his niece, the niece couldn't imagine why you would dance with anyone besides your date at the prom—one dance partner for the whole of the evening. To date someone came to mean a long-term relationship rather than a competitive system for establishing popularity.

These changes were brought about in part as a result of the demographics of the time period. During World War II, there were suddenly fewer men available to date, and so fewer men to cut in on the women on the dance floor and to date in general. In this new social reality, there was more incentive to hold on to the few men who were available through going steady. Demographically, this points to the importance that the gender makeup of a society (the relative number of marriageable or fertile women and men) can have on dating and marriage practices. Another explanation for the shift toward going steady is the important fact that dating throughout its brief history has been thought of as training wheels for marriage or for life in general. Dating-rating-mating was perceived as good training into the highly competitive world of adulthood. Beginning in the early part of the 20th century, the idea of sexual compatibility in marriage began to gain in importance. Today, the assumption that a healthy sex life is an important part of a successful marriage is fairly widespread, but that's a relatively recent idea, dating back to the period between the two world wars in the 1930s, when advice books began to emphasize the importance of good sex to a successful marriage (Hawkes, 1996).

Going steady, although it initially made many parents very nervous, was partly a response to these changes in the demands of marriage. If being able to have a healthy sex life was now going to be important to marriage, it made sense for couples to be able to have some sexual experiences prior to marriage. Although that's not to say sexual encounters didn't occur under the rating-dating-mating complex, if sex were commonplace under that system, it would have led to quite a bit of sex with multiple partners. And in a period before the widespread use of the birth control pill among women, that could have had some interesting consequences in unplanned pregnancies. But if a boy and a girl went steady for long periods of time and experimented sexually with a long-term partner, the behavior looked a lot more like what we expect from marriage today. Sex is retained, at the very least, as something that happens within the context of a long-term, monogamous relationship—even if it's not quite marriage, per se. The ideal during this period was still that young people should wait to have sex until marriage, and especially that *women* do so. But going steady made some sexual experimentation prior to marriage OK, as good practice for the sexual compatibility that would be necessary in marriage.

What does dating mean now, and what gender implications does the current system of dating have? When you say you're dating, are you more likely to mean you've been seeing one person exclusively in a long-term relationship or that you're going out on dates with several different people in succession or, perhaps, at the same time? When we have these conversations in my gender classes, there often seem to be competing and sometimes contradictory definitions of what dating means. This is in part because many of the historical meanings of dating have stuck with us; for some students, it's still more like dating-rating-mating, while for others, it's more about going steady. In addition, we have to throw in the myriad terrain of contemporary slang to describe the complicated relationships between young women and men in high schools and on college campuses. A brief sample tells us that you can be "talking," "talking to," "hanging out," "seeing," "hooking up," or as is increasingly popular, "friends with benefits."

Hookups and Friends With Benefits

The hookup may very well be the next phase in the evolution of courtship. Hookups are probably familiar to many students on college campuses; a survey study of students at 55 northeastern colleges and universities in 2000 found that greater than three fourths of respondents had experienced at least one hookup (Paul, McManus, & Hayes, 2000). What exactly is a hookup? In this particular study, **hookups** were defined as a sexual encounter, usually lasting only one night, between two people who were strangers or brief acquaintances. For the purposes of this study, a hookup includes some physical sexual interaction, although not necessarily sexual intercourse. Hookups are generally spontaneous sexual experiences, something that just happens, and although the hookup itself is often planned, the particular individual with whom a person eventually hooks up is unknown. Generally, hookups occur with a stranger or acquaintance and do not lead to a continuing relationship; in fact, hookup partners may never see each other again. This general definition has received some corroboration from college students themselves (Rodberg, 1999; Vanderkam, 2001).

So far, hookups have been studied largely within the context of colleges, which are already characterized by more sexual permissiveness relative to other parts of society as well as patterns of sexual activity that may include multiple partners (Paul et al., 2000). Casual sex on or off college campuses is not completely new, but is the pattern of hooking up something different than mere casual sex? Some evidence suggests that hooking up is on the increase at the same time that dating is on the decline. One study reported that only half of college women had been on six or more dates during college, while a third had been on no more than two (Vanderkam, 2001). This is certainly different from the pattern of dating in the United States prior to World War II, but does this mean hooking up is really replacing dating?

One observer of social life and culture among Princeton University students, David Brooks (2001), argued that college students no longer date because they simply no longer have the necessary time. This generation of college students, Brooks argued, is the result of overscheduled childhoods in which they were shuttled from soccer practice to music lessons to scout meetings and so on. In the ultracompetitive world of trying to get into elite colleges like Princeton University, these students spent their high-school years engaged in every extracurricular activity they could find to pad their college admission applications. In college, these same kids continue to create busy schedules for themselves, now to gain a competitive edge in getting into medical school, law school, graduate school, or a first job in the competitive corporate world. Brooks's argument is that these young people don't have the time it takes to date, but they still have sex drives. If sexual experiences can be efficiently gained outside of dating through the system of hooking up, why waste time with dating?

Brooks's (2001) argument is based on casual conversations with a very elite group of college students, and not on actual social scientific research. Current research on the nature of hookups is contradictory. In general, researchers have found little difference in how often women and men hook up on college campuses and a great deal of agreement among men and women on what constitutes a typical hookup. Men report more hookups that involve sexual intercourse, while women report more hookups that involve sexual experiences outside of intercourse (Paul et al., 2000). In one study, a common theme in women's accounts of their worst hookups was experiencing pressure to engage in unwanted sexual behavior; these stories sometimes focused on the aggressiveness of their male partner, but also on their own as well as their partner's use of alcohol or drugs, gendered peer pressure, and their own low self-esteem as contributing to their experiences (Paul & Hayes, 2002).

Another gender difference in this study suggested that although both men and women sometimes regret hookup experiences, the reasons for this regret break down along gender lines. Women experienced regret that centered on shame and self-blame, often connected to not knowing their partner or to the lack of further contact with their hookup partner. Men, on the other hand, expressed regret about the bad choice of a hookup partner, usually because the partner was unattractive or undesirable because of a reputation for promiscuity. These findings suggest that although both men and women are participating in the hookup culture on college campuses, their subjective experiences of these encounters may be quite different.

In contrast, researchers in one study expected, as a result of the double standard, that women engaging in hookups would experience lower self-esteem than men who engaged in hookups. The double standard in our society tells women that they will be judged more harshly for having more sexual experiences, as well as for having sexual experiences that are outside the context of marriage, a close intimate relationship, or that do not at all involve feelings of affection and love. In other words, romantic love and sex are supposed to be more strongly tied together for women. But in this study, there were no significant differences in the self-esteem of women and men who participated in hookups, suggesting that women felt no worse about themselves in general for having participated in this casual sex behavior (Paul et al., 2000).

This may be an example of what sociologists of sex call **sexual convergence**, or the trend for the norms and ideals surrounding women's sexuality to become increasingly similar to the norms and ideals of men's sexuality. Expectations for sexual behavior between the two genders have begun to converge and become more similar. This convergence is reflected in ideals about virginity and masturbation, as well as in actual behaviors. The percentage of high-school boys who were having sexual intercourse stayed roughly the same from the 1940s to 1991, while the percentage of high-school girls who were sexually active increased from 20% in the 1940s to 60% in 1991 (Lewin, 2003). This convergence can be explained in part by the emphasis by many of those involved in the women's movement on sexual empowerment as an important goal for women. Many feminists emphasized the important connection between sexuality and power, and so taking control of their own sexuality was an important goal of feminism for many women. Another important factor leading to sexual convergence is the increasing availability, social acceptance, and safety of various methods of birth control for women, among the most important being the birth control pill and access to legal abortion. Although various contraceptive methods have been around for a long time, these new methods have made preventing pregnancy much easier and more socially acceptable for women. Women can have sex without having to worry about the consequences in ways that had only been possible for men in the past. Thus, hooking up can be said to reflect women's gains in equality that have been achieved since the women's movement and the changing norms surrounding women's sexuality.

Is sexual convergence a good thing for women? For society as a whole? Can you think of disadvantages of women's sexuality becoming more like men's? Is there a value to a coupling of romantic love and intimacy with sex?

The final new pattern of intimacy between men and women in the United States is the "friends with benefits" phenomenon. Googling this phrase produces a proliferation of articles and Facebook groups, including quite a few advice columns that help their readers think through the advantages and disadvantages of this new type of relationship. One of the first studies conducted on friends with benefits, or the FWB phenomenon, was among Michigan State University students, and it indicated that 60% of the students reported having had at least one relationship that fit the criteria of friends with benefits (Carey, 2007).

Friends with benefits is generally defined as having sex in a nonromantic friendship. It is in many ways the realization of many of the tensions we discussed earlier in the chapter surrounding both cross-gender friendships among heterosexuals and same-gender friendships in gay communities. How exactly do you draw the line between a nonromantic relationship that qualifies as friends with benefits and someone you're dating, and therefore someone in whom you have a romantic interest? In the study at Michigan State, only one tenth of the friends with benefits relationships went on to become actual romantic relationships. In another third of the relationships, the sex eventually ended while the friendship continued. One in four of the relationships broke off both the friendship and the sex, while the rest of the respondents continued with both the sex and the friendship. Women and men in this study seemed motivated to enter into these relationships because they were perceived as a safer relationship, free of the level of commitment assumed in a typical romantic relationship. In most other ways, the relationships truly resembled friendships more than any other type of relationship.

Assuming that hooking up and friends with benefits are, in fact, the rising new trends in sexuality and intimacy in places like the United States, what exactly do these behaviors say about our changing culture? Certainly, they reflect changing norms around both sexuality and marriage in the United States. Although it is true that most people in the United States will get married, the number of people who are not married recently surpassed the number of people who *are* married in the population as a whole. It is also true that people in the United States are putting off marriage until a later age, although we will discuss these figures in historical perspective in Chapter 8.

If marriage is something that is presumed to happen later in life, if at all, for young people in college, then it makes sense that courtship activities among these groups are no longer really about courtship. Neither hooking up nor friends with benefits seem to be activities that are oriented toward the task of finding a life partner. This is especially true given the other piece of the puzzle: the increasing liberalization of sexual norms during the particular historical period, starting at the beginning of the 20th century, during which calling transformed into dating. Sexual convergence means the norms around the sexual behavior of women, specifically, have changed in the last 100 years, and the prohibitions against sex before marriage have generally loosened—although some abstinence movements seek to reverse this trend. Hooking up and friends with benefits are practices that seem to have few rules relative to the complicated and highly gendered systems of calling and dating. That these behaviors seem, at least for now, to have few gender implications may suggest a growing similarity in the sexual and romantic experiences of women and men.

What about love, though? Does romantic love have any place within hooking up and friends with benefits? The evidence so far suggests otherwise, but the evidence is still rather sparse and sometimes contradictory (Paul et al., 2000). Although anecdotal evidence suggests that friends with benefits is a practice that continues beyond college, it is unclear whether either of these patterns apply to many of the young people in the United States who do not attend college. With current research, it's still uncertain whether these trends apply to a great many colleges or just a few. Is hooking up more likely to occur at big campuses or small campuses? Does it happen at colleges and universities that are not primarily residential, with large populations of students living on campus?

One of the most important correlates of hookup behavior researchers have uncovered is the consumption of alcohol, and this is especially true for hookups that include sexual intercourse (Paul et al., 2000). This suggests that hookups would be less prevalent on campuses with a social life that is centered around activities beyond drinking (assuming a few such places exist in the United States). Perhaps most importantly, there is little research on whether these trends apply outside of college campuses at all, both for young people who are not in college and for older people, who also have romantic and sexual lives. Time will tell whether hookups and friends with benefits are here to stay, but it is also important to keep in mind, as we discussed previously, that many college students still have romantic relationships. And among heterosexuals, most people will end up married. Perhaps as that event happens increasingly later and seems less permanent than it has in the past, hooking up and friends with benefits become easy ways to fill the time spent in between with minimal hassle.

Radha and Krishna

Romeo and Juliet are two of the most famous lovers in the tradition of Western cultures, but Radha and Krishna embody love for many of the 837 million followers of Hinduism, making up 13% of the world's population. Unlike Romeo and Juliet, Radha and Krishna may or may not have actually gotten married. They may or may not have had sex, depending on which particular tradition in Hinduism you follow. And Radha, although she was in love with Krishna, may or may not have had a husband, a husband who was not Krishna. This story helps us think about the question of romantic love in cultures outside of the Western world because, very much unlike Romeo and Juliet, Radha and Krishna are both figures who are worshipped and venerated in Hindu society, and their love for each other has much greater religious significance than does the love of Romeo and Juliet.

Krishna is 1 of the 10 avatars of Vishnu, one of three gods who make up the Hindu trimurti, or a trinity—Vishnu, Shiva, and Brahma (Waterstone, 2005). An avatar is, roughly, an incarnation, and so Krishna is one particular manifestation of the god Vishnu. In his incarnation as Krishna, Vishnu is raised as a human (in other avatars, he's a boar, a fish, and a tortoise) and grows up among a group of female goat-herders called *gopis*. Radha and Krishna grow up together, and Krishna is depicted as seducing or, at the very least, inspiring some rather strong desire in all of the gopis, who are sometimes numbered as high as the thousands. But Radha is the most special among them, and it is the love between Krishna and Radha that is celebrated in Hindu art, literature, and popular culture, including contemporary Bollywood soap operas and films.

Is the love between Radha and Krishna that is celebrated through religious rituals, offerings, and festivals the same as the romantic love depicted between Romeo and Juliet? In Bollywood films, world famous for their depictions of romantic love, Radha and Krishna are used as models for the great, passionate love affair, passionate being defined in the world of Bollywood movies that do not depict kissing, let alone any of the more involved sexual acts displayed rather casually in American films. But their relationship is also viewed as a metaphor for the love between devotee and god, or of the quest for union with the divine. In fact, in many aspects of Hinduism as a religious system, passion between male and female, including sex itself, is perceived as deeply spiritual and connected to the divine union of male and female that is represented in the divine.

Romantic Love in Cross-cultural Perspective

The story of Radha and Krishna instructs us in two things to consider in looking at the relationship of cultures outside of Anglo-European societies to romantic love. First, romantic love certainly is an important cultural feature of societies outside of the Western world. But second, these traditions are, not surprisingly, somewhat different in the ways they depict romantic love. In the Christian religious traditions that currently dominate much of the United States, it is hard to imagine Jesus having a passionate love affair with one of the women in the Bible. In fact, much controversy has been raised by suggesting that such a thing ever happened. Romantic love in this view is largely incompatible with the spiritual world. But in Hindu traditions, romantic love can represent the highest forms of spiritual love, and love stories are important parts of religious teachings.

Many cultures outside of the developed world may have strong traditions, myths, and stories about romantic love. Scholars also argue that with the increasing influence of Western cultures through globalization, more and more cultures are beginning to follow and aspire to ideals of romantic love as they dominate in places like the United States. In places like Saudi Arabia, strict rules enforce an almost complete and total separation of women and men, and only recently among some more liberal Saudis are a bride and groom allowed to see each other or talk to each other before they are married (Slackman, 2008; Zoepf, 2008). Young people in Saudi Arabia generally support these strong religious strictures that make anything that vaguely resembles dating almost impossible.

But both young women and young men in these countries wonder about romantic love and how exactly it fits into their lives. Young girls debate whether it is appropriate under the strict Wahabi sect of Islam enforced in Saudi Arabia to talk to boys on Facebook, or even have them as Facebook friends, as well as wondering about what to talk to their future husband about during their one or two opportunities to speak with them before the wedding. Saudi boys have the theme song from *Titanic* as their cell phone ring and describe themselves as very romantic as they dangerously push the boundaries of acceptable behavior as it is defined between men and women by trying to get phone numbers from women they see in public. For a generation of young people who grew up watching Oprah and Dr. Phil, as well as seeing American movies like *Pride and Prejudice,* it becomes difficult to reconcile their own experiences with courtship with those they see depicted in popular culture.

How do other cultures mix their own views of romantic love with the particular systems of courtship and marriage that are native to their own society? Among the Igbo of Nigeria, the concept of romantic love is not a new tradition that has been imported from Anglo-European societies. Notions of romantic love have existed in Igbo society, although interestingly they were more likely to occur through institutionalized extramarital affairs. An old Igbo proverb notes that "Sweetness is deepest among lovers," suggesting that romantic love is more likely to exist outside of marriage than within. In his study of romance, courtship, and marriage among the Igbo, Daniel Jordan Smith (2001) noted that the important question for those researching romantic love in cross-cultural perspective is not whether such a thing exists, but how exactly romantic love is integrated into the fabric of social life.

Among the Igbo, marriage selection in the past was based on more of a collectivistic orientation. In individualistic societies, the needs of the individual outweigh the needs of any collective. The needs of the collective are perceived as more important than individual needs in a **collectivistic** culture, and in these types of societies, marriage is less likely to be a matter of individual choice. Marriage in collectivistic societies is viewed less as a coming together of two individuals based on free choice, and more as a strategic alliance that should provide some important benefits to larger groups, including families, larger kinship groups, tribes, or villages. As we'll explore in Chapter 8, this particular view of marriage is much more common both historically and cross-culturally than the more individualistic approach that dominates Anglo-European societies today. Since World War II, individual choice in marriage has become increasingly common among the Igbo. In addition, the importance of conjugality in marriage has also increased. **Conjugality** is a term used to describe the personal relationship between husband and wife, so an increasing emphasis on conjugality means that marriage is increasingly perceived as about this relationship—between husband and wife. What else would marriage be about, you might ask, as this seems to be the only thing in Anglo-European society? But in a collectivistic orientation, the personal relationship between husband and wife is secondary to the collective interests of larger groups. So an emphasis on conjugality is part of the transition away from collectivistic orientations toward a more individualistic view of marriage.

But the changes taking place within Igbo culture are not as simple as a straightforward transition from collectivistic to individualistic. While Igbo men and women increasingly choose their marriage partners and base these decisions primarily on romantic love and conjugality, they still do so within certain parameters that are set by the influence of older traditions. Smith (2001) told the story of Chinyere and Ike, two Igbo young people who met while Chinyere (an Igbo woman) was attending a teacher's college and Ike (an Igbo man) was a building contractor in the same town. Chinyere was educated, as well as beautiful, and chose Ike from among many potential suitors. Chinyere and Ike dated for two years, during which time they often went out for dinner and dancing, before deciding to marry. The couple went to social events together, where they were acknowledged as a couple, and started sleeping together before they married.

When the two agreed to get married, only then did they involve their families, in part to begin arranging the traditional Igbo marriage ceremony, the *igba nkwu*. Chinyere and Ike's story differs a great deal from those told by Igbo men and women older than 60 years of age, for whom mate selection involved their mother or father pointing to someone and saying, "That will be your spouse" (Smith, 2001, p. 134). Older Igbo who had married under this system had their difficulties, as many told stories of women who ran away to avoid marriage or men who married the woman of their own choosing in defiance of their parents. These stories demonstrate that romantic love did happen under the old system of marriage—it just wasn't always perceived as something that happened within marriage.

Contemporary Igbo couples choose their own mates, like Chinyere and Ike, but they do so within constraints. Traditional Igbo society dictates that marriages be between men and women from different villages as a way of fostering important alliances and connections among different groups of people. Thus, Igbo marriages were generally between a man and a woman from two different villages, but the two villages were generally close to each other and had longstanding ties between them. Marrying someone from a neighboring village helped to maintain peace between the two villages, created potential allies in the event of warfare with a third village, and facilitated networks of trade and other potentially profitable relationships (Smith, 2001).

With the increasing geographic mobility of young people like Chinyere and Ike in Igbo society, and the growing prevalence of individual choice and romantic love in marriage, there is no longer any guarantee that marriages will occur between closely neighboring villages. Families and village members have practical reasons to encourage these alliances between geographically close villages. At a distance, it is more difficult for the woman's family to make sure the marriage is going well, and the couples' families will not become as familiar and involved in the lives of children from such a marriage. This lack of involvement means that the wife's family will not as easily be able to benefit from any resources the children of the marriage could eventually provide. The family of the husband also feels potentially cut off from the newly formed family, as well as unable to appeal to their new relatives through marriage for support or assistance. Smith (2001) suggested that in Igboland, these boundaries of geography are stronger than those of ethnicity because it is generally more difficult for men and women to marry someone from a distant village than from a different ethnic group.

This doesn't mean that people from distant villages don't fall in love and decide on their own to get married. But family still has a strong role to play in the final steps of the process of getting married because, eventually, couples talk to their families and need their cooperation to plan and organize their *igba nwku,* or marriage ceremony. When a woman and man from distant villages approach their families, their families often try to talk them out of the match, usually by relying on stereotypes about the people in the distant village. Often, the woman's family in these situations exerts its control by repeatedly putting off the delegates from the man's family when they come to negotiate bridewealth, or the amount of money that the husband's family will pay to the wife's family. Families sometimes use this tactic of delay until the man gives up and withdraws his suit or the woman herself tells her suitor she's no longer interested.

Although individual choice, romantic love, and conjugality are increasingly important among the Igbo, there are still real constraints on the power of romantic love. At first, this may sound strange and foreign to those in cultures like the United States. But the idea of a man or woman's family being upset that their grandchildren will end up living very far away from them isn't really that hard to translate across cultures. A family in the United States may not fight quite as hard to prevent the marriage, but there are other constraints placed on romantic love in Anglo-European cultures. All the rules we discussed in the section on friendship apply to whom we court and love as well. We generally can't fall in love with someone whom we never have the opportunity to meet, and in our highly segregated society, this means we're less likely to fall in love with someone who is very different from us in terms of age, social class, race, or ethnicity. Romantic love, although we believe it transcends all boundaries, has real limits and barriers placed on it in all cultures.

THE GENDER OF LOVE

Having acknowledged that the ideal of romantic love may not be quite the *Titanic*-inspired, "I'm the king of the world," kind of experience for everyone, it becomes easier to observe how identities like gender can have an impact on something as personal as how we love. As with friendship, researchers have suggested that there are masculine and feminine styles of romantic love. These styles bring us back to our discussion of sex or gender role theory, which posed that the male gender role was characterized by instrumentality, while the female role was expressive. The instrumental masculine gender role is focused on accomplishing tasks, while women's expressive role is more oriented toward interacting

Theory Alert!
Sex Role Theory

with others. How does this play itself out when it comes to love? Researchers focused on long-term, heterosexual love in the contemporary United States have argued that there are masculine and feminine styles of love, as well. **Instrumental love** is expressed through doing things for the other person by providing material help and practical assistance. When a husband takes his wife's car to be serviced, this is a good example of instrumental love. **Affective love** is being emotionally and affectionately connected through communication and the sharing of feelings. A husband who sits down with his wife and tells her how his day was is an example of affective love. Can you guess at this point which style of love is generally perceived as masculine and which is generally perceived as feminine?

These differences line up pretty well with the masculine and feminine styles of friendship we discussed already. Just as men in their friendships are perceived as more likely to do things together, men in heterosexual relationships are perceived as more likely to provide assistance and help as ways of demonstrating love. Women who share more in their friendships also express love more through communication and conversation in their romantic relationships. As with friendship, love is typically defined more in terms of affective love, as a relationship that involves sharing and communication of feelings. In her own discussion of love and gender, Francesca M. Cancian (1986) argued that men get something of the short end of the stick in this formulation. Both among researchers and the general public, love is considered to be almost exclusively affective, and therefore, men's particular style of loving is not acknowledged. This can lead to a great deal of tension and misunderstanding within romantic, heterosexual relationships. Cancian gives the example of one man in a study who is encouraged to increase his affectionate behavior toward his wife. He goes out and washes her car and is surprised when neither the researcher nor his wife views that as an affectionate act (Cancian, 1986). Thus, many heterosexual couples' different love styles leave them working at cross-purposes in their relationships.

In explaining these apparent differences between feminine and masculine styles of love, many of the same explanations used for men's and women's friendship styles are applied to romantic relationships. Psychoanalytic theory is used to explain women's increased ease with emotional attachment and feeling connected relative to men. An additional explanation for masculine and feminine patterns in both friendship and romantic love takes more of a historical view, focusing attention on large-scale changes in society. This explanation relies on the historical development of the doctrine of separate spheres. Prior to widespread industrialization in places like the United States and Europe, the family was an integrated economic unit. The tasks of raising children and making a living all happened within the physical realm of the home. This was because a much larger percentage of the population were farmers, and on a farm, work is a family affair that all takes place within the realm of home. Even for those individuals who were not farmers, shops were generally run out of a living establishment; the blacksmith or printer generally located his shop in the downstairs of his home. Family members were expected to all work together as a productive unit. There was no separation between work and home in the way we now conceive of it.

This changed with industrialization when family members began to move off the farm and to take jobs in factories. Work became something that increasingly took place outside of the home, and although it was not initially or inevitably the case, paid work outside the home became something that men did. The doctrine of separate spheres physically separated women and men, but it also created a set of expectations about the gendered aspect of each place. The home became gendered in a way it had not been before, and women as the primary caretakers of this domain came to be associated with a whole set of characteristics considered to be important to the physical realm of the home. Women became the domestic experts, and they were expected to be good at emotional expression, love, nurturance, and compassion—the skills perceived as necessary to taking care of a family. Men moved into the world of factories and offices, where emotional expressiveness was not considered to be appropriate to this competitive, rational landscape. The division between men as instrumental and women as affective is therefore not a product of any developmental process that takes place in early childhood but of a structural rearrangement of the organization of the social lives of women and men. The doctrine of separate spheres encouraged women to become experts at emotions and expressiveness, while it taught men to shut those feelings down and emphasize rational competition.

If you believe the explanation that men's and women's styles of friendship and love are a result of the historical doctrine of separate spheres, how would you go about erasing these differences?

This historical take on men's and women's styles of intimacy provides a very different explanation for these patterns than one based in psychoanalytic theory, but it still predicts that with the separation of spheres, women will by and large be the romantic ones, while men will be focused more on rational matters. From this historical perspective, this particular situation may be a recent development, and it certainly does not describe the situation across many different cultures, but the separation of spheres still firmly crowns women as the winners in the game of romance. Is this actually true? Some research confirms these differences, as in findings that women place greater importance on sex occurring within the context of a caring, committed relationship (Carroll, Volk, & Hyde, 1985; Cohen & Shotland, 1996; Hill, 2002; Oliver & Hyde, 1993; Sprecher, 1989).

Another study found that men are more likely to fall in love based on qualities like physical appearance; men also say that they are more easily attracted to members of the opposite sex (Hatfield, 1983). But as with other areas of sex difference research, there is also evidence that men and women are generally looking for similar things in relationships. One survey of 738 people found the following similarities in men and women's approach to relationships: Both value the provision of support as a central element of close personal relationships; both value supportive communication skills in a wide range of relationships; both men and women agree on what counts as sensitive, helpful support; and both respond in similar ways to support efforts (Barnett & Rivers, 2009). Roughly the same percentage of women and men (61% and 62%, respectively) report resolving conflicts in their relationships by compromising with their partners, suggesting that men and women have similar patterns within relationships. One meta-analysis found that women do disclose slightly more than men in relationships, but the difference was small (Barnett & Rivers, 2009). As with many areas of sex difference research, your perspective depends a great deal on whether you're more interested in similarities or differences.

SUMMING UP

In the end, it's probably safe to say that almost all cultures have some basic idea of something like what we call love and intimacy. Indeed, it's difficult to imagine what a rewarding human life would look like without these important experiences. It's also safe to say that the particular ways in which different cultures define exactly what constitutes love and intimacy vary a great deal. The level of intimacy that is expected among a married couple in the United States may seem grossly inappropriate for a married couple in Saudi Arabia, where marriage is largely a political and economic arrangement that has little to do with the emotions of the couple.

In the United States, many people could probably come up with a kind of hierarchy of love and intimacy, ordering the types of love that are most and least important to their lives. Love for family might be most important, or closely tied with the romantic love one feels for a partner. Next might be the love one experiences for friends, or for extended family such as aunts and grandparents. This hierarchy would probably roughly correspond with levels of intimacy as well. But what we have learned is that this particular hierarchy would probably be unique to the culture of the United States, and not necessarily shared by those in other parts of the world. In traditional Indian society, the relationship between a mother and a son would probably have much greater importance than the relationship between a husband and a wife (Kanin, 1970). Among the Bangwa in the Cameroon, male friends perform funeral rites, and this friendship is trusted above the family, where relatives are perceived as scheming and self-interested in ways that friends are not (Brain, 1976). The ways in which a society provides for the basic human needs of love and intimacy vary a great deal, and it's not for us to say which particular system might be better or worse. They probably all have their advantages and disadvantages.

What we can also say is true about these patterns of love and intimacy is that they are deeply influenced by each culture's notions of gender. Friendship, for most of its history, was assumed to be a sex-segregated phenomenon, a relationship that made sense only between two women or two men, and this is still true in many parts of the world. This fact powerfully demonstrates for us the ways in which gender affects our everyday lives. Although the particular paths that lead individuals into institutions like marriage vary globally, some including love and some not, in all of these paths gender plays an important role. Although we might like to think of love as something that is untouched by the influences of culture or as something that exists in a realm beyond gender, or class, or race, this is simply not the case. The particular types of people for whom we feel affection, the way we express that affection, and the meaning of that affection are all deeply affected by gender.

BIG QUESTIONS

- The social network theory of gender as applied to friendship suggests that when women and men have similar networks, they will orient toward the world in similar ways, implying that gender will become less important. Some research on men's and women's friendships seems to bear this out. What evidence could you use to argue against this perspective? If you believe this theory of gender, how do you go about reducing gender inequality?
- In Chapter 5, we discussed the history of Anglo-European categories of sexuality and the emergence of the concepts of homosexuality and heterosexuality. How do this history and these concepts connect to our discussion in this chapter of the history of friendship in Anglo-European society? How are the two histories connected? What does this reveal about the connections between sex category, gender, sexuality, and relationships in general?
- From the perspective of many Anglo-European societies, it seems wrong not to be able to choose your own romantic partner or friends. Is it fair to judge these practices as wrong or unjust? If you take the perspective of collectivistic cultures, in which friends and romantic partners may be chosen for you, how can you see some possible advantages of this system?
- If you believe the research on feminine and masculine styles of love and follow it through to one logical conclusion, would it make more sense for people to be in romantic and sexual relationships with people of the same gender? If men have similar styles of intimacy, wouldn't romantic relationships between men be easier than heterosexual relationships? Why or why not?
- Each era in the history of courtship both shaped gender beliefs and was shaped by the gender beliefs of the time. What do the current practices of hooking up and friends with benefits reveal about our current gender beliefs? How are they shaping the way we think about gender today and in the future?

GENDER EXERCISES

1. Much of the research on the gender of friendships is contradictory. Try testing out some of the findings with your own survey or interviews on gender and friendship. Keep in mind Walker's (1994) findings about the differences between global questions about friendship and questions about behaviors within one specific friendship. What do your findings seem to suggest about gender and friendship? Are there important differences based on factors besides gender, such as age, social class, and race?
2. Social network theory of gender places importance on the different size of social networks among men and women in childhood and continuing into adulthood. Think about your own networks of friends when you were a child. Did the friend networks of boys seem larger than

the friend networks of girls? What about now? Make a diagram of your own friendship network, and then ask some of your friends to do a similar diagram. Include friends of both genders. How do these networks compare both in their size and in the type of people in the networks (kin or nonkin)?

3. Much of Beth Bailey's (1988) history of courtship in the United States is drawn from magazines and advice books. Do your own content analysis of popular magazines or dating advice books. What norms for courtship do they seem to prescribe? Are they different for women and men? What do these sources tell you about the directions courtship is going and the changes that are taking place?

4. What makes people attracted to each other, as we discussed in this chapter, is a difficult subject to study. Investigate some aspect of attractiveness for yourself. You might set up an experiment in which you show subjects pictures of different types of people and ask them which they find attractive. Or you might ask people to make a list of all the people they've been in some kind of relationship with (friendship, romantic, or sexual) and ask them to describe what was attractive about those people. Or you might look at descriptions in online dating sites like Match.com or HotorNot.com.

5. Dating websites like Match.com and eHarmony.com are increasingly popular ways of meeting people around the world. Do an Internet investigation of different kinds of dating sites and how they differ. For example, you might explore JDate.com, a dating site for Jewish people; IndianDating.com for people from India; or MuslimFriends.com for people of the Islamic faith. How many different kinds of dating sites can you find, and how do they differ in their representations of attractiveness, romance, dating, and gender?

TERMS

nuclear family

kin

sociability

friendship

"face-to-face"

"side-by-side"

social segregation

residential segregation

fictive kin

bridewealth

romantic love

individualistic

calling

rating-dating-mating complex

going steady

hookups

sexual convergence

friends with benefits

collectivistic

conjugality

instrumental love

affective love

SUGGESTED READINGS

Relationships in general

Allan, G. (1990). *Friendship: Developing a sociological perspective.* Boulder, CO: Westview Press.

Brain, R. (1976). *Friends and lovers.* New York, NY: Basic Books.

Hendrick, C. (1989). *Close relationships.* Newbury Park, CA: Sage.

Wood, J. (1986). *Gendered relationships: A reader.* Mountain View, CA: Mayfield Press.

Friendship

Nardi, P. M. (1991). *Men's friendships.* Newbury Park, CA: Sage.

Nardi, P. (1999). *Gay men's friendships.* Chicago, IL: University of Chicago Press.

O'Connor, P. (1992). *Friendships between women: A critical review.* New York, NY: Guilford Press.

Rubin, L. (1985). *Just friends: The role of friendship in our lives.* New York, NY: Harper and Row.

Walker, K. (1994). Men, women and friendship. *Gender & Society, 8* (2), 246–265.

Werking, K. (1997). *We're just good friends: Women and men in nonromantic relationships.* New York, NY: Guilford Press.

Courtship and romantic relationships

Bailey, B. (1988). *From front porch to back seat: Courtship in twentieth-century America.* Baltimore, MD: Johns Hopkins University Press.

Cancian, F. M. (1986). The feminization of love. *Signs, 11* (4), 692–709.

Dion, K., & Dion, K. (1993). Individualistic and collectivistic perspectives on gender and the cultural context of love and intimacy. *Journal of Social Issues, 49,* 53–69.

Jankowiak, W., & Fischer, E. (1992). A cross-cultural perspective on romantic love. *Ethnology, 31,* 149–155.

Kanin, E. T., Davidson, K. R., & Scheck, S. R. (1970). A research note on male-female differentials in the experience of heterosexual love. *Journal of Sex Research, 6* (1), 64–72.

Paul, E. L., & Hayes, K. A. (2002). The causalities of "casual" sex: A qualitative exploration of the phenomenonology of college student hookups. *Journal of Social and Personal Relationships, 19,* 639–661.

Paul, E. L., McManus, B., & Hayes, A. (2000). "Hookups": Characteristics and correlates of college students' spontaneous and anonymous sexual experiences. *Journal of Sex Research, 37,* 76–88.

Slacman, M. (2008, May 12). Young Saudis, vexed and entranced by love's rules. *New York Times.* Retrieved June 12, 2008, from http://www.nytimes.com/2008/05/12/world/middleeast/12saudi.html?_r = 1&ref = generation_faithful

Smith, D. J. (2001). Romance, parenthood, and gender in a modern African society. *Ethnology, 40,* 128–151.

Stone, L. (1988). Passionate attachments in the West: A historical perspective. In W. Gaylin & E. Person (Eds.), *Passionate attachments: Thinking about love* (pp. 15–26). New York, NY: The Free Press.

Turner, J. S. (2003). *Dating and sexuality in America.* Santa Barbara, CA: ABC-CLIO.

Waller, W. (1937). The rating and dating complex. *American Sociological Review, 2* (5), 727–734.

Zoepf, K. (2008, May 13). Love on girls' side of the Saudi divide. *New York Times.* Retrieved June 12, 2008, from http://www.nytimes.com/2008/05/13/world/middleeast/13girls.html?ref = generation_faithful

WORKS CITED

Allan, G. (1990). *Friendship: Developing a sociological perspective.* Boulder, CO: Westview Press.

Antonucci, T., & Akiyama, H. (1987). An examination of sex differences in social support in mid and late life. *Sex Roles, 17,* 737–749.

Bailey, B. (1988). *From front porch to back seat: Courtship in twentieth-century America.* Baltimore, MD: Johns Hopkins University Press.

Barnett, R., & Rivers, C. (2009). Men and women are from Earth. In E. Disch (Ed.), *Reconstructing gender: A multicultural anthology* (pp. 226–230). New York, NY: McGraw-Hill.

Birdsey, T. (n.d.). *Steak and BJ bay.* Retrieved April 28, 2009, from http://www.steakandbjday.com

Brain, R. (1976). *Friends and lovers.* New York, NY: Basic Books.

Brooks, D. (2001, April). The organization kid. *The Atlantic Monthly, 287,* 40–54.

Brown, L. S. (2005). Outwit, outlast, out-flirt? The women of reality TV. In E. Cole & J. H. Daniel (Eds.), *Featuring females: Feminist analyses of media* (pp. 71–84). Washington, DC: American Psychological Association.

Cancian, F. M. (1986). The feminization of love. *Signs, 11* (4), 692–709.

Carey, B. (2007, October 2). Friends with benefits, and stress, too. *New York Times.* Retrieved from http://www.nytimes.com/2007/10/02/health/02sex.html?scp = 1&sq = friends % 20with % 20benefits&st = cse

Carroll, J., Volk, K., & Hyde, J. (1985). Differences between males and females in motives for engaging in sexual intercourse. *Archives of Sexual Behavior, 14* (2), 131–139.

Cohen, L., & Shotland, R. (1996). Timing of first sexual intercourse in a relationship: Expectations, experiences and perceptions of others. *Journal of Sex Research, 33* (4), 291–299.

Cohen, T. F. (1992). Men's families, men's friends: A structural analysis of constraints on men's social ties. In P. M. Nardi (Ed.), *Men's friendships* (pp. 115–131). Newbury Park, CA: Sage.

Cohen, T. F. (2001). *Men and masculinity: A text reader.* New York, NY: Wadsworth.

Dion, K., & Dion, K. (1993). Individualistic and collectivistic perspectives on gender and the cultural context of love and intimacy. *Journal of Social Issues, 49,* 53–69.

Fischer, C. (1982). *To dwell among friends.* Chicago, IL: University of Chicago Press.

Fischer, C., & Phillips, S. (1982). Who is alone? Social characteristics of people with small networks. In L. Peplau & D. Perlman (Eds.), *Loneliness: A source-book of current theory, research and therapy* (pp. 21–39). New York, NY: Wiley-Interscience.

Galliano, G. (2003). *Gender: Crossing boundaries.* Belmont, CA: Thomson-Wadsworth.

Goldberg, H. (1976). *The hazards of being male: Surviving the myth of masculine privilege.* New York, NY: Nash.

Harris, S. R. (1997). Status inequality and close relationships: An integrative typology of bond-saving strategies. *Symbolic Interaction,* 20 (1), 1–20.

Hatfield, E. (1983). What do women and men want from love and sex? In E. Allegier & N. McCormick (Eds.), *Changing boundaries: Gender roles and sexual behavior* (pp. 106–134). Mountain View, CA: Mayfield Press.

Hawkes, G. (1996). *A sociology of sex and sexuality.* Maidenhead, England: Open University Press.

Hill, C. A. (2002). Gender, relationship stage, and sexual behavior: The importance of partner emotional investment within specific situations. *The Journal of Sex Research, 39,* 228–240.

Inman, C. (1986). Closeness in the doing: Male friendship. In J. Wood (Ed.), *Gendered relationships: A reader* (pp. 95–110). Mountain View, CA: Mayfield Press.

Jankowiak, W., & Fischer, E. (1992). A cross-cultural perspective on romantic love. *Ethnology, 39,* 149–155.

Johnson, F. (1996). Friendships among women: Closeness in dialogue. In J. Wood (Ed.), *Gendered relationships* (pp. 79–94). Mountain View, CA: Mayfield Press.

Kanin, E. T. (1970). A research note on male-female differentials in the experience of heterosexual love. *Journal of Sex Research, 6* (1), 64–72.

Kilmartin, C. (1994). *The masculine self.* New York, NY: Macmillan.

Lewin, T. (2003, May 20). One in five teenagers has sex before 15, study finds. *New York Times.* Retrieved from http://www.nytimes.com/2003/05/20/us/1-in-5-teenagers-has-sex-before-15-study-finds.html?scp = 1&sq = one % 20in % 20five % 20teenagers % 20has % 20sex % 20before % 2015&st = cse

Lewis, R. (1978). Emotional intimacy among men. *Journal of Social Issues, 34* (1), 108–121.

Nardi, P. (1992). "Seamless souls." An introduction to men's friendships. In P. M. Nardi (Ed.), *Men's friendships* (pp. 1–14). Newbury Park, CA: Sage.

Nardi, P. (1999). *Gay men's friendships.* Chicago, IL: University of Chicago Press.

O'Connor, P. (1992). *Friendships between women: A critical review.* New York, NY: Guilford Press.

Oliver, M., & Hyde, J. (1993). Gender differences in sexuality: A meta-analysis. *Psychological Bulletin, 114,* 29–51.

Owens, C. (1987). Outlaws: Gay men in feminism. In A. Jardine & P. Smith (Eds.), *Men in feminism* (pp. 219–232). New York, NY: Methuen.

Paul, E. L., & Hayes, K. A. (2002). The causalities of "casual" sex: A qualitative exploration of the phenomenon-ology of college student hookups. *Journal of Social and Personal Relationships, 19* (5), 639–661.

Paul, E. L., McManus, B., & Hayes, A. (2000). "Hookups": Characteristics and correlates of college students' spontaneous and anonymous sexual experiences. *Journal of Sex Research, 37,* 76–88.

Pleck, J. (1976). The male sex role: Definitions, problems, and sources of change. *Journal of Social Issues, 32,* 155–162.

Powers, E., & Bultena, G. (1976). Sex differences in intimate friendships of old age. *Journal of Marriage and the Family, 38,* 739–747.

Roberto, K., & Scott, J. (1989). Friendships in later life: Definitions and maintenance patterns. *International Journal of Aging and Human Development, 28,* 9–19.

Rodberg, S. (1999, May 6). *Culture front online.* Retrieved May 6, 1999, from http://www.culture-front.org

Rose, S. (1995). Women's friendships. In J. C. Chisler & A. H. Hemstreet (Eds.), *Variations on a theme: Diversity and the psychology of women* (pp. 79–105). New York: State University of New York Press.

Rubin, L. (1985). *Just friends: The role of friendship in our lives.* New York, NY: Harper and Row.

Seiden, A., & Bart, P. (1975). Woman to woman: Is sisterhood powerful? In N. Glazer-Malbin (Ed.), *Old family/ new family* (pp. 189–228). New York, NY: Van Nostrand.

Sherrod, D. (1989). The influence of gender on same-sex friendships. In C. Hendrick (Ed.), *Close relationships* (pp. 164–186). Newbury Park, CA: Sage.

Shulman, N. (1975). Life-cycle variations in patterns of close relationships. *Journal of Marriage and the Family, 37,* 813–821.

Slackman, M. (2008, May 12). Young Saudis, vexed and entranced by love's rules. *New York Times.* Retrieved June 12, 2008, from http://www.nytimes.com/2008/05/12/world/middleeast/12saudi.html?_r = 1&ref = generation_faithful

Smith, D. J. (2001). Romance, parenthood, and gender in a modern African society. *Ethnology, 40,* 128–151.

Sprecher, S. (1989). Premarital sexual standards for different categories of individuals. *The Journal of Sex Research, 26* (2), 232–248.

Tannen, D. (Director). (2001). *He said, she said: Gender, language, communication* [Motion Picture]. Los Angeles, CA: Into the Classroom Media.

Tognoli, J. (1980). Male friendship and intimacy across the life span. *Family Relations, 29* (3), 273–279.

Tschann, J. (1988). Self-disclosure in adult friendships: Gender and marital status differences. *Journal of Social and Personal Relationships, 5,* 65–81.

Turner, J. S. (2003). *Dating and sexuality in America.* Santa Barbara, CA: ABC-CLIO.

Vanderkam, L. (2001, July 25). Hookups starve the soul. *USA Today.* Retrieved September 2, 2001, from http:// www.usatoday.com/news/opinion/2001-07-26-ncguest1.htm

Walker, K. (1994). Men, women and friendship. *Gender & Society, 8* (2), 246–265.

Waller, W. (1937). The rating and dating complex. *American Sociological Review, 2* (5), 727–734.

Waterstone, R. (2005). *India: The cultural companion.* New York, NY: Barnes & Noble.

Wright, P. (1998). Toward an expanded orientation to the study of sex differences in friendship. In D. Canary & K. Dindia (Eds.), *Sex differences and similarities in communication* (pp. 41–63). Mahwah, NJ: Erlbaum.

Zoepf, K. (2008, May 13). Love on girls side of the Saudi divide. *New York Times.* Retrieved June 12, 2008, from http://www.nytimes.com/2008/05/13/world/middleeast/13girls.html?ref = generation_faithful

CHAPTER 7

How Does Gender Matter for How We Think About Our Bodies?

The Gender of Bodies and Health

What does it feel like to have a body? How do you think about your body? Do you like the way it looks and the way it feels? What would you change about your body if you could? How do you think gender affects the particular way you feel in your body? Is there a gender dimension to this division between the body and the mind, or the body and the soul? Does the idea of the body, in an abstract sense, have a gendered dimension to it? Who's supposed to feel closer to their bodies, women or men? Who's supposed to feel more comfortable in their bodies, men or women? Who seems to take better care of their bodies, and who seems unafraid, by and large, to put their bodies at risk? Is there a difference between the dangers posed to our bodies based on whether we're a man or a woman? How do you know the difference between a "good" body and a "bad" body, or between a healthy body and an ill body? How is gender a part of how we make those distinctions? How do those dangers and the feelings we have about our bodies intersect with all the other identities we occupy? Does gender play a part in who gets to say what happens to bodies, and how do other identities impact the amount of power we have over our own bodies? And finally, is there even such a thing as a body, or can we make the argument that the body itself is largely socially constructed, and how weird would that be to tell you that the very flesh and blood you occupy is a social construction?

These are some of the questions we'll ask in this chapter, which explores the gender of the body. As you can see through these questions, there's a lot to explore in relation to gender and the body. We've already begun to realize the ways in which bodies are important to gender. If you think about many of the controversial issues we've already touched on, bodies are often the terrain where these battles take place. In the case of intersexed individuals, their bodies serve as the battleground for debates about what sex and gender are. What aspect of one's body makes one male, and what aspect of one's body makes one female? Sexuality necessarily involves the body, as do the considerations of physical attractiveness that we touched on in relation to dating and love. Sociobiology and evolutionary psychology as theoretical approaches to gender firmly locate many of our gendered behaviors in the body, whether that means brain structure, hormones, or the content of our DNA. Issues related to the body are important throughout this book, and almost every topic related to gender can be explored through a focus on the body. In this chapter, we will explore several specific topics that bring our bodies into focus while acknowledging that bodies are important throughout our discussion of gender.

A BRIEF HISTORY OF BODIES

Most gender courses are likely to include some conversation about the body, and it's no coincidence that bodies come up often in the context of discussing gender. This frequent connection between gender and bodies has to do with what we might call a history of bodies and, specifically, with how that history plays out in many parts of the developed world, where very specific ways of thinking about the body have historically emerged. Many scholars have identified the mind–body dualism as one of the ideas most fundamental to Western thought, perhaps the very first of many dualisms that form the basic structure of our ways of thinking. The **mind–body dualism** is a belief, expressed in many different forms, in a split between the physical body and the nonmaterial entity we call mind (or spirit, soul, thought, etc.), where the mind is perceived as superior in many ways to the inferior body. This mind–body dualism is expressed in sayings like "The spirit is willing, but the flesh is weak." In this simple formulation, which is repeated in many shapes throughout Western cultures and dates back to Plato and Aristotle, the spirit represents the higher selves to which we aspire; the body is the force that prevents us, through its weakness, from achieving those higher goals. The mind–body dualism is a very old tension in Anglo-European cultures, one that is still being debated as neuroscientists try to locate many of the functions of the mind in the physical stuff of our brains—while other scientists argue that even the most complicated technologies for looking into the brain will never allow us to see the nonmaterial workings of the thing called "mind."

This dualism might be interesting enough in and of itself, and it has certainly generated libraries worth of scholarship and debate. It is interesting to our specific examination of gender because this fundamental dualism has its own gender dynamic. The mind is associated with masculinity; it is rational, aspires to the best efforts of the self, is closer to god, and is working toward ultimate self-realization (Bordo, 2003). The body is feminine and is the heavy drag on all the higher aspirations of the mind. It's difficult to say whether the particular male–female dualism came before the mind–body dualism or whether the gender aspect of the mind–body dualism was laid atop preexisting notions about gender differences. Dorothy Dinnerstein (1976) speculated that the gendered dynamic of the mind–body dualism can be traced to women's primary responsibility for taking care of human bodies as infants. Regardless of its genesis, the distinctly gendered nature of the mind–body dualism means that the negativity associated with the body is also associated with women and femininity. Like the body, women came to represent "distraction from knowledge, seduction away from God, capitulation to sexual desire, violence or aggression, failure of will, even death" (Bordo, 2003, p. 5). This is, needless to say, quite a lot of baggage to be located in one body.

The mind–body dualism positions women as more closely associated with their bodies than are men. Given all the negative associations of the body we just listed, this connection between women and their bodies raised important questions about the nature of bodies for feminists of the second wave in the global North. What are the effects of women's strong association with their bodies? What does it mean, exactly, to position women farther away from the mind and closer to the physical realm of the body? Certainly, important connections to sexuality emerge. As we discussed in Chapter 5, some societies view women as hypersexual beings, as temptresses whose bodies are made to lead men away from the path of spiritual or intellectual strivings. The Adam and Eve story in the Christian Bible, in most interpretations, is a classic example of this dynamic. Eve, after being led astray by Satan in the form of the serpent, tempts Adam with knowledge in the form of an apple from the Tree of Knowledge and, thus, in many traditions, is blamed for humanity's fall from grace. Eve's sexuality literally gets humanity thrown out of the Garden of Eden and, therefore, permanently removed from the presence of God. In what other ways does the mind–body dualism affect both men and women in relation to their bodies?

In your experience, does the gender of the mind–body dualism hold true? Are women perceived as closer to their bodies than men? What evidence might you use to contradict this closer connection between women and bodies?

Whose body is more important to them, men's bodies or women's bodies? A lot depends on exactly what we mean by important, but certainly in recent history in the United States, women have seemed to show more concern about their bodies and specifically about what we call body image, although men may be gaining ground. **Body image,** as we use the term here, involves the perception and evaluation of one's own bodily appearance. What does my body look like, and how do I feel about what my body looks like? Although the mind–body distinction positions the body as the less important part of the dualism, body image has been demonstrated to be connected to self-esteem and emotional stability, suggesting the ways in which our bodies can have important implications for the state of our minds (Wykes & Gunter, 2005).

The importance of our bodies to how we think and feel about ourselves is nothing new, and it is certainly nothing unique to Western cultures. All over the world, bodies reflect a person's status in his society and are used to project important identities. Among the Tiv of Nigeria, beauty is achieved among both women and men through scarification, or creating complex designs on the body through cutting it to create permanent scars (Burton, 2001). In the United States, we have surgery and invest in beauty products that promise to hide or remove the tiniest scars or imperfections on our bodies. Among the Tiv, these scars tell complicated stories about gender and the complex ways in which a particular Tiv individual fits into his or her patrilineal (descent through the male line) system of family and inheritance. On a Tiv woman's abdomen, straight lines and concentric circles represent the history of her lineage and the way it will continue to grow (Burton, 2001, p. 84).

Among the Tiv, scarification on a body tells about gender, age (many scars are added to women only after puberty), and how that person fits into the historical and present social structure of Tiv society. What does the absence of any scars or imperfections on a person in the United States tell us? A face free of scars in a place like the United States is probably a good sign of social class because the surgeries and products necessary to remove or reduce scars can be expensive. In addition, people in higher classes generally have jobs and lives that are healthier and less dangerous, therefore, reducing the chances that they will pick up scars and imperfections on their faces in the first place. Women are also more likely to be concerned about scars and blemishes on their faces than are men because the "rugged" (in moderation) is considered a sign of masculinity. Scars in the United States and scars among the Tiv tell us very different things, but in both places, they demonstrate how the body serves as an important marker of our social status.

What are examples in your own culture of the ways in which a person's status is marked on their bodies (not just gender, but statuses such as race, social class, ethnicity, age, nationality, or religious background)? How much can you tell about a person by looking at his or her bodies?

What about the person in the United States who, for whatever reason, cannot avoid having scars? In a society that promotes perfection in our faces and our bodies, the inability to live up to those standards necessarily affects the way an individual thinks about himself or herself. This deeply social aspect of bodies is how we understand the connection between our bodies, our self-esteem, and our emotions. This synergy between bodies and our image of ourselves has always been true historically and across a wide range of cultures. But when feminists first began their examination of the body in relation to gender, they began to argue that the impacts of the social norms about the body were much greater on women than they were on men. Perhaps as a result of the gendered nature of the mind–body dualism, feminists argue that the body is, in fact, one of the main methods used for controlling women and maintaining male power. This does not mean that gender norms don't also impact men and their bodies, as we will explore later. Feminists are not the only group interested in unpacking many of the dominant cultural ideas about bodies, but we will start with a feminist perspective and then explore other voices on body issues.

THE BEAUTY MYTH

One of the main ways feminists argue that the body is used to maintain the power imbalance between women and men is through ideals of beauty. This concern with how beauty ideals affected women dates back to some of the very first writings by women on the question of gender inequality. In 1792, Mary Wollstonecraft addressed the important role of beauty in women's subordination in her famous essay, "A Vindication of the Rights of Woman." Echoing later feminist scholars, Wollstonecraft wrote the following:

> Genteel women are, literally speaking, slaves to their bodies, and glory in their subjection . . . women are everywhere in this deplorable state. . . . Taught from their infancy that beauty is a woman's scepter, the mind shapes itself to the body, and, roaming round its gilt cage, only seeks to adorn its prison. (1792/1988, p. 82)

In 1914, one of the rights listed on the agenda of the first Feminist Mass Meeting in the United States was "[t]he right to ignore fashion." One of the first public demonstrations during the second wave of the feminist movement in the United States was a protest of the Miss America pageant, in part because of the ways in which the institution of beauty pageants enforces certain beauty ideals on women (Bordo, 2003). The concern with beauty as a weapon used against women is nothing new to feminist social movements or scholarship.

Is Mary Wollstonecraft's (1792) statement about women's relationship to their bodies still true of women today? Are women slaves to their bodies because of beauty ideals? Do women contribute to their own subjection through following the beauty ideal? Are women's bodies literally a cage for their own imprisonment?

A great deal of scholarship and a number of personal essays have explored the myriad ways beauty impacts women's lives. Naomi Wolf's (1991) book, *The Beauty Myth,* catalogues in great detail the ways in which cultural ideals of beauty create an unrealistic standard that is impossible for any real woman to achieve. The **beauty myth**, according to Wolf, is the belief in a quality called beauty that is real and universal and that women, as a result of biological, sexual, and evolutionary factors, should want to embody, while men should desire the women who embody that ideal of beauty. The important idea Wolf sets out to demonstrate in her book is that all of these ideas are part of a myth and are, therefore, not at all *actually* true. There is no universal and real measure of what beauty is because different cultures and historical time periods have completely different and often contradictory ideas about what makes someone beautiful.

In addition, there is no biological reason why women pursue beauty ideals; rather, as Wollstonecraft (1792) and other feminists have repeatedly argued, women pursue beauty because culture strongly reinforces and rewards this pursuit. The beauty myth is not true, but its existence is a powerful force in keeping women focused on the pursuit of beauty and in providing both women and men with a way to judge and limit women by their physical appearance. Beauty, Wolf (1991) contended, is politically and economically determined, and the beauty myth serves as the "last, best belief system that keeps male dominance intact" (p. 12). This is especially true, Wolf argued, since the second wave of the feminist movement, when the beauty myth became even more powerful as a kind of backlash against women's gains in power and equality with men.

Beauty and Gender Inequality

How exactly does the beauty myth become so important to upholding gender inequality? One way the beauty myth contributes to male dominance is by channeling much of women's energies toward the pursuit of beauty. When women fail to achieve the goals set up by the beauty myth, which they inevitably will (because there is no such thing as real, objective, and universal beauty), it takes a psychic and emotional toll on women everywhere. Most research demonstrates that women are more dissatisfied on the whole with their bodies than are men (Ambwani & Strauss, 2007; Cash & Henry, 1995; Garner, 1997; Muth & Cash, 1997) and that rates of dissatisfaction with bodies among both men and women are increasing over time (Feingold & Mazzella, 1998; Jefferson & Stake, 2009). In 1985, 30% of women in the United States said they were unhappy with their overall appearance, while by 1993 the figure had increased to 48%, or nearly one in two (Cash & Henry, 1995). Among college women, 80% report body dissatisfaction (Spitzer, Henderson, & Zivian, 1999), while 76.8% of adolescent girls report wanting to be thinner (Ricciardelli & McCabe, 2001).

Why do women seem to be becoming even less satisfied with the appearance of their bodies? Consistent with Wolf's (1991) concept of the beauty myth, the ideals about what exactly a perfect female body looks like are always changing. In just the past century, the ideal female body type in the United States has gone from the cinched-waist ideal of the 1900s to the flat-chested, straight-bodied flapper of the 1920s, to the full-chested, hourglass figure of the 1950s, to the skinny, yet large-breasted supermodels of today's fashion industry (Shields & Heinecken, 2002, p. xii). The body type most commonly depicted in today's popular culture is a body type attainable to roughly 4% of the female population—as these models are on average 9% taller and 16% thinner than the average woman (Jhally & Kilbourne, 2000; Zones, 2005). No amount of dieting or exercise will make that body type attainable for the other 96% of the female population, despite the fact that images of women with these body types are those with which both men and women are bombarded on a daily basis. Although more than half of *Vogue*'s readers wear a size 14 or larger, most of the images in the magazine are of models who are a size 6 or smaller (Zones, 2005).

The media and the images of beauty displayed in popular culture are part of the reason women are becoming increasingly less satisfied with their bodies. When the ideal body type held up for women is so far outside the normal range of women's bodies, the possibility of ever attaining that ideal becomes further and further out of reach for most women. When we consider the ways in which beauty is not just highly gendered, but also racialized, classed, and sexualized, achieving beauty becomes forever out of reach for most women in the United States, as well as around the world. Just as images of beauty have changed over time and vary across cultures, there has never been one dominant ideal of beauty in the United States. Dominant ideals of beauty always compete with those that exist in specific racial, ethnic, and class subcultures.

Yet although there may be different ways of thinking about beauty among Latinos or Italian Americans, these localized beauty ideals are always in competition with the dominant ideals that are legitimated through popular culture. In this dominant ideal, to be beautiful as a woman, you must also be white (or perhaps light-skinned at the very least), able-bodied, young, and thin (Zones, 2005). Women who exist in any way outside of these norms can never conform to this ideal, although there are a myriad of beauty products and advertisers who might attempt to convince them that they can. Women with disabilities are rarely ever held up in popular culture as images of beauty, and the experiences of many women with disabilities indicate that, with visible disabilities, the sign of disability often overwhelms any other impression individuals might have of a disabled person. One college student in a survey said, "I think the visual impact of a person sitting in a chair with wheels on it is so great as to render all other impressions, such as dress or grooming, virtually insignificant" (Kaiser, Freeman, & Wingate, 1984).

CULTURAL ARTIFACT 1: FAIR AND LOVELY—SKIN COLOR AND BEAUTY

What color is beautiful skin? For much of Anglo-European history, women aspired to have skin that was as pale as possible. Freckles, sun spots, a tan, or any sign of exposure to the sun was to be avoided at all costs. Think of Scarlett O'Hara in *Gone With the Wind* being cautioned to wear a hat and not expose her bosom to the afternoon sun. In the United States, being "tan" for white people became fashionable in the 1960s. Today, people with fair skin in the United States spend millions of dollars on tanning beds and other products to make their skin look darker. Ten percent of the U.S. population uses some kind of indoor tanning every year (Indoor Tanning Association, 2005). Despite the dangers posed by indoor and outdoor tanning, for many people in Anglo-European societies, darker skin (on white people) is beautiful. In India, women, and more recently men, spend millions of dollars to become lighter skinned. Products sold to lighten the skin make up 60% of skin care sales in India, bringing in $140 million a year for the maker of the most popular brand, Fair and Lovely (Leistikow, 2003). Recently, the same company introduced a skin lightener cream targeted for men, Fair and Handsome. In one commercial for Fair and Lovely, two women sit in a bedroom having an intimate conversation. The lighter skinned woman has a boyfriend and is happy, while the darker skinned woman lacks a boyfriend, and is not. The solution? Buy some Fair and Lovely to wash away the dark skin that's keeping the men from flocking. In another ad, a woman lands a well-paying job as a flight attendant to support her father after using the skin lightening cream. In a Fair and Handsome commercial, a man becomes a successful movie star, complete with a bevy of beautiful women, only after he uses the skin lightening cream. On a popular Indian talk show, a movie director suggests that darker skinned women are viewed as the women you take to a hotel, while the lighter skinned women you take home to your mother. This suggests that in places like India, skin color is believed to signal not just beauty but also sexual availability. How might you account for the variations in skin color and attractiveness across historical periods and cultures? How does this fit with the idea of the beauty myth? How are other ways of defining beauty also connected to assumptions about sexuality and to other identities like race and ethnicity?

Exporting the Beauty Myth

In regard to body type and the beauty myth, the ideal in the United States of thinness as the ultimate goal for all women has begun to spread globally, displacing beauty norms in many cultures that venerate and view as normal and desirable a body type that would now probably be considered overweight by U.S. standards. In 2001, a Nigerian woman won the Miss World title for the first time after the producer for the local Nigerian beauty pageant instructed his judges to pick a Nigerian representative to the Miss World pageant based on international, rather than local, standards of beauty. The difference? The ideal body shape for women for many Nigerians is "Coca-Cola bottle voluptuousness," and so the Nigerian winner of the Miss World pageant, who was 6 feet tall and skinny, was ironically not perceived as particularly beautiful by many local Nigerians.

Nigerians, especially those older than the age of 40, did not find their pageant winner particularly attractive. This is a country in which festivals are held to celebrate large women, and in the Niger region, women eat livestock feed or vitamins to bulk up, rather than slim down. Among certain groups in Nigeria, brides are sent to fattening farms before their weddings, where caretakers feed them huge amounts of food until the wedding, when the big brides are let out and paraded in the village square. But in the year since their Miss World pageant victory, and with the increasingly widespread reach of Western media images, ideals of feminine beauty have begun to shift in Nigeria. Thin girls like the Miss World winner are called *lepa,* a Yoruba word that means *thin,* but which has just recently been applied to people. Songs and movies are being made about *lepa* girls, and among many of the younger generation in Nigeria, thin is in and voluptuousness is out (Onishi, 2002).

A similar transition seems to be taking place in Brazil, where the ideal female body type of the past was guitar-shaped, which meant having an ample waste, hips, and bottom. Brazil's new international supermodel, Gisele Bundchen, is tall, thin, and busty—far from guitar-shaped. Young Brazilian girls used to play with Susi, a doll whose body reflected Brazilian norms. Susi was darker skinned than Barbie, her counterpart in the United States, as well as being fleshier in the hips, waist, and fanny. In the 1970s, Barbie arrived in Brazil, and by the mid-1980s, production of the Susi doll had ended, leaving Brazilian girls with the blond and skinny Barbie as a model of beauty. This shift in cultural norms is not without consequences; the percentage of the Brazilian population taking appetite suppressant pills nearly doubled between 2001 and 2005. Brazilians recently became concerned about eating disorders; in 2007, six young women died of anorexia in quick succession, among them a 21-year-old model (Rohter, 2007). In less urban areas of Brazil, the norms favoring fleshy women continue to be relatively strong, and researchers have demonstrated that Brazilian men by and large still prefer women who are *popozuda,* or fleshy in the rear.

The problems surfacing in places like Brazil in the form of eating disorders and the use of diet pills are not new to women in the United States. Cultural norms that venerate thinness as the ideal body type are important contributors to the prevalence of a wide range of eating disorders such as anorexia and bulimia in the United States. Both anorexia and bulimia affect women disproportionately over men, and men who do suffer from anorexia or bulimia tend to be involved in activities that have rigid weight requirements, such as dancing, modeling, and wrestling (Bordo, 2003). By now, most Americans are familiar with the signs and symptoms of anorexic and bulimic behavior, as well as the personality type and family backgrounds that may contribute to the development of these diseases. Both disorders emerged in the medical and psychological literature as recently as the 19th century, and for much of their history, they were disorders most likely to occur among upper- or middle-class white women.

Can the emphasis on other body types in places like Brazil and Africa also have negative implications for women? The emphasis on thinness is often linked to eating disorders. What negative implications might an emphasis on more full-figured body types for women have? Is the best goal to remove an ideal body type or just to change that type?

Hysteria and Eating Disorders

In her book, *Unbearable Weight,* Susan Bordo (2003) examined eating disorders such as anorexia through a cultural, feminist perspective that focused on the importance of the relationship between gender and bodies to understanding these diseases. She pointed to the importance of the historical and cultural context of the emergence of anorexia as a disease and drew parallels between anorexia and another predominantly female "disease" of the 19th century: hysteria. **Hysteria** was a disease that began to be diagnosed among primarily upper-class, and sometimes famous and prominent, white women in the United States and Europe in the late 19th century. The symptoms of hysteria included headaches, muscular aches, weakness, depression, menstrual difficulties, indigestion, as well as an overall fatigue and debility that was believed to require constant rest (Ehrenreich & English, 1978).

Among the prominent women who were diagnosed with hysteria during their lives were the writer and feminist Charlotte Perkins-Gilman, the urban reformer and activist Jane Addams, suffragette Elizabeth Cady Stanton, and the founder of Planned Parenthood and advocate for birth control, Margaret Sanger (Bordo, 2003). The primary treatment for hysteria, fictionally portrayed by Charlotte Perkins-Gilman in her famous short story, "The Yellow Wallpaper," was a regime of total rest and isolation from all of the very public activities in which many of these women were engaged. Perkins-Gilman's famous doctor, S. Weir Mitchell, who wrote a famous book outlining the new disorder, instructed Perkins-Gilman to "live as domestic a life as possible. Have your child with you all the time. . . . Lie down every hour every day after every meal. Have but two hours intellectual life a day. And never touch pen, brush or pencil as long as you live" (Bordo, 2003, p. 158).

The epidemic of hysteria among these middle- and upper-class women was explained physiologically as both originating in the uterus and as a result of the general frailty of specifically upper-class women. Although doctors by and large did not concern themselves with the health of working-class women or women of color at the time, their health situation was in actuality considerably worse than that of upper-class women. Nonetheless, the popular ideology of the time felt that these women, located farther down on the ladder of civilization and evolution, were better suited for their suffering, much of which doctors believed they had brought on themselves by having too many children. Upper-class women were by their nature frail and delicate, and hysteria was just one particular manifestation of this female disposition. Medical knowledge of the time also dealt in the **psychology of the uterus,** or the belief that women's whole persona was dictated by her uterus, and therefore, any medical or psychological problems encountered by women were traced back to some dysfunction in this particular organ. This explanation gives hysteria as a disorder its name because *hysteria* is the Greek word for uterus. Hysteria, then, was commonly conceived of as a disorder of the uterus.

Hysteria was considered to be a medically legitimate, widespread disease in the 19th century. Books were written about the disease, and many treatments were developed, ranging from removing women's ovaries to the complete rest and isolation suffered by Charlotte Perkins-Gilman to hydrotherapy, in which doctors used water sprayed directly onto their patient's vulva. Electrical stimulation was also commonly used to treat hysteria, and the first vibrators to be privately sold were marketed to women as a means of treating their hysteria at home, rather than paying for the same treatment in the doctor's office (Maines, 1999). How many women do you know today who have been diagnosed with hysteria? Unless you go to a very strange doctor, you've probably never heard of a woman being diagnosed as having hysteria, although you might have been told or heard someone else being told not to get *hysterical* or, in the more colloquial version, not to *have a hissy fit*. This language is what remains of the commonly recognized disease of hysteria, which most doctors, scientists, and scholars now look back on as nothing but a reflection of the particular social situation in which upper-class white women found themselves during the 19th century. Although hysteria was considered to be very much a real disease during this particular time period, it is now placed along with melancholia and being phlegmatic in the long list of diseases that are perceived as never having really existed.

Although the women who were diagnosed with hysteria were certainly suffering, the modern-day interpretation is that they were suffering more from their social and psychological conditions than they were from any underlying physiological condition. Hysteria is largely explained now as a reflection of the shifting gender norms of the period. On the one hand, middle- and upper-class women had been rendered largely useless except for their reproductive role. Charlotte Perkins-Gilman compared the affluent wife to the dodo as an evolutionary anomaly. She did not do any productive work outside of the home, and most of the tasks to be done within the household were done by servants. Her primary job was to give birth to the heirs of businessmen, lawyers, or professors (Ehrenreich & English, 1978). At the same time, some nondomestic work opportunities were opening up for women outside of the home (Bordo, 2003). More women were pursuing education at the new women's colleges, places such as Smith (which opened in 1875), Wellesley (1875), and Bryn Mawr (1885) or older institutions such as Cornell and Harvard, which began to open their doors to women (Ehrenreich & English, 1978). In 1848, the Seneca Falls Convention served to launch the suffrage movement in the United States, and in 1889, the Women's Franchise League was formed.

The particular group of women who were understood to suffer from hysteria were precisely those women who were poised to benefit from these opportunities, but they were also still expected to uphold the ideals of their class backgrounds—that a proper middle- or upper-class woman does nothing. Sigmund Freud also treated hysterics, as women suffering from hysteria were called, and published a whole book on the treatment of the disease. Although he acknowledged that the women he treated for hysteria tended to be "unusually intelligent, creative, energetic, independent, and, often, highly educated," as well as endowed with "gifts of the richest and most original kind," Freud didn't make the connection between the cultural restrictions placed on these talented women and their "afflictions"

(Bordo, 2003). Hysteria, many scholars argue, was a physiological manifestation of the social and psychological tensions experienced by these women.

What does the case of hysteria reveal about the social construction of disease, health, and medicine? Can you think of "new" diseases or disorders that have emerged recently? Did these diseases and disorders always exist and just hadn't been identified, or is there another possible explanation?

What does all this have to do with anorexia? Bordo (2003) argued that the emergence of anorexia was also grounded in the specific historical time period: the period during the 1950s and 1960s when Betty Friedan identified the feminine mystique and women of that generation had just been fired en masse from the jobs they had occupied during World War II. The new postwar ideology celebrated women back into their roles as wives and mothers at the same time that the second wave of the feminist movement was experiencing its very first signs of life. The images of ideal women during this time period were very different from what subsequent scholars have called the "tyranny of slenderness." Jane Russell, an actress of the time period, represented the fuller figure in fashion during a time that celebrated voluptuous, large-breasted women (Bordo, 2003).

Against these cultural backdrops, Bordo (2003) saw anorexia partly as a method of resisting these dominant ideals of femininity or as a response to the particular cultural pressures women began to face during the 1950s and 1960s, continuing into today's culture. Among the common themes in studies of anorexics is their reluctance to acquire fully female bodies through the experience of puberty. One anorexic explained her desire to stay thin in the following way: "I would never have to deal with having a woman's body; like Peter Pan I could stay a child forever" (Bordo, 2003, p. 155). As Bordo pointed out, this girl not only wants to keep her prepubescent body, but in choosing Peter Pan as an example, she expresses a desire to be a boy literally and have a boy's body. In other studies, anorexics recounted childhood fantasies and dreams of growing up to be boys. The family dynamics of many anorexic girls often included a submissive mother who gave up a career to care for her children and husband. Some researchers and therapists interpret anorexic women's resistance to becoming a grown woman as a fear of becoming their mothers. In this way, Bordo argues that anorexia must be understood as a disease firmly grounded in a culture that still sometimes provides limited options and contradictory messages for young girls about the proper way to be a woman.

Other scholars, including Bordo (2003), point to the difficulty in drawing the fine line between medically diagnosable disorders such as anorexia and bulimia and the way most women in places like the United States think about eating, exercise, health, and their bodies. An example from the medical literature is the case of BIDS, or **Body Image Distortion Syndrome** (Bordo, 2003). BIDS was first described as a "disturbance in size awareness" (Bordo, 2003, p. 55) and has been considered an indicator of anorexia both among clinicians and doctors responsible for making diagnoses and in popular culture depictions.

Body Image Distortion Syndrome is perceived as an important way of distinguishing between anorexia as a disease or disorder and what is "normal" in relation to how girls and women perceive their bodies. BIDS was conceived of as a visuospatial problem, rooted perhaps in impaired brain function or in defective processing of body experiences related to developmental problems. Either of these explanations made BIDS a clearly medical or psychological problem. The classic image of BIDS is the anorexic girl or woman looking in a mirror; the girl is emaciated and thin, but the image she sees projected back to her is grossly obese and overweight. BIDS, in this sense, prevents anorexics from correctly perceiving what their body looks like.

The problem with BIDS as a way of distinguishing between normal body perception and the distorted body images of anorexics is that large numbers of women seem unable to perceive their own bodies correctly. In a 1984 survey of 33,000 women conducted by *Glamour* magazine, 75% of women considered themselves too fat, when only one quarter of the women could be deemed overweight by standard weight tables (Bordo, 2003, p. 56). Thirty percent of the women in this survey, based on their reported height and weight, would actually be considered underweight, and yet they still saw themselves as too fat. Another study found that 95% of women among a total of 100 who were "free of eating-disorder symptoms" (Bordo, 2003, p. 56) overestimated their body size; on average, their own estimates were one fourth larger than their actual body size. These studies seem to indicate that significant proportions of women in the United States suffer from BIDS, even though the rates of anorexia are obviously not consistent with the numbers of women who demonstrate distorted body images.

CULTURAL ARTIFACT 2: GOOD HAIR, BAD HAIR, AND RACE

In 2012, Gabby Douglas, the "Flying Squirrel," became America's darling by helping the U.S. women's gymnastics team win their first team gold medal since 1996. So why did it seem like the only thing some people noticed was her hair (Marques, 2012)? On Twitter and other social media, fans seemed more interested in the state of Douglas's hair than they were in her ability to vault or tumble. Similar conversations take place around the hairstyles of first lady Michelle Obama. The conversation brought to the forefront long-standing debates about beauty, race, and the definition of "good hair." As Chris Rock explored in his 2009 documentary, "Good Hair," Black women's hair is a multimillion-dollar industry. Many African American women are comfortable with keeping their hair "natural," which often translates into a kinky or curly hairstyle. But in a society where definitions of beauty are deeply racialized, "good hair" is defined as straight hair—the kind of hair that most white women supposedly have. As Norell Giancana writes, "'The good hair' is what it is, an open secret, something we all recognize without a textbook definition. Intuitively, we know what it means, what it looks like, and who has it" (Giancana, 2005, p. 211). To get straight hair, many African American women must either treat their hair with chemicals or buy expensive hair exentensions. Women of color with "good hair" are often considered to be more attractive—by other men and women of color as well as by society in (Continued) general. Straight hair has meanings beyond attractiveness, as well. The particular combination of images used when Don Imus referred to the Rutgers University women's basketball team as "nappy-headed hos" is no coincidence (Desmond-Harris, 2009). Natural hair in Black women is associated with both sexuality and political beliefs. When *The New Yorker* magazine set out in a cartoon to depict the First Lady as a country-hating, Black radical, they drew her with an Afro. Thus, when Black women straighten their hair, they as perceived as more attractive, pure, and compliant. As you watch television, surf the Web, or go to the movies, pay attention to the hair of the women of color you see. Do any of them have natural Afros, or is their hair straight?

The Problem With Bodies

The argument Bordo (2003) and other researchers of body image, eating disorders, and gender are making is not that anorexia and bulimia do not exist as real problems with serious health consequences for increasing numbers of women. Rather, their purpose is to draw attention to anorexia and bulimia as one end of a continuum of problematic relationships women in the United States have with their bodies. Not every woman in the United States suffers from anorexia or bulimia, but surveys indicate that 84% of women have dieted to lose weight, compared with 58% of men (Stinson, 2001). One survey of young women in the United States between the ages of 13 and 18 found that 8% of the girls had vomited during the past year to lose weight, 2% had used diuretics, and 17% had taken diet pills (Wykes & Gunter, 2005). Another survey reported that 20% of young female college students claimed to have starved themselves to lose weight (Pyle, Neuman, Halvorson, & Mitchell, 1990). Are all these women anorexic or bulimic? Is that the most important question, or is it safe to say, as feminist scholars like Bordo would point out, that our culture makes women's relationship to eating and to their bodies a difficult one to navigate for all women?

Another attempt to draw this line between normal and abnormal behavior in women focuses on a more general concern with body image and is called **body dysmorphic disorder**, or BDD. Indications of BDD include "frequent mirror checking, excessive grooming, face picking and reassurance seeking" (Jeffreys, 2009, p.167). Applying excessive amounts of makeup, buying excessive amounts of hair products, as well as engaging in hair removal to excess are all signs of BDD. The question, then, is the following: What constitutes "excessive" behavior in these instances? How many hair products do you have to have in your shower or bathroom before you're in danger of being diagnosed with BDD? If you shave or wax your arms, as some women do, do you have BDD? Another guide is that people with BDD "actively think about their appearance for at least an hour a day" (Jeffreys, 2009, p. 167). Are we to believe that if we think about what's wrong with our appearance for just 30 or 45 minutes every day, we're okay and the extra 15 minutes makes all the difference?

The importance of beauty takes its toll on women in the form of actual eating disorders, as well as in the range of negative perceptions of their bodies and eating that characterize large numbers of the female population. Excessively cruel beauty practices such as foot-binding among upper-class Chinese women (in which women's feet were tightly wrapped in cloth from a very young age, distorting the growth of their feet and making it almost impossible for many women to walk) may have disappeared, but some feminists argue that makeup and other beauty regimes are still detrimental to women's health and well-being. Cosmetics are increasingly marketed to younger and younger girls, while the health effects of using makeup are not fully regulated by any government agency in the United States. One researcher found that a woman engaging in a typical daily beauty ritual exposes herself to 200 synthetic chemicals before breakfast, many of which have been identified as toxic by the U.S. Environmental Protection Agency. Studies have suggested a link between women who use hair dyes and chromosomal damage because hair dyes often contain carcinogens such as benzene, xylene, naphthalene, phenol, and creosol (Jeffreys, 2009).

Eyelids and Empowerment: Cosmetic Surgery

For some women, cosmetics, as toxic as some researchers suggest they might be, are still not enough to alter their appearance in ways that bring them in line with dominant ideals of beauty. For many scholars who study the beauty myth, as well as people in general, cosmetic surgery is considered to be a more extreme step in the alteration of one's body, an invasive procedure that is qualitatively different from exercising, dieting, hair styling, using makeup, or other techniques aimed at body alteration. Those who choose to have cosmetic surgery appear to many observers to have placed so much importance on their physical appearance as to put their very lives at risk by undergoing surgery (Gimlin, 2002). Nonetheless, much of the stigma surrounding cosmetic surgery will probably continue to fade thanks to the increasing numbers of women and men who elect to have cosmetic surgery. Plastic surgeons performed 2.2 million procedures in 1999, a 153% increase in procedures in the seven short years since 1992. The most common cosmetic surgery procedure is liposuction, and this procedure alone was performed 230,865 times in 1999; this was an increase of 264% since 1992, and it reflects the obsession not just with being thin, but with being free of any and all flab, softness, or imperfections on the body. Breast augmentation is the second most common cosmetic surgery, and thus, it's not surprising that fully 90% of cosmetic surgery operations are performed on women (Gimlin, 2002).

Although the practice of cosmetic surgery has certainly improved over time so that many procedures can be done on an outpatient basis, without ever having to spend the night in a hospital, any surgical procedure carries with it a set of risks. Cosmetic surgery is no exception. Liposuction can cause pain, numbness, bruising, and discoloration, sometimes for up to six months after the operation. Face-lifts can leave the recipient's face permanently numb. More serious potential complications from cosmetic surgery include fat embolisms, blood clots, fluid depletion, and potentially, death. Breast augmentation carries with it a 30% to 50% risk of serious side effects, including loss of feeling in the breasts, painful swelling or congestion of the breasts, and hardening of the breasts that make lying down or lifting the arms painful and difficult. Possibly the worse outcome from breast augmentation is encapsulation, in which the body reacts to the breast implants as foreign material and forms a capsule of fibrous tissue around the implants. In these cases, doctors must either massage the coverings away manually or remove the implants altogether; either process can be extremely painful to the patient (Gimlin, 2002). Those who argue that cosmetic surgery is qualitatively different from frequent dieting or obsessive exercising point to these facts about the dangers of cosmetic surgery.

Race and the Beauty Myth

Perhaps a figure that might be surprising to many people is the third most popular cosmetic surgery operation, blepharoplasty. Blepharoplasty is eyelid surgery, and this procedure was performed on 142,033 patients in 1999, 85% of them women (Gimlin, 2002). What exactly is eyelid surgery, and why is there such a high demand for this type of surgery? The procedure is marketed by plastic surgeons as reversing signs of aging and fatigue by helping patients get rid of the "droopy eyelid" look and the appearance of bags under the eyes. In this sense, any woman or man seeking to erase the signs of aging might elect to have this procedure, as part of the beauty ideal in the United States is someone who looks eternally young. But a specific type of blepharoplasty is called double eyelid surgery, or Asian blepharoplasty (American Society of Plastic Surgeons, n.d.; Kaw, 1993).

In her research on Asian American women and plastic surgery, Eugenia Kaw (1993) found that most of the Asian American women who sought plastic surgery at the specific site where she conducted her research were seeking double eyelid surgery. Statistics show that in 1990, 20% of all individuals seeking out cosmetic surgery were Asian American, Latino, or African American. What exactly is double eyelid surgery? Many people of certain Asian ancestries are born with a different anatomical structure to their eyelids that results in what some refer to as being *single-lidded* rather than *double-lidded,* as many people of non-Asian ancestry are. This anatomical difference is what makes the eyes of many (but not all) Asian individuals look different from what is considered normal among white people in the United States. Note that we should be careful about defining double-lidded as anything approaching normal; 1 in 4 people on the planet are Chinese or of Chinese ancestry, which means that huge shares of the world population are likely to be single-lidded, and this is only including Asians of Chinese ancestry.

Nonetheless, in the United States, whites and whiteness continue to be important axes for defining the norm, regardless of figures projecting that in the near future, white people will no longer be the statistical majority in the United States. This means that double-lidded eyelids are perceived as the norm, and as desirable by many Asian American women. Double eyelid surgery creates this double-lidded look for Asian American men and women. The Asian American women in Kaw's study used double eyelid surgery and other types of cosmetic surgery to improve their appearance, but as Kaw pointed out, there is a very specifically racial dimension to these women's pursuit of beauty. When white women elect for liposuction or a face-lift, they are seeking to conform to norms of beauty that center around thinness or age, but they are not trying to alter features that are specific to their racial background. This is because in the United States, white features are still generally perceived as the standard against which beauty is measured.

That many Asian American women would then want to make themselves appear more like white women is not particularly surprising against this cultural background. How do these women make sense of their decisions? In Kaw's (1993) in-depth interviews with Asian American women, she found that all of the women were proud to be Asian American and made no claims to having any desire to look white. Instead, respondents talked about the negative connotations of the facial features that are typically associated with many Asian faces. A 21-year-old Chinese American woman considered double eyelid surgery to avoid the stereotype of the "Oriental bookworm" who is "dull and doesn't know how to have fun" (Kaw, 1993, p. 79). Another woman in Kaw's study talked about how when she looks at other Asian American women who have not had eyelid surgery, she sees their eyes as "slanted and closed" and thinks about how much better they would look with their eyes more "awake" (Kaw, 1993, p. 79).

Kaw argued that these women were articulating many of the stereotypes that are associated with Asian Americans in general, and Asian American women more specifically. Asian American women are perceived as docile, passive, slow, and unemotional, and the women in Kaw's study came to associate their characteristically Asian features with these negative qualities. By removing the visible symbols of these qualities from their faces, they are increasing their chances for success in a society that generally does not value these qualities. One woman who had double eyelid surgery explained that she would encourage her daughter to have the surgery as well because she believes this will help her daughter in the future in finding better jobs. Because beauty is associated with success for many women, becoming more beautiful by the dominant standards is a means to improving their chances for these women.

Kaw (1993) also pointed out that the cosmetic surgeons whom these women consult about their surgery largely confirm these negative associations between characteristically Asian facial features and personality traits. These doctors add the power of the medical institution to their "diagnoses," creating a sense that having this set of facial features is in itself "abnormal" and a medical problem that needs to be corrected. In the following quote, Dr. Gee, one of the cosmetic surgeons interviewed by Kaw, described his sense of the importance of double eyelid surgery for Asian Americans: "I would say 90% of people look better with double eyelids. It makes the eye look more spiritually alive. . . . With a single eyelid frequently they would have a little fat pad underneath [which] can half bury the eye and so the eye looks small and unenergetic" (Kaw, 1993, p. 81). When these beliefs are presented to Asian American patients in the official medical setting of the doctor's office, complete with lab coat and technical equipment for measuring facial features and imagining postoperative appearances, these views of a woman's facial features as in need of correction take on a kind of objective truth.

In the consumer-oriented society of the United States, both the doctors and patients in Kaw's (1993) study argue that Asian American women who elect to have cosmetic surgery are actually asserting their agency and independence in their decision to alter the appearance of their face to conform more closely to white ideals. Cosmetic surgery in this interpretation is not an attempt to conform to any set standards of beauty, but rather a way for individuals to use their bodies and the latest technology to sculpt their flesh in unique ways that reflect their own personality and desire. This perspective raises an interesting argument within the literature on beauty and bodies. Early in feminist explorations of the relationship between gender and bodies, the body was considered to be one of the primary tools of patriarchy. The beauty myth outlined by Naomi Wolf (1991) was perceived as a method for maintaining gender inequality. Keeping women focused on their appearances was a clear way to keep women in their place and, therefore, to stifle efforts by women to increase their status and standing in society.

Can you think of other ways in which the pursuit of beauty and attractiveness can carry risks or disadvantages, or other ways in which people (women and men) might hurt themselves in an effort to be attractive—historically, cross-culturally, or in your own culture?

Is Beauty Power?

But as feminist critiques evolved, and as the second wave of feminism gave way to the shifts of the third wave, more feminists began to argue that the pursuit of beauty could in some ways be considered a means of empowerment for women, rather than just a primary tool of their oppression (Bordo, 2003; Jeffreys, 2009). In her book, *Lipstick Proviso,* Karen Lehrman (1997) argued that women should not have to sacrifice their femininity to achieve gender equality, and those aspects of femininity that should be maintained include the right to wear lipstick and short skirts. There are several ways in which to view beauty practices as a means of affirming women rather than contributing to their subordination, and the first of these we will discuss focuses on the ways in which beauty practices can be important foundations on which a unique women's culture and female bonding is built.

Women sharing tips and advice about how to remove body hair can create unique spaces for the expression of what are primarily feminine concerns and issues. In her interviews with young women about their experiences with bodily changes at puberty, Karin A. Martin (1996) discovered one of the few positive aspects of puberty for many young girls was learning how to shave. This was partly because of the bonds between girls and older women, whether mothers or older sisters, that learning how to shave created. In her study of dieting culture, Kandi Stinson (2001) noted the co-optation of feminist messages in the world of weight loss programs, which are composed primarily of women and emphasize control over eating and weight loss as one means of female empowerment. Women in these settings bond together in their collective quest for weight loss. In our earlier discussion of women's friendships, we noted that working out and doing aerobics was a common activity women share together. All of these instances, some feminists argue, suggest that the pursuit of beauty can serve as a source of cohesion and bonding for women as a group.

The second kind of argument made in support of beauty as being about more than just women's subordination draws attention to the potentially creative and subversive nature of many beauty practices. For example, a brief history of makeup in the 20th century reveals that initially the production, marketing, and selling of makeup was largely controlled by women (Jeffreys, 2009). These businesses served as an entrance for women into the predominantly masculine world of the paid economy and allowed early female entrepreneurs to employ other women. Many of the central figures in the early history of the beauty industry in the United States were working-class women, ethnic women, and women of color. These women were able to take control of the politics of appearance and funnel the profits from their efforts back into their own communities. Women involved in the early history of the beauty industry were able to exert a great deal of control over the ideals of beauty and femininity and to make the pursuit of beauty through consumption a respectable endeavor. Prior to this period, *makeup* in the United States was more commonly called *paint* and was associated with prostitutes and women in the theater, neither of whom were considered to be appropriate models for proper women.

Although the beauty industry itself was gradually taken over by large corporations run primarily by men in the 1930s, some historians argue that the spread of makeup usage helped enable women in the early 20th century to begin entering public life in large numbers. As we discussed in Chapter 6 in relation to dating, public life was relatively off limits to "respectable" middle- and upper-class women in Europe and the United States until the latter parts of the 19th century and the beginning of the 20th century. Job markets were also opening up for middle-class women in this period through occupations such as office worker and teacher. The growing beauty industry helped women move into the public world of face-to-face interactions and into a new marriage market that emphasized sexual freedom. Kathy Peiss (1998), a historian who focuses on this early history of makeup in the United States, wrote the following about the women of this period:

> Moving into public life, they staked a claim to public attention, demanded that others look. This was not a fashion dictated by Parisian or other authorities, but a new mode of feminine self-presentation, a tiny yet resonant sign of a larger cultural contest over women's identity. (p. 55)

From this point of view, makeup was a tool with which women armed themselves to encounter the new world of public life.

In her book, Susan Bordo (2003) argued that bodies in cultures like those of the United States are increasingly viewed as plastic. In a wide variety of ways, people in places like the United States can use their bodies like canvases on which to reflect whatever unique manifestation of their personality or beliefs they choose to express. Is this newfound ability a realization of our increasing freedoms, or just a new evolution of ways in which we feel compelled to make our bodies conform to cultural norms and ideals? Research suggests that some women who have cosmetic surgery are, in fact, more satisfied with their bodies and their lives after having the surgery (Gimlin, 2002). Liz Frost (1999) argued that "doing looks" for women can be a potential source of pleasure and power, even positively contributing to women's overall mental health. Regardless of the dangers discussed previously, does the improvement in self-esteem outweigh those risks? If Asian American women really do increase their prospects for career success by altering certain facial features to look more stereotypically white, are their decisions not about increasing their sense of empowerment? These are important questions to explore, and they begin to suggest the difficulty of making the simple mind–body distinction that seems so basic to many modes of thinking. If altering our bodies does in fact make us better people or people who are better able to navigate the world successfully, can any kind of body alteration really be wrong?

PILLS AND POWER TOOLS: MEN AND BODY IMAGE

When we focus on body issues in places like the contemporary United States, the mind–body dualism seems to handicap women in relation to their bodies in ways that, perhaps until recently, were not as true for men. We will consider in a moment how many of the pressures about bodily appearance experienced by women have their equivalencies among men, but it is important to remember that across the globe, male bodies are also used as markers of gender, status, and group membership. In Nazi Germany, circumcision as a way of marking group membership could have life or death consequences. For men, being Jewish was marked by circumcision of the penis, and historically this has often been an important marker of Jewish identity. In the story of a historical encounter between Australian explorers and a previously "undiscovered" group in New Guinea, the explorers were not recognized as humans because of their light skin but also because all the men in this particular group covered their penises with a woven sheath and decorated their nostrils with pig tusks; from their perspective, other humans who lacked these characteristics were clearly not human, and certainly not men (Nagel, 2003).

That both gender and other kinds of status are inscribed on men's bodies, just as they are on women's, is not subject to much debate. The question that arises in a discussion of gender and bodies is whether there are important differences in the ways in which gender is inscribed on men's bodies as compared with women's bodies. This is a debate that commonly emerges in my gender courses when we enter into our discussion of body image, eating disorders, and the beauty myth. Men in the classroom, many of whom can recount their own lifetime of struggles with issues of being overweight or underweight, and therefore too fat or too skinny to be considered appropriately masculine, point out that they face their own set of issues in relation to bodies. Although the mind–body dualism clearly equates masculinity with the world of the mind, we all know that men have physical bodies. How does the mind–body dualism affect the relationships that men have to their bodies in ways that are different from those faced by women? And how are some of the body issues faced by men very much similar to those faced by women?

Appearance and physical attractiveness, although they may be expressed differently based on gender, are important determinants of the way people react to us and therefore how we feel about ourselves. Those who are considered to be more physically attractive, regardless of whether they are male or female, are believed to possess more socially desirable personality traits and are expected to lead happier lives (Zones, 2005). This has important impacts across a wide range of social situations because research demonstrates that cute babies are cuddled more than less attractive babies, attractive toddlers are punished with less frequency, teachers pay more attention to better looking students, and jurors are more sympathetic to attractive victims in the courtroom. For those who are fortunate enough to be considered attractive, gliding through a world in which people generally show you more affection, expect more of you, and endow you with many positive qualities creates a self-fulfilling prophecy; the end result is that people who are considered to be attractive do, in fact, tend to be more socially competent and accomplished (Zones, 2005).

The Importance of Being Tall

Attractiveness is important for both genders, but it is often perceived differently for women and men. Height can be an especially important component of attractiveness for men, in ways that may not be the case for women. Height is important in modern, Anglo-European cultures like the United States in the sense that tallness in men is perceived as more attractive than being short. Studies have demonstrated that taller men are more likely to be trusted and that tall men enjoy an advantage in hiring and promotions relative to their shorter counterparts (Gascaly & Borges, 1979; Gieske, 2000). Height has often been demonstrated to be viewed as a sign of leadership potential, including its potential importance in places like West Point to promotion through military ranks (Mazur, Mazur, & Keating, 1984; Stogdill, 1974). Other research demonstrates that height is a consistent marker of status and rank across many different cultures, and it is one aspect of a person's body we assess in our first encounter with a new person. Erving Goffman (1988) noted the importance of height in the positioning of figures in advertisements; a person positioned above someone else, regardless of whether that person was actually taller, was a clear indication of someone with power.

Theory Alert! Doing Gender Theory

Given these associations, the way in which height becomes an important component in heterosexual coupling is not really all that surprising. From the doing gender perspective, the strong cultural norm among heterosexual couples dictating that the man should be taller than his female counterpart is a prop that allows men and women to perform gender across many different contexts and situations. Specifically, these height differences allow heterosexual couples to act out gender dominance; if height indicates power and status, it makes sense that, in heterosexual couplings, men should be the taller partner. It is true that on average, men are slightly taller than women. But when we speak of averages, we must keep in mind that this does not mean that a significant number of women are not, in fact, taller than a significant number of men. It is also true, as an interesting side note, that the average differences between the heights of men and women have been decreasing historically according to the records we have available (Schiebinger, 2000). Although our height is determined in part by genetics, nutrition is also an important factor in determining eventual height; some scholars speculate that women are closing the height gap because girl children in many places around the world are no longer deprived of valuable food resources that in the past would have been given to boys instead. As of now, men still maintain a slight height advantage on average over women, but this in itself does not lead to the logical outcome that all tall men should pair themselves with shorter women. Although it's difficult for us to imagine, the norms could just as easily be for tall women to pair off with shorter men.

In fact, that men should tower over women has not always been the norm in Anglo-European cultures. The reaction for many people of seeing a heterosexual couple whose heights' violate these assumed norms is similar in some ways to the reaction to interracial couples, or other couples who clearly violate a basic norm we have about the proper way of pairing off. These couples seem to represent a fundamental reversal of our basic gender assumptions; the taller woman is attempting to "be the man" by dominating her partner, while the man is relinquishing his power, but also demonstrating his specialness as someone who was "chosen" despite his height. In the 18th century among the European aristocracy, there were no norms governing the height of heterosexual couples. A woman could be taller than her husband in her natural state, but even if she wasn't, the fashion norms of the time often made women appear to tower over their mates. The fashion among aristocratic women included high heels and sometimes towering wigs, and both of these tended to give women a distinct height advantage over the men accompanying them.

What are other gender norms for heterosexual couples like the height differential? Are there other appearance differentials (regarding weight, hair color, attractiveness, etc.), social class differentials, racial differentials, or age differentials? What do these other norms say about gender and heterosexual relationships?

Sabine Gieske (2000), in her history of height and coupling, argued that these gender norms changed because of shifting ideas about social class in the 18th century. During this period in Europe, the middle class was on the rise, and the middle class sought to define themselves in opposition to the aristocracy. Gender distinctions among the aristocracy were less important than class distinctions, and aristocratic women were often able to obtain a certain level of financial independence, education, and political influence. As the emerging middle class defined a new set of values, the differences between women and men became important components of this new culture. As such, cartoons and caricatures from this transitional period showed the height and body differences between men and women among the aristocracy as dangerous to the social order and unnatural. The bodily differences between women and men, previously considered unremarkable, became the focus of science and popular writing. The small differences in stature between women and men became important justifications for a relatively new set of gender relations, which considered women to be the "weaker sex." Advice from experts on the proper height matching between couples continued to be contradictory, and up until 1889, scientists and scholars wrote that the choice of a spouse should be largely independent of stature (Gieske, 2000). By the 20th century, the norm of a taller man with a shorter woman was firmly entrenched, thus, enacting predominant notions of men's physical superiority. This historical case is an interesting example of intersectionality, as we discussed in Chapter 2. Remember that intersectionality directs our attention to the ways in which gender intersects with other identities such as social class, race, ethnicity, and nationality. Here, gender differences are used as a tool to create and reinforce important class differences in 19th-century Europe.

Theory Alert!
Intersectionality

In today's societies, these assumed height differentials can create problems for "vertically challenged" men, as well as for "vertically advantaged (or disadvantaged, all depending on your perspective)" women. There are real advantages for men in being taller, both in terms of attracting potential mates and in the sense of the benefits that accrue to those who are viewed as attractive in general. As early as the 18th century, various tools promised men the ability to either appear or actually become taller. Historians of gender and the body note that the top hat, popular in the 19th century, was one relatively simple means of allowing men to hide any height deficiencies (Gieske, 2000). Other advertisers promised muscle and ligament stretching techniques in the 19th century, while today hormones are sold for some of the same purposes.

Naked Bodies

Surely the problem of being taller than your partner is less of a concern for gay men and lesbians who are not necessarily enacting dominance in the same way as heterosexual couples. But what other set of body issues might these groups face? Much of the literature and research on men and body images involves gay men and body images within gay communities. In her book on men and body images, *The Male Body,* Susan Bordo (1999) argued that the increasing number of images in advertisements that offer up men as sexual objects are probably less a result of advertisers' desire to appeal to women and more a result of their desire to sell their products to gay men. Recently, heterosexual men who demonstrate a level of concern with their appearance that is considered outside the norm for masculinity have been labeled **metrosexuals**. This name evokes the urban context (metro) where some men are more invested in their looks, but it also suggests a strong link between concern with one's appearance and a man's sexuality. Is metrosexual meant to suggest that the sexuality of a man who cares about his appearance is somehow different from the sexuality of a "normal" heterosexual man? Or is the term meant to suggest that a metrosexual acts like homosexual men in his concern for what he looks like, but he is distinctly heterosexual? What is this connection between men's body images and sexuality about?

It is difficult in general to separate out issues of sexuality from a discussion of the body, and it is certainly the case that sex is at play in the relationship between women and their bodies. Part of the lingering effects of the mind–body dualism is to perceive that women, who are more associated with the body end of the dualism, are more sexualized because of their closer connection to their bodies. For those living in modern, Anglo-European cultures, it is much easier to perceive a naked woman's body as a sexual object than it is a man's, and this is true regardless of one's gender. Women in the United States do not escape the overall tendency toward evaluating and judging women's bodies, including their own bodies and the bodies of other women. What this means is that all of us, women and men, gay and straight, are accustomed to looking at women's bodies as sexual objects, regardless of whether we actually have any desire or experience having sex with women.

The history of men's bodies being held up as objects of desire and images of beauty has a complicated history in places within the Anglo-European cultural tradition. Under the one-sex model of sex differences, a model that persisted well into the 18th century, men's bodies were considered to be the ideal representation of beauty and perfection. As we discussed in Chapter 6, this one-sex model originated with ancient Greece and held that rather than two separate and distinct sexes of male and female, females were a lesser and inferior version of males. Obviously, the ancient Greeks knew something of anatomy and were aware of the differences in the bodies of men and women, but they saw all of women's sexual organs as internalized counterparts to male organs. Into the 18th century, what we call women's ovaries were still called testes, reflecting the belief that ovaries were, in fact, internalized versions of men's external testes. The vagina, according to this line of thinking, was considered to be an inverted penis.

Within the hierarchy of the one-sex system, it was the male body that was venerated and held up as representing the ultimate ideal of beauty. This is reflected in much of Greek art, as well as in the origins of Olympic sport. In the gymnasium, the location where male, Greek citizens gathered to practice sports, all the men were naked, and the first Olympic events were as much about admiration for the male form as they were about actual athletic competition. Athletes competed naked and rubbed olive oil on their bodies, much as today's body builders do, to emphasize and accentuate the beauty of their physical form. Yet when you think of an art class today sitting down to paint a nude figure, whose nude figure do you more commonly imagine them using? Probably a woman's body, and this tendency seems to reflect a general belief in the superior beauty of the female form. How did we make this almost complete turnaround in the gender of beauty, and how are we starting to move in a very different direction more recently?

Beginning as early as the 15th century in Europe, artists began to use live women as artistic models (Bernstein, 1992). These women were often female servants who performed double duty as sexual partners for the painters. Painters used female models in their private studios, but most of the art academies did not provide female models until as late as the 19th century. Since roughly the 19th century, the female nude has become perhaps the preeminent subject for generations of artists. French impressionist painters such as Édouard Manet, Pierre-Auguste Renoir, and Edgar Degas took up the subject of the female nude, as did post-Impressionists such as Henri Matisse. Pablo Picasso continued the tradition in the modernist, Cubist style, and Georges-Pierre Seurat drew pointillist female nudes. By the 20th century, anyone describing a collection of "nudes" was generally presumed to be describing a collection of nude women, rather than nude men.

That female bodies were increasingly considered to be appropriate subjects for artists is partly explained by the mind–body dualism to which we continually return. Art beginning in the Renaissance increasingly aspired to portray life as it really was, rather than the stylized depictions popular during the Middle Ages. This included representations of nature in the form of landscapes, a development that coincided with the rise of science and empirical observation during the Renaissance. Leonardo da Vinci, as one of the earliest Renaissance thinkers, provides a good example of how science and art were fused; da Vinci used his anatomical examinations of the human body to aid his art, while his art, in the form of sketches and paintings, also helped to illustrate his scientific discoveries. Because in the mind–body dualism, women are equated with the body, and therefore perceived as closer to nature; therefore, the difference between painting natural landscapes and painting women's bodies is negligible. Both are considered objects of beauty worthy of being depicted in art.

The Rules of Manhood

Feminist scholars and art historians during the second wave feminist movement in the United States and Europe began to identify how these cultural traditions in Anglo-European society have conditioned us to be very comfortable with looking at the nude female body, or at female bodies in general. Although many feminists have raised interesting questions about the boundaries that are drawn between female nudity in high art and female nudity in pornography, both create a culture of women and men who are well accustomed to looking at naked women. Feminist film critic Laura Mulvey (1975) identified this phenomenon in the context of film as the **male gaze**. Mulvey argued that in films, women are generally considered to be objects of the gaze (of both the audience and characters in the film) because control of the camera is shaped by the assumption of a heterosexual, male audience. The male gaze, then, helps describe the ways in which women become sexual objects in film and other media (Mulvey, 1975). Until more recent trends in art and advertising, the same has not been true of naked men. In the 1970s in the United States, artists, including many gay male artists, began to put together exhibitions that included fully nude males. The response of many art critics to these artistic works shows how far away the Western art world has moved from the veneration of the male body by the Greeks; one art critic wrote in *The New York Times*, "Nude women seem to be in their natural state; men, for some reason, merely look undressed. . . . When is a nude not a nude? When it is a male?" (Bordo, 1999, p. 178). Another critic wrote that, "there is something disconcerting about the sight of a man's naked body being presented as a sexual object" (Bordo, 1999, p. 178).

For Bordo (1999) and Mulvey (1975), these statements reflect the unfamiliarity and newness of the naked male body to modern viewers, and especially depictions of men's bodies that seem to portray men as sexual objects; the male gaze is heterosexual and therefore objectifies women's bodies, not men's. As we will discuss in more depth in Chapter 10, women's bodies are ubiquitously used as sexual objects to sell a wide range of products and services across a broad range of popular culture venues. Although scantily clad men are not completely unknown in popular culture, and certainly the history of male sex symbols goes back to the earliest history of the film industry in the United States, Bordo (1999) argued that it is only recently that men have begun to be depicted specifically as sex objects in ways similar to women's depictions. The difficulty to be overcome in perceiving men as sexual objects has to do with the ways in which hegemonic masculinities are constructed. One crucial theme of the hegemonic masculine role in the United States as identified by Brannon (1976) is "**No Sissy Stuff**," indicating that real men don't do anything that carries with it the least suggestion of femininity. Demonstrating too much concern for your appearance or allowing yourself to be admired for your appearance are both activities or positions that contain that suggestion of femininity. Allowing oneself to be admired and gazed on is especially dangerous for men because of its suggestion of vulnerability. Being the person who is judged based on appearance is to give power to the person doing the judging. This is especially true when the gaze and judgments have a sexual component to them. Because being vulnerable and passive are qualities associated with femininity, these activities are considered threatening to masculinity.

Theory Alert! Hegemonic Masculinities

Are there situations in which men's bodies are admired without the threat of vulnerability, passivity, and femininity? What about male body builders or male athletes in general? How are they perceived as different from male models?

Bordo (1999) demonstrated these dominant views about masculinity by drawing on a particular advertising campaign used by Haggar to sell men's pants. One print ad for this campaign depicted a man in boxer shorts with the copy, "I'm damn well gonna wear what I want. Honey, what do I want?" (Bordo, 1999, p. 194). This ad attempted to sell men's fashion to men while emphasizing that men aren't really that good at making fashion decisions largely because they're not really concerned with or interested in fashion. The ad, like others in the campaign, firmly established a gender dichotomy between fashion-conscious women (the "honey" who will help the man figure out what to wear) and fashion-clueless men, as well as reinforced the heterosexuality of the man (who had a presumably female "honey" to consult). A television commercial in the Haggar campaign reinforced this message of hegemonic masculinity. It began with a man sleepily pulling on a pair of Haggar pants and then going outside to get the paper. The basic script for the commercial, as described by Bordo (1999), read as follows:

> "I am not what I wear. I'm not a pair of pants, or a shirt." (He then walks by his wife, handing her the front section of the paper.) "I'm not in touch with my inner child. I don't read poetry, and I'm not politically correct." (He goes down a hall, and his kid snatches the comics from him.) "I'm just a guy, and I don't have time to think about what I wear, because I've got a lot of important guy things to do." (Left with only the sports section of the paper, he heads for the bathroom.) "One-hundred-per-cent-cotton-wrinkle-free khaki pants that don't require a lot of thought. Haggar. Stuff you can wear." (p. 195)

The voice-over featured the voice of John Goodman, who played the large and gruff husband on *The Roseanne Show,* and it sends a clear message about hegemonic masculinities. Real heterosexual men cannot be bothered to think about what they wear, and they are therefore certainly not aspiring to become sexual objects for women or other men to gaze at.

Calvin Klein's controversial ad campaigns for his jeans and line of underwear were some of the first popular culture images to question these formulations of hegemonic masculinity. Klein's innovative advertising style echoed much of the art, and especially photography, of many gay male artists emerging during the 1970s, and his advertisements were directly influenced by his experiences in the gay male community in New York during the 1970s. At the Flamingo, a popular gay male club in New York City, Calvin Klein observed a "vision of shirtless young men with hardened torsos, all in blue jeans, top button opened, a whisper of hair from the belly button disappearing into the denim pants" that inspired his own line of jeans and later, underwear (Bordo, 1999, p. 180). Although Jockey had broken ground in the depiction of men's bodies by introducing an advertisement with baseball player Jim Palmer in a pair of briefs, Klein took the imagery of scantily clad men to a whole new level in 1981.

Calvin Klein placed a 40 foot by 50 foot photograph of an Olympic pole-vaulter, Tom Hintinauss, on a billboard in New York City. Unlike previous underwear advertisements, including the Jockey ad with Palmer where the underwear was airbrushed to remove any indication of underlying genitalia, the outline of Hintinauss's considerable penis was clearly visible through the cloth of his briefs. The Calvin Klein ad suggested no fictional reason for why he was being depicted in his underwear, as in previous advertisements that suggested men were captured in the middle of dressing or undressing. Hintinauss is clearly just on display, perhaps sunbathing, but leaning backward in a passive stance that was new to most advertising images of men. Most previous advertisements selling men's clothing generally depicted men—clothed and not—in a pose Bordo (1999) described as **"face-off masculinity"** (p. 186).

In this pose, men look directly into the camera and at the viewer, conveying a sense of being "powerful, armored, emotionally impenetrable" (p. 186). In these ads, the male models stare down their audience in a way that is reminiscent of the hostility and power of looking at someone that is familiar in the animal world and across a wide range of human cultures. The body posture of these men is often aggressive and powerful as well, with their legs planted firmly and far apart and their hands on their hips. Despite the fact that large parts of the men's bodies are exposed, the ads are still somewhat traditional depictions of masculinity in that they fall in well with a second theme of hegemonic masculinity, **"Be a Sturdy Oak"** (David & Brannon, 1976). Face-off masculinity is consistent with this theme of masculinity as it reinforces the idea that men should never show vulnerability or weakness, even when they are in a potentially vulnerable position (being without their clothes). Men are calm and reliable in a crisis and certainly are not supposed to show their emotions.

The advertisements for many Calvin Klein products for men, beginning with the Hintinauss ad described previously, began to depict men who violated the face-off masculinity of many previous depictions. Men in these advertisements did not look into the camera to stare down the person looking at their picture. In one of the first ads of this type, the man's eyes were averted, his hair falling into his eyes. As Bordo (1999) pointed out, the man in the advertisement was not in the typical wide, aggressive stance, but he had his hip cocked in an S-curve typical of the way women were posed in ads. His eyes were downcast, and although he was not naked (the ad is, after all, supposed to be for underwear), the outline of his penis is again clearly visible through his underwear. The ad seems to say, "Feast on me, I'm here to be looked at, my body is for your eyes," and this is a somewhat unprecedented message for a man's body to send.

In discussing Calvin Klein's underwear ads and the shifts in the depictions of male nudity in general, Bordo (1999) had two main points to make related to our discussion of gender and bodies. The first was that, at least as masculinity has been constructed in contemporary times in the United States, it has been difficult for men's bodies to be common objects of people's gazes as well as sexual objects. Regarding the underwear ad mentioned previously, Bordo pointed out that part of the difficulty in men being perceived as sexual objects is that these men are then subject to being judged by someone else, and although women are accustomed to being judged in this way, it is a sign of weakness and passivity not usually associated with masculinity. To look at and judge someone's appearance is active, and therefore masculine, as with the male gaze. To be the object of that gaze and judgment is passive and therefore feminine. Bordo described this tendency with a phrase that captures a visual rule in both paintings and contemporary advertisements: "men act and women appear" (p. 196). Femininity is structured such that women are supposed to enjoy being looked at and admired, to seek out situations in which they are looked at and admired, and to gear their behaviors and attitudes toward an assumption that they *will* be looked at and admired. Men, in both their artistic depictions and in real life, are the ones doing the looking and admiring, or whatever other type of action is called for in the situation.

Masculinity, then, is structured in ways that tend to discourage the objectification of male bodies. Movies such as *The Full Monty* and *Boogie Nights,* which include some male nudity, are partially about the anxieties that men experience when they become the objects of gazing eyes. The flip side of this equation, as Bordo (1999) pointed out, is that women themselves are not used to viewing men's bodies as sex objects. Research from the famous Kinsey Institute on male and female sexual responses to depictions of nudes shows that 54% of men were erotically aroused by nudity versus 12% of women (Bordo, 1999, p. 177). This statistic, which indicates that men are as much as four times more likely to be aroused by nudity than women, is roughly analogous on a large scale to the pornography industry, which caters almost exclusively to men.

Nude pictures of men that appeared in the magazines geared toward female audiences, *Viva* and *Playgirl,* were eventually dropped from *Viva* because the readers didn't much like them and the editor of the magazine herself said she found them "slightly disgusting" (Bordo, 1999, p. 177). Such studies are used to suggest that women are biologically hard-wired not to be particularly aroused by nude women or men. Bordo counters that perhaps culturally, the use of men's bodies as sex objects is so rare that women have simply not learned to respond to them sexually. Many heterosexual women do get turned on by nude male bodies, with their own lists of particular parts that get them going. A more recent study suggests that changes might be in the making; 30% of women between the ages of 18 and 44 said they found watching a partner undress to be very appealing, while only 19% of women aged 45 to 59 did so. Men are still more likely to enjoy a partner's strip tease (50% and 40%, respectively, among men), but the age gap suggests younger women are becoming more accustomed to doing the gazing rather than being the object of the gaze.

GENDER AND HEALTH: RISKY MASCULINITY AND THE SUPERMAN

That masculinity is constructed in ways that make demonstrating any vulnerability difficult has implications beyond the objectification of male bodies. Men may generally possess more of a sense of agency in their bodies, but this doesn't necessarily mean that gender can't have dangerous effects on the ways in which men relate to their bodies. Don Sabo (2009) argued that traditional ideas of masculinity can, in fact, pose a real danger to men's health. When you consider the mortality of men as compared with women, men are actually the weaker sex in a wide variety of ways. From the womb, men's chances of dying during the prenatal stage of development are about 12% greater than women's, and during the neonatal (newborn) state, men's chances of dying are 130% greater than women's. This is in large part a result of a host of neonatal disorders that are common to male babies but not to females. The gap between male and female infant mortality rates has closed over time, but a gap still exists that favors female infants. Most people are probably aware that women, on average, live longer than men in most societies across the globe. Worldwide, the life expectancy for males is 62.7, while for females, the figure is 66 years, giving women worldwide three extra years of life on average (Rosenburg, 2007). In places like the United States, the gap in life expectancy between women and men widened for much of the 20th century. Women's life expectancy increased because of the decrease in maternal mortality, or fewer women dying during or after childbirth—once a risky endeavor for many women. Men's life expectancy decreased during the same time period because of increases in deaths from lung disease and coronary heart disease, both a result of men's higher rates of smoking. In the latter part of the 20th century, the gap seemed to be closing again between how long women and men could expect to live. This narrowing is again more a result of social and cultural factors because women have begun to catch up with men in their rates of smoking.

Thinking about life expectancies begins to reveal some of the underlying explanations for these differences in women and men's overall health, as well as demonstrates the incredible power of gender in our lives. We've already explored how gender is important for a wide range of behaviors and attitudes, but here we see how gender becomes a part of bodies, ultimately determining both the quality and the length of the gendered lives we lead. Men around the world live shorter lives than women in part because of the choices they make, like choosing to smoke. Why would men on average be more likely to smoke than women? The answer for many who study masculinity and health helps to explain more than just men's relatively higher rates of smoking. Many hegemonic masculinities, as they are constructed in the developed world, encourage risk-taking behaviors.

This certainly varies across the many different types of masculinities that exist across the wide spectrum of society, but the information on death and other health-related behaviors seems to indicate that masculinity is an important force leading men to take more risks with their bodies on average than women. For example, morbidity, as opposed to mortality, addresses the prevalence of disease, illness, and injury. Women on average are more likely to suffer from a host of chronic illnesses than are men, a point we will return to in a moment (Sabo, 2009). These chronic illnesses do not necessarily lead to death but just to a potentially uncomfortable life. The area of morbidity where men do exceed women is in their tendency to sustain injuries. What do smoking and being more prone to being injured have in common? Both can be considered risk-taking behaviors. Men are more likely to be injured through their attempts to demonstrate masculinity in reckless ways; through their higher levels of involvement in contact sports like American football, rugby, and boxing; and as a result of the types of jobs many men are more likely to work, especially working-class men, whose jobs can put them in a great deal of danger.

In interviews with men from working-class backgrounds in Australia, R.W. Connell (1995/2005) documented the high incidence of violence in these men's encounters with power in many forms. The stories these men told seemed, to Connell's eyes, filled with fights, beatings, brawls, assaults, bashings, and sometimes knives. The men were both the perpetrators and the victims of violence, whether being caned by cruel teachers in their schools or beaten by their fathers or other family members (including sisters). These experiences led to a general code regarding violence among these men, namely, that violence is OK when it is justified. The main way in which violence was always justified was when the other man started it. Among these working-class young men, their violent encounters with the authority of school figures eventually led to hostility toward other authority figures in general, including the police. When these boys left school, some eventually found themselves in jail and encountered the violence that is inherent in those institutional settings as well. Among the other risky behaviors these men engaged in were speeding in cars, trucks, and bikes (three of the young men owned motorbikes and another two were passionate about biking). At least one of the young men Connell interviewed had also been involved in a police chase, complete with a roadblock and a serious crash.

Can you think of other examples of risky behaviors that men seem more likely to participate in than women? Is this true of all men? What types of men do you think are more or less likely to be risk takers? What kinds of things do you think would make women more likely to be risk takers?

Is Masculinity Bad for Your Health?

Maybe these sound like men you know, and maybe not. Certainly not all men lead lives quite as dangerous and as filled with violence as the group of men Connell (1995/2005) interviewed. But some of the most basic masculine activities for the "typical" man in the United States carry with them a great deal of potential danger and bodily harm, including the all-American activity of playing football (American football rather than soccer). American football is not *the* most dangerous sport. Those would be boxing and auto racing, both sports that also predominantly involve men. Some boxers are killed from injuries during or immediately after a match, while others, like Muhammad Ali, have their health and lifespan compromised by the long-term effects of competing. But far fewer men in the United States box or auto race than the numbers who play football at many competitive levels, and many would argue that football has replaced baseball as *the* American sport. It is the most profitable professional sport in the world, and its prime-time games in the United States win the ratings competition over every other sport (Ozanian, 2007). One research study estimated that an average of 13 high-school football players die in the United States every year, half from injury and half from overexertion. Other estimates put the numbers as high as an average of 40 deaths every year. Every year, football produces approximately 30 catastrophic injuries such as permanent brain damage or paralysis and a grand total of 600,000 other types of injuries. In an average college or professional football season, usually at least one player suffers an injury that leads to paralysis (Kilmartin, 1994). Football is a fairly dangerous sport, and yet for many men in the United States, it represents the very epitome of what it means to be a man.

In regard to the connection between masculinity and men's bodies, this veneration of football is no coincidence. Football players are expected to sacrifice their bodies for victory and the team, to feel no pain. Players frequently play through injuries, and they violate an essential code if they use an injury as an excuse for a poor performance. At the core of the culture of American football is a belief in the value of bodily sacrifice and the denial of any desire for self-preservation and safety (Kilmartin, 1994). The masculine body exists to be put at risk, and this may be one of the core ways in which gender norms contribute to health outcomes for men. Other risky behaviors in which men are more likely to engage than women include habitual drinking to excess, drunk driving, high-speed driving, drug dealing, sharing hypodermic needles, using firearms, engaging in gang violence, and working in dangerous jobs (Kilmartin, 1994). Among those arrested for alcohol and drug-abuse violations, more than 90% are men.

All of these statistics demonstrate an important contradiction about the relationship between many men and their bodies. Relative to women, they may experience a much greater degree of agency in their bodies. Football, dangerous and violent as it may be, surely can also provide a feeling of empowerment and accomplishment. But masculinity also seems to encourage men to act in their bodies in ways that frequently place their bodies at great risk of physical harm. Certainly the most extreme examples of these truths relate to the rather grim statistics regarding homicide and suicide among men. Although women in the United States are more likely to attempt suicide than men, men make successful suicide attempts three times more often than do women (Kilmartin, 1994). *Successful* here somewhat ironically refers to suicide attempts that actually lead to death, and men are better at obtaining this outcome in large part as a result of the more violent means they employ in attempting to end their lives. Men are more likely to use guns or to hang themselves, as opposed to women's more common use of pills (Sabo, 2009). Homicide is the second leading cause of death among men between the ages of 15 and 19 in the United States, and men between the ages of 15 and 34 make up almost half of all homicide victims in the United States (Sabo, 2009). Many studies have demonstrated that men are also disproportionately the perpetrators of homicide; one study of 600 murders in the United States found that men were the perpetrators in 95% of the cases (Barash, 2002).

One way of understanding these risk-taking, aggressive, and sometimes violent behaviors in men is through another essential feature of Anglo-European hegemonic masculinity as outlined by Robert Brannon (1976). The first two we already touched on, "No Sissy Stuff" and "Be a Sturdy Oak." A third component of manhood as defined by Brannon is "Give 'em Hell," and it is certainly well suited to explaining many of the behaviors we've discussed in relation to masculinity. **"Give 'em Hell"** dictates that real men exude an aura of "manly daring and aggression," encouraging men to "go for it" and "take risks" (David and Brannon, 1976, p. 199). This aspect of the dominant way of defining manhood is positive in that it certainly can encourage acts of physical bravery and courage. But as we've seen, this imperative of manhood can also pose great risks to men's overall health.

Can you think of other ways in which the "Give 'em Hell" imperative is demonstrated for men? What other norms or expectations serve as examples of the "Give 'em Hell" ideal?

The way in which this Give 'em Hell ideology contributes to men's increased levels of violence directed toward themselves and others is not surprising given the way that connections between masculinity and violence are institutionalized in many different cultures. The most common institutionalization of male violence is through the military and war. Although men's frequent obligation to perform military service can also be considered a means of controlling male bodies, men also gain great prestige from going to war if they survive; their service reinforces the masculine norm that men's bodies are meant to be risked and sacrificed in the service of some crucial set of values. Theodore Roosevelt made this connection between masculinity, sports, and war explicit. Roosevelt encouraged the spread of "manly" sports such as football because he was concerned men were becoming physically feeble under industrialization. Men working office jobs would not be able to defend our country come wartime because of the inactivity of their lifestyles. Football could help men continue to develop not just the physical abilities necessary for warfare but the principles of toughness and self-sacrifice as well.

Dangerous Masculinity in Palestine

In a particularly interesting example of the way in which male bodies are considered to be material to be sacrificed for the larger goal of establishing masculinity and protecting some core cultural values (usually patriotism or nationalism in the case of military service), Julie Peteet (1994) explored how among Palestinians during the Intifada (Palestinian term for an uprising or *shaking off*) of 1987, being beaten, imprisoned, or tortured by Israeli soldiers became a new kind of ritual transition into manhood for many young Palestinian men. In Israel and the occupied territories, control is exercised by the Israel government in large part through control of Palestinian bodies, often through the use of myriad checkpoints that Palestinians must negotiate on a daily basis. In her ethnography among Palestinians in Israel, Peteet found that during the period of this particular intifada, beatings became much more public, and most Palestinian men had either been beaten themselves or knew someone who had been beaten. Although the Israeli soldiers saw the beating of Palestinians as a way to control and end the uprising, the meaning given to being beaten or imprisoned among Palestinians themselves was quite different.

The evidence of violence on the bodies of Palestinian men is used as evidence, especially for Western eyes, of a history of subordination and powerlessness at the hands of Israelis but also, and importantly, of the Palestinians' continued resistance to that subordination within their continuing nationalist struggle. Peteet demonstrated this interpretation of physical violence through the instance of one Palestinian mother who brought her son in to show the scars on his body from being beaten and then shot by Israeli soldiers when he got up to run away after being accused of throwing rocks at soldiers. Being beaten or imprisoned by Israelis also brought a kind of status to young Palestinian men that reversed the usual privileged status accorded to older men in this Arab society. Young men who had been released from prison earned the right to negotiate community disputes and to occupy the central position in male gatherings that were traditionally reserved for high status, older men. Peteet argued that in this particular time period and culture, being beaten or imprisoned became a new ritual marking transition to manhood. This made Palestinians not very different from many other cultures that mark men's entrance into adulthood through some physical trial of courage. In the case of Palestinians, the honor of having endured a beating in the cause of nationhood became this trial, and a dangerous one at that.

Masculinity, Health, and Race

The case of Palestinian young men demonstrates the important point that because masculinity takes many different forms across cultures, masculinity can pose more or less risk for different types of men. To demonstrate your masculinity through being shot or beaten is considerably more dangerous than playing football in the United States. Among different men in the United States, bodies can be impacted in different ways by their culture and can have very different meanings placed on them. Class, race, and masculinity intersect to make the simple act of making a living a hazardous proposition for many men. Immigrant populations in the United States have always been forced into the occupations least desired by dominant groups, and those undesirable jobs are often the most dangerous. Slaughterhouses have a long history of using male immigrant labor, initially in the large cities of the Midwest like Chicago and, more recently, in rural areas in Iowa. Upton Sinclair documented the horrors of slaughterhouse work for the human workers, the animals, and the consumers who would eventually eat the meat (Sinclair, 1906). Today's slaughterhouses are not a considerable improvement over those of the past, with an injury rate three times greater than in any other kind of factory in the United States. Each year, about a quarter of the 40,000 men and women working in these jobs suffer a work-related injury or illness that requires medical attention beyond first aid (Schlosser, 2002). In some slaughterhouse facilities, 50% of workers are Hispanic, often including illegal workers who have less ability to fight back against unsafe occupational practices. Daily cleanup of the slaughterhouse is the most dangerous job, as well as being the most poorly paid. The number of Hispanic immigrants in this occupation is even higher.

Working dangerous jobs is just one factor among many that contributes to dramatic discrepancies in the health and welfare of men of color and working-class men as compared with white and middle-class men in the United States. The long history of discrimination faced by Native Americans, African Americans, and Latinos in the United States becomes literally etched onto their bodies. The average life expectancy of a Black man in the United States is 6 years less than his white male counterpart (69.5 and 75.7 years, respectively), while for Native American men, the difference is 5 years (71 years) (Arias, 2007). Native Americans as a group have the highest rates of obesity (36%) and cigarette smoking (38%) (Gallagher, 2007). For Black men living in Harlem, which is 96% African American, the survival rates beyond the age of 40 are lower than those for men living in Bangladesh, casting some doubt on development models that believe places like the United States are healthier. African American men have higher rates of alcoholism, infectious diseases, and drug-related conditions than their white counterparts, but the health care they receive is likely to be inferior. The rate at which Black men older than the age of 13 contract AIDS is almost five times higher than the rate for white males. These figures have led some scholars to describe especially young African American men as an endangered species (Sabo, 2009).

Being in an African American male body is a dangerous proposition. But in an interesting irony, the bodies of Black men are viewed as dangerous to others in situations as mundane as walking down a city street. In an essay, Brent Staples (2009), an editor for *The New York Times,* described some of the reactions he commonly encounters as a Black man, as well as what it feels like to be the one causing those reactions. The "hunch posture" is how he describes the body language of women he encounters after dark in Brooklyn: "They seem to set their faces on neutral and, with their purse straps strung across their chests bandolier style, they forge ahead as though bracing themselves against being tackled"

(Staples, 2009). Being perceived as dangerous largely because of the symbolism attached to the body you occupy takes its own psychological toll. Staples also told a story about being mistaken for a burglar in his own place of employment. He was chased around the building by security until he reached his editor's office and the safety of someone who knew him to be employed at the newspaper.

For many Black men, the power of their bodies as images of danger and violence lead them to monitor very carefully their actions in what many white people would consider a routine, if unwanted, encounter like being pulled over by the police. Articles and books have been written counseling Black men on how to behave in their encounters with the police, with good reason. In America's 10 largest cities, the percentage of victims of police shootings who are Black exceeds the percentage of Black people in the total population for that city, suggesting that African Americans are victims of police violence at a much higher rate than their representation in the population. This perception about the danger of Black male bodies makes being an African American man dangerous, and it is tied to a long history of shifting images of race and gender in the United States. At various points in this history, Black bodies have been depicted as more durable (the strong body of the African American slave woman as compared with the feeble and sickly white, upper-class woman in the 19th century) or as sickly (in the early part of the 20th century, medical scientists predicted African Americans were an inferior race that would naturally die out without the protection of slavery) or as super-human (as in some current racial ideology about the superior biological and physical athletic abilities of Black men and women). In all these different moments, we observe how the gender of bodies is always complicated by the intersection of gender with other identities, such as race, class, nationality, and sexuality.

Theory Alert!
Intersectionality

That masculinity itself leads to more risk-taking behavior and more dangerous lives in a variety of ways is the first way in which being a man can be perceived as hazardous to your health. The second way in which masculinity might affect the health of men is through their likelihood of seeking help for their health-care problems. Namely, men may be less likely to admit they need help or to seek out help when it comes to health issues. Men fail to perform the kinds of behaviors necessary for maintaining health (Kilmartin, 1994). Men are more likely to stop taking medications that manage chronic and life-threatening diseases such as high blood pressure. They are less likely to take time off from work when they are sick or injured. Although prostate and colorectal cancers are common diseases among older men (affecting nearly 75% of men older than 50 years of age), 50% of the male population does not know the symptoms of these cancers or of prostate enlargement, a precursor to prostate cancer. Some research demonstrates that women experience more acute illnesses and some chronic illnesses than men, but the type of chronic illness is important. Acute illnesses include respiratory conditions, infective and parasitic conditions, and digestive system disorders, and in these categories, women exceed men. Among chronic illnesses, women are more likely to suffer from anemia, migraine headaches, arthritis, diabetes, and thyroid disease (Sabo, 2009).

But among the most life-threatening of the chronic diseases, men tend to take the lead over their female counterparts (Kilmartin, 1994). For example, the male-to-female ratio for heart disease as a cause of death among men and women is 101:100, which means that for every 100 women who die of heart disease, 101 men also die of heart disease. This seems like a relatively small difference, but when you look specifically at age groups, a different picture emerges. Among those between the ages of 24 and 44, the ratio of male-to-female deaths from heart disease is 283:100. Thirty-five percent of the population in the United States will eventually die of heart disease at some point in their life, but men are more likely to die much earlier from the disease than are women. Men also die at greater rates than women from every type of cancer except for breast cancer.

At first glance, these statistics seem to suggest again that perhaps men are genetically weaker than women and less equipped to survive the barrage of chronic and acute diseases that afflict many of those in the global North. But a more likely explanation again has to do with the particular demands masculinity places on men in places like the United States. To admit you are sick and to seek help is a sign of vulnerability and weakness. Studies of testicular cancer, a common disease among men, suggest that denial is an important barrier to treatment. Many men do not know about testicular cancer, and those who do may recognize symptoms but still not go to a doctor to seek treatment (Sabo, 2009). These tendencies suggest that many of the differences in overall health reported between women and men may reflect men's reluctance to identify themselves as sick as much as they suggest actual differences in the rates at which women and men suffer from illnesses. Women report higher rates of digestive disorders, but perhaps this is merely because women are more likely to admit that something is wrong with them and to, therefore, show up in these reports. According to official statistics, there seems to be a gender difference in rates of depression, with women more likely to suffer from this disease than men.

But Terrence Real (1997) pointed out that the way classic depression is defined in the medical community misses the ways in which the symptoms of depression among men can be quite different. As a result of their gender socialization, women tend to internalize their emotional and psychological pain, while men externalize these feelings. Although male psychiatric patients are more likely to be violent, female psychiatric patients are more likely to engage in self-mutilation. Because depression is considered to be a feminine disease by both sufferers and clinicians, men are far less likely to seek help for their depression or to recognize their problems as stemming from depression. Depression is perceived as a wimpy disease even though it is a disease that can turn deadly in the form of suicide. Depression falls under the category of "No Sissy Stuff" in the definition of modern manhood, and it is therefore not something real men have. In this way, the masculine reluctance to admit vulnerability and to seek help is another factor potentially contributing to men's shorter lives, thereby making masculinity a somewhat dangerous prospect.

Women, Doctors, Midwives, and Hormones

When you go to your doctor, do you think about her or his gender? Did gender matter when you first picked your doctor, assuming you had the privilege of picking your own doctor or that you are lucky enough to have a doctor to call your own at all? When you think of all the medical doctors you've encountered over the course of your life, have they been mostly women or men? Do you think men or women tend to be better at certain medical specialties? Do women, for example, make better obstetricians and gynecologists, while men make better urologists or surgeons? Should doctors specialize in treating one gender or the other, or do the best medical models disregard gender? Just as we saw previously, the ways in which gender becomes internalized in the bodies of men and women, affecting how long they live, what diseases they get, and how everyone around them reacts to their bodies, gender is also important to the question of how to take care of bodies—to the practice of medicine and health care. Gender has been important in determining who occupies the jobs that involve taking care of bodies, as well as in influencing the ways in which narratives about feminine and masculine health are constructed. The mind–body dualism becomes important again when doctors think about a whole host of diseases and syndromes that are unique to women and connected to women's supposedly closer relationship to the body.

What happened when people became ill before there were doctors? And exactly how long have doctors been around? The answers to these questions all depend on exactly what you mean by *doctor*. Many cultures designate some person or group of people as responsible for healing. But is healing what doctors actually do? In the early history of the United States, there were people who called themselves doctors, but most of the population didn't necessarily go to these people for help when they were sick, injured, or about to give birth. The tradition of medicine that primarily male, American doctors were attempting to re-create in the United States was based on some training in Latin and on reading the works of Greek and Roman philosophers. This training involved almost no experience with actual sick people and was based on a theory of illness that held that air, water, and light were the main causes of illness. Those acquainted with these theories lived in fear of getting wet or being subjected to a breeze, rarely bathed, and kept their houses closed off to any kind of outside ventilation. Doctors knew that blood circulated, but they had no idea how, and the search was for the one disease that was believed to cause all illnesses. This was the period when bloodletting and purging (inducing vomiting) were the main tools at doctors' disposals, and one of the most commonly used drugs was calomel, a mercury salt. This poisonous substance was used for maladies ranging from diarrhea to teething pain, and long-term use caused the gums, teeth, tongue, and the entire jaw to erode and eventually fall off (Ehrenreich & English, 1978).

These doctors of the time were mostly male, and it's little wonder that most people did not use them. Up into the 18th century, when people in the United States became sick or injured, they turned largely to women as healers. Like their European counterparts, many of whom had been prosecuted and murdered as witches in the 15th and 16th centuries, women healers in the United States learned their craft through centuries of accumulated wisdom. Healers in the United States added the knowledge they gained from Native Americans and African slave communities to their repertoire of healing tools. Goldenseal powder and pennyroyal as an herbal remedy came from Native Americans, while the use of cayenne pepper was learned from transplanted Africans in the West Indies. Female healers, like their male medical counterparts, had nothing resembling what we would now call a rational theory of disease.

But unlike male doctors, female healers used gentle cures that largely encouraged the natural abilities of the body to heal itself. Male doctors employing "heroic medicine," such as bloodletting and purging, often did more harm than good in their quest to appear to be doing something rather dramatic to heal the patient. Bloodletting generally involved bleeding an individual until she or he fainted or the pulse stopped, whichever happened first. This was dramatic, but not generally as effective as the remedies of female healers, whose more subtle methods did less harm to the sick patient. In one account, a patient dying of a fever was visited by a female healer. The woman opened the doors and windows previously ordered closed by the doctor (to keep out the dangerous, disease-causing air) and bathed the patient with cold water, still a common method for alleviating fever. The patient, assumed to be at death's door, felt better within five minutes and eventually recovered (Ehrenreich & English, 1978).

Part of the reason male doctors engaged in the dramatic and dangerous practice of heroic medicine was to convince their patients that their services were in fact worth the money to be paid. Female healers were generally less likely to ask for money in return for their services because healing was considered to be a system of reciprocity and community building. In addition, the services provided by female healers were much more holistic than what the program male doctors had in mind. Early doctors in the United States had to convince people that healing was something real and tangible that could be provided in exchange for money; they had to figure out how to **commodify** healing, to turn it into something that could be bought and sold.

This is a problem that still faces the medical community, but it was especially acute against the historical dominance of female healers. Healing involved more than just a particular drug or technique; it included the many kindnesses and encouragements and the knowledge that the healer had at her disposal about the strengths and fears of her patient. Midwives often moved in with the family of a pregnant woman long before the birth and stayed with her until she fully recovered from childbirth. Healing was something deeply embedded in a web of human relationships, and the new field of medicine had to convince people that healing could be plucked out of that complex network as something to be bought and sold (Ehrenreich & English, 1978).

The complete control of doctors and the medical field over healing in the United States was not fully accomplished until the beginning of the 20th century, and when it was finished, women, as well as anyone but the most economically privileged classes, were almost completely shut out of medicine. Science triumphed within medicine over folk healing, and scientific medicine required the presence of a laboratory. Although under a more flexible system of training for doctors that existed up until the 1900s there were many women, as well as working-class and African American doctors, the need to fund laboratories pushed the cost of medical training far beyond the reach of these groups. In their book on the history of expert advice for women, *For Her Own Good,* Barbara Ehrenreich and Deirdre English (1978) argued that this was no mere coincidence. Male doctors purposefully sought to carve out a niche for themselves through the development of medicine as a profession and, therefore, as the exclusive domain of a limited set of people, mainly upper-class, white men. Prior to this consolidation, there were 10 exclusively female medical schools as well as 7 Black medical schools. All but 3 of the female medical schools closed, while African Americans lost all but 2.

The last bit of territory still claimed by women was, not surprisingly, childbirth. As recently as 1900 in the United States, 50% of babies were delivered by midwives. These midwives were responsible not just for delivering the baby but also often for all the rituals that still surrounded childbirth (and potentially, death) for many of the lower- and working-class individuals whom the midwife served. The male medical profession was not concerned with midwives so much because of the competition they presented, as they had been in their previous crusade to eliminate female healers and nonscientific forms of medicine. Few doctors at the time specialized in what we would now call gynecology and obstetrics. Rather, the midwives stood in the way of the scientific development of medicine. Young doctors needed experience observing live patients, including live women giving birth. If a disproportionate number of women used midwives to give birth, how would these generations of young doctors ever be trained?

From this dilemma, we gain the historical relationship between training hospitals and charity hospitals that exists to this day in the United States. Those who were wealthy during the time period would certainly never consent to have a room full of male doctors observing the most intimate process of childbirth. The poor, who had no choice, were the best options for the training of new doctors, and so hospitals began to attach themselves to the nearest charity hospital to provide *material* in the form of the sick bodies of the poor. The overall campaign against midwives specifically portrayed them as dirty and un-American. Midwives were common among many of the new immigrant groups of the time period, and eradicating midwifery became synonymous with the task of assimilating these new groups into the "American" way of life. Although some midwives were dirty, as in less than hygienic in their practices, a solution could have been a training program for midwives, which was what England *did* elect to do with their midwives. Nonetheless, between 1900 and 1930, midwives were effectively eliminated in most of the United States, through legal means or general harassment. This removed the last vestiges of feminine control over the health of bodies in the United States and made medicine a distinctly male profession for roughly the next 30 to 40 years. That this is hardly inevitable is demonstrated by the Russian case; as a result of different historical circumstances, Russian women began entering the medical field during a time when there was a dire shortage of doctors. Perhaps as a direct result, doctors in today's Russia are 70% female, compared with the United States, where female doctors in 2006 made up only 28% of the total population of physicians (American Medical Association, 2008; Ehrenreich & English, 1978).

What are the ideal qualities you think of for a doctor? Are those qualities gendered? Do the ideal qualities differ by the type of doctor or specialty, and does the gendered nature of those qualities change as well?

The Gender of PMS

Perhaps this masculine dominance in medicine helps explain the perspective toward women's bodies that developed in 19th and 20th centuries among doctors and other medical professionals. Feminists who study gender in the medical world have argued that women's bodies are often viewed as flawed or diseased in some way. Hysteria itself is a thing of the past, but there are other feminine hormonal disorders to take its place. **Premenstrual syndrome**, or PMS, broke into the popular imagination and discourse in the 1980s, in part because of a well-publicized case in Britain where two women used PMS as a defense in a murder trial (Rittenhouse, 1991). The "syndrome," originally called *premenstrual tension,* was first identified in medical literature as early as the 1920s, but it was added to the *Diagnostic and Statistical Manual,* or *DSM,* as a disorder officially recognized by the American Psychiatric Association in the 1990s, amid much debate.

The *DSM* is important because it is what doctors and insurance companies use to designate officially the disorder for which patients are being treated. The *DSM* legitimates a disorder or disease as something objectively real and therefore treatable by the medical community. The debates about whether to include premenstrual syndrome in the *DSM* involved gender in important ways because PMS would be a disorder from which only women could suffer. On the one hand, the medicalization of PMS gave legitimacy to women's claims to be suffering real symptoms that seemed to be associated with the onset of menstruation. Before the medical recognition of PMS, a woman who complained about some distress or discomfort leading up to her period was told that these symptoms were all in her head (Markens, 1996). With the official recognition of PMS, these symptoms were perceived as real, with a discernible cause and possibilities for medical treatment.

That women's experiences be taken seriously and that medical research be devoted to investigating premenstrual syndrome seem like positive developments. The problem that many feminists raised with perceiving PMS as an actual psychological or medical disorder is the extent to which this reinforces a view of women's bodies and all their biological processes as inherently abnormal and pathological. For most of medical history, male bodies have been defined as the norm while many of women's *normal* biological processes have been viewed as diseases, syndromes, or disorders. This view is well encapsulated in a quote from the president of the American Gynecological Association from 1900:

> Many a young life is battered and forever crippled in the breakers of puberty; if it crosses these unharmed and is not dashed to pieces on the rock of childbirth, it may still ground on the ever-recurring shadows of menstruation and lastly upon the final bar of menopause ere protection is found in the unruffled waters of the harbor beyond the reach of sexual storms. (as quoted in Fausto-Sterling, 1986)

This quotation suggests that women spend most of their lives suffering as a result of the normal reproductive functions of their bodies. This strong connection between women and their bodies has often been used to control women and to limit their abilities to compete with men. In the late 1800s, during the same period in which hysteria had its golden age, physicians argued that women were biologically ill-disposed to receive an education. Part of their arguments rested on the belief that education, and the ensuing independence that resulted from education, inhibited women's reproductive capacity. Education would literally cause women to become infertile, and ironically, they weren't far off. Education does, in fact, reduce women's fertility but not through any biological process. Across the globe women who have access to an education choose to have fewer children because of the increased opportunities that are available to them. In addition to education rendering women infertile, 18th century doctors also argued that women were just not as intelligent as men and that menstruation made women too sickly and unfit for the hard work of earning an education. For one week out of every month during the very best years of their lives, women were reduced to complete invalids, according to the experts in the medical field.

Against this historical backdrop, the proposition of a new syndrome related to women's menstruation that requires medical treatment understandably made some feminists a bit nervous. The idea that menstruating makes women emotionally unstable or incapable is something that is very much still a part of our culture in the United States. Most recently, images appeared on the Internet during Hillary Clinton's campaign to be the Democratic nominee for president of the United States that reinforced the dangers perceived by many in putting women in positions of power. One image depicted Clinton's face against the backdrop of a nuclear mushroom cloud with the text: "This is what you want once a month?" Women's bodies may no longer be used to prevent them from getting an education, but this image still uses menstruation as an argument for keeping women out of powerful positions. Women are still legally prevented from occupying certain jobs because of the health hazards it would pose in the form of birth defects, while no similar consideration is made for men, who contribute half of the genetic material for a child and are therefore just as likely to contribute to birth defects as are women.

In the research that has been conducted about premenstrual syndrome, no definitive cause has been identified, and researchers are generally unclear about exactly how to define what premenstrual syndrome is. Many women experience various symptoms leading up to menstruation, and certainly some of these are probably severe for some women. But the research indicates no clear baseline for what the "normal" experience of premenstruation is like for women and, therefore, no way to determine what would be abnormal in the case of premenstrual syndrome. One scientist suggests that all women's experiences of menstruation in the modern world are, in fact, abnormal, as in our primitive hunter-gatherer past, short life spans and the lack of fat in the diet meant that women might menstruate no more than 10 times in their whole lives. Modern women's average of 400 total menstrual cycles over the course of a lifetime *is* a physiologically abnormal status.

Menopause in Cross-cultural Perspective

Had the people responsible for the Internet image of Hillary Clinton been more thoughtful, it might have occurred to them that Clinton, at 51 years of age, might be postmenopausal, and therefore no longer menstruating (the typical age range in which menopause occurs is between 45 and 55). Menopause is the event that marks the last menstrual cycle experienced by women and, therefore, the end of a woman's fertility, assuming that she ever was fertile (many women are not). Postmenopausal women no longer experience menstruation. The effects of menopause itself, since about the 1960s in the United States, have been defined as its own disease. Menopause, a process that has occurred naturally to women who were lucky enough to survive for that long throughout our human history, began to be defined as a hormone deficiency. Menopause was conceived of as a process that deprived women of the necessary "sex hormone," estrogen, and therefore turned them into something like lifeless zombies, "walking stiffly in twos and threes, seeing little and observing less. . . . The world appears as through a grey veil, and they live as docile, harmless creatures missing most of life's values" (Fausto-Sterling, 1986, p. 110). Menstruation may make women dangerous, but according to this perspective, the end of menstruation puts women in a virtual coma-like state. The solution to what one doctor described as the "vapid cow-like negative state" (Fausto-Sterling, 1986, p. 110) of postmenopausal women was to replace the missing estrogen with hormone replacement therapy, or HRT. Many point out the happy coincidence in the fact that the doctor who first identified menopause as a problem of estrogen deficiency was also tied to a pharmaceutical company that produced estrogen in the form of a drug, Premarin. By 1975, 6 million women in the United States were taking Premarin, making it the fourth or fifth most popular drug in the country.

In the 1970s, research began to suggest a link between estrogen treatment and uterine cancer. Although many women and their doctors stopped using hormone replacement therapy to treat the symptoms of menopause, the controversial practice of using estrogen continued until a definitive study was released in 2002. This large-scale research project conducted by the Women's Health Institute actually terminated part of the study because of the dangers women were being exposed to as a result of taking hormones. The study found statistically significant increased risks of breast cancer, coronary heart disease, strokes, and pulmonary emboli among women taking estrogen. Currently, hormone replacement therapy, now called hormone replacement treatment, is recommended for women only in the lowest possible dose and for the lowest possible amount of time. After the report was released in 2002, the number of women receiving HRT dropped by almost half.

What exactly was hormone replacement theory supposed to treat, and how was it supposed to work? The answers to these questions reveal some of the flaws in the kind of thinking that led to hormone replacement therapy and highlight some of the ways in which gender assumptions affect what are supposedly scientific, and therefore objective, pursuits. The symptoms that are probably most commonly associated with menopause, besides turning women into vapid cows, are depression, osteoporosis (weakening and loss of bone mass), hot flashes, and vaginal dryness. One study found that 75% of women reported no remarkable menopausal symptoms, which suggests that most women go through menopause experiencing none of the common symptoms listed here. Cross-culturally, one study of Navajo women found that no word exists in the Navajo language for the event we call menopause (Wright, 1982). Although women who were more familiar with Anglo culture could discuss the symptoms and experience of menopause, women more isolated from Anglo culture did not really mark menopause in any important way, which is reflected in the lack of any word in their language to describe this "event." Many Navajo women in general were happy about the end of their menstrual cycle, as it meant a new level of freedom and opportunities open to them. Navajo women follow a set of rituals that isolate women during their menstrual cycle, and so the end of menstruation meant increased freedom and, for some women, that they could now pursue a new life as a medicine woman, which was not open to women who were still menstruating.

This research within Anglo society and among the Navajo suggests that menopause is not really a condition that calls for medical treatment among large numbers of women who *do not* experience menopause as an inconvenient or even notable event. But if we use the study that indicates that 75% of women experience no remarkable menopausal symptoms, that leaves roughly 25% of women who do. Hot flashes and vaginal dryness can be uncomfortable and annoying, while depression and osteoporosis are symptoms with potentially more serious consequences. Without hormone replacement therapy, hot flashes will eventually go away because they are a temporary effect of menopause. Vaginal dryness, for those women who experience this symptom, is a more permanent condition than hot flashes, although continued sexual activity helps, as well as the use of vaginal creams, jellies, or lubricants. These two symptoms are the only two for which hormone replacement therapy has definitively proven effective. The 2002 study of HRT found that there was a reduction in the rates of hip fracture, which is commonly associated with osteoporosis, but other studies suggest that the effects of estrogen treatment are temporary. Hormone replacement therapy works for a short while in arresting osteoporosis, but it eventually ceases to be effective. In addition, other, less risky methods of preventing osteoporosis are available.

For the final symptom of menopause, depression, no research has ever established a relationship between serious depression and the onset of menopause. The bodily changes experienced by some women during menopause may cause some women to lose sleep and be irritable, but even the assumption that menopause as a time of crisis for women can lead to depression seems to lack much support. Like Navajo women, the stage of life during which menopause is likely to occur for many women can be a fairly positive period; assuming women at this stage are without small children, these women are less depressed, have higher incomes, and have an increased sense of well-being on average when compared with women with small children.

If menopause is an event that is definitively linked only to the symptoms of hot flashes, vaginal dryness, and osteoporosis, all of which can be dealt with easily in the absence of hormone replacement therapy, why did Premarin become such a popular drug for so many women? How did an event that barely registers in the lives of most women come to be defined as a deficiency in need of treatment? Part of the answer is in the opportunism of the medical field and the pharmaceutical industry. In the 1940s, scientists had begun perfecting the ability to identify and purify **sex hormones** such as estrogen, and pharmaceutical companies were searching for something to do with these hormones. One researcher argued that drug companies turned to women and menopause because they were more accessible as patients than men through their gynecologists, for which there is no male equivalent. Gynecology allowed drug companies to market hormone replacement therapy easily to women (Oudshoorn, 1994). But the cultural logic of hormone replacement therapy lies in a gendered understanding of the role of hormones that dates back to the beginning of the 20th century. The idea that estrogen needs to be replaced in postmenopausal women because it is no longer being produced makes a kind of sense to scientists, doctors, and patients. Estrogen, as a sex hormone, is an important part of what gives women their sex, and to lose this hormone would certainly mean losing important parts of what it means, biologically speaking, to be female.

There are several important flaws in this line of thinking. The first is that to identify menopause as primarily about the loss of estrogen is a gross oversimplification. There are at least six different hormones that are involved in monitoring a woman's menstrual cycle (including testosterone), and three of those hormones are different types of the broad class of hormones identified as estrogen (Fausto-Sterling, 1986). Menopause is a complex process involving not just a woman's reproductive organs but also her pituitary gland and adrenal gland. After menopause, estrogen does not disappear from a woman's body; rather, estrogen production shifts toward different types of estrogen than are present in premenopausal women. The body shifts from estrogen produced by the ovaries to estrogen produced by the adrenal glands, and this in part explains the postmenopausal evening out of the monthly ups and downs of hormone production that predominate during menstruation. Hormone replacement therapy is based on something of a flawed logic, then, in replacing estrogen—a hormone that has not technically disappeared from the postmenopausal woman's body.

The second flaw in the line of thinking represented by hormone replacement therapy is contained in the idea of sex hormones themselves. Neither male nor female sex hormones are present exclusively in the bodies of the corresponding sex. Scientists in 1932 were aware of this disturbing fact, and they had also discovered that the sex hormones they were so interested in affected many bodily processes that had nothing to do with reproduction. Estrogen could stunt growth, produce fat deposits, increase the breakdown of the thymus gland, and decrease kidney weight (Fausto-Sterling, 2000, p. 185). Other tissues of the body that could be affected by sex hormones included the bones, nerves, blood, liver, and heart. Regardless of these scientific discoveries, the idea of sex hormones was codified in two conferences in the 1930s. A female sex hormone was a hormone capable of affecting the estrus of a female, or her menstrual cycle, while a male sex hormone was a hormone responsible for the expression of male secondary sex characteristics. Female sex hormones were defined in relation to reproduction, and male sex hormones in relation to the expression of sex characteristics such as size and the decorative comb of roosters.

The study of the effects of these hormones that were unrelated to either reproduction of secondary sexual characteristics was largely dropped. The belief became well-established within the scientific community and eventually the public that sexual dimorphism, the existence of two distinct, discrete sex categories, was also reflected at the chemical level in our biology. The fact that sex hormones were not really about sex in the sense of being present only in male or female bodies, *or* in exclusively affecting only *sexual* bodily processes, was proven false at the beginning of the 20th century, but the belief in sexual dimorphism shaped what happened to this scientific evidence, so that the idea of sex hormones is still very much with us today. Strong social constructionists would point to this as another example of how our social values and beliefs in sexual dimorphism shape the way in which we perceive, and therefore categorize, biological reality. Our culture leads us to believe in two types of sex hormones—male and female—when the real picture is much more complex.

Theory Alert!
Strong Social Constructionist Perspective

Anne Fausto-Sterling (2000) argued that this is probably no coincidence; much of the research on sex hormones took place during a historical period characterized by debates about gender, class, race, and reproduction. Understanding hormones was perceived as important to understanding reproduction, and many people during the early part of the 20th century were concerned with controlling reproduction, especially among the lower classes and certain minority groups. Funding for much of the research into hormones came from the Committee for Research in Problems of Sex, an organization that included several strong supporters of the eugenics movement—which brings us to our final exploration of the body as it relates to reproduction and sexuality.

Eugenics, Sterilization, and Population Control

"The struggle for reproductive self-determination is one of the oldest projects of humanity; one of our earliest collective attempts to alter the biological limits of our existence," wrote Linda Gordon (1976, p. 403) in her book, *Woman's Body, Woman's Right,* on the history of birth control in the United States. There are few issues that generate as much emotion and controversy as reproductive rights. The right to have children is not covered in the Bill of Rights in the United States, although the right to found a family is covered in the U.N. Universal Declaration of Human Rights. It's not something you'll probably learn about in your U.S. history class, but there's a long history of attempts to control the reproduction of women and men in the United States, continuing until fairly recent times. The collision of rights, governments, reproduction, and bodies can result in some tragic consequences, as in the instance of the eugenics movement in England and the United States.

The word *eugenics* was coined by Sir Francis Galton from Greek roots, and its etymology suggests *good race,* as in a race of people (Greer, 1984). Galton was Charles Darwin's cousin, and some of the basic ideology of the eugenics movement can be traced back to Darwin's theory of natural selection, as applied to human populations. **Eugenics** came to mean the study or practice of improving the human race through selective breeding and, sometimes, restrictive immigration policies. Eugenics developed in England in response to concerns about the overpopulation among the lower classes, while in the United States, eugenics was applied to the lower classes and racial minorities. As a science, which eugenics claimed to be, eugenics is potentially harmless enough. But eugenics became a social movement that involved large numbers of people in the United States and England, including many people who were powerful enough to make the ideals of eugenics a reality.

Eugenics as a practice involved preventing undesirable people from reproducing while encouraging those with desirable genetic material to have more children. In its more benign forms, this could include "Better Baby and Fitter Families" contests at state fairs sponsored by the American Eugenic Society in the 1920s. But the same group, the American Eugenic Society, also advocated using the newly developed IQ test to determine who should and should not reproduce. The eugenics movement had considerable influence in the United States and England during the beginning of the 20th century; its adherents included Charlotte Perkins-Gilman, the writer, activist, and victim of hysteria we discussed previously, as well as Margaret Sanger. John D. Rockefeller donated large amounts of money to funding the Eugenics Research Office, while Alexander Graham Bell wrote and delivered a paper on eugenics to the American Breeding Association (Greer, 1984). Margaret Sanger was a feminist and advocate of birth control in an age when most of the existing methods of birth control were illegal in the United States, but her connection to the eugenics movement demonstrates the frequent tensions that emerge around reproductive rights. This powerful group of eugenicists was able to lobby the government in support of the 1924 National Origins Act, which severely limited immigration to the United States from all countries outside of northern Europe (Solinger, 2007). Racial science of the time viewed non-Nordic Europeans as inferior races, and this included Italians, Irish, Polish, and Russians.

Where eugenics gained influence, involuntary sterilization often became a legal practice, and a practice that was most often used against women. The case of Carrie Buck became a national story because her case went all the way to the U.S. Supreme Court. Buck was born in Virginia to a mother who had apparently been a prostitute and contracted syphilis. Buck's mother was assigned to a home for epileptics and the feeble-minded, the rough historical equivalent of an insane asylum. Buck was sent to live with a family, where she proved to be a good student and housekeeper. In 1923, she became pregnant, and upon being informed, a Virginia peace officer immediately began the process of having Carrie Buck declared feeble-minded so that she could be sent to the same home where her mother had gone. It was common during this time period for authorities to take women who had been labeled *promiscuous* and officially classify them as *morons* or as *feeble-minded*. These were legal statuses somewhat equivalent to being ruled legally insane today.

When Buck gave birth to a baby girl in the home to which she had been assigned, that baby was labeled *subnormal,* thereby implying that the parent was socially inadequate. This evidence was used to rule that Carrie Buck should be "sexually sterilized . . . and that her welfare and that of society will be promoted by her sterilization" (Solinger, 2007, p. 93). Buck's case went all the way to the Supreme Court, where that body upheld the right of a state to engage in eugenic sterilization. The famous Justice Oliver Wendell Holmes delivered an opinion in support of Buck's sterilization, saying, "It is better for all the world, if instead of waiting to execute degenerate off-spring for crime, or to let them starve for their imbecility, society can prevent those who are manifestly unfit from continuing their kind. . . . Three generations of imbeciles is enough" (Solinger, 2007, p. 94). In 1918, there were 22 states with sterilization laws on the books, although the extent to which they were actually invoked varied.

Carrie Buck's case sounds deeply unjust to our more modern perspective, like something out of a less enlightened past. But attempts to control women's fertility through sterilization and other means persist in the United States and other parts of the world. As recently as the 1970s, doctors working with the Indian Health Service in the United States sterilized Native American women, often without their consent or knowledge. In one case, two 15-year-old girls were surgically sterilized while they were hospitalized for appendectomies (Lawrence, 2000). The racial science of eugenics developed in the United States during the early part of the 20th century and was later put to catastrophic use by the Nazis in Germany; they began their program to eradicate Jews and other undesirable races with sterilization programs. In 1974, India's prime minister, Indira Gandhi, declared a state of emergency and used the opportunity of suspended democracy to engage in large-scale forced sterilizations of poor people, women and men, in her country. The widespread practice of hysterectomies in the United States reveals a similar pattern based on race and class determining what kind of women have their uterus removed. Working-class women and women of color are more likely to have hysterectomies, thereby rendering them sterile and incapable of reproducing (Elson, 2004).

Against this backdrop, it's not surprising that women around the world are suspicious of attempts to curb overpopulation through what has come to be known as family planning. Population growth has been perceived as a real crisis globally since the days of Malthus and his predictions about reaching the limits of the Earth's carrying capacity, or our ability to sustain ourselves within resource limits. Certainly, the rate of population growth globally continued to speed up well into the 20th century because it took 14 years (1960 to 1974) to go from 3 billion to 4 billion people, and 13 years to add another billion in 1987, and only 12 years to get to 6 billion in 1999 (Bell, 2004). As of 2013, the world population has now reached, which indicates a slight slowing in global population growth. The problem is that most of that population growth will happen in less developed countries and, therefore, in places with the least amount of resources available to their governments and their people.

The eugenics movement on a small scale sought to limit the reproduction of poor people in the United States and England because they were labeled undesirable. Although gaining control over population growth is beneficial to a society, trying to limit the fertility of the less developed world can be perceived as having disturbing parallels to the eugenics movement. An important fact to keep in mind in this picture is that the United States represents 5% of the world's total population but consumes 25% of the world's resources. Meanwhile, the U.S. government is the largest donor to population programs around the world (Dixon-Mueller, 1993). As many population experts and environmentalists argue, perhaps population growth is less of a problem than is inequality in the worldwide distribution of resources. The important lesson is that attempts to control reproduction have gendered implications in that they are often exacted on women's bodies, but they also reverberate with other systems for distributing power nationally and globally.

THROWING LIKE A GIRL

Theory Alert! Biosocial Perspective

Theory Alert! Sex Differences

On one level, it seems perfectly natural to think about the connection between gender and bodies. If you subscribe to a biosocial approach to gender, it is our bodies that in part give us gender. Many social meanings are written onto the physical differences in the bodies of men and women, but in the end, some real differences between males and females are embedded in our bodies. Of course, then, bodies are important to the study of gender. On another level, leaving aside the assumed biological differences between women and men, the ways in which gender becomes ingrained in our bodies is amazing and sometimes disturbing. How many times have you heard someone use the expression, "throwing like a girl"? To tell someone, regardless of their gender, that they throw like a girl is hardly a compliment. In translation, to throw like a girl means that you don't really know how to throw, and in this sense, the phrase represents one of the central ethics of male sports, and masculinity in general. Whatever you do, don't do it the way a girl would do it—"No Sissy Stuff," in other words. There's a distinct component of sexism in the phrase, but there is also an observable difference in the throwing styles of many people, and they sometimes do seem to run along gendered lines. What's going on here? Why do many women and men seem to throw differently?

Bodies seem to be the best place to look, and many people might guess that there's actually something different about the bodily structure of men and women. Something about the structure of their shoulders or the rotator cuff. But they'd be wrong. Men and women have the exact same anatomy in their shoulders and arms. From there, those who are very wedded to the idea of biological explanations might turn to sex hormones, or differences in strength or muscle mass. But it's a stretch to argue that any hormones could have such an effect, and the fact is, some men "throw like girls" and some women do not. Female softball players use the same throwing action on the field as do male baseball players. If you ask a man with a correct throwing motion to throw with his other hand (assuming he's not ambidextrous), chances are he'll "throw like a girl." Actor John Goodman, who is right handed and played baseball in his youth, had to learn how to throw left handed for his role as Babe Ruth in a movie. Goodman found he had to teach himself all over again how to throw left handed, and in the beginning, his left-handed throw was nothing to brag about (Fallows, 1996).

So why do so many girls and women throw in a very different style than so many boys and men? The answer lies in the way that, leaving aside any differences rooted in our biology, gender deeply shapes our bodies and, in this case, the way we move. Girls are far less likely to be taught by someone how to throw, and they are far less likely to be motivated to figure out an athletic throwing motion on their own. In addition, some have suggested that the motion of athletic throwing runs counter to the very ways in which women are taught to move their bodies. Women take up less space when they are sitting. When they walk, they take shorter steps and their steps are closer together, as if they are walking a tightrope compared with men whose steps are generally farther apart. The subtle differences in the way women and men move are part of what allow us to make **gender attributions**, to make guesses about who is male and who is female, and this is why many male-to-female transsexuals take classes from dance instructors to learn how to move like a woman. Given all these ways in which our every move is gendered, is it really surprising that when we engage in the simple act of picking something up to throw it that gender is there?

Throwing is just one of the ways in which gender mediates the relationships we have to our bodies. As we have discussed, gender impacts the things we can do with our bodies (like throwing a ball or walking down a street without causing fear). It dictates the way we feel about our bodies, through body image and the beauty myth. Gender influences the ways in which we care for our body, as in the relationship between masculinity and health. And gender often dictates the amount of control we have over what is done to our bodies, as in the case of hysterectomies and other reproductive rights. In all these ways, our bodies become an important site for the complexities of living gender in our everyday lives.

BIG QUESTIONS

- At the beginning of this chapter, we suggested that strong social constructionists view the body as a blank screen onto which societies project their ideas about gender. Do you agree with this perspective? What evidence might you use to argue against this point of view?
- Many feminists argue that the mind–body dualism is a key feature of gender inequality and that dismantling gender inequality would have to involve rethinking this division. Based on what you've read in this chapter, do you agree that the mind–body dualism is key to understanding gender inequality? How would society be different without this particular way of thinking about bodies and gender?
- The mind–body dualism is generally constructed with women positioned as representing the body, and with men representing the mind. Can you think of evidence that contradicts this gendered take on the mind–body dualism? Are there ways in which men might be positioned as closer to their bodies than women?
- In Chapter 2, we discussed the doing gender approach and the resources that are available to help us accomplish an accountable performance of gender. In this chapter, what are ways in which the body is a resource for doing gender?

- Many of the topics in this chapter have in common a concern with how people with different amounts of power in society have more or less control over what happens to their bodies. Based on what you've read, does gender affect how much control people have over their bodies? Do other categories such as social class, race, and nationality?
- In Chapter 5, we discussed the complex relationship between gender and sexuality. How is the gender of sexuality connected to the gender of the body? How does the mind–body dualism discussed in this chapter line up with our discussion of feminine and masculine sexuality in Chapter 5?
- In this chapter, we suggested that controlling bodies in various different ways (through body image, reproduction, or health) is a way of maintaining power over people in general. Do you believe bodies are an important way of maintaining power in society? Can you think of other examples where this is true?

GENDER EXERCISES

1. If you belong to a local gym or sports club, or have access to a place where people play sports or exercise, do some observation, paying close attention to the gendered nature of physical activity. Are there spaces in your location that are more masculine, feminine, or gender neutral? What's the gender makeup of people engaged in different types of activities in these locations? Does the orientation of women and men to these activities appear to be different? What are similarities you observe in the behavior of men and women in these settings? How does your observation connect to the discussion of body image in this chapter?
2. Visit a local art museum or art gallery and pay attention to the gender story being told in the works of art. What is the predominant gender of the artists themselves? What is the gender breakdown of the people depicted in paintings, photographs, and sculptures? Are any of these depictions genderless, and what does that look like? What are differences in how women and men are portrayed in these images? Do there appear to be differences in the way men and women are portrayed based on the gender of the artist? How do your observations line up with the gender of bodies discussed in this chapter?
3. Investigate the gender of current products designed to help people alter their bodies in some way. You might begin by making a list of all the products that fall into this category (if you're creative, this could be a fairly long list). Then think about which of these products seem to be aimed primarily at men, which at women, and which at both sexes. Look at the advertisements for these products in magazines, television ads, or on the Internet. What gender messages are being sent in these advertisements? What does your investigation suggest about men's and women's feelings about their bodies?
4. What is the current medical and cultural ideology surrounding menstruation and menopause? Find books, magazine articles, advice columns, or websites that are targeted toward women and address menstruation and menopause. Do these sources seem to treat menstruation and

menopause as problems or disorders? Look for alternative health advice on these topics. Is the approach to menstruation and menopause different for these different sources?

5. Many cultures, like the Navajo discussed in this chapter, have menstruation taboos, dictating behaviors women can and cannot engage in when they are menstruating. Use online resources and your library to do some research on how different cultures think about menstruation and the norms they have regarding this biological process. How do these practices compare with those in your own culture?

6. Interview or survey women and men on their feelings about their bodies. You might focus on issues of body image, behaviors they engage in to alter their bodies, or their perspectives on the mind–body dualism. Are there gender similarities in how men and women answer these questions? Are there gender differences? How do other identities such as age, race, social class, and nationality matter?

TERMS

mind–body dualism
body image
beauty myth
hysteria
psychology of the uterus
Body Image Distortion Syndrome (BIDS)
body dysmorphic disorder (BDD)
metrosexuals
male gaze
"No Sissy Stuff"
"face-off masculinity"
"Be a Sturdy Oak"
"Give 'em Hell"
commodify
premenstrual syndrome
sex hormones
eugenics
gender attributions

SUGGESTED READINGS

Bodies and gender

Arthurs, J., & Grimshaw, J. (1999). *Women's bodies: Discipline and transgression*. London, England: Cassell.

Bordo, S. (1999). *The male body: A new look at men in public and in private*. New York, NY: Farrar, Straus and Giroux.

Bordo, S. (2003). *Unbearable weight: Feminism, western culture, and the body*. Berkeley: University of California Press.

Burton, J. W. (2001). *Culture and the human body: An anthropological perspective*. Prospect Heights, IL: Waveland Press.

Fausto-Sterling, A. (2000). *Sexing the body: Gender politics and the construction of sexuality*. New York, NY: Basic Books.

Schiebinger, L. (2000). *Feminism and the body.* Oxford, England: Oxford University Press.

Silliman, J., & Bhattacharjee, A. (2002). *Policing the national body: Sex, race and criminalization.* Cambridge, MA: South End Press.

Wolf, N. (1991). *The beauty myth: How images of beauty are used against women.* New York, NY: William Morrow.

Body image

Cash, T., & Henry, P. (1995). Women's body images: The results of a national survey in the U.S.A. *Sex Roles, 33,* 19–28.

Gimlin, D. L. (2002). *Body work: Beauty and self-image in American culture.* Berkeley: University of California Press.

Jeffreys, S. (2009). Making up is hard to do. In E. Disch (Ed.), *Reconstructing gender: A multicultural anthology* (pp. 165–185). Boston, MA: McGraw-Hill.

Jhally, S. (Director), & Kilbourne, J. (Writer/Director). (2000). *Killing us softly III* [Motion Picture]. United States: Media Education Foundation.

Muth, J., & Cash, T. (1997). Body-image attitudes: What difference does gender make? *Journal of Applied Social Psychology, 27,* 1438–1452.

Peiss, K. (2001). On beauty . . . and the history of business. In P. Scranton (Eds.), *Beauty and business. Commerce, gender, and culture in modern America* (pp. 7–23). New York, NY: Routledge.

Shields, V. R., & Heinecken, D. (2002). *Measuring up: How advertising affects self-image.* Philadelphia, PA: University of Pennsylvania Press.

Stinson, K. M. (2001). *Women and dieting culture: Inside a commercial weight loss group.* New Brunswick, NJ: Rutgers University Press.

Wykes, M., & Gunter, B. (2005). *The media and body image: If looks could kill.* Thousand Oaks, CA: Sage.

Bodies, health, and science

Dixon-Mueller, R. (1993). *Population policy & women's rights: Transforming reproductive choice.* Westport, CT: Praeger.

Ehrenreich, B., & English, D. (1978). *For her own good: 150 years of the experts' advice to women.* New York, NY: Anchor/Doubleday.

Elson, J. (2004). *Am I still a woman? Hysterectomy and gender identity.* Philadelphia, PA: Temple University Press.

Gordon, L. (1976). *Woman's body, woman's right: A social history of birth control in America.* New York, NY: Penguin.

Greer, G. (1984). *Sex and destiny: The politics of human fertility.* New York, NY: Harper and Row.

Maines, R. P. (1999). *The technology of orgasm: Hysteria, the vibrator and women's sexual satisfaction.* Baltimore, MD: Johns Hopkins University Press.

Markens, S. (1996). The problematic of "experience": A political and cultural critique of PMS. *Gender & Society, 10* (1), 42–58.

Oudshoorn, N. (1994). *Beyond the natural body: An archeology of sex hormones.* London, England: Routledge.

Rittenhouse, C. A. (1991). The emergence of premenstrual syndrome as a social problem. *Social Problems,* 38 (3), 412–425.

Sabo, D. (2009). Masculinities and men's health: Moving toward post-Superman era prevention. In E. Disch (Ed.), *Reconstructing gender: A multicultural anthology* (pp. 585–602). Boston, MA: McGraw-Hill.

WORKS CITED

Ambwani, S., & Strauss, J. (2007). Love thyself before loving others? A qualitative and quantitative analysis of gender differences in body image and romantic love. *Sex Roles, 56,* 13–21.

American Medical Association. (2008). *Physician characteristics and distribution in the United States, 2008.* Chicago, IL: Author.

American Society of Plastic Surgeons. (n.d.). *Eyelid surgery.* Retrieved June 18, 2008, from http://www.plasticsurgery.org/patients_consumers/procedures/Blepharoplasty.cfm

Arias, E. (2007, December 28). United States life tables, 2004. *National Vital Statistics Reports, 56* (9). Retrieved May 5, 2009, from http://www.cdc.gov/nchs/data/nvsr/nvsr56/nvsr56_09.pdf

Barash, D. P. (2002). Evolution, males and violence. *The Chronicle of Higher Education, 37,* B7–B9.

Bell, M. M. (2004). *An invitation to environmental sociology.* Thousand Oaks, CA: Pine Forge Press.

Bernstein, J. G. (1992). The female model and the Renaissance nude: Durer, Giorginone and Raphael. *Artibus et Historiae, 13* (26), 49–63.

Bordo, S. (1999). *The male body: A new look at men in public and in private.* New York, NY: Farrar, Straus and Giroux.

Bordo, S. (2003). *Unbearable weight: Feminism, western culture, and the body.* Berkeley: University of California Press.

Burton, J. W. (2001). *Culture and the human body: An anthropological perspective.* Prospect Heights, IL: Waveland Press.

Cash, T., & Henry, P. (1995). Women's body images: The results of a national survey in the U.S.A. *Sex Roles, 33* (1–2), 19–28.

Connell, R. (1995/2005). *Masculinities.* Berkeley: University of California Press.

David, D., & Brannon, R. (1976). *The forty-nine percent majority: The male sex role.* Reading, MA: Addison-Wesley.

Desmond-Harris, J. (2009, Sept. 7). *Time Magazine.* Retrieved May 14, 2013, from Why Michelle Obama's Hair Matters: http://www.time.com/time/magazine/article/0,9171,1919147-2,00.html

Dinnerstein, D. (1976). *The mermaid and the minotaur: Sexual arrangements and human malaise.* New York, NY: Harper and Row.

Dixon-Mueller, R. (1993). *Population policy & women's rights: Transforming reproductive choice.* Westport, CT: Praeger.

Ehrenreich, B., & English, D. (1978). *For her own good: 150 years of the experts' advice to women.* New York, NY: Anchor/Doubleday.

Elson, J. (2004). *Am I still a woman? Hysterectomy and gender identity.* Philadelphia, PA: Temple University Press.

Fallows, J. (1996, August). Throwing like a girl. *The Atlantic Monthly.* Retrieved from http://www.theatlantic.com/magazine/archive/1996/08/throwing-like-a-girl/6152

Fausto-Sterling, A. (1986). *Myths of gender: Biological theories about women and men.* New York, NY: Basic Books.

Fausto-Sterling, A. (2000). *Sexing the body: Gender politics and the construction of sexuality.* New York, NY: Basic Books.

Feingold, A., & Mazzella, F. (1998). Gender differences in body image are increasing. *Psychological Science, 9* (3), 190–195.

Frost, L. (1999). "Doing looks": Women's appearance and mental health. In J. Arthurs & J. Grimshaw (Eds.), *Women's bodies: Discipline and transgression* (pp. 117–136). London, England: Cassell.

Gallagher, C. A. (2007). *Rethinking the color line: Readings in race and ethnicity.* Boston, MA: McGraw-Hill.

Garner, D. (1997). The body image survey. *Psychology Today, 30,* 32–84.

Gascaly, S., & Borges, C. (1979). The male physique and behavioral expectancies. *Journal of Psychology, 101* (11), 97–102.

Giancana, N. (2005). No fairy tale. In A. Byrd & A. Solomon (Eds.), *Naked: Black women bare all about their skin, hair, hips, lips and other parts* (pp. 211–215). New York, NY: Perigee.

Gieske, S. (2000). The ideal couple: A question of size? In L. Schiebinger (Ed.), *Feminism and the body* (pp. 375–394). Oxford, England: Oxford University Press.

Gimlin, D. L. (2002). Cosmetic surgery: Paying for your beauty. In D. Gimlin (Ed.), *Body work: Beauty and self-image in American culture.* Berkeley: University of California Press.

Goffman, E. (1988). *Gender advertisements.* New York, NY: Harper and Row.

Gordon, L. (1976). *Woman's body, woman's right: A social history of birth control in America.* New York, NY: Penguin.

Greer, G. (1984). *Sex and destiny: The politics of human fertility.* New York, NY: Harper and Row.

Indoor Tanning Association. (2005). *About indoor tanning.* Washington, DC: Author. Retrieved May 22, 2009, from http://www.theita.com/?page = Indoor_Tanning&hhSearchTerms = about + and + indoor + and + tanning

Jefferson, D. L., & Stake, J. E. (2009). Appearance self-attitudes of African American and European American women: Media comparisons and internalization of beauty ideals. *Psychology of Women Quarterly, 33* (40), 396–409.

Jeffreys, S. (2009). Making up is hard to do. In E. Disch (Ed.), *Reconstructing gender: A multicultural anthology* (pp. 165–185). Boston, MA: McGraw-Hill.

Jhally, S. (Director), & Kilbourne, J. (Writer/Director). (2000). *Killing us softly III* [Motion Picture]. United States: Media Education Foundation.

Kaiser, S. B., Freeman, C., & Wingate, S. B. (1985). Stigmata and negotiated outcomes: Management of appearance by persons with physical disabilities. *Deviant Behavior, 6* (2), 205–224.

Kaw, E. (1993). Medicalization of racial features: Asian American women and cosmetic surgery. *Medical Anthropology Quarterly, 7* (1), 74–89.

Kilmartin, C. (1994). *The masculine self.* Boston, MA: Macmillan.

Lawrence, J. (2000). The Indian health service and the sterilization of Native American women. *American Indian Quarterly, 24* (3), 400–419.

Lehrman, K. (1997). *Lipstick proviso: Women, sex and power in the real world.* New York, NY: Doubleday.

Leistikow, N. (2003, April 28). Indian women criticize "fair and lovely" ideal. *Women's eNews.* Retrieved May 22, 2009, from http://www.womensenews.org/story/the-world/030428/indian-women-criticize-fair-and-lovely-ideal

Maines, R. P. (1999). *The technology of orgasm: Hysteria, the vibrator and women's sexual satisfaction.* Baltimore, MD: Johns Hopkins University Press.

Markens, S. (1996). The problematic of "experience": A political and cultural critique of PMS. *Gender & Society, 10* (1), 42–58.

Marques, M. (2012, August 10). *Huffington Post.* Retrieved April 30, 2013, from Gabby Douglas' Hair: US Olympic Gymnast Gets Heat Via Twitter Over Her Hairstyle (POLL): http://www.huffingtonpost.com/2012/08/01/gabby-douglas-hair_n_1730355.html

Martin, K. A. (1996). *Puberty, sexuality and the self: Girls and boys at adolescence.* New York, NY: Routledge.

Mazur, A., Mazur, J., & Keating, C. (1984). Military attainment of a West Point class: Effects of cadets' physical features. *American Journal of Sociology, 90,* 125–150.

Mulvey, L. (1975). Visual pleasure and narrative cinema. *Screen 16* (3), 6–18.

Muth, J., & Cash, T. (1997). Body-image attitudes: What difference does gender make? *Journal of Applied Social Psychology, 27,* 1438–1452.

Nagel, J. (2003). Race, ethnicity, and sexuality: Intimate intersections, forbidden frontiers. New York, NY: Oxford University Press.

Onishi, N. (2002, October 2). Globalization of beauty makes slimness trendy. *New York Times.* Retrieved August 15, 2006, from http://www.nytimes.com/2002/10/03/world/lagos-journal- globalization-of-beauty-makes-slimness-trendy.html?scp = 1&sq = globalization%200f%20beauty%20makes%20slimness%20trendy&st = cse

Oudshoorn, N. (1994). *Beyond the natural body: An archeology of sex hormones.* London, England: Routledge.

Ozanian, M. K. (2007, September 13). Cowboys top list of NFL's most valuable teams. *MSNBC.com.* Retrieved May 1, 2009, from http://www.msnbc.msn.com/id/20763666

Peiss, K. (1998). *Hope in a Jar: The Making of America's Beauty Culture.* Philadelphia: University of Pennsylvania Press.

Peteet, J. (1994). Male gender and rituals of resistance in the Palestinian "Intifada": A cultural politics of violence. *American Ethnologist, 21* (1), 31–49.

Pyle, R. L., Neuman, P. A., Halvorson, P. A., & Mitchell, J. E. (1990). An ongoing cross-sectional study of the prevalence of eating disorders in freshman college students. *International Journal of Eating Disorders. 10* (6), 667–677.

Real, T. (1997). *I don't want to talk about it: Overcoming the secret legacy of male depression.* New York, NY: Scribner.

Ricciardelli, L., & McCabe, P. (2001). Dietary restraint and negative affect as mediators of body dissatisfaction and bulimic behavior in adolescent girls and boys. *Behavior Research and Therapy, 39,* 1317–1328.

Rittenhouse, C. A. (1991). The emergence of premenstrual syndrome as a social problem. *Social Problems, 38* (3), 412–425.

Rohter, L. (2007, October 2). In the land of bold beauty, a trusted mirror cracks. *The New York Times.* Retrieved October 5, 2007, from http://www.nytimes.com/2007/01/14/weekinreview/14roht.html?_r = 1&scp = 1&sq = in%20the%201and%200f%20bold%20beauty&st = cse

Rosenburg, M. (2007, August 19). Life expectancy. *About.com.* Retrieved June 23, 2008, from http://geography.about.com/od/populationgeography/a/lifeexpectancy.htm

Sabo, D. (2009). Masculinities and men's health: Moving toward post-Superman era prevention. In E. Disch (Ed.), *Reconstructing gender: A multicultural anthology* (pp. 585–602). Boston, MA: McGraw-Hill.

Schiebinger, L. (2000). Introduction. In L. Schiebinger (Ed.), *Feminism and the body* (pp. 1–21). Oxford, England: Oxford University Press.

Schlosser, E. (2002). *Fast food nation: The dark side of the all-American meal.* New York, NY: Harper Collins.

Shields, V. R., & Heinecken, D. (2002). *Measuring up: How advertising affects self-image.* Philadelphia, PA: University of Pennsylvania Press.

Sinclair, U. (1906). *The jungle.* New York, NY: Doubleday.

Solinger, R. (2007). *Pregnancy and power: A short history of reproductive politics in America.* New York, NY: New York University Press.

Spitzer, B., Henderson, K., & Zivian, M. (1999). Gender differences in population versus media body sizes: A comparison over four decades. *Sex Roles, 40,* 545–565.

Staples, B. (2009). Just walk on by: A black man ponders his power to alter public space. In E. Disch (Ed.), *Reconstructing gender: A multicultural anthology* (pp. 204–207). New York, NY: McGraw-Hill.

Stinson, K. M. (2001). *Women and dieting culture: Inside a commercial weight loss group.* New Brunswick, NJ: Rutgers University Press.

Stogdill, R. (1974). *Handbook of leadership.* New York, NY: Free Press.

Wolf, N. (1991). *The beauty myth: How images of beauty are used against women.* New York, NY: William Morrow.

Wollstonecraft, M. (1792/1988). A vindication of the rights of woman. In A. Rossi (Ed.), *The feminist papers* (pp. 55–57). Boston, MA: Northeastern University Press.

Wright, A. (1982). An ethnography of the Navajo reproductive cycle. *American Indian Quarterly, 6* (1–2), 52–70.

Wykes, M., & Gunter, B. (2005). *The media and body image: If looks could kill.* Thousand Oaks, CA: Sage.

Zones, J. S. (2005). Beauty myths and realities and their impact on women's health. In M. B. Zinn, P. Hondagneu-Sotelo, & M. A. Messner (Eds.), *Gender through the prism of difference* (pp. 65–80). New York, NY: Oxford University Press.

PART III

How Is Gender an Important Part of the Way Our Society Works?

CHAPTER 8

How Does Gender Impact the People We Live Our Lives With?

The Gender of Marriage and Families

What kind of family did you grow up in? Do you think your family was fairly normal compared with other people's families? Is there such a thing as a normal family, and what would that look like? How was gender a part of the family in which you grew up in? Do you think your family followed the gender norms for family life pretty closely? What did your family teach you about gender, and how much of what you believe and know about gender comes from your family? Could you have families without gender, and what would those families look like? What makes a "good" marriage and a "bad" marriage, and how might gender be an important part of that answer? Are some marriages more or less gendered? How do different people define what makes a marriage and a family, and are there any similarities across different places and time periods? How have marriage and families changed over time, and how have those changes impacted the way we think about gender? What are the changes taking place in marriage and families right now, and how do those changes influence the ways in which we experience gender? How have the many transformations in our ways of thinking about gender impacted the institution of marriage? How do marriage and families become important sites for the acting out of our gender identities, and does marriage have to involve gender in order to work? Are families the best way to figure out who does what in a society? How do most families make those decisions, and how does gender become an important part of the division of tasks in a family?

These are all questions we will explore in this chapter on marriage and families, two institutions that are both intimately familiar and, as we will explore, inherently strange and alien. Almost everyone has a family of some sort, or has been a part of a family of some sort—even if that family isn't made up of people to whom you are biologically related.

Marriage sometimes, but not always, helps to form the basis of families, and in the United States, we often view marriage as the basic building block of family. Most people in the United States, despite warnings about the decline of marriage, will get married at some point in their lives. The familiarity of marriage and families is easy to grasp. In this chapter, we'll be exploring the strange and alien part of marriage and families, especially as it relates to gender.

In many places around the world, gender is deeply influenced by the institutions of marriage and family. From the theoretical perspective of an institutional approach to gender, it is institutions such as marriage and the family that produce gender, and without these institutions, gender as we know it does not exist. Would it even be possible for us to talk about gender in the same way without such things as marriage and families? The history of these two institutions, as we have seen already throughout this book, is deeply interwoven with the history of gender. And if you want to be able to tell something about the way gender works in a particular culture, looking at the ways in which it structures marriage and families is a good place to start. Whole classes are taught about the sociology of marriage and families, so we certainly can't cover everything there is to say about this topic, but we'll try to make the very familiar institutions of marriage and family a little bit more strange and alien by questioning exactly what these two institutions are and exactly how they work (or don't).

Theory Alert! Institutional

A BRIEF HISTORY OF MARRIAGE

"Mawage. Mawage is wot bwings us togeder tooday. Mawage, that blessed awangement, that dweam wifin a dweam. . . ." So says the Impressive Clergyman in the movie classic, *The Princess Bride,* as incomprehensible as it may be; and so we begin our own brief overview of the history of marriage. This particular wedding ceremony is a good place to start this history because, oddly, it reflects some of the more important aspects of marriage in historical perspective. First, in *The Princess Bride,* marriage and love are two very separate things—things that are actually at odds in the movie. The Princess Bride Buttercup's marriage is exactly the thing from which Westley has to save his beloved. As we discussed in Chapter 6, this incompatibility of love and marriage is accurate historically and still cross-culturally in many places. People fell in love, but as with Buttercup and Westley, the people they fell in love with were probably not the same people they married. The idea of finding love in marriage is only about 200 years old in Anglo-European cultures. Love and marriage, in fact, have *not* always gone together like a horse and carriage, and you might think about this pairing as a somewhat recent historical experiment. Later in the chapter, we'll explore our sense of exactly how this experiment has worked out so far.

The second historical lesson we can learn from the satirical fairy tale of *The Princess Bride* is that marriages were much more likely to be about political power than about love and much more likely to lead to war than to anything like the happy ending ride into the sunset at the end of the movie. In *The Princess Bride,* neither the bride nor the groom standing in front of the Impressive Clergyman is there because of love. They both have ulterior motives; Buttercup, the bride, believes she is saving the life of Westley, the man she actually loves. Prince Humperdinck, the groom, is plotting to murder his wife and start a war with the sworn enemy of the country he rules. Humperdinck's motives are pretty accurate for much of the history of marriage—and especially marriage among the most powerful in early societies. Marriage involved political intrigue, murder, family feuds, and incest to an extent that led Stephanie Coontz (2005), in her history of marriage, to compare these stories with real-life soap operas, with sometimes large-scale consequences.

Antony and Cleopatra: The Real Story

The story of Cleopatra and Marc Antony is a classic example. Billed as a classic love story, the real tale has much more to do with the politics of Egypt and Rome than with passion or lust. During this particular period in the long history of marriage, marriage had become a system for gaining power or forging alliances among the elite groups in a society. Daughters and sisters were often used in these power exchanges, and they often came out on the losing end of these deals. Women were often sent far away from their homes to marry men they had never met. In the first century BC, one Chinese princess from the Chu'u family was sent to marry the ruler of Wusun in Central Asia. Liu Jieyu was far from home, and when the first ruler she married died, she remarried his cousin, who served as regent. When this husband died, she was forced to marry her stepson, and with him, she had an heir. Her third husband was murdered, and her son eventually succeeded to the throne and became ruler of a large segment of the kingdom.

Only when Liu Jieyu was 70 years old was she allowed to return to her home, where she was rewarded for her considerable service with houses, estates, and slaves. The fact that these women were often married off to represent their father's or brother's interests sometimes also gave these women access to power and made them more than mere pawns in a dangerous chess game being played by men. A ruler married to the daughter or sister of a more powerful ruler had to tread carefully not to anger or slight her, lest he ignite the anger of her father or brother. In pre-Columbian Mexico, war was waged and an entire kingdom destroyed when a less powerful ruler insulted the sister of the Aztec emperor. Wives represented the power and interests of their in-laws, and they often plotted to displace their royal husbands in order to replace them with their heir. Even the sons of these marriages were sometimes not safe because a son might still have strong ties to his father's family and could therefore be considered "an obstacle to be removed" (Coontz, 2005, p. 59) in the quest for power. For this reason, some male rulers took commoners as wives or married their own sisters or half-sisters. Commoners had no powerful in-laws to worry about, while sisters or half-sisters shared the same family and, therefore, had no competing family loyalties. At least, in theory, this was supposed to be the case—which brings us back to the story of Cleopatra and Marc Antony.

In the romantic version of the story, Cleopatra, queen of Egypt, goes out to meet Antony, ruler of Rome, on a golden barge. Her beauty overwhelms him, and together they unite to fight a war against the Roman emperor, Octavian. When they lose, Antony commits suicide believing that Cleopatra is already dead. He lives long enough to die in her arms (as the tales of her death were greatly exaggerated), and she follows him, committing suicide by poisonous asp bite to the breast. That's one version of the story, but the reality—as far as historians and classicists understand it—is a bit more complicated. Cleopatra was the daughter of the Egyptian emperor, and when he died, he designated Cleopatra, then 17, and her 10-year-old brother, Ptolemy XIII, as co-heirs and left instructions for them to marry. Having his children marry each other, the Egyptian emperor hoped, would rule out the possibility of a power struggle between various sets of Cleopatra's and Ptolemy's in-laws.

But his children did not marry, and instead they went to war for sole control of Egypt. Cleopatra was a smart woman, well-versed in several languages, and when Julius Caesar, the then ruler of Rome, came to try to broker a peace deal between her and her brother, she understood the value of an alliance with Rome. The primary way to forge alliances in those days was through marriage or children, and whether the affair also involved love or passion is less important than the fact that Cleopatra and Julius Caesar had an affair and that she bore him a child, Caesarion. Several marriages later, Cleopatra and her son became involved in the affairs of Rome as a result of the threat that Caesarion posed to the unstable rule of Octavian, the adopted son of Julius Caesar. Cleopatra was a powerful woman in charge of a large army, and Marc Antony was involved in a power struggle with Octavian for control of the whole of the Roman Empire. Antony was looking for an ally in this struggle, and although the two may have had genuine affection or passion for each other, the relationship was helped along a great deal by the mutual benefits to each of an alliance.

As alliance meant marriage and children, Antony and Cleopatra did both. Roman law didn't recognize marriage to foreigners like Cleopatra, and even if it had, it probably wouldn't have prevented Antony from also marrying Octavia, the sister of Octavian, in order to cement a deal to split up rule of the Roman Empire. Eventually, Marc Antony and Cleopatra went to war with Octavian in support of Caesarion as the rightful heir to the Roman Empire, a gamble through which they would have won a kingdom. When they lost this gamble, they committed suicide probably to avoid being paraded through the streets of Rome as the spoils of victory, as was common for the defeated enemies of Rome. The story of these two tells a fairly typical tail of incest, murder, war, and marriage—all in pursuit of power.

Marriage, as this story demonstrates, was the historical equivalent of the peace treaty, nonaggression pact, or trade agreement. The exchange, usually of women, sealed the deal, and sometimes literally took the form of a sister for a sister or a daughter for a daughter. Some women, like Cleopatra, used this power to become the rulers themselves, rather than the mere currency of exchange. During the Christian era of the Roman Empire, the emperor's sister, Pulcheria, was able to gain control of the Roman Empire and rule partly by declaring herself a sacred virgin, foreshadowing the strategy used by Queen Elizabeth I, the Virgin Queen, in England.

Do any of these historical qualities of marriage persist today? Are power and money at all factors in how people in modern Anglo-European societies decide whom to marry?

Although women were more often the method of exchange in these power systems, men also attempted to use marriage to gain power for themselves. The marriage gradient worked somewhat differently in these time periods. The **marriage gradient** is a pattern in which women tend to "marry up" while men "marry down" in mate selection, meaning that women usually marry men who are older, richer, and generally of higher status. In these royal intrigues, men were often motivated to marry up to gain the power and wealth of their wife's family. In China, low-level scholars with exceptional talents might be given a wife from a noble family in the hopes that he would rise through the ranks and this would eventually benefit the noble family through the alliance.

So What Is Marriage, Then?

These historical accounts demonstrate that marriage, regardless of your gender, was a means to gain power and resources, rather than a joining of soul mates. If we traveled farther back into human history, to the hunter-gatherer groups that predominated before humans began to engage in the sedentary lifestyle required by agriculture, we'd find a slightly different version of marriage. Hunter-gatherer societies were much more egalitarian in general than agricultural societies, where surpluses of food allow more inequality and specialization to emerge. Marriage involved little of the scheming described among royalty and elites because no one really acquired enough valuable stuff to scheme over. Against the backdrop of these societies, anthropologists and other scholars have competing theories about exactly why marriage seems to have developed across a wide range of cultures around the world. In fact, in a survey of all the cultures of which we have knowledge, there's only one we know of where marriage is not a central component of the organization of social and personal life—the Na people of China (Coontz, 2005). This certainly does not mean that marriage has looked the same everywhere, as we'll discuss later. Given the many variations in how different societies define marriage, it's difficult to come up with a standard definition of marriage that actually covers all the forms this institution takes around the world (DeCoker, 2001).

Marriage as Protection

The two competing theories for why marriage in all its different forms evolved have very different implications for our understanding of gender and marriage. The first, and probably more familiar, story about where marriage came from and why it evolved argues that marriage is an institution that allows for the protection of women. We'll call this the **protection theory of marriage**. This theory builds on some of the ideas we discussed related to the anthropology of gender in Chapter 2—specifically the concept of man the hunter. Among early human societies, this theory goes, marriage evolved as a way for women to trade exclusive sexual access in exchange for the meat provided by a man. The woman could also offer some of the nuts and berries she gathered during the day, but this basic unit of male as provider and the female as keeper of the family is, according to these theorists, the "root of the truly human society" (Coontz, 2005, p. 36).

In Chapter 2, we already discussed some of the problems with the model of man the hunter. One problem is that in the contemporary hunter-gatherer societies anthropologists have studied, women's foraging is not an add-on, but it contributes the bulk of food for the group. Hunting, rather than foraging, can be perceived as the add-on to the food supply of the group. The use of the word *group* rather than *family* suggests another problem with the first theory of marriage. The first theory imagines a family in the form of what we now call the **nuclear family**, a family grouping that consists of a mother, father, and their children. Anthropologists and others who study early human societies can't know for sure exactly what the social organization of these groups looked like, but they have several different theories.

None of those theories describes a group divided up into nuclear family units because it would have been impossible for hunter-foragers to have survived had they distributed food only within a nuclear family. In addition, hunting as it first evolved did not include only men, and it certainly did not include one individual man engaging in a hunt by himself. The hunting of large animals by early humans originally involved driving the animals off of a cliff, an activity that included everyone in the group. In other contemporary hunter-gatherer societies, women sometimes go hunting as individual hunters or in all-female groups. The primary problem with the protection theory is less about who did the hunting than the basic idea that sharing was crucial to the survival of these early groups, and sharing meant across the whole of the group, rather than within any smaller unit that looked like a nuclear family. There was certainly a division of labor between males and females but not one man who provided for one woman and her children. In this social context, marriage probably did not evolve for the protection of women.

Marriage as Exploitation

The second theory on why marriage emerged is influenced by feminist thinking about marriage. Marriage, rather than an institution that emerged for the protection of women, is an institution that primarily involves the exploitation of women in a system of exchange, and so we'll call this the **exploitation theory of marriage**. The famous anthropologist Claude Levi-Strauss said of marriage alliances that they were "not established between men and women, but between men *by means of women*" (Coontz, 2005, p. 42). Women, according to this theory, were used like currency, in some ways no different from the exchange of an animal or a clay pot. Women needed to be controlled through marriage because, according to this theory, women had important roles in both the production and the reproduction of the group. Women probably played an important role in the development of agriculture because their experiences with gathering plants would have given them the knowledge necessary to begin cultivating certain plants rather than just gathering what was already growing. Women also had potential power within the group as those responsible for the reproduction of the group itself, through pregnancy and childbirth. Marriage evolved to give men a means of controlling these potential sources of power for women. Women were coerced into marriage through a variety of sometimes violent means, while fathers gained status by giving their daughters to young men in exchange for loyalty or goods.

The protection theory of marriage is sometimes used to reinforce the idea that men as protector and women and children as protected is built into our hereditary past. The exploitation theory of marriage also has modern-day consequences because some feminists argue that marriage, even in its contemporary form, is still primarily a system for exchanging women and for gaining control over their labor. From this perspective, marriage is "the cornerstone of patriarchal power" and the primary way in which men benefit from the labor of women (Coontz, 2005, p. 42). This theory has some support in the historical case of the Blackfoot Indians of North America. Prior to the introduction of the horse, the gun, and the fur trade by Europeans, everyone in the tribe took part in hunting (by driving animals into traps or off of cliffs) and the results of the hunt were distributed among the tribe. The introduction of the horse and the gun allowed men to kill many more buffalos than through previous methods, creating much more work to be done in tanning and hiding.

The fur trade also created a greater incentive to kill many more buffalo than could be eaten by the tribe in order to make money. Men suddenly needed more labor, and women were often responsible for tanning and hiding. To fill this need for labor, there was a sharp increase in the number of wives per husband among the Blackfoot, as well as a drop in the average age at marriage among women. The practice of multiple wives was most common among groups who engaged in the fur trade, and the restrictions placed on wives among these groups became much more severe compared with those not involved in the fur trade. This case seems to suggest that in a system where intensive labor suddenly became necessary, men used marriage and therefore the labor of their wives to fill that need, and this resulted in a loss of power and autonomy among women.

But marriage does not seem to always involve women's loss of power. In many societies, women may go to the village or households of their husbands, but often in smaller societies, women could return to their parent's home for protection if needed. Other evidence that contradicts the exploitation theory of marriage comes from societies in which men are the ones who leave their parent's house and go to live with their wives' family upon marriage. This is the case among the Minangkabau of Indonesia, and they refer to a husband as "the borrowed man" (Coontz, 2005, p. 43). Among the Hopi of North America, a ceremonial presentation of cornmeal to the family of the groom is understood as a way of paying for him. In these societies, men, rather than women, are exchanged through marriage, and some evidence suggests that the exchange of men in marriage may have been much more common in the past. Even in societies where women are exchanged through marriage, the amount of power they have once they are married varies a great deal.

What both of these theories on the origins and purposes of marriage have in common is that, to some degree, they reflect modern ideas about what marriage is and what it should (or should not) be. The protection theory of marriage reflects beliefs that emerged only as recently as the 19th century with the doctrine of separate spheres. These ideas about marriage are still familiar to us in the 21st century but at the time were relatively new. They reflected the idea that men and women were inherently different (again, a relatively new idea) and that women, as the weaker sex (also relatively new), needed male protection within the household. The exploitation theory of marriage reflects many of the critiques of marriage developed by feminists during the second wave of the women's movement. Emerging from the historical period of the 1950s, when women were confined to the household and marriage was held up as the only appropriate state for women and men, it's not surprising that many feminists saw marriage as an inherently oppressive institution for women.

Marriage as Cooperation

A third theory for the evolution of marriage that is a bit less grounded in contemporary ideologies is that a pairing off between women and men did make good sense in light of sexual companionship, child rearing, and organizing the daily tasks of survival. This practice gained strength because marriage also became a useful way to establish many ties that would facilitate the sharing of resources within and across groups. Rather than using marriage to monopolize resources, marriage was primarily a way to share resources and therefore to help ensure the survival of greater numbers of the kin or tribal group (Coontz, 2005). Marriage survived and thrived across many different societies not because it was about protection or exploitation but because it was about cooperation and circulation, and so we can call this last theory the **cooperation theory of marriage**. Marriage turned "strangers into relatives and enemies into allies" (Coontz, 2005, p. 44).

What do the exploitation theory and cooperation theory explanations for the origins of marriage tell us about the ways in which our current social reality affects how we understand the past? Can you think of other potential examples of projecting our future social assumptions onto historical reality?

The early role of marriage in establishing relationships of cooperation and sharing later evolved into marriage as a method for establishing alliances and gaining access to the resources of other groups. In either scenario, marriage was never about love, and it certainly was not about the choices of individual people. Marriage was too important to leave up to the whim of two individuals making a decision based on something as potentially unstable as love or affection. This means that most of the things we now assume to be true about marriage in places like the United States have only been true about marriage for roughly the past 150 years, and they are still not true in all parts of the world. You might think about the **love marriage** as a fairly recent experiment, and that's important to keep in mind when we begin to discuss the most recent conversations about the decline or end of marriage. When many people bemoan the coming death of marriage as an institution, they may be right. A **social institution** is something that gives us a set of clear-cut rules about how to go about accomplishing something in society. One problem we'll explore is that marriage and families in recent years have become less institutionalized, and more like something we make up as we go along than a set of established rules to follow. This shift has its own set of advantages and disadvantages, and it has interesting implications where gender is concerned.

Defining Marriage

We still haven't really answered the question about what marriage is, though, and it's a more difficult problem than you might suspect when you take into consideration all the variations in marriage across history and culture. One definition for marriage proposed by anthropologist George Peter Murdock in 1949 is "a universal institution that involves a man and a woman living together, engaging in sexual activity, and cooperation economically" (Coontz, 2005, p. 26). This sounds fairly reasonable, except that husbands and wives don't always live together. Among both the Ashanti of Ghana and the Minangkabau of Indonesia, tradition dictates that men live with their mothers and sisters even after they marry, while only the main meal is shared between husbands and wives. In many Islamic countries, women and men dwell in separate parts of the same dwelling unit (Deaver, 1980; Kimball, 1980). Even within the history of marriage in Anglo-European societies, servants in many European countries as well as in the United States did not live with their husband or wife who served in a completely different household. Sojourner immigrants to the United States from places like China were very often married men who left their wife and children behind and sometimes never saw them again, although they stayed married and sent their income back home for the rest of their lives. Neither is the economic cooperation component of this definition of marriage universal, as many African societies maintain separate resources for wife and husband after marriage. In response to these problems, a better definition was proposed by the Royal Anthropological Institute of Britain: "a union between a man and a woman such that children born to the woman are the recognized legitimate offspring of both partners" (Coontz, 2005, p. 27).

CULTURAL ARTIFACT 1: I LOVE YOU, BUT I DON'T WANT TO LIVE WITH YOU—LIVING APART TOGETHER

Married couples who didn't live with each other in the past often did so as a matter of necessity rather than as a matter of choice. Increasingly, some couples in contemporary Anglo-European societies are not living together after marriage or in long-term committed relationships as a matter of choice. Demographers have labeled this new trend *living apart together* (LAT), and those who choose to follow this trend are called LATs (Bennett, 2007; Newman, 2007). According to 2006 data from the U.S. Census Bureau, 3.8 million married couples do not live under the same roof (Newman, 2007). In Great Britain, demographers estimate that 2 million couples, despite being in a committed relationship, do not live together, while in Sweden, the number of LATs has risen from 6% of the population in 1993 to 14% in 2002 (Bennett, 2007). Sasha Roseneil, a professor of sociology and gender studies, found through interviews with LATs in Great Britain that most had made a conscious decision to maintain separate domestic lives. Their reasons for not living together included the breakdown of a previous cohabiting relationship and a reluctance to impose a new partner on children from a previous relationship. A few LATs in Roseneil's study were "regretfully apart," as a result of work or family responsibilities, while a third of the group were "undecidedly apart," finding themselves in relationships that were committed but that weren't necessarily leading to cohabitation (see Bennett, 2007). In her personal account of LAT life, Judith Newman cited the expense of finding a living space large enough for both of their households in New York City (including her husband's multiple pianos), as well as the fact that she and her husband had little in common beyond loving each other and their children, a fact that might make living together difficult (see Bennett, 2007). Famous couples who are married or committed but don't live together include the actress Helena Bonham Carter and her director husband, Tim Burton, as well as author Arundhati Roy and her filmmaker husband, Pradip Krishen. Is the LAT lifestyle the wave of the future, or just a fad involving artistic, Hollywood types? What does this phenomenon do to attempts to define what marriage is, and what does it suggest about how marriage is changing?

This definition still does not cover all the examples of the institution identified as marriage across the globe. West African societies, such as the Igbo, recognize the ability of a woman to become a "female husband" (Amadiume, 1987). Among the Igbo, these women received the title *ekwe,* and the women who were formally their co-wives became their actual wives. *Ekwe* women wore the same string anklet as men and were eligible for many of the same privileges as men in Igbo society, including the right to the labor and to the children of their wives (Amadiume, 1987). They did not have sexual relations with their wives, but when their wives had sex with other men and bore children, the *ekwe* woman was viewed as the father of these children. Other societies have acknowledged male–male marriages, while intersexed individuals have married throughout history, leading to some real legal and religious dilemmas in working out whether these constituted a marriage or not and exactly which laws, if any, had been broken (Fausto-Sterling, 2000). These instances of female–female, male–male, and intersexed marriage all violate the basic idea that marriage is between a woman and a man. Chinese and Sudanese societies have a tradition of marriages between women and ghosts or spirits. In Chinese societies, these marriages were arranged by families and married a live son or daughter to the dead son or daughter of another family.

This second definition is also problematic in that many marriages across the world and historically are not between just *one* man and *one* woman. **Polygamy**, or marriage between one man and several women, is a form of marriage that is common in many cultures around the world, including among Native Americans and in Europe, Africa, China, and India, to name a few. Polygamy was common in medieval Europe, where the problems it sometimes posed for an orderly succession among kings may be part of the reason these rulers began to turn to Christianity. Christianity was a historical oddity in both the way early Christianity had a less than favorable view of marriage and in its strict rules against polygamy, a norm that was rare in the historical context of the Middle Eastern world in which Christianity first developed. Both early Judaism and Islam, as the two other religions that originated in the Middle East, allowed for polygamy. **Polyandry**, a marriage that involves one woman and multiple men, also occurs in some cultures, although not as frequently as does polygamy.

Long before contemporary populations of gay and lesbian individuals began demanding the right to participate in marriage, then, marriages sometimes involved gender combinations that contradict the conception of marriage as a relationship between one woman and one man. The current debate about gay marriage often draws on arguments about the true purpose of marriage, and perhaps this is a solution to the problem of defining marriage. Marriages, regardless of the format they take, serve some basic function in society. What might that basic function be? Economic cooperation and pooling of resources is already out of the picture because this does not happen in every kind of marriage. Having sex with each other and bearing children together might be a basic function of marriage, although not all married couples in contemporary societies perform both of these functions. Historically, there have been many heterosexual relationships that involved both sex and the birth of children but were not acknowledged as marriages.

If we argue instead that the primary function of marriage is to distribute legal rights to both goods and services, as well as people (in the form of children), there are still exceptions. A whole history could be written on the idea of illegitimacy, and it would reveal that ideas about whether there are significant differences between children born inside and outside of wedlock vary a great deal. Until very recently in places like the United States, children born outside of marriage were entitled to fewer rights of inheritance and child support than were their "legitimate" brothers and sisters. But in other cultures, and even some inside the United States, whether or not a child is conceived and born within the institution of marriage is relatively unimportant. For many years in European cultures, the most common pattern among commoners was for a woman to become pregnant and then marry, rather than the other way around. This was also true of women and men in Appalachian communities in the United States in the late 19th century (Waller, 1988). Sex outside of marriage was common during the colonial period among the Montagnais-Naskapi Indians of North America. A French missionary was shocked by what he viewed as the promiscuous behavior of the women and warned a Naskapi man that he might never know for sure which children his wife bore were biologically his. The Naskapi man was shocked by the missionary, and he replied, "You French people love only your own children; but we love all the children of our tribe" (Coontz, 2005, p. 29).

You may have grasped the point by now that figuring out exactly what marriage is, and what it does or should do, is not at all an easy thing. You might argue that some of these examples of marriage that don't fit attempts at standard definitions are just weird exceptions that prove the rule. That's all good and fine as long as your particular version of marriage isn't defined as the weird one. For our purposes, the lack of any of these consistencies is exactly the point to keep in mind, especially in light of the contentious arguments about marriage and gender. For example, arguments about the roles of women and men in "traditional" marriages depend on exactly what you think a traditional marriage is. Given the historical and cross-cultural variations we've just briefly explored in the institution of marriage, you could make the argument that traditional marriage is, in fact, polygamy, or that in marriages that were considered traditional throughout much of human history, infidelity on the part of the husband was expected and widely tolerated. If you want to make an argument about the traditional marriage as the ultimate basis for determining how men and women should think and behave, we should at least be moved to ask how you decided which particular form of marriage is, in fact, the traditional one.

How would you define marriage? How is your definition different from the criteria we've discussed here? Would your definition fit all of the people you know who are married?

What we *can* say is true about marriage is that it is, in itself, a method of creating categories important to the working of societies—the married and the nonmarried. How those distinctions are made and the meanings attached to them vary, but marriage is one way of marking an important boundary in all cultures, often the boundary between childhood and being an adult. We can also say that marriage, regardless of the particular form it takes, is an important engine for shaping the way people think about gender. This is true no matter what the particular gender breakdown is of the people involved in a marriage. Male–male marriages among many Native American groups still enforce gender roles, in that one of the men still must take on the roles associated with being a woman in that society; this is what makes this man a *berdache,* or two-spirit, as we discussed in Chapter 5. Feminists may not be correct in arguing that marriage is the cornerstone of patriarchy and male power, but it is true that if you want to find out something about the way gender works in a particular society, an examination of how marriage works is a good place to begin.

THE DEMOGRAPHICS OF MARRIAGE

Defining marriage is a difficult task to undertake, but if we narrow our focus to the assumptions about what marriage is and how it works within particular time periods and particular cultures, then we can begin to understand the ways in which gender has important impacts on the institution of marriage. For cultures that by and large assume that marriage is an institution that involves one man and one woman, demographics can have an interesting part to play. **Demographics** are a way to describe the basic population characteristics of a group, including the age distribution, the sex ratio, how long people live, the composition of different racial and ethnic groups, and so on. In discussing the average life expectancies among men and women in Chapter 7, we were discussing demographics.

Demographics can also tell us the relative mixture of women and men within a given population. The **sex ratio** is usually expressed as the number of males to females in a given society, so worldwide, a sex ratio of 1.01 means that there are 101 men for every 100 women. The sex ratio varies by age group, so that when you look at the population worldwide older than the age of 65, there are now 79 men for every 100 women. The sex ratio becomes important to marriage in a society where women mostly marry men because there need to be enough of each group available to marry each other. Too few women or too few men produces a **marriage squeeze**, or a shortage of one sex or the other in the age group in which marriage generally occurs (Guttentag & Secord, 1983). Marriage becomes like a game of musical chairs, and in that there are too few partners available to pair off with, someone is going to be left standing without a chair at the end.

A marriage squeeze happens only when the sex ratio is off among people who under normal circumstances would be marrying each other. Having too few males or females among the population younger than the age of 10 in most societies isn't going to create a marriage squeeze, although it might when that population reaches marrying age. Too few of one gender or the other among those between the ages of, say, 20 and 29, and you might have a problem. The potential for a marriage squeeze is also affected by the marriage gradient that we discussed earlier, and this is where gender becomes important. In a group of women and men between the ages of 20 and 29, not just any man will marry just any woman. If the marriage gradient dictates that women should marry up while men marry down, then a marriage squeeze can exist even with a 1-to-1 sex ratio among men and women within a particular age group. This is because men generally marry women who are younger than them while women generally marry men who are older. A marriage squeeze can also be created by too many men with lower educational attainment relative to women in the same group or too many women making more money relative to the same group of men.

The term *marriage squeeze* was first coined by demographers in the United States in the wake of the post–World War II baby boom to describe the problem faced by women during this period (Glick, 1988). Remember that women marry up in terms of age, so the first wave of baby boom women had as their marriage pool a group of men who had been born before them and, therefore, before the advent of the baby boom. The sex ratio among boomers themselves was fine, but for the women of this generation, there were fewer older, marriageable men. This may have helped along the reemergence of the women's movement in the 1960s because the marriage squeeze may have led to more unmarried women who pursued education and careers. The discrimination they faced in colleges, universities, and the workplace may have helped to develop some of the ideology that became part of the second wave feminist movement.

How do the concepts of a marriage gradient and a marriage squeeze connect to ideas of compulsory heterosexuality?

The marriage gradient and a marriage squeeze can have some interesting and fairly significant impacts on a society. A country that lacks an adequate supply of women or men may experience an out-migration of one sex or the other. This partly explains trends in Irish immigration to the United States during the 18th century, which included many single Irish women. These women came to the United States and often worked as maids because economic conditions in Ireland made finding a marriageable man difficult. A marriage squeeze can also cause a group to reconsider exactly who is and who isn't marriageable. Throughout this discussion, we've used the somewhat ambiguous terminology of the sex ratio within a particular group. That group can include a whole country, but often even within those boundaries, some groups are considered more or less marriageable.

Race and the Marriage Squeeze

African American women within the United States are often identified as facing a considerable marriage squeeze that might help explain the lower rates of marriage within African American communities in the United States. This is especially true for Black women with higher levels of education; Black men achieve college degrees at much lower rates than their female counterparts, and fewer African American men with high levels of education makes it harder for an educated Black woman to marry up in terms of education (Crowder & Tolnay, 2000). All of this assumes that, within the United States, African Americans marry only African Americans; the tendency (sometimes enforced by norms or rules) to marry only others within the same social group is called **endogamy**. The pressure of a marriage squeeze on African American women could, in theory, increase the incidence of interracial marriages. But the marriage gradient is a good example of the complex ways in which race and gender can intersect. Within the dominant racial hierarchy, women of color in the United States can marry up by marrying white men. Yet in interracial marriages between African Americans and whites, it is more common for Black men to marry white women than for white men to marry Black women.

If we strictly followed the marriage gradient rules for marrying up, there would be more interracial marriages between Black women and white men, but this is not the case. In fact, demographers argue that the large number of Black men marrying white women makes the marriage squeeze for highly educated African American women even worse. This points us to an important point to be made about exactly what it means to marry up or marry down; these terms are, not surprisingly, socially constructed. Exactly what it means to marry up depends on who and where you are. It's not clear exactly why Black women do not marry white men or whether this trend is explained by reluctance on the part of African American women or white men. Some researchers suggest that Black women are reluctant to marry white men because they feel it would be betraying their sense of loyalty to their own racial identity and to their racial group as a community (Dalmage, 2000). Nonetheless, the marriage gradient can become complex, and this is easily demonstrated transnationally when groups attempt to solve a marriage squeeze by defining the boundaries of a particular group in more global terms.

Hung Cam Thai's (2004) research demonstrated the very specific effects of globalization on the marriage gradient among Vietnamese both in Vietnam and in the diaspora. A **diaspora** refers to the worldwide scattering of a nationality or ethnic group. When we refer to the Vietnamese diaspora, we mean all the Vietnamese living in Vietnam, as well as the many emigrants living in the United States, Canada, France, or Australia, just to name a few of the more common destinations for Vietnamese emigrants. The Vietnamese face what Thai identified as a double marriage squeeze. Within Vietnam itself, there are more women relative to men, as a result of the high male mortality rate during the Vietnam war and the high numbers of Vietnamese men who migrated after the war. Because of the higher rates of male migration, there are more men relative to women outside of Vietnam, in places like the United States and Australia. In Vietnam, among those between the ages of 30 and 34, there are 92 men for every 100 women. In the United States, among Vietnamese Americans between the ages of 25 and 29, there are 129 men for every 100 women. Too many women in Vietnam and too many men outside of Vietnam is a problem that in our era of globalization seems to have an easy solution—move the people around. After all, this large-scale global movement of people is a significant part of what globalization is all about.

Transnational Marriage

Yet we still need to keep in mind the marriage gradient, one way in which gender reasserts itself into the equation in a very tangible way. In an era of globalization, what exactly does it mean to marry up and marry down? In her interviews with couples involved in **transnational marriages**, Thai (2004) found that the marriage gradient was still a very important ideal to both the women in Vietnam looking for husbands and the men in the United States and Canada searching for wives. To outline the complexities faced by these women and men, she focused on the specific case of Minh and Thanh. Thanh was a 32-year-old single lawyer in Vietnam whose education and age made her marriage prospects more complex. As she described her situation, men in Vietnam with levels of education and wealth similar to her own wanted to marry down, which meant marrying a woman considerably younger than Thanh. This made it difficult for her to marry up within Vietnam, although she described the alternative, marrying a man with less education or who made less money than her, as "very unappealing" (Thai, 2004, p. 279).

In addition, Thanh believed that any Vietnamese man, although especially one of lower social status than herself, would want to follow traditional gender norms about Vietnamese marriage by trying to control her emotionally or dominating her physically. For Thanh, marrying a Vietnamese man who migrated to the United States was marrying up, first, because he lived in the United States and, second, because having been influenced by American values, an American husband would be less invested in the gender norms of traditional Vietnamese marriage. Thanh and women like her assumed that the standard of living for Vietnamese in the United States was far better than their own in Vietnam and that the more egalitarian gender norms of American culture would have become part of the value system of Vietnamese American men.

The problem Thai (2004) identified in her research was that both of these assumptions on the part of Vietnamese women about their counterparts in the United States were likely to be misguided. Thanh, like the members of many wealthy and successful families in Vietnam, was able to achieve that level of prosperity in large part through the system of remittances. **Remittances** are the money and income that immigrants send back to their families in their country of origin. A Vietnamese immigrant in the United States like Minh may not earn a lot of money by American standards, but the money he sends back to his relatives in Vietnam can considerably raise their standard of living. Thanh's family reaped the benefits of remittances by an uncle in the United States, and Minh remitted $500 of his $1,400 monthly paycheck to his family in Vietnam. This system partly explains the often reduced standard of living of immigrants like Minh relative to their families back in their native country. Five hundred U.S. dollars can buy much more in Vietnam than it does in the Seattle suburb where Minh worked as a deep fryer and assistant cook at a Chinese restaurant. Low-paying jobs are the other part of the picture explaining how Vietnamese American men are often in worse shape than their families back in Vietnam; Minh had completed three years of engineering school in Vietnam before he emigrated to the United States, but without knowing English, there was no way for him to translate these skills into a high-paying, high-status job in the United States.

Like many immigrants, Minh experienced a considerable drop in his social status in his move from Vietnam to the United States, although his family back in Vietnam benefitted greatly. Many Vietnamese American men are not, in fact, economically better off than their female counterparts in Vietnam. In addition, their loss of status in the destination country makes preserving their control within marriage even more important than it might be to a man in Vietnam. As Minh said about his life in the United States, "If you have money, everyone will pay attention [to you], but if you don't, you have to live by yourself" (Thai, 2004, p. 281). Vietnamese American men like Minh look for wives in Vietnam because they want a woman who will show them the amount of respect assumed to be normal in traditional Vietnamese marriages, respect they do not receive in their everyday lives as immigrants in the United States and which they notice is not the norm among many marriages in the United States. Men like Minh are seeking from their Vietnamese wives exactly the kinds of things that Thanh was expecting to escape by marrying a Vietnamese American man.

In this situation, marrying up comes to have a very strange meaning. As Thai (2004) described this couple, Minh spent the bulk of his life working in the service industry in a small Chinese restaurant in suburban, Middle America. He hadn't read a book in recent memory, lived with his fellow Vietnamese restaurant workers in a modest three-bedroom apartment, and counted among his possessions a used Toyota Tercel. Thanh, on the other hand, was a successful lawyer who lived in urban Saigon, used Chanel perfume, and wore Ann Taylor shirts. Thanh was more comfortable speaking English than Vietnamese, while Minh used Vietnamese in his interview with Thai. Thanh counted *The Great Gatsby* as one of her favorite books, took aerobic classes, and wore her hair dyed and permed. The social divide between the worlds of these two people were great, and not at all in the direction you would expect given the assumptions many people make about the developed world of the United States as compared with the developing world of Vietnam.

The expectations each brought to their marriage were almost completely at odds, and yet both interpreted their pairing as following a marriage gradient and as a completely logical solution to the double marriage squeeze. What will happen when Minh and Thanh actually enter married life? Thai presented several possibilities, including a happy or an unhappy adaptation on the part of either half of the couple. Perhaps Minh might be persuaded to give up on his ideas of having control in his marriage, or perhaps Thanh will be content with a more traditional Vietnamese marriage. Marital conflict is likely in either case, and perhaps many of these marriages will end in divorce or, worse, physical or emotional abuse. Their story demonstrates both the ways in which the basic idea of the marriage gradient can become flexibly defined in different social contexts, as well as the ways in which gender norms in the form of the marriage gradient can make the actual business of being married a fairly difficult prospect.

Social class is an important part of the marriage gradient, and defining social class varies across time and place. Demographers suggest that the lack of available, age-appropriate men to marry for baby boom women may have contributed to the women's movement. In turn, some demographers suggest that the women's movement may have created a new marriage squeeze, especially for men with little education. Rates of marriage among men across all racial groups in the United States without a college degree have been decreasing. In 2006, about 18% of men between the ages of 40 and 44 with no college degree had never married. This number is up from only 6% within that category in 1980 (Porter & O'Donnell, 2006). These statistics are part of a general trend we'll discuss later in the chapter, in which many groups in the United States are deciding not to marry or are spending more of their lives outside the marital state. But the trend is most marked among men without a college education, and two possible explanations are the declining wages available to men who do not have a college degree and the increasing independence of women. Whereas, in the past, these men might have paired up with women who felt they could not support themselves on their own, more women today may feel perfectly able and willing to do so. In the language of the marriage gradient, more women have been pushed up the ladder ahead of men in terms of women's higher rates of college graduation and better job prospects. If both men and women continue to follow the marriage gradient, there are fewer women below these men to marry down.

Which leaves us with a question that may have occurred to you already: If the marriage gradient often becomes difficult to follow, as well as potentially creates the problems we discussed among transnational Vietnamese couples, why do people continue to follow it? As we saw among powerful royal families historically, there were distinct advantages to men's ability to marry up and improve their family's social status. The marriage gradient has not always been enforced or assumed in the same form we see today. Some people do defy the marriage gradient in its current form; sometimes women marry younger men or men marry women who make more money or have more education than they do. Why don't more people follow their lead? There are several different explanations you might offer up, but all of them would have to do with the power of gender. We've discussed the existing gender norms about height differences between women and men in Chapter 7, and doing gender theory argues that these size disparities are part of the resources women and men use to do gender cooperatively. A woman standing next to a man who is taller than her is a very simple visual message about who is in charge and who has more power.

Theory Alert! Doing Gender Theory

The marriage gradient can be viewed in a similar light; when women choose men who are older, richer, or more powerful, they are reinforcing some of our basic ideas about who should be in charge—even within intimate relationships such as marriage. An institutional explanation might point to the exploitation theory of marriage. If we agree that marriage evolved as a tool for the subordination of women, then it makes sense that a married woman should always be less powerful in some way than her husband. This is the main point of marriage. Even if you don't believe that keeping women in their place is the main purpose of marriage, an institutional approach would still point to the marriage gradient as a way in which this particular institution creates a set of rules to follow that reinforce gender norms and gender inequality. It may not have ever been the main purpose of marriage, and it certainly may not be the main purpose now. Nonetheless, this is the way institutions work; rules get set up and followed, becoming beyond question regardless of what their actual consequences are. In this way, the marriage gradient demonstrates the way in which gender and marriage are intertwined even before anyone walks down the aisle.

Theory Alert! Institutional

WHO DOES WHAT? THE GENDERED DIVISION OF LABOR

Gender is a large part of the process that helps lead certain types of people, in certain kinds of combinations, to the altar—or to whatever particular ritual signifies that a marriage has commenced in that culture. Gender is also crucial to dictating what happens once the particular combination of people is officially married. Although it might be difficult to agree on one universal aspect of marriage, figuring out who does what is generally an essential function that is served by marriage in a wide variety of societies. In many places and times, marriage is a matter of survival simply because one person, woman or man, could not perform all of the tasks necessary for their survival on their own, let alone all the tasks necessary to the survival of two people plus any possible offspring.

This was true in medieval Europe among serfs, or the peasants who farmed land for a lord. Most people in medieval times were serfs, and marriage was important for this class of people because a farm simply could not be run single-handedly (Coontz, 2005). Some tasks were separated along gender lines, and some were not; the husband plowed the field, spread manure, dug peat for fuel, harvested crops, threshed the grains, turned the hay, and sometimes hired himself out to work in the fields of larger landowners. The wife milked the cows, made butter and cheese, fed the chickens and ducks, cleaned and carded wool, prepared flax, brewed beer, carried water, took any surplus products to market, washed clothes, and had grain pounded at the mill. Both husband and wife helped with the harvest, gleaned the fields, and collected firewood, and wives, like husbands, sometimes also hired themselves out as farm workers (Coontz, 2005, pp. 110–111). The term husbands and wives used for each other for much of this period in history, "yokemates," makes sense given the workloads shared by both. Some tasks overlapped, while some were viewed as more appropriate to one gender or the other. But the idea that one or the other person in the marriage did more or less than the other, or that one person did nothing at all, would have sounded largely absurd to these medieval couples.

Sexual division of labor is one of the oldest and most common ways of distributing tasks in a society, with age as the only other close competitor in terms of dividing up who does what. The **sexual division of labor** is simply a sense that certain tasks are more appropriate to one gender or the other. Brewing beer among medieval serfs was a woman's task, while plowing the fields was what men did. In many African societies, it is a man's job to clear fields for cultivation, but a woman's responsibility to grow the crops used to feed the family (Beneria & Sen, 1997). When we talk about the sexual division of labor, we're talking about all forms of labor a family performs, regardless of whether anyone gets paid for those tasks. Caring for children is a form of labor, although it is a task that is rarely part of the paid labor market, unless in the form of a babysitter or nanny.

Every society has its own particular sexual division of labor that tells individuals within a marriage or family unit who should do what, but as we've already discussed, there's variation in how tasks are assigned based on gender (Du, 2000). Farming in many Asian societies is reserved strictly for men, while in many African societies, farming is what women do (Beneria & Sen, 1997). Child care is a task that many of us would assume is associated with women, and generally, most societies do designate some group of women as primarily responsible for child care. But the particular women who do child care in any given society may or may not be the biological mothers of the children, and they may or may not be biologically related to the children at all. One survey of nonindustrial societies suggests that in 8 out of 10 such societies, children between the ages of 18 months and 5 years old have spent less than half of their time with their own mothers (Coltrane, 1996). Often, older women or older children do much of the child care, but some research suggests that half of all societies also have close father–child relationships. For example, among the Aka pygmies in Africa, fathers sleep with their infants, feed and soothe them, and generally show them a great deal of affection, even though mothers still do the bulk of child care (Coltrane, 1996).

The Mbuti, another African tribe, have a relatively flexible sexual division of labor. Both men and women gather food in the forest, and hunting is most often a joint activity between women and men. Women and men both make hunting nets, and both men and women help to build and maintain the huts in which they live. Husbands and wives have roughly equal amounts of say in issues affecting the family and the larger group. The role of Mbuti men in child care is exemplified in the way in which the Mbuti wean children off of breast milk and onto solid food. At a certain point, when Mbuti children are ready, their father picks them up and fondles them affectionately just as their mother would. Mbuti fathers take the child to their breast and hold it there, as a mother would to breastfeed the child. Being as familiar with its father as the child is with its mother, the child begins to explore for milk, but instead it is given its first taste of solid food. In this way, men become a "kind of mother" (p. 185) to Mbuti children in giving them their first solid food and playing a large role in their care.

The Mbuti stand in stark contrast to the Rwala Bedouins of the Middle East and to their lives in the north Arabian desert before their exposure to Anglo-European influences. The Rwala enforced a strict sex segregation and strict sexual division of labor between women and men. Women were expected to serve men at meals, and the tent dwellings were strictly divided into men's and women's spaces. The important works of entertaining and conducting business were conducted in the male spaces of the tent, and women were strictly forbidden from participation. Unlike the Mbuti, Rwala women had few legal rights, no political role, and were viewed largely as the property of their husbands (Coltrane, 1996).

The Sexual Division of Labor and Gender Inequality

Theory Alert! Radical Feminist Theory

These two cases begin to suggest why the sexual division of labor in any given culture is an important part of gender and marriage. The beliefs about gender as well as the relative degree of gender equality within a society are greatly influenced by the particular sexual division of labor that serves as the norm in that society. For Marxist and socialist feminists, the sexual division of labor is the main source of gender norms and gender beliefs in a society. It is this primal division of tasks along the lines of sex that gives rise to everything we view as gender, according to a Marxist feminist approach. How does this happen? As we've already discussed, there are a wide variety of ways in which the sexual division of labor is formulated. What seems universal so far is that, even in societies like the Mbuti where men have a relatively large role to play in the care of children, women still do most of the child care throughout the world. According to Marxist feminists, it is this allocation of child care to women that results in the devaluation of women as a gender. Child care is perceived as less important to the survival of early groups of humans than the supposedly male tasks of hunting and providing. Women, as the group engaging in this devalued task, become the gender that is less valued as a consequence.

You can probably already think of some problems with this particular explanation of the sexual division of labor as the root cause of gender inequality. We know that hunting has often been a task that involved both women and men, as it does among the Mbuti. We also know that even when men were primarily hunting and women gathering, the gathering of women contributed the bulk of sustenance for early human societies. Under either of these scenarios, it's difficult to understand why hunting would be valued so greatly over gathering, or why hunting would even be perceived as a primarily male task. We must also ask whether there is some reason why taking care of children should universally be perceived as a less important task than providing food through hunting. In contemporary societies like the United States, we would at least probably claim that we feel raising children is an important task, although as we'll discuss, some of our actions, beliefs, and policies contradict that privileging of child care. It would be incorrect to say that people in all times have not cared about their children and the welfare of their children, but the particular forms that concern may have taken have varied greatly from culture to culture. Children have been important because they were a means to continue a line of descent, as with kings and queens, or because they added much needed labor to a family unit. The care of children is important at a very basic level because no surviving children signals the end of the group's very existence.

Whether child care in and of itself is perceived as devalued, regardless of the gender of the person doing the task, is a difficult question to answer because we have no cases where someone besides women have done most of the child care. The implications of this argument, though, are that if we were able to create such a society, and some people in contemporary society seem to be moving in that direction, a more equal division of child care would result in more gender equality. In fact, in his book on fatherhood, Scott Coltrane (1996) argued that the particular form that masculinity takes in various cultures is directly tied to the degree to which men in those cultures are or are not involved in the act of child care. When men do a great deal of child care, masculinity tends to be less aggressive, violent, and dominant. The way to change our ideas about gender and the degree of gender inequality is to change our behaviors in regard to who does child care.

If you know men who do a great deal of child care, do they seem to exhibit masculinity that is less aggressive, violent, and dominant? Does being responsible for child care seem to change the way these men express masculinity?

From a different perspective, it is not child care itself that leads to women being devalued relative to men. Rather, child care is devalued as a kind of labor precisely because it is done everywhere by mostly women. From this perspective, the sexual division of labor by itself is not enough to explain gender inequality or the existence of gender. The sexual division of labor is just one particular expression of gender inequality, and it is perceived as a result of gender inequality rather than as its main cause. Some evidence supporting this particular point of view is that even if we leave aside the particular task of child care, in almost every society, the set of tasks women do are valued less than the set of tasks men do. Farming, an activity performed by both men and women depending on the particular culture in which you find yourself, is more valued when it is perceived as men's work than when it is perceived as what women do. This suggests that the actual content of the tasks done by women and men is less important than androcentrism—the idea that men and masculinity are better than women and femininity. The exact origins of this gender inequality are left unexplained, but this perspective suggests that changing the gender of child care alone is not enough to undo gender inequality. It certainly wouldn't hurt for women and men to take more of an equal role in the job of raising children, but the sexual division of labor is not the magic key to ending gender inequality for all time.

This second perspective, that the work women do is devalued precisely because women are doing the work, is a view we will discuss again in the next chapter when we turn to inequality within the world of paid work. The sexual division of labor applies to tasks in a society regardless of whether those tasks are part of the formal economy. For example, in many of the African cultures where women do much of the agricultural labor, women's farming is subsistence agriculture while men are more likely to do paid agricultural labor. **Subsistence agriculture** is farming that's done to feed one's self or one's family, rather than to produce goods to sell to someone else to consume. Housework and child care are other kinds of labor that are often unpaid and exist outside the formal economy. When you were a child, the person who gave you a bath, got you to school, and bandaged your knee was probably not paid for any of those services, unless you had a nanny or other type of servants. Yet growing food for a family and bathing children are both forms of labor that are essential to the welfare of family members, regardless of whether anyone gets a paycheck at the end of the day.

For much of history, and for some societies still today, there was little need to make a distinction between paid work in the formal economy and unpaid work that happened outside the bounds of the formal economy. In our example of the medieval wife, her responsibility for taking care of ducks and chickens could be formal or informal, depending on what she did in the end with her ducks and chickens. If she used them only to feed her family, this is unpaid work, but if she took her surplus ducks, chickens, or eggs to market, that became part of a formal economy. This distinction becomes important when people in a society begin to work for wages rather than to support themselves or to produce surplus for the person in charge, such as serfs for the lord of the manor. Industrialization and the system of paying people wages brought a whole new gender dimension to the sexual division of labor, the repercussions of which have become so widespread and taken-for-granted in many places that we've completely forgotten that things ever used to be any different. Nonetheless, it's important to remember, as we'll discuss, that the definition of *traditional family* that most people casually throw around is dependent on this new division of labor. Historically and to some extent globally, what we now view as traditional is somewhat abnormal in the grand sweep of history and across the wide range of other cultures.

THE DOCTRINE OF SEPARATE SPHERES

For most of human history, the home was the factory and family members were the workers. This meant everyone in the family—children, wife and husband, grandparents, nephews, and so on—was expected to do whatever the work of that particular family unit was. Most of that work took place inside the home or close by, in the fields, or in a shop downstairs or next door. This is still true for many families in the United States and around the world. Among groups like the Amish in the United States, everyone's labor is necessary to sustain the family. Small family farms usually call on husband, wife, and children to do the work necessary to feed themselves and others. Many immigrant families in the United States start businesses like restaurants or hotels, living above the restaurant or in the hotel while all members of the family help to make the business run. In places like the United States, these situations that wed work and home seem like exceptions, but for most of history and for many people, the main point of marriage was to form a productive, self-sustaining, and successful unit.

This function of marriage often took a much greater precedence over what we now often think of as a primary reason to marry—having children. In France of the 1720s, historians discovered a spike in the marriage rate immediately after a plague in 1720. From our perspective, we might assume more people married after the plague to replenish the population by having children. But taking a closer look at the records revealed that in fact, most of those getting married were well past their childbearing years. Their motivation for marrying was not, then, to have more children, but to take advantage of new prospects for inheritance and to fill the empty slots among farmers and shopkeepers upon the death of husbands and wives. In other words, marriages increased as a result of the economic opportunities brought about by deaths from the plague, which is not the way we are generally accustomed to thinking about our decisions to marry (Coontz, 2005).

Modern Marriage

Marriage, in the form with which many in the United States are most familiar, is an institution that began to take shape about 150 years ago in Europe and its colonies. This particular form of marriage, which we'll call **modern marriage**, is an institution freely chosen on the basis of love and compatibility and composed of a sole male breadwinner plus economically dependent wife and children. The idea that people should get to choose whom they marry as individuals, largely free from the constraints placed on them by family and community, evolved in part as a result of the rise of wage labor and the simultaneous emergence of Enlightenment ideals about political and philosophical rights. Wage labor suddenly gave young people more freedom from their parents than had existed in the past; if you could go out on your own and get a job, you became less dependent on your parent's inheritance or their influence in finding you an apprenticeship or other way of supporting yourself. This meant there was less pressure on young women and men to make a marriage match that was satisfactory to their parents. At the same time, Enlightenment philosophers began to extol the virtues of individual rights while challenging the right of institutions such as churches and the family to dictate who could marry whom. Marriage, according to Enlightenment ideals, should be freely chosen based on reason, and at least for Enlightenment thinkers, that meant marriages chosen on the basis of love.

This radical change in marriage took some time and brought about a great deal of anxiety on the part of many people. For years, the expression of mild affection among married couples was looked down on and perceived as potentially disruptive to the real business of marriage as an institution for making society work. Those with the time and inclination to think about the implications of such a transformation had a long list of worries on their minds. Would people be able to choose a spouse wisely if love were involved? If marriage was expected to make people happy, what would hold marriages together when things went for worse rather than for better? In a marriage of intimacy between men and women, would women begin to expect equality with their husbands? And if parents, churches, and states no longer made the decisions about who married whom, how could society ensure that the right people married and had children or that the wrong people did not (Coontz, 2005, p. 150)? This long list of anxieties probably sounds somewhat familiar to us today, even though they were being articulated as early as the 18th century. Yet the love marriage is now considered the norm across many parts of the world, and where it is not, the ideology of marriage for love exerts an increasingly powerful influence through popular cultural images from the United States and other parts of the world.

In your opinion, did any of the fears of this time period about the dangers of love marriage turn out to be true? Do people choose their spouses wisely and stick together? If you feel some of their fears did come true, what does this say about the state of modern marriage?

The fact that modern marriage is freely chosen and based on love carried with it some gender implications, as reflected in the set of anxieties listed previously. If you can trust women to make their own decisions about marriage, how do you continue to exert control over them within that marriage? This anxiety reflects the assumptions that were true about the institution of marriage for much of its history; men and women may have done the same amount of work and formed a cohesive productive unit, but that by no means meant that women were considered equal to men. Women had few rights within previous forms of marriage, and they were perceived as lesser versions of men. This meant that women were not considered to be fundamentally different from men, but that they were also not considered to be in any way worthy of being equals to men in any meaningful capacity.

But it is the other component of modern marriage that many would argue had the most important gender implications for men and women living in places like Europe and the United States, and that is the idea of a sole male breadwinner with a dependent wife and children. With the advent of industrialization and the concept of the modern marriage, the ideal increasingly became for women and children to stop working, inside or outside of the household. It is important to note that this was an ideal, which is to say it became what many families aspired to, but what many more families were never actually able to achieve. This ideal image of the modern marriage, where the husband alone works to support his wife and children, is what gives us the doctrine of separate spheres—a spatial and gendered separation that came to have far-reaching impacts on the way we think about gender still today.

In the 19th century, in places like the United States and Europe, the process of industrialization was well under way. This shift to industrialization, along with the accompanying trend of urbanization, happened in a fairly short amount of time. In 1871, two thirds of Americans were still self-employed, meaning they were generally farmers and artisans who produced goods or provided services. These people worked for themselves rather than being paid by someone else for their labor, and usually everyone in the household was a part of that productive unit. The husband/father was in charge of this productive unit, and you might think of him as like a factory foreman, where his workers also happened to be his relatives and any other apprentices or farmhands that were added into the mix. In this system, husbands/fathers were thought of less as part of the family than they were the person in charge of running the family. The few advice manuals written for husbands or fathers sounded similar to the directions one might give for breaking a horse; the ideal was that a husband should achieve complete dominance over his wife, as well as the rest of his family (Coontz, 2005, p. 118). By the turn of the century, most people in the United States depended on wage labor to support their families, and this generally meant someone was working outside of that household unit. Families transitioned from working side by side in the fields or in their shop to sending someone out to the new factories to sell their labor for wages.

Decisions had to be made about exactly who was the best candidate to be sent out to work for pay in these new factory settings, and that it would be male children or husbands was far from a given. If you've read much Charles Dickens, you know that in places like England and the United States, children commonly worked long and dangerous hours in the early days of industrialization. For families who had formerly farmed, using children as a source of labor was nothing new, although the conditions within these factories were probably far worse than anything they would have faced on a farm. In the United States, it was common for farming families to send their unmarried daughters to work in the mills for the extra income. Sons, so the logic went, were more valuable for the physical labor they could provide on the farm, while more return on the labor of daughters could be gained from sending them to work in the factory. As we will discuss in more detail in the next chapter, that husbands be the sole member of the family to work outside of the home was not an inevitable development. The number of families who were actually able to achieve this ideal was also fairly limited.

Gender and the Doctrine of Separate Spheres

Nonetheless, among the limited numbers of the white, middle-class families in places like the United States and Europe, the ideal family became one in which the father worked to support a woman who stayed home and was responsible for taking care of home and children. This physical separation of the lives of women and men came to be known as the *doctrine of separate spheres*. According to the **doctrine of separate spheres**, it is a man's duty to take care of his family by being a breadwinner and protector, while a woman's duty is to be a good wife and mother. Women, through their association with the home and child care, are supposed to be kind and gentle, while bold and aggressive men pursue success in the competitive world of business and the economy (Coltrane, 1996, p. 25). The doctrine of separate spheres is a complicated and far-reaching ideology, and it accomplishes several different tasks related to gender. First, the doctrine of separate spheres marks two different realms of our social lives as distinctly gendered. The place now emphasized as home, and as a safe harbor from the stresses of the outside world, is associated with women and femininity. Women were responsible for creating a "haven in a heartless world" (Bonvilliain, 2007, p. 184). The economic sphere where people were paid for their labor in the formal economy became an inherently masculine realm of social life. This realm was associated with lower moral standards, as men were expected to work in a world characterized by the dog-eat-dog attitude of the business world.

Where the home was associated with sentiment and comfort, the world of work was associated with rationality and competition. Many scholars identify this as an important separation between the public and the private spheres. The **public sphere** is the world of market relations and productive behavior, a world that is labeled as distinctly masculine with the separation of spheres. This includes the world of paid labor but also all the other institutions that intersect with the concerns of the market, including politics. The doctrine of separate spheres dictates that all of these institutions are best suited for men. The **private sphere** is anything that's not public, and it is primarily the home and family life; this becomes defined as the place where women belong. As an institution, the doctrine of separate spheres had a radical effect against the long backdrop of history in suddenly defining the family in relation to women and femininity and the public world of work in relation to men and masculinity.

Make a list of all the qualities you associate with home and all the qualities you associate with work. Are these qualities gendered, and do they line up with the doctrine of separate spheres? Are the qualities associated with home more feminine and the qualities associated with work more masculine?

The doctrine of separate spheres first achieved a gendering of the physical spaces and institutions of the home and work. We'll discuss how this matters in the world of work in more detail in the next chapter. The second change achieved by the doctrine of separate spheres was to transform the very way in which people thought about what it meant to be gendered. In a reciprocal kind of relationship, gendering the institutions of work and home in turn affected the content of gender roles for femininity and masculinity. The doctrine of separate spheres changed the way people conceived of what it meant to be a man or a woman in ways that are now difficult for us to understand because they have become so deeply embedded in our own gender ideologies.

The doctrine of separate spheres gives us the idea that a woman's place is in the home, and everything that ideology implies about gender. The most important component of this shift is the idea that men and women are to be perceived as fundamentally different from each other. These differences surface in a variety of ways, but most importantly in a shift from perceiving women to be lesser versions of men to perceiving women to be an entirely different species from their male counterparts. In the past, a woman could have some independent roles as a widow or unmarried woman and was considered to be a kind of "deputy husband" (Coontz, 2005, p. 170). Remember that under the previous system, there was some overlap between the tasks men and women performed, and women were perceived to be yokemates to their husbands, as members of the productive labor force. In this worldview, women certainly weren't as good as men, but they really weren't that different, either. With the doctrine of separate spheres, women were no longer allowed to act in any way that would even begin to suggest that they were like men.

Some specific examples help demonstrate how the separation of spheres radically transformed the meanings attached to being a woman or a man, creating this sense of fundamental differences between the two. Most traditional views of gender roles that emerged with the separation of spheres viewed mothers as the natural family member charged with the job of educating and caring for children. The qualities that make women well qualified for this role are their ability to provide maternal love and their superior feminine virtue (Laslett & Brenner, 1989). This "superior feminine virtue" originates in women's sense of morality as superior to that of men because women's place in the home shelters them from the worldly struggles for success and achievement. Explanations for this moral superiority were often biological, as Charles Darwin himself believed that women were "naturally more tender and less selfish than men" (Sayers, 1987). This superior moral compass didn't actually translate into real power outside the home for women, except as they were able to exert an influence on their husbands. Some scholars suggest the rewards of this domestic feminism were a better bet for many women than the abstract and uncertain promises of actual equality in the public realm. But the idea that women were possessed of a better sense of right and wrong than men was a relatively new concept, and especially the idea that this made women better suited to the job of educating and disciplining children.

Before the separation of spheres, women were viewed as too overly indulgent to be given the responsibility of encouraging sound reasoning in their children and of helping their offspring to restrain their sinful urges; only a father, the stronger of the two parents, was truly capable of the harsh discipline that was considered necessary to successfully raising a child to adulthood (Coltrane, 1996, p. 29). Although women did much of direct caretaking of infants and children in earlier periods, before the 19th century, fathers were thought to have greater responsibility for children and to exert greater influence in this area. Child rearing focused on the transmission of patriarchal authority and religious doctrine, and the idea that love and caring were necessary to these tasks would have seemed absurd to families before the doctrine of separate spheres. This was reflected in the tendency during this period for children to go with their fathers after a divorce, a practice that almost completely reversed in the 20th century. After the emergence of the doctrine of separate spheres, mothers rather than fathers came to be viewed as the expert parents and the family member most qualified and, therefore, largely responsible for the socialization of children.

A second way in which the separation of spheres influenced basic ideas about gender was in the connection between masculinity and domesticity. All things associated with the home became the exclusive realm of femininity, so any virtues that had once been associated with masculinity but stayed in the home were redefined. In the centuries leading up to these transformations in marriage and family, domesticity was an ideal toward which both men and women were assumed to aspire. The labor contributed by men and women was considered to be equally important to household functioning (Bonvilliain, 2007). Saving money, working hard, and keeping order are certainly not behaviors that would be particularly likely to be characterized as feminine today, but when men and masculinity were separated from the home, the meaning of domesticity itself changed. With it, the idea that a man should aspire to demonstrate domesticity also disappeared.

In these and many other ways, the doctrine of separate spheres created a fairly rigid gender system in which some very basic components of human behavior became strictly divided along gender lines. Moral virtue belonged to women, while the ability to pursue rational and intellectual interests became the territory of men. Women as mothers were perceived as responsible for "watching over the hearts and minds of our youthful citizens" (Matthaei, 1982, p. 113), and this was their role in forming national character. Along with moral virtue, women won a monopoly on sexual purity, while men were condemned to a titanic struggle with their unseemly physical urges. In this way, the ability to have a real sexuality was divvied up along gender lines. Women's role in the family became to love and nurture, and this was the meaning given to the tasks that were once considered to be part of the productive power of the household itself, while men's role became one of economic provision. Women continued to grow food for the family, tend animals, cook, repair household implements, and make clothing, but these tasks were now viewed not as economic activities but as acts of love (Coontz, 2005, p. 155). The doctrine of separate spheres gave us a new word, *housework,* to describe these same activities that women had always performed. Whereas once *all work* was housework, as all productive work was done within the household, now a new word defined what would, until the end of the 20th century, by and large be considered women's work.

Separate Spheres in Global Perspective

The doctrine of separate spheres should sound familiar to those of us in places like the United States because it is the basic content of the "traditional" family, the decline of which causes a great deal of worry for many people. In a broad survey of the history of families and the roles played by men and women in them, a period in which women were in theory expected to do nothing that was recognized as work is something of an anomaly. But this became the exact situation for some middle- and upper-class white women in the United States, some of whom were able to escape even housework through the use of servants and nannies. The fact that they had servants and nannies available to do housework for them means that those women (the servants and the nannies) *were* working outside of their homes and, therefore, that the ideal of the separation of spheres could never be realized for all families. Working-class families were less capable of achieving a family characterized by a male breadwinner, as working-class wives often worked to supplement family income. Immigrant women and African American women were also often unable to follow the doctrine laid down by the separation of spheres, and they continued to work in factories, as agricultural workers, and as servants. Although the rates of women in general participating in the paid labor force increased throughout the end of the 19th century and into the 20th century, the number of white married women working outside of the home in 1900 was less than 5% (Coltrane, 1996).

CULTURAL ARTIFACT 2: THE DIVISION OF HOUSEHOLD LABOR IN GLOBAL PERSPECTIVE

The doctrine of separate spheres developed in Anglo-European society, and although its ideology has spread to other parts of the world, it is still, in the end, one particular idea about how labor should be divided along the lines of sex category and gender. The amount of global variation in both the types of tasks that need to be done and the ways in which sex category and gender are used to make those divisions is vast. One study that used the collection of ethnographies from 185 different societies included in the World Ethnographic Sample found that two thirds of the 50 tasks identified

(Continued)

(Continued)

exhibited varying degrees of flexibility in regard to gender, in that the same task was perceived as primarily masculine in one society and primarily feminine in another (Murdock & Provost, 1973). What would happen if we shifted the language we used to describe the basic ways in which people in different societies decide who does what? As Shanshan Du (2000) suggested in her review of the extensive literature on this topic, the language of division suggests just that—a way of dividing women and men. In the new language Du suggested, what would a system of allocating tasks look like that was not necessarily based on division but instead on unity? Du found just such a system among the Lahu, an ethnic group in southwest China. Reinforced by their myths and religious beliefs, the Lahu believe that women and men share equal roles in fertilization and conception and that this joint responsibility should extend throughout child rearing and other tasks a household must accomplish together. This is exemplified in the strong role played by the husband during pregnancy and childbirth. The husband is expected to take over much of his wife's work during her pregnancy, and he participates in the pregnancy by monitoring his wife's experiences and bodily changes. During childbirth, the husband serves as midwife, assisting his wife in all phases of giving birth to their child and then taking care of both mother and infant during a set postpartum period. The Lahu principle of child rearing is consistent with this theme of unity in stating that the joint efforts of father and mother in child rearing "do not divide you and me" (Du, 2000, p. 535). Lahu parents take their infants and children with them into the fields to work, and in all facets of their life, they share responsibility for the care of children. Although mothers breastfeed, fathers are responsible for bathing infants, cleaning them up after urination and defecation, and often for holding the babies during the night to keep them from crying. In sustenance tasks, the Lahu live by the principle "work hard to eat," and here also they believe this is a task to be shared jointly by wife and husband. In her study, Du found that 51 out of the 63 households in the Lahu village she studied had an equal share of labor, both inside and outside the household. The cases where this labor was not shared equally were a result of illness or other social responsibilities that interfered with this general pattern. What does the case of the Lahu reveal about different ways of understanding the division of labor and how such a system is bound to lead to division? What does it say about the role of culture and ideology in shaping households in relation to gender? How would the household division of labor look differently under an ideology that suggested unity and joint effort, rather than division?

The doctrine of separate spheres presents an interesting case for the ways in which identities such as race, social class, and nationality can intersect with gender. The doctrine of separate spheres, based on the statistics mentioned previously, did seem to drive many white married women, and especially white married women with children, into the exclusive sphere of the home. But the doctrine of separate spheres has always been a *proscriptive* ideal, rather than a *description* of the way large numbers of people were really living. This was true during the time period in which the separation of spheres developed and is still true today. African American, Latino, and Native American families in the United States face discrimination that often prevents them from achieving the level of economic success necessary to conform to the male breadwinner model of family. This model assumes a man with a job that pays well enough to support everyone in the family, and this has never been universally accessible to all men in the United States.

Racial discrimination and blocked economic opportunities interact with native African cultural traditions imported to the United States to create family structures in some African American communities that are much more flexible than the doctrine of separate spheres would demand. In many of these communities, cultural norms contradict this ideology by suggesting that locating sole responsibility for the mothering of children in one person may not be wise or possible (Collins, 1990). **Othermothers** are women who may or may not be biologically related to children but play a crucial role in assisting bloodmothers with the task of caring for and raising children. Women-centered networks made up of grandmothers, sisters, aunts, or cousins can act as othermothers. Grandmothers became especially important to the welfare of children in many African American urban communities during the epidemics of drug and crime experienced during the 1980s and 1990s; as mothers and fathers fell victim to the ravages of drug use or imprisonment, many Black grandmothers took over the raising of their children's children. Carol Stack (1974) described these women-centered networks as *fictive kin* because many Black communities extended the boundaries of those responsible for the care of children beyond the limited pool of biological relatives.

Transnational Motherhood

Latino families in the United States have faced many of the same barriers of racism and reduced economic opportunities. Just as many Caribbean and African American women have left their families down South or in their native country to seek work in the North, Mexican and Latin American women now leave their children at home in their countries of origin to take jobs in the United States, creating a new arrangement some scholars call **transnational motherhood** (Hondagneu-Sotelo & Avila, 2005). Latina women are not immune to the cultural dictates of the doctrine of separate spheres originating in the world of white, middle-class families—an ideology that filters into Latin America through the institutions of industrialization, urbanization, and globalization. Within their own cultural traditions, Latina women experience pressure as a result of the rural legacy of women who in the past were more easily able to combine child care with productive agricultural work. In addition, the Catholic images of the Virgin Madonna create a strong tension between the ideals of motherhood and employment. From all these sources, Latina women face pressures to conform to the separation of spheres by dedicating themselves solely to the task of mothering and caring for the household.

But the economic circumstances of many Latino families make the possibility of a mother who stays home and dedicates all her time to child care beyond reach for many families. In their survey of 153 domestic workers in Los Angeles, Pierrette Hondagneu-Sotelo and Ernestine Avila (2005) found that 40% of these women had at least one of their children "back home" in their country of origin. This separation between biological mothers and their children, which sometimes lasts for decades, necessitates a redefinition of what it means to be a mother, given the ideal that biological mothers are meant to be the primary and most important caretakers of their children. Transnational Latina mothers make an important distinction between their own situations and those of mothers who may have become estranged from their children, abandoned them, disowned them, or put them up for adoption, all behaviors of which they generally disapprove. They emphasize that they are able to "mother" their children at a distance by talking to them on the phone or sending them letters and other gifts. One Salvadoran mother described this connection as "I'm here, but I'm there" (Hondagneu-Sotelo & Avila, 2005, p. 313).

Motherhood for these women is transformed to include aspects of the typically male, good provider role; mothers become breadwinners. But rather than replacing the typical caregiving role that characterizes motherhood, transnational mothers redefine breadwinning as a kind of caregiving while still seeking to maintain what limited aspects of traditional caregiving are available to them. In addition, many of these Latina mothers feel their work helps not only to improve their children's economic situation but also to shelter their children from the kind of discrimination and racism that they face in the United States. Keeping their children in their country of origin means they never have to experience the disdain many of these women feel from their employers and others in the United States. One mother who did bring her young daughter with her to the United States was given permission by her employer for her daughter to live with her and the employee's family. This was generous, but the employer also quarantined the daughter until she saw her vaccination papers and then would not allow her to interact with her own children. The Latina mother in this case felt there were emotional costs to being with her daughter under these circumstances.

Transnational Fatherhood

What about the husbands and fathers of these transnational families? If many women from the global South are becoming the breadwinner, how does that affect their husbands? Although a great deal of research has been done on women who migrate for work, there is less information about what happens to the men who stay at home. In their research on transnational families in Vietnam, Lan Anh Hoang and Brenda S. A. Yeoh (2011) investigate "left-behind" husbands. In Vietnam, the husbands of female migrants struggle with the implications of their wives' absence and their role as primary breadwinners. In interviews, the men often emphasized that their wives' decision to migrate was made only with their permission, asserting the Confucian view of men as the "pillar of the household" (p. 722). Among the 23 left-behind fathers the researchers interviewed, only one did not still do paid work. Although the fathers struggled with balancing their work with the demands of caring for children, they all felt it was important to continue working for pay to avoid being perceived as a man who was sponging off his wife.

In places like Sri Lanka and the Philippines, female relatives are the primary caregivers in families with a migrant mother (Save the Children, 2006). Studies on Mexican transnational families have similarly demonstrated a preference for maternal grandmothers to take over the care of children in their mother's absence (Dreby, 2010). But in Vietnam, survey research indicates that 71% of children in families with a mother who had migrated were cared for by their fathers; only 18% were cared for by their paternal grandmothers. Some of the men in Hoang and Yeoh's (2011) study had been involved in parenting even before their wives migrated. The researchers found that Vietnamese men had a flexible view of masculinity that allowed for them to prioritize the presence of a parent in their children's lives—even if that parent was the father. In some cases, men had left their own jobs overseas when their wives decided to migrate to come home and care for children. Although ideas about masculinity sometimes made it difficult for left-behind men to negotiate their roles, they did not prevent men from becoming primary caregivers in their wives' absence.

On a global scale, the experiences of these transnational families are probably more typical than a family that perfectly conforms to the doctrine of separate spheres. Where women around the world do not engage in paid work in the formal economy, they still work to feed and clothe their children. Women are also much more likely, especially in the developing world, to work in the informal economy; the most recent UN report indicated that 60% of women in the developing world work in the informal economy (Braun, 2005). Informal economic activities can include illegal activities such as prostitution or selling and manufacturing drugs, which are not included in the formal calculations of economic activity because they are off the books. But they can also include women who sell their goods at a market, or people who are paid for legal services off the books. Some estimates suggest that more than half of agricultural work worldwide is done by women. These statistics are important to keep in mind to demonstrate the ways in which the doctrine of separate spheres, assuming it is a desirable or optimal form of family organization, is feasible for a very small percentage of the world's population. Yet the power of its ideology is exerted within the United States and around the world in its pressure on determining how families should divvy up the important tasks of who does what.

THE DIVISION OF HOUSEHOLD LABOR

The legacy of the doctrine of separate spheres is still very much with us today when we think about what it means to be in a family and what the respective roles of women and men are within this institution. The separation of spheres sets the terms for the current debates about how two of the most important institutions in our lives—work and family—should interact. For men, the separation of spheres made it very difficult to do anything but work for money outside of the home. Some sociologists have identified this as *the good provider role* for men. The **good provider role** dictates that a man's primary obligation to his family is to work so as to meet his family's material needs. A good father, according to the good provider role, does not need to perform the day-to-day tasks of caring for his children or keeping a household running. Unlike in previous centuries, he is not responsible for the religious education of his children or for ensuring their moral education. The good provider role narrows a man's obligation to his family to its economic value and, in this sense, sets up a tension between the world of work and the world of home for men. A man who is fulfilling his main obligation to his family according to this role will spend as much time and energy as possible trying to make money or increase the amount of money he can contribute to his family. In this way, the good provider role puts a great deal of stress on men who are not able to fulfill this financial obligation to their family, and it causes stress for men who are able to do so about the prospect of losing that ability. In a sense, the separation of spheres and the good provider role encourage men to focus more of their time and energy away from their family, as well as potentially intensifying the stress they feel about their work lives.

This effect of the doctrine of separate spheres on masculinity helps explain another element of contemporary Anglo-European manhood. We discussed the first three elements in the last chapter: "No Sissy Stuff," "Be a Sturdy Oak," and "Give 'em Hell." The fourth is **"Be a Big Wheel,"** and it emphasizes the way in which masculinity is measured by power, success, wealth, and status (Brannon, 1976). The expression, "He who dies with the most toys wins," is a good encapsulation of the "Be a Big Wheel" ideal for masculinity. In the good provider role, economic success is specifically linked to the ability to provide for a family, while the "Be a Big Wheel" dictate is largely unconnected from the family. You might consider "Be a Big Wheel" to be the good provider role on steroids; exactly how much money do you have to make to be a good provider? This transformation represents a general shift in masculinity in the United States from ideals that were based on men's role in the family, to an increasing emphasis on men's success in the marketplace.

Can you think of other examples of the cultural imperative for men to "Be a Big Wheel"?

During the early colonial and revolutionary period in the United States, men like George Washington, Thomas Jefferson, and Paul Revere represented the ideals of manhood for that time period. Men like Washington and Jefferson were refined, elegant intellectuals who would not have considered it amiss to admit to a bit of casual sensuousness (Kimmel, 2009). Their identities didn't depend on their success in the market or on making lots of money but on their status as landowners. This ideal of the **Genteel Patriarch,** represented by men like Jefferson and Washington, doted on his family and spent much of his time in their company or supervising his estate; think of Washington's wish to be able to return to his home, Mount Vernon, after serving as President. For men who did not own large plantations, there was the **Heroic Artisan** ideal. Men like Paul Revere and yeoman farmers embodied physical strength but also dedication to their country and the budding ideals of democracy. The Heroic Artisan was also a family man, teaching his son his trade or craft. But like the Genteel Patriarch, the ideal represented by the Heroic Artisan was dedication to others and to economic autonomy, and not necessarily to overwhelming success in the marketplace. For both of these models of manhood, service to others (family and community) was more important than individual success, and acquiring the most toys was not at all an important way of measuring manhood.

How did we end up with the imperative on men to "Be a Big Wheel"? With the increasing industrialization of the United States, **Marketplace Manhood** increasingly replaced the ideals of the Genteel Patriarch or the Heroic Artisan. Under Marketplace Manhood, men derived their identity from their success in the capitalist economy in accumulating wealth, power, and status. Unlike the Heroic Artisan and the Genteel Patriarch, whose status was based on stable sources (land and a craft), the Marketplace Man constantly had to prove his masculinity in the winner-take-all world of the marketplace. The Marketplace Man also differed from previous ideals of manhood in that his masculinity was based firmly in the world of the market and other men rather than in his role within the family. To achieve success within the market, the new man had to spend most of his time away from his family, his home, and his estate. Marketplace Manhood follows the doctrine of separate spheres by effectively detaching what it means to be masculine from the context of the family.

For women, the doctrine of separate spheres tells them not just that the home is their natural environment but also that any desires they might have to explore the world outside of the home are a violation of femininity. Early in the development of the separation of spheres, women who were not able to stay in the home were considered to be immoral, degenerate, "fallen" women, and the line between housewife and harlot was a thin one indeed (Coontz, 2005, p. 169). Doctors of the time period worried that women who sought an education had "masculine brains" that would prevent them from being able to fulfill the task of motherhood (Bonvilliain, 2007). In the late 19th century and early 20th century, the doctrine of separate spheres was so powerful that any woman who intentionally violated it could simply not be a real, biological woman. Certainly, we no longer question the biological femaleness of women who work outside of the home, but the idea that taking care of children and the home is an important duty, and perhaps something that women are more inclined to want to do because of their natural instincts, is still very much with us today.

The Second Shift

If you question whether this is necessarily the case, you only need to look at the consensus of research social scientists have done in the past 60 years or so on the division of household labor. The division of household labor is a concept that makes sense only after the doctrine of separate spheres, in that it makes a distinction between all the labor that happens outside of the house and the labor that occurs within the household. In this sense, the **household division of labor** is the way in which the tasks necessary to the care and running of a household are distributed. The content of these tasks varies depending on exactly who is in the household and what the household is like, but in places like the United States, these tasks generally include child care in families with children, as well as cooking, cleaning, laundry, yard work, household maintenance, financial accounting, kin-keeping (maintaining relationships with extended family), and car maintenance. As more married women with children began entering the labor force in the United States in the 1960s, social scientists became increasingly interested in how families handled the household division of labor. All of the studies have reached consensus on two facts, using a wide variety of research methods. First, women do a larger share of household labor of all types than do men (Blair & Johnson, 1992; Coltrane, 1996; Lewis, 1992).

In her famous book on this topic, Arlie Hochschild (1989) labeled this the *second shift*. The **second shift** is the extra burden of child care and housework added on for women who also do paid work outside of the home, and in Hochschild's study, this second shift meant that women worked an average of 15 hours longer per week than men, adding an extra months' worth of a 24-hour workday each year to a working woman's life. The second finding from this body of research is that women and men perform distinctly different types of tasks within the household. This is true even for **dual-earner couples**, or couples where both spouses are working, and for those couples who are deeply committed to a more egalitarian division of household labor. A recent article in *The New York Times* on families who are attempting to achieve equality in their division of labor described classes and workshops families can attend to help them in this goal, including complicated spreadsheets to track the amount of time each spouse spends in various tasks (Belkin, 2008). Why is creating an equal division of who does what in the house so difficult that couples seem to need to take classes to achieve this goal?

One interesting way to try to answer this question as to why an equal division of household labor seems to be so difficult to achieve is to consider what happens when we take gender out of the equation, assuming for the moment that such a thing is possible. Certainly, in today's society, we can't observe a family of genderless people raising children; gender identities may be shifting in very interesting ways, but no one can yet claim to be genderless. What we can look at are situations in which the gender of the adult couple that forms a family is the same. Gay and lesbian families provide an interesting case where two people of the same gender form a household and are faced with the same set of decisions about who exactly will do what in this particular family situation. Regardless of your sexual identity, there's garbage to be taken out, meals to be provided, clothes to be washed, and some level of cleanliness and order to be maintained. When you add the increasing numbers of gay and lesbian couples who have children through artificial insemination, adopt children, or who already have biological children from previous heterosexual relationships, you add to the mix of household responsibilities the task of caring and providing for children. In these situations, there are no gender norms to provide a guide for who should do what. How do gay and lesbian families decide who makes the bed, cleans the toilet, changes diapers, or puts the children to bed?

In theory, many scholars have argued that the division of household labor among gay and lesbian couples should be more egalitarian than it is among heterosexual couples. In the absence of gender, no one partner is expected to do more of the household labor or to take on certain types of tasks rather than others. This happens in heterosexual couples because of the power of gender as the easiest way of making the difficult decisions about who should take responsibility for all the tasks associated with being a family. There are competing theories about exactly how this works, drawing on the different gender theories we discussed in Chapter 2.

Theory Alert!
Institutional

An institutional approach to explaining the gendered household division of labor points to the ways in which marriage as an institution comes with its own set of norms about how a family should work. What individuals want or desire in a particular situation is less important than the restraints placed on them by their position within an existing institution (Dalton & Bielby, 2000). Institutions like marriage and family provide social scripts for actors that serve as guides for appropriate action. Since the separation of spheres within the family, these social scripts influence the personal attributions of couples, their habits, and the practical actions they take (Dalton & Bielby, 2000). In the context of the family, this includes who should be the provider, how household labor should be divided, and who should organize family dinners, just to name a few. Marriage and family as gendered institutions give us easy answers to these questions; men should be the providers, women should do child care while men do outdoor household chores, and women should be in charge of organizing family dinners. Escaping this prescribed division of household labor is difficult, from an institutional perspective, because the desire of any individuals to do things differently always encounters the power of institutions to guide our actions. Even our attempts to violate the norms of an institution will still be responses to those guidelines. For example, gays and lesbians who marry are in many ways offering a challenge to the institution of marriage and seeking to alter that institution in important ways. But in posing this challenge, most gay and lesbian couples still follow the institutional norm of engaging in a wedding or commitment ceremony. Their challenge to the institution of marriage is also conditioned by its norms. From this perspective, wives and husbands as well as mothers and fathers find escaping the prescribed household division of labor difficult because institutions exert a powerful influence on our behaviors. Although institutions may give us some wiggle room, we remain largely within the power of their grip. Another theoretical perspective often used to explain the persistent power of the gendered household division of labor is doing gender.

Theory Alert! Doing Gender Theory

From the doing gender perspective, the most important location for gender is within our day-to-day interactions. Marriage as an institution may place some constraints on our behavior, but women and men follow gender norms about the household division of labor primarily because this is a way to produce an accountable performance of gender. Gender is performed in the ways in which our behaviors are understood by someone else as representing femininity or masculinity, thereby indicating that someone is male or female. This is true regardless of whether the person actually is male or female, as the case of transsexuals and cross-dressers demonstrates; we can make assumptions that someone is biologically female even if they are not. The household division of labor is an important means within family systems for the performance of gender. A man can produce an accountable performance of gender by working to support his family, placing a priority on his role as a breadwinner, and demonstrating less of a sense of responsibility for child care and housework. An accountable performance is not necessarily a performance that follows gender norms perfectly; fathers who are the primary caregivers for their children are not mistakenly assumed to be female. But their behaviors are always understood within the framework of gender expectations about the division of labor; men who do not work outside the home but take care of their children full time are congratulated for their actions, especially by women. A woman who stays home full time to take care of her children is not likely to be praised for such an activity because it is what is expected of her. According to a doing gender perspective, the household division of labor remains in place because of its importance as a resource for the performance of gender.

Gay and Lesbian Households

What happens to these theories in the context of families formed by same-gender couples? Some evidence suggests that the division of household labor among gay and lesbian couples is more egalitarian than in heterosexual couples (Blumstein & Schwartz, 1983; Dalton & Bielby, 2000; Kurdek, 1993; Kurdek, 1995). Other research indicates that there are differences between the division of labor in gay male versus lesbian couples; gay men seem to be more likely than lesbian couples to follow a traditional division of labor, in this case based on a division between breadwinner and caretaker (McWhirter & Mattison,

1983; Peplau & Cochran, 1990). In these couples, the division of labor mirrors the ideal in heterosexual couples, where one partner works and is viewed as supporting a second partner, who is viewed as responsible for taking care of the household. The main difference is that in the case of gay men, both of these partners are men. These differences between lesbian and gay male couples may be traceable to lesbian couples being more invested in ideals of fairness and in maintaining their independence than are gay male couples (Blumstein & Schwartz, 1983; Clunis & Green, 1988).

Much of this research focuses on gay men and lesbians in households without children, and there seem to be important differences between what happens in households with and without children. The disparities in the amount of household work that women do in families relative to men is most pronounced during the early years of child rearing, as caring for infants and small children creates an intense new level of household work for parents (Suitor, 1991). What happens in gay and lesbian families with children? Maureen Sullivan's (1996) study of lesbian co-parents explored the possible criteria that might be used as a basis for structuring the division of household labor in the absence of parents of a different gender. For the lesbian co-parents in Sullivan's study, as well as for many gay and lesbian families, this division could potentially be based on biological parenthood rather than on gender. This could be true both for gay and lesbian couples who have children together, as well as for gay and lesbian families that were formed by the children of one partner from a previous heterosexual relationship. In either case, one partner is biologically related to the child or children, while one partner is not. In Sullivan's study, this was not the case even for lesbian mothers who had carried and given birth to the child; the biological mother did not automatically become the primary caregiver. Another criterion that might be used to divvy up household tasks is the characteristics of either partner's job. For example, perhaps the partner with the more flexible schedule comes to do a larger proportion of housework and child care. Or the partner who contributes more to the household income and has the higher salary could be exempted from more of the household work because she or he is contributing in different ways to the family.

Sullivan (1996) found that most of the lesbian co-parents in her study did not use any of these possible criteria to create an unequal division of household labor but took conscious steps to ensure a more egalitarian household. For some of these couples, this meant that both partners, rather than just one, reduced their work hours to accommodate child-care responsibilities and share them equally. For other couples, one partner did much of the child care during the week, while the second partner did most of the housework and child care on the weekends. Sullivan described this as a strategy of taking the second shift and dividing it into **quarter shifts**, smaller portions to be distributed more equally among household members. Other couples balanced the division of labor through what Hochschild (1989) identified as a **reduction of needs**. In this strategy, one partner who is "unwilling or unable to maintain the home according to certain standards simply lower household expectations by denying that work needs to be done" (Sullivan, 1996, p. 758). In heterosexual couples, this is often a strategy used by husbands to exempt themselves from housework; a husband whose vacuuming is not up to his wife's standards may argue that it's not really important that the vacuuming be done to meet those standards or may simply refuse to do the vacuuming at all because he cannot do it well. In Sullivan's research, lesbian co-parents used this strategy to manage housework with apparently little tension. Among this sample of lesbian co-parents and their children, it seems that most families are able to escape the need to perform gender through the household division of labor. In their family's lives, these couples seem to be able to undo rather than do gender.

Returning to our earlier question as to why the traditional household division of labor seems so difficult to violate, the case of these lesbian co-parents suggests that gender is, in fact, the important variable. In the absence of gender, married couples and families are better able to create a more equal division of labor. Heterosexual couples, on the other hand, are compelled to use housework and child care to create accountable performances of gender. But the picture is not quite so simple. Sullivan (1996) found some of the lesbian co-parents in her study followed what she described as a "Rozzie and Harriet" (p. 756) pattern, in which they conformed to the traditional, heterosexual ideal of family with one, sole breadwinner and a dependent partner who was completely responsibly for home and children. This often meant that one partner gave up her career completely to stay home and take care of the child or children. These situations sometimes created tensions similar to those experienced by heterosexual couples who follow this model. The partner who serves as the primary caregiver seems to have less power in the relationship as a result of her lack of economic contribution to the family and, therefore, often less power to negotiate for time off from her child-care responsibilities. This is in part because these families follow the dominant division of household labor, including the idea that the work of child care and housekeeping is not really valued as work in the same way as is work outside of the home. Thus, the job of primary caregiver is perceived as not as important as that of financial provider.

Sullivan (1996) suggested that the small group of lesbian couples in her study who followed this more traditional model of family life did so because of their desire to provide their children with what they often called a "sense of family" (Sullivan, 1996, p. 763). The ability of one parent to stay home full time helps to create parent–child bonding that is crucial to the development of a strong sense of family for these couples. This sense of family is important for these couples in part as a future resource for their children in dealing with the homophobic oppression that they will surely someday face. Thus, they view themselves as arming their children against future discrimination and marginalization by providing a good family background.

What is interesting about this argument, as Sullivan (1996) pointed out, is that it is a largely heterosexual form of family structure that these couples use to arm their children against future homophobia. The breadwinner/caregiver model of traditional heterosexual families, it is hoped, will provide the sense of family necessary to facing a homophobic world. The insistence on the importance of instilling a strong sense of family through following these norms also lends support to an institutional approach to explaining the household division of labor. When these lesbian couples seek to create good families for their children, the institution of family provides a ready-made guide for the proper way to do so. In a good family, our culture tells us, one person works and one person stays home to take care of the children. Just because some of these lesbian couples were not following the family norm that says one of these persons should be male and one should be female does not mean they were not motivated to follow these other norms related to the structure of family life. Rozzie and Harriet couples suggest the power of family as an institution for everyone in society, dictating how we make decisions about the division of tasks on a daily basis.

How do you imagine the household division of labor might be different in different family structures? In a single-parent household? An extended family household (with grandparents, parents, and children all living together)? In a "family" composed of friends living together?

Power and the Household Division of Labor

You might ask yourself at this point exactly what's so bad about this separation of the world into public and private, good providers and housewives. Is it really so bad for a woman to want to take care of her children and her house, while her husband works to support the family? Aren't these both pretty important tasks that need to be accomplished? These are all good questions, and they lead us to one of the central questions surrounding gender issues in general. Is it OK for men and women simply to want to do different types of things? If we assume that gender equality is something that we want, does equality also mean that everyone does the same thing? Can you have equality and still acknowledge that maybe women are somehow better or more inclined to do things like take care of children? Why does that necessarily have to make women and men unequal?

These are big questions, without any definite answer. In the specific case of the separation of spheres and the division between paid work and housework, our culture places a great deal more value on paid work. Research suggests, as in the case of the Rozzie and Harriet couples mentioned previously, that the amount of power a woman has in a family is related to the income she contributes (Hochschild, 1989; Juber, 1988; Kamo, 1988). In India, many women who are part of dual-earner couples are able to negotiate a reduced share of housework for themselves because of their economic contribution to the household. Women who do paid work outside of the home seem to have a greater say in important decisions that impact the entire family, such as buying a home, moving, changing jobs, how to educate children, and who will do housework (Baxter, 1992; Coleman, 1988). Within the family itself, then, women's role as caretakers does not seem to be viewed as equally important to the role of economic provider; the person responsible for primary caregiving and housework loses out within the power dynamics of the family unit.

Researchers and feminists suggest that this loss of power experienced by women who work inside the home translates to a lack of power outside of the family as well. In this argument, the household division of labor is an important part of explaining women's inequality across society, beyond the borders of the family as an institution. The tasks of taking care of a home and children are devalued within families and in the wider society at large. In one study that evaluated the amount of skill necessary to do different jobs (and the amount of pay appropriate for the job), evaluators ranked attendants at dog pounds and parking lots over nursery school teachers and ranked zookeepers more highly than day care workers (England, 1992). These evaluations suggest that taking care of animals and parking lots is perceived as having greater value and requiring more skill than taking care of small children. That women are of the gender more likely to be engaged in these tasks means that women themselves come to be devalued. Paid child-care workers in day care centers or those who work in private homes as nannies are some of the lowest paid workers, demonstrating the monetary value that is placed on these tasks when they are done by persons outside of the family.

The United States specifically provides very little government support for parents who would like to take time off from the labor force to do child care; the United States and Australia are the only industrialized nations that provide no federally mandated paid parental leave. The lack of any policy that would make child care easier for working parents can be perceived as an indication of the low priorities placed on the act of caring for children, especially when compared with the government support provided for other services that are important to the society as a whole, such as military service. Although prior to the doctrine of separate spheres much of the work done by women and men within the household was considered to be productive labor, in today's world, most of the work that is done inside the home is not considered real work at all. The hard task of keeping a household running and caring for children is something that becomes associated with the private, feminine sphere. Tasks that were once considered essential components of a productive and stable household, such as preparing meals, are now viewed as reflections of women's emotional and nurturing nature, existing outside the rationality of the public sphere. The area where women's traditionally primary role as caregiver perhaps has the greatest effect outside the family is in the world of work, and we will explore these connections in more detail in the next chapter.

Do you agree that the household division of labor and the doctrine of separate spheres are important sources of gender inequality? Are these differences that necessarily lead to inequalities?

For now, we can say that one of the most important end results of the doctrine of separate spheres was the creation of not just a difference in how we think about what men and women do but also a hierarchy in how those tasks are valued. The separation of spheres created important splits between what we view as masculine and what we view as feminine, both within individual people and in dividing specific areas of society. Just as the home became feminine and the workplace became masculine, all the qualities associated with home became the exclusive territory of women and all the qualities associated with the workplace were distributed to men. If these two life areas—public and private, work and home—were truly valued equally, the consequences would not be as great. But in cultures like the United States, we believe that the important stuff of the world takes place in the public realm, and defining that world as masculine puts women at a distinct disadvantage.

FAMILIES IN TRANSITION

You may have noticed that in many of today's societies, people seem to be pretty worried about both the state of marriage and the related state of what they often call *the family*. At this point, a phrase like *the family* might make you a little nervous. What exactly do people mean when they talk about *the family?* Do they mean your family? Do they mean families in Italy or on Native American Indian reservations in the Southwest? Is there something called the family that covers all those situations? Unfortunately (or fortunately, depending on your point of view), family is no easier to define than is marriage. The U.S. Census needs a definition of family that they can use to go out and measure the status and characteristics of this thing called family. According to the Census, a family is a group of individuals who cohabit and are related by blood in the first degree, marriage, or adoption. This definition is limited in that it excludes gay and lesbian couples, as well as any group of people who are not related by blood, marriage, or adoption but think of themselves very much as family—what social scientists are now calling **families of choice** (Weston, 1991).

The idea of families of choice originated within gay and lesbian communities, where individuals were often cut off from their biological families and therefore formed new family relationships of their own choosing with people not necessarily related to them by blood. The idea has spread beyond gay and lesbian communities to include urban people in general who form tight-knit groups of friends, neighbors, or coworkers who serve many of the functions normally reserved for family. If this is hard to imagine, you need only think of television shows such as *Friends* and *Sex in the City,* in which the relationships among friends are the focus much more so than the relations within a biological family; when the characters on these television shows need help or support in any form, it is their friends to whom they turn rather than their biological family members.

If the modern family most closely resembles the nuclear family we described previously, families of choice represent what Judith Stacey (1996) called the *postmodern family*. Remember, the nuclear or modern family consists of a husband and wife (one male and one female), living in the same house with their biological children, where the father is expected to provide financial support while the mother takes care of the home and children. The **postmodern family**, in contrast to the modern family, is a changing, unsettled, recombination of different family forms. If you remember our discussion of postmodernism from Chapter 3, you'll remember that postmodern thinking emphasizes the problems in meta-narratives, in one story that seeks to explain all the diverse experiences that exist in the real world.

The idea of the postmodern family acknowledges the difficulty in figuring out exactly what makes a family—precisely because the form that real families take is always changing. In addition, any attempt to come up with one definite, all-inclusive definition, like the one used by the U.S. Census, is automatically going to exclude some people who seem pretty family-like in important ways. The postmodern family, then, can include the families of gay and lesbian couples with biological or adopted children, as well as the family communities formed by gays and lesbians, as well as heterosexuals who consider close groups of friends to be more family-like than their biological families. The postmodern family also helps to account for the experiences of transnational families, like the Latina workers in Los Angeles we discussed previously, who often work to maintain the ties of family across great distances and with increasing mobility (Zinn, 2000).

The postmodern family certainly seems to be a better way of making sense of the actual reality of what family life increasingly looks like in places like the United States and around the world. As an interesting experiment, you might go around the room or take a mental census of your group of friends, classmates, or coworkers and count how many of them are the product of a modern family. That is, how many people do you know who grew up in a family with a mother and father who were married to each other, are still married to each other, and have only ever been married to each other, where mom stayed home and dad worked, and all the children were biological? Assuming you know anyone who fits this profile, what are the chances that they'll marry only once, stay married, have biological children, and conform to the gender norms of mom staying home and dad working? Statistics tell us that the chances that you know someone who fits this particular family profile are fairly slim. In the United States, only 44.6% of all married couples with children younger than the age of 18 (between 2000 and 2003) were living in the same house with those children.

For the first time, according to census records in 2005, married couples became the minority category among all households in the United States, meaning that most American households do not include a married couple (Roberts, 2007). Fifty-one percent of women in the United States live without a spouse, which means they either don't have a spouse at all or their husband is no longer living with them. Only 68% of children in the United States in 2007 lived with two, married parents, as compared with 77% in 1980. For African American children, only 35% live with two parents who are married to each other (Winerip, 2007). The percentage of children living only with their mothers in 2007 was 23%, while 3% of children live only with their fathers and 3% live with two unmarried parents (Federal Interagency Forum on Child and Family Statistics, 2008). In 2006, 38% of all births were to unmarried women. Although the overall divorce rate in the United States has fallen (currently, 3.6 divorces per 1,000 people compared with 5.3 in 1981), the rate of divorce after remarriage has decreased relative to rates in the 1950s, and currently, 43% of all marriages will end in divorce within 15 years (Coontz, 2005). Both women and men are waiting longer to marry, and increasing numbers of people in the United States and Europe are living together and having children together without ever marrying, including large numbers of heterosexual couples who *could* marry if they so chose (unlike gay and lesbian couples in many places who could not legally marry) (Kiernan, 2002). Experts estimate that 50% of children in the United States will spend at least part of their lives in a household that does not contain both of their married, biological parents (Coontz, 2005). All these statistics add up to fewer and fewer people who are raised in the modern, nuclear family and to fewer and fewer people who are choosing to form these kinds of families themselves.

If we remember that part of the modern family also includes the ideal of the male sole breadwinner, dependent children, and mother as caretaker and homemaker, the percentage of families conforming to this ideal becomes even smaller. In 2005, only 1.5% of the couples with children younger than the age of 18 in the United States had a sole male breadwinner, meaning that only the father was employed (U.S. Census Bureau, 1999). Among women with children younger than the age of six, 62% are in the labor force, with considerable variation among racial and ethnic groups. Among African American women, 89.6% of these women are in the paid labor force, compared with 61.1% of white women, 57.9% of Asian women, and 51.0% of Hispanic women (U.S. Bureau of Labor Statistics, 2005).

In the late 1990s, some statistics seemed to indicate that the labor force participation of women with children had slowed and was perhaps beginning to decrease; in 2000, the labor force participation of women with infants less than one year of age went down instead of up for the first time in a quarter century (Coontz, 2005). Media outlets such as *The New York Times Magazine* and *Time* magazine used these numbers to claim that increasing numbers of women, especially highly educated women, were deciding to forego their careers to stay home with their young children. Yet women with children aged one and older now have the exact same rates of labor force participation as childless women. This is true even among mothers with increasing levels of education, as more than two thirds of women with college degrees and three quarters of all women with graduate or professional degrees who have children younger than the age of six were in the paid labor force in 2002 (Coontz, 2005). Rather than the father as sole breadwinner, in 30% of households in the United States, the working wife earns more than her husband, and in more than 2 million American households, working fathers provide primary child care for their children while their wives work (Coontz, 2005).

This is a small slice of the extensive mountain of statistics that can be assembled to demonstrate the increasing rarity of the modern family. Often, these statistics are assembled to sound a warning bell about the demise of marriage and the family. In this story, the modern, nuclear family is quickly becoming extinct relative to its predominance in some mythical past. We've already discussed that this story about the demise of marriage and the family all depends on what your particular historical reference point is. Even if we take the 1950s, post–World War II United States as our starting point for the decline of marriage, the story doesn't hold true. Behind the façade of the happy nuclear family of the 1950s, 20% of all couples considered their marriage unhappy and between one quarter and one third of all marriages of the time ended in divorce (Coontz, 1992). Still, it might be a difficult prospect to give up completely on the idea that marriage and families weren't simpler or better at some point in the past. We might take some comfort, though, from thinking about the many different forms that these two institutions have taken over time and across different places. There is no one way to do marriage or family successfully; much depends on how these institutions fit into the larger context of the society or culture in which they exist.

The modern family, as we will discuss in the next chapter, is a current impossibility, in part because of the economic structure of our societies. It is simply not possible for many families around the world to survive on only one income, and this has been impossible for many families for a very long time. It is also probably no longer possible in places like the United States to return to marriages that are not based on love and choice; immigrant groups with a tradition of arranged marriages soon assimilate to a model of love marriage, and you'd be hard-pressed to convince most Americans to marry someone chosen for them by their parents. On the other hand, despite the many changes and variations in the exact meaning of marriage and families, these institutions continue to persist across all cultures. What we call a family in the 21st century may look very different, but we still place a great deal of importance on having a group of people who serve those particular functions. Despite the increasing age at first marriage and the divorce rate, most people in the United States will marry at some point. What marriage means to them may change, but tying the knot seems still to be a fairly popular thing to do.

In relation to gender, the case of marriage and families demonstrates the complex ways in which institutions become important sites for establishing gender differences and hierarchies. We can debate whether the primary purpose of marriage is to oppress women, but when we examine the household division of labor, it does seem that families contribute to the maintenance of existing systems of inequality. Changes to these institutions have important gender implications for society as a whole. The separation of spheres radically transformed family life in ways that still influence us today and had deep-seated implications for the way we think about gender. Some changes to marriage and families as institutions have helped to lessen gender inequality. No-fault divorce laws made it much easier for women to escape abusive relationships. But marriage and families are also some of the most personal and therefore most controversial sites for making gender distinctions. The gay marriage debate is in part a debate about whether marriage is an institution that is inherently gendered or whether it can be separated from the idea of woman and man. In this way, discussions of marriage and families go to the heart of questions about whether the project of ending gender inequality is also a project of ridding ourselves of all gender differences.

BIG QUESTIONS

- In this chapter, we explored the difficulty of defining basic concepts like marriage and family. What makes coming up with one definition important? Why do we look for definitions, and what kind of power do definitions have?
- Based on what you read in this chapter, how would gender be different in a society with no marriage and no family? What would marriage and family be like in a society with no gender?
- In this chapter, we explored many of the historical changes in ways of understanding marriage. How might this discussion inform current debates about marriage and, specifically, regarding gay marriage? How would this change, allowing gay men and lesbians to marry, change the ways we think about marriage, and how would those changes compare with changes that have already occurred in relationship to marriage?
- From the institutional approach discussed in this chapter, institutions such as marriage have the power to dictate our behaviors even when we don't necessarily agree with the values reflected in those institutions. Do you agree that institutions have this kind of power? How else might you explain the persistence of a social phenomenon like the unequal household division of labor even among couples who hold egalitarian values?
- In this chapter, we discussed the important ways in which gender, marriage, and the family are all connected. How would you describe this as a causal relationship? Does gender affect the shape of marriage and families, or does the structure of marriage and families affect the particular gender system in a given society? How could you determine which is the proper direction for this relationship?
- Why do you believe many people hold up the idea of a traditional family or the modern marriage as a benchmark against which families or marriages are judged? Are certain forms of family better than others, and why would this be so? How is gender part of the arguments over what makes the ideal family form or the ideal marriage?

- The sexual division of labor and the division of household labor are both good examples of gender differences that seem to lead to gender inequalities. Does this have to be the case? Is a division of labor necessary to society, and what might a division that is not based on sex category or gender look like? How could you imagine these differences existing without leading to inequality?

GENDER EXERCISES

1. Write an essay in which you describe what the division of household labor was like in your own house growing up. Who did what, and how did everyone seem to feel about it? Were there tensions over who did what tasks? What would be your ideal division of household labor if you were to form your own household?

2. Interview some couple about their division of household labor. You might interview both together or each separately. Come up with an extensive list of all the tasks that are involved in maintaining a household, and ask your respondents who does each of these tasks on a regular basis. You might also ask them what they feel the ideal division of household labor is and how close their own situation comes to that. How does your interview line up with the research in this chapter?

3. Make a list of all the tasks that are essential to the running and maintenance of a household. Include a household with children and one without children. Then investigate how much it would cost to have all of these tasks done by someone in the paid labor force. For example, for laundry, find out how much it would cost to have laundry done by a laundry service, and for cooking, how much it would cost to hire a personal chef. Based on your calculations, how much are the services of someone who labors inside the household worth?

4. What's the relationship between weddings as ceremonies and what we believe about marriage as an institution? Does what happens at weddings reflect our expectations about what will happen when people are married? Do some research on wedding ceremonies. This might involve observing at actual wedding ceremonies, doing content analysis of wedding magazines, or gathering information about wedding ceremonies in cross-cultural perspective. What are the gendered implications of wedding ceremonies? Can you use wedding ceremonies to make presumptions about what couples believe about marriage itself?

5. You could argue that one way to measure how much a society values child care and child rearing is in how much support their government provides for these activities. Do some research on government policies in different countries toward families, marriage, and child care. What are the benefits provided by different governments to encourage certain kinds of behavior related to these institutions? For example, in the United States there are tax benefits to be gained through marriage, while in Holland, the government provides a yearly subsidy for all parents for each child they have. Which governments seem more or less family, marriage, and child care friendly?

6. How does popular culture line up with the idealization of modern marriage and the nuclear family? In this chapter, we discussed some basic statistics reflecting how many people in the

United States conform to these two ideals. Do your own research on these demographics. Then, do a content analysis of some popular culture format, focusing on their representation of families and marriage. Do the media accurately reflect the changing reality of marriage and family types?

TERMS

marriage gradient
protection theory of marriage
nuclear family
exploitation theory of marriage
cooperation theory of marriage
love marriage
social institution
polygamy
polyandry
demographics
sex ratio
marriage squeeze
endogamy
diaspora
transnational marriages
remittances
sexual division of labor
subsistence agriculture
modern marriage
doctrine of separate spheres
public sphere
private sphere
othermothers
transnational motherhood
good provider role
"Be a Big Wheel"
Genteel Patriarch
Heroic Artisan
Marketplace Manhood
household division of labor
second shift
dual-earner couples
quarter shifts
reduction of needs
families of choice
postmodern family

SUGGESTED READINGS

Marriage and families in general

Coltrane, S. (1996). *Family man: Fatherhood, housework, and gender equity.* New York: Oxford University Press.

Marriage and families in cross-cultural perspective

Amadiume, I. (1987). *Male daughters, female husbands: Gender and sex in African society*. London, England: Zed Books.

Blumstein, P., & Schwartz, P. (1983). *American couples.* New York, NY: William Morrow.

Crowder, K. D., & Tolnay, S. E. (2000). A new marriage squeeze for black women: The role of racial intermarriage by black men. *Journal of Marriage and Family, 62,* 792–807.

Du, S. (2000). "Husband and wife do it together": Sex/gender allocation of labor among the Qhawqhat Lahu of Lanchang, Southwest China. *American Anthropologist, 102* (3), 520–537.

Hondagneu-Sotelo, P., & Avila, E. (2005). "I'm here, but I'm there": The meanings of Latina transnational motherhood. In M. B. Zinn, P. Hondagneu-Sotelo, & M. A. Messner (Eds.), *Gender through the prism of difference* (pp. 308–322). New York, NY: Oxford University Press.

Lewis, S., Izraeli, D. N., & Hootsmans, H. (1992). *Dual-earner families: International perspectives.* London, England: Sage Ltd.

Stack, C. (1974). *All our kin: Strategies for survival in a black community*. New York, NY: Harper and Row.

Thai, H. C. (2004). For better or worse: Gender allures in the Vietnamese global marriage market. In L. Richardson, V. Taylor, & N. Whittier (Eds.), *Feminist frontiers* (pp. 275–286). Boston, MA: McGraw-Hill.

Marriage and families in historical perspective

Coontz, S. (1992). *The way we never were: American families and the nostalgia trap*. New York, NY: Basic Books.

Coontz, S. (2005). *Marriage, a history: From obedience to intimacy, or how love conquered marriage*. New York, NY: Viking Press.

Glick, P. C. (1988). Fifty years of family demography: A record of social change. *Journal of Marriage and the Family, 50,* 861–873.

Stacey, J. (1996). *In the name of the family: Rethinking family values in the post-modern age*. Boston, MA: Beacon Press.

The household division of labor

Hochschild, A. (1989). *The second shift: Working parents and the revolution at home*. New York, NY: Viking Penguin.

Kurdek, L. (1993). The allocation of household labor in homosexual and heterosexual cohabiting couples. *Journal of Social Issues, 49,* 127–139.

Suitor, J. J. (1991). Marital quality and satisfaction with the division of household labor across the life cycle. *Journal of Marriage and the Family, 53,* 221–230.

Families of choice and postmodern families

Clunis, M., & Green, D. (1988). *Lesbian couples.* Seattle, WA: Seal.

Dalton, S. E., & Bielby, D. (2000). "That's our kind of constellation": Lesbian mothers negotiate institutionalized understandings of gender within the family. *Gender & Society, 14,* 36–61.

Kiernan, K. (2002). Cohabitation in Western Europe: Trends, issues, and implications. In A. Booth & A. C. Crouter (Eds.), *Just living together: Implications of cohabitation on families, children and social policy* (pp. 3–31). Mahwah, NJ: Lawrence Erlbaum.

Kurdek, L. (1995). Lesbian and gay couples. In R. D'Augelli & C. Patterson (Eds.), *Lesbian, gay and bisexual identities over the lifespan: Psychological perspectives* (pp. 243–261). New York, NY: Oxford University Press.

McWhirter, D., & Mattison, A. (1983). *The male couple: How relationships develop.* Englewood Cliffs, NJ: Prentice-Hall.

Sullivan, M. (1996). Rozzie and Harriet?: Gender and family patterns of lesbian coparents. *Gender & Society, 10* (6), 747–767.

Weston, K. (1991). *Families we choose: Lesbians, gays, kinship.* New York, NY: Columbia University Press.

WORKS CITED

Amadiume, I. (1987). *Male daughters, female husbands: Gender and sex in African society*. London, England: Zed Books.

Baxter, J. (1992). Power, attitudes and time: The domestic division of labor. *Journal of Comparative Family Studies, 23* (2), 165–182.

Belkin, L. (2008, June 15). When Mom and Dad share it all. *The New York Times.* Retrieved July 1, 2008, from http://www.nytimes.com/2008/06/15/magazine/15parenting-t.html?_r = 1&scp = 1&sq = when%20 mom%20and%20dad%20share%20it%20all&st = cse

Beneria, L., & Sen, G. (1997). Accumulation, reproduction and women's role in economic development: Boserup revisited. In N. Visvanthan, L. Duggan, L. Nisonoff, & N. Wiegersma (Eds.), *The women, gender and development reader* (pp. 42–51). London, England: Zed Books.

Bennett, R. (2007, May 12). Couples that live apart . . . stay together. *TimesOnline.* Retrieved May 25, 2009, from http://property.timesonline.co.uk/tol/life_and_style/property/article1779831.ece

Blair, S. L., & Johnson, M. P. (1992). Wives' perceptions of the fairness of the division of household labor: The intersection of housework and ideology. *Journal of Marriage and the Family, 54,* 570–581.

Blumstein, P., & Schwartz, P. (1983). *American couples.* New York, NY: William Morrow.

Bonvilliain, N. (2007). *Women and men: Cultural constructs of gender.* Upper Saddle River, NJ: Prentice-Hall.

Brannon, R. (1976). The male sex role: Our culture's blueprint of manhood and what it's done for us lately. In D. David & R. Brannon (Eds.), *The forty-nine percent majority* (pp. 1–14). New York, NY: Random House.

Braun, N. (2005, August 31). Press release. *UNIFEM.* Retrieved July 4, 2008, from http://www.unifem.org/news_events/story_detail.php?StoryID = 295

Clunis, M., & Green, D. (1988). *Lesbian couples.* Seattle, WA: Seal.

Coleman, M. T. (1988). The division of household labor: Suggestions for future empirical consideration and theoretical development. *Journal of Family Issues, 9,* 132–148.

Collins, P. H. (1990). *Black feminist thought.* New York, NY: Routledge.

Coltrane, S. (1996). *Family man: Fatherhood, housework, and gender equity.* New York, NY: Oxford University Press.

Coontz, S. (1992). *The way we never were: American families and the nostalgia trap.* New York, NY: Basic Books.

Coontz, S. (2005). *Marriage, a history: From obedience to intimacy, or how love conquered marriage.* New York, NY: Viking Press.

Crowder, K. D., & Tolnay, S. E. (2000). A new marriage squeeze for black women: The role of racial intermarriage by black men. *Journal of Marriage and Family, 62* (3), 792–807.

Dalmage, H. (2000). *Tripping on the color line: Black-white multiracial families in a racially divided world.* New Brunswick, NJ: Rutgers University Press.

Dalton, S. E., & Bielby, D. (2000). "That's our kind of constellation": Lesbian mothers negotiate institutionalized understandings of gender within the family. *Gender & Society, 14,* 36–61.

Deaver, S. (1980). The contemporary Saudi woman. In E. Bourguignon (Ed.), *A world of women: Anthropological studies of women in the societies of the world* (pp. 19–42). New York, NY: Praeger.

DeCoker, G. (2001). Japanese families: The father's place in a changing world. In T. F. Cohen (Ed.), *Men and masculinity: A text reader* (pp. 207–218). Belmont, CA: Wadsworth.

Dreby, J. (2010). *Divided by borders: Mexican migrants and their children.* Berkeley: University of California Press.

Du, S. (2000). "Husband and wife do it together": Sex/gender allocation of labor among the Qhawqhat Lahu of Lanchang, southwest China. *American Anthropologist, 102* (3), 520–537.

England, P. (1992). *Comparable worth: Theories and evidence.* New York, NY: Aldine de Gruyter.

Fausto-Sterling, A. (2000). *Sexing the body: Gender politics and the construction of sexuality.* New York, NY: Basic Books.

Federal Interagency Forum on Child and Family Statistics. (2008). America's child in brief: Key national indicators of well-being, 2008. *Childstats.gov.* Retrieved July 23, 2008, from http://www.childstats.gov/americas children/famsoc.asp

Glick, P. C. (1988). Fifty years of family demography: A record of social change. *Journal of Marriage and the Family, 50,* 861–873.

Guttentag, M., & Secord, P. F. (1983). *Too many women? The sex ratio question.* Beverly Hills, CA: Sage.

Hoang, L. A., & Yeoh, B. S. (2011). Breadwinning wives and "left-behind" husbands: Men and masculinities in the Vietnamese transnational family. *Gender & Society, 25* (6), 717–739.

Hochschild, A. (1989). *The second shift: Working parents and the revolution at home.* New York, NY: Viking Penguin.

Hondagneu-Sotelo, P., & Avila, E. (2005). "I'm here, but I'm there": The meanings of Latina transnational motherhood. In M. B. Zinn, P. Hondagneu-Sotelo, & M. A. Messner (Eds.), *Gender through the prism of difference* (pp. 308–322). New York, NY: Oxford University Press.

Juber, J. (1988). A theory of family, economy and gender. *Journal of Family Issues, 9* (1), 9–26.

Kamo, Y. (1988). Determinants of household division of labor: Resources, power and ideology. *Journal of Family Issues, 9,* 177–200.

Kiernan, K. (2002). Cohabitation in Western Europe: Trends, issues, and implications. In A. Booth & A. C. Crouter (Eds.), *Just living together: Implications of cohabitation on families, children and social policy* (pp. 3–31). Mahwah, NJ: Lawrence Erlbaum.

Kimball, L. A. (1980). Women of Brunei. In E. Bourguignon (Ed.), *A world of women: Anthropological studies of women in the societies of the world* (pp. 43–56). New York, NY: Praeger.

Kimmel, M. S. (2009). Masculinity as homophobia: Fear, shame, and silence in the construction of gender identity. In T. E. Ore (Ed.), *The social construction of difference and inequality: Race, class, gender and sexuality* (pp. 132–149). Boston, MA: McGraw-Hill.

Kurdek, L. (1993). The allocation of household labor in homosexual and heterosexual cohabiting couples. *Journal of Social Issues, 49,* 127–139.

Kurdek, L. (1995). Lesbian and gay couples. In R. D'Augelli & C. Patterson (Eds.), *Lesbian, gay and bisexual identities over the lifespan: Psychological perspectives* (pp. 243–261). New York, NY: Oxford University Press.

Laslett, B., & Brenner, J. (1989). Gender and social reproduction: Historical perspectives. *Annual Review of Sociology, 15,* 381–404.

Lewis, S. (1992). Introduction: Dual-earner families in context. In S. Lewis, D. N. Izraeli, & H. Hootsmans (Eds.), *Dual-earner families: International perspectives* (pp. 1–18). London, England: Sage Ltd.

Matthaei, J. (1982). *An economic history of women in America.* New York, NY: Schocken.

McWhirter, D., & Mattison, A. (1983). *The male couple: How relationships develop.* Englewood Cliffs, NJ: Prentice-Hall.

Murdock, G. P., & Provost, C. (1973). Factors in the division of labor by sex: A cross-cultural analysis. *Ethnology, 12,* 203–225.

Newman, J. (2007, November 1). For some couples, distance is key to closeness. *MSNBC.* Retrieved May 25, 2009, from http://www.msnbc.msn.com/id/21369007

Peplau, L. A., & Cochran, S. (1990). A relational perspective on homosexuality. In D. McWhirter, S. Sanders, & J. Reinisch (Eds.), *Homosexuality/heterosexuality: Concepts of sexual orientation.* New York, NY: Oxford University Press.

Porter, E., & O'Donnell, M. (2006, August 6). Facing middle age with no degree, and no wife. *The New York Times.* Retrieved December 10, 2007, from http://www.nytimes.com/2006/08/06/us/06marry.html?scp = 1&sq = facing%20middle%20age%20with%20no%20degree%20and%20no%20wife&st = cse

Roberts, S. (2007, January 16). 51% of women are now living without spouse. *The New York Times.* Retrieved July 22, 2008, from http://www.nytimes.com/2007/01/16/us/16census.html?scp = 1&sq = 51%%200f%20 women%20are%20now%20living%20&st = cse

Save the Children. (2006). *Left behind, left out: The impact on children and families of mothers migrating for work abroad.* Sri Lanka: Author.

Sayers, J. (1987). Science, sexual difference and feminism. In B. Hess & M. M. Ferree (Eds.), *Analyzing gender* (pp. 68–91). Newbury Park, CA: Sage.

Stacey, J. (1996). *In the name of the family: Rethinking family values in the post-modern age.* Boston, MA: Beacon Press.

Stack, C. (1974). *All our kin: Strategies for survival in a black community.* New York, NY: Harper and Row.

Suitor, J. J. (1991). Marital quality and satisfaction with the division of household labor across the life cycle. *Journal of Marriage and the Family, 53,* 221–230.

Sullivan, M. (1996). Rozzie and Harriet? Gender and family patterns of lesbian coparents. *Gender & Society, 10* (6), 747–767.

Thai, H. C. (2004). For better or worse: Gender allures in the Vietnamese global marriage market. In L. Richardson, V. Taylor, & N. Whittier (Eds.), *Feminist frontiers* (pp. 275–286). Boston, MA: McGraw-Hill.

U.S. Bureau of Labor Statistics. (2005). *Women in the labor force: A databook.* Washington, DC: Author. Retrieved July 24, 2008, from http://www.bls.gov/cps/wlf-databook2005.htm

U.S. Census Bureau. (1999). *Current population survey: Definitions and explanations.* Retrieved July 21, 2008, from http://www.census.gov/population/www/cps/cpsdef.html

Waller, A. L. (1988). *Feud: Hatfields, McCoys, and social change in Appalachia, 1860–1900.* Chapel Hill: University of North Carolina Press.

Weston, K. (1991). *Families we choose: Lesbians, gays, kinship.* New York, NY: Columbia University Press.

Winerip, M. (2007, December 9). In gaps at school, weighing family life. *The New York Times.* Retrieved July 21, 2008, from http://www.nytimes.com/2007/12/09/nyregion/nyregionspecial2/09Rparenting.html?scp = 1&sq = in%20gaps%20at%20school&st = cse

Zinn, M. B. (2000). Feminism and family studies for a new century. *The ANNALS of the American Academy of Political and Social Science, 571,* 42–56.

CHAPTER 9

How Does Gender Affect the Type of Work We Do and the Rewards We Receive for Our Work?

The Gender of Work

What did you want to be when you grew up? Do you think your answer to that question or the way people reacted to your answer was connected to your gender? Did anyone ever tell you that you couldn't be what you wanted because of your gender? How do certain jobs get defined in more or less gendered ways? Assuming that a job is where you do something called *work*, exactly what is work? How is gender important in determining what kind of work people do and how much work they do? What criteria do we use to reward people for the work they do, and how are gender, sexuality, race, and class all important to those decisions? How important is the amount of money you get paid for a job to your decision about whether it's something you'd like to do or not? Do you think the importance of money varies by gender? Are men more invested in a high-paying job, and why would that be? What are the differences in what women and men are paid? How have those differences changed over time, and how do we explain those differences? Does gender influence the types of jobs for which we think men or women are better suited? How does gender impact how well people do in the jobs they have, including their likelihood of being promoted or getting a raise? Does gender have an impact on who gets hired for a job? What would a workplace without gender look like?

These are the questions we'll be exploring in this chapter on gender and the world of work, another important institution where gender is enacted and that helps to create and reinforce our ideas about what gender is. As the preceding questions reveal, determining exactly what we're talking about with a focus on work is not an easy task. In the last chapter, we discussed the important historical development within Anglo-European societies of the separation of spheres. The separation of spheres is important to our consideration both of family and work because it sets up a fundamental division between these two realms of social life, and this has important gendered implications for the world of work. The separation of spheres affected families by removing some aspects of productive work from within the house. The doctrine of separate spheres affected work in Anglo-European societies by creating a whole new sphere that largely did not exist before: the workplace as separate and distinct from the home. In this way, the separation of spheres forever changed the way many people in the world thought about work. Work became something done largely by men, outside of the house, for pay. What happened to all of the many other tasks that are done by women, inside the home, or for no pay?

WHAT IS WORK?

Take the example of Haruko, a wife and mother in Bessho, a rural farming community in Japan (Conway-Turner & Cherrin, 1998). Haruko described herself as "just a housewife" who had a lot of free time, but anthropologist Gail Lee Bernstein observed Haruko, like many other women in her village, laboring side by side with men in the rice paddies weeding, transplanting, and harvesting (Bernstein, 1993). Haruko called this work "helping her husband" rather than "work" because Haruko was not paid for her labor in the field. As far as the abundance of free time Haruko had, Bernstein (1993) described the daily worklife of this Japanese housewife as follows:

> Haruko's daily round of chores began at six o'clock. Every morning she prepared a breakfast of *mososhira,* boiled rice, and green tea. After sending the children off to school at eight o'clock in a flurry of last-minute searches for clothing and books and hastily delivered instruction, Haruko ran a load of wash in the washing machine and hung it out to dry. As the woman of the house, Haruko also had several community obligations that were impossible to shirk.
>
> In addition to being a homemaker, Haruko was the family's chief farm worker. Haruko also worked with her mother-in-law and husband on a neighborhood team husking rice. One of Haruko's principal farm chores was feeding pigs. Twice a day, once in the morning and once at night, the couple fed the ninety pigs and cleaned the pigsty.
>
> By early November, Haruko usually looked for part-time wage-paying jobs. Haruko was taken on as a *dokata,* or construction worker (literally a "mud person"), and she worked on a team with two other women and three men. In most other wage-paying jobs women and men worked apart, at distinct kinds of work, but on the construction teams they worked side by side. The female *dokata's* work was physically demanding: women hauled heavy boulders, climbed down into trenches to lay irrigation pipes, constructed bridges over irrigation ditches, and shoveled snow from steep mountain slopes. (pp. 226–229)

Some of the things that Haruko did may seem somewhat foreign to women outside of her particular cultural context. If you asked most people in the United States to guess the gender of a construction worker (which is essentially what a *dokata* is), their safest bet would probably be male. In fact, some people in the United States might tell you that construction work is a job that men are inherently better suited for based on their belief in men's greater overall strength. Yet Haruko, and many women like her around the world, move boulders, lay irrigation pipes, and shovel snow—largely without the aid of the modern implements that reduce the amount of physical strength involved in these tasks in places like the United States. The case of Haruko begins to reveal the ways in which our sense of how certain jobs are gendered or better suited to one gender or the other may be a product of our own cultural biases.

The labor for which Haruko is *not* paid may sound more familiar to the experiences of those from Anglo-European societies. Certainly, many women prepare breakfast and do laundry, even if fewer women (and people in general) are responsible for many of the farm tasks included in Haruko's round of daily chores. Are doing the laundry, making sure children are prepared to go to school, feeding pigs, or husking rice somehow less important than the work Haruko's husband does in the paid labor force? Why is Haruko's work not really considered work, even by Haruko herself, who describes herself as *just a housewife?* Her story is typical of women around the world, not necessarily in the specific tasks she performs (not everyone has pigs to feed), but in the amount of work she does, the many different work roles she occupies, and in the devaluation of all her tasks as not really work (Conway-Turner & Cherrin, 1998). We begin with Haruko's case to make an important point in our ongoing examination of work. Many ways of accounting for work and discussing work within the social sciences have historically ignored the lives of people like Haruko. If we want to examine the relationships between gender and work, we must first make sure that we become aware of the gender biases that are already there as soon as we use the word *work*.

Can you think of other examples of hidden work that is performed by women and not usually perceived as work?

MEASURING THE WORLD'S WORK

The difficulty in deciding exactly what constitutes work makes it hard to do very simple things like measure who works around the world and exactly how much work they do. On a global level, organizations like the United Nations are interested in measuring the amount and kinds of work done by people in different countries, and the statistics they produce can have important implications for decisions about who receives much-needed international aid. Some evidence indicates that there is an important link between the economic growth of a country or region and the rates of employment of women, and this makes measuring the amount of work done by women an important task (International Labor Office, 2008). Initially, many of the statistics collected by international organizations like the United Nations and the International Monetary Fund simply excluded the work women did outside of the paid labor market. This meant that most of the work Haruko did on a daily basis was not accounted for in these numbers. If her construction work included some kind of legal contract, an hourly wage or salary, or some kind of paycheck from which taxes were taken (meaning the income was reported to the government), then this is **formal paid labor**. But much of the labor that takes place around the world is **informal labor**, which means there is no legal contract for the work, not necessarily a set hourly wage or salary, and none of the income is reported to the government by the employer or employee. Women are far more likely worldwide to be employed in the informal labor market than are men, so there's a strong chance that Haruko's part-time position as a construction worker still did not include her in the official statistics once collected by the United Nations.

In the wake of the 1995 Beijing Fourth World Conference on Women held by the United Nations, the United Nations began to reconsider its methods for measuring work and employment. The current United Nations usage of "economically active" persons includes those who are own-account workers and those who are contributing family workers (International Labor Office, 2008). **Own-account workers** are those who are self-employed with no employees working for them. Cottage industry work is often an example of own-account work, and it involves individuals making or producing some product in their home to sell for cash. If Haruko grows her own pigs to sell eventually on the market, she is an own-account worker, and is counted as "economically active" under new UN measures. **Contributing family workers** are a kind of own-account worker who works without pay in an establishment operated by a related person living in the same household. If Haruko's husband is the one who gets the money from raising and selling the pigs, while Haruko receives no pay in return for her part of the labor, she becomes a contributing family worker. Both these categories help account for much of the work done by women and children that is outside the formal labor market but still important to the economic functioning of families as well as nations as a whole.

When the category of economically active people is used, worldwide there are less than 70 (66.9) economically active women for every 100 economically active men (International Labor Office, 2008). This rate of economic activity has been fairly stable in the 10 years since 1997, and it indicates the large share of the world's work that is done by women, excluding any of the domestic work they do that does not directly contribute to family income. Making breakfast, sending children to school, and doing laundry are still not included in these calculations of what it means to be economically active. The rate of economically active women is highest in Europe and the developed world, where there are 82 women working for every 100 men. Northern Africa and the Middle East present some of the lowest rates, with 35 and 39 economically active women for every 100 men, respectively.

Vulnerable Employment

Expanding the ways in which we define work to include a wide range of activities is important in that it begins to dismantle our very gendered notions of work as something that is primarily done by adult men. In fact, much of the world would come to a grinding halt if we tried to survive solely on the labor of adult men. But on a global scale, there are important differences in the kinds of economic activity in which women and men are engaged. The types of economic activity women are more likely to be engaged in worldwide are labeled by the United Nations as *vulnerable employment.* More than 60% of women who work worldwide are employed in the informal sector, including jobs like own-account worker and contributing family member (United Nations Development Fund for Women, 2005). These include the work Haruko does to help her husband in the rice fields and to care for the family pigs. Worldwide, women's work in the informal sector includes women like Cherifa, a woman in Tunisia who began selling pastries to friends and family members when her husband squandered her family inheritance (El Amouri Institute, 1993). Cherifa makes the pastries out of her own home, and although one of her sons has set up a shop selling the same pastries and paying taxes, Cherifa neither counts her money nor reports income from what is her own business. In Colombia, women like Auntie Donkor and Madam Faustina participate in the informal economy and support their families by illegally selling foodstuffs like fish and produce at a market in the city (Okine, 1993). They cannot afford the taxes and fees required to sell food legally at the market, so they sell after 4:00 P.M., when the police are less likely to see them, and establish a network of customers who know where to find them. Like Cherifa, both of these women support their families through the income they gain from these informal activities.

The activities of these women are considered **vulnerable employment** for several reasons, including the very basic fact that these jobs are lower paid overall relative to jobs in the formal economy. One household member working a job in the informal economy will find it difficult to raise his or her family out of poverty on that one income, and this is an especially troubling fact in much of the developing world, where fully 50% to 80% of non-agricultural employment is in the informal sector (United Nations Development Fund for Women, 2005). When most jobs are in the low-paying informal economy, moving vast numbers of families out of poverty in the global South becomes a nearly impossible task. These types of employment make women and their families vulnerable to falling into or remaining in poverty, but because they are in the informal economy, they also make women vulnerable in other ways. Those working in the informal economy are less likely to have access to basic social services while being more likely to be fall prey to illness, property loss, disability, and death. Their lack of access to social services, because they are not *officially* employed in the informal economy, means that they have a reduced ability to deal with contingencies such as illness, property loss, and disability; workers in the informal economy are not eligible for or cannot afford the very things that help to buffer the effects of these life events, like health insurance, home insurance, life insurance, and disability insurance.

Not surprisingly then, those in the informal economy have lower levels of health, education, and longevity relative to their counterparts in the formal economy. They experience fewer of the rights and benefits that come with employment and are therefore often cut out of the very decision-making processes in government or marketing institutions that might help to improve their situation. For example, organizing a union in the world of the informal labor market is difficult, and this leaves these workers with limited options for improving their working conditions or pay. Within the government, informal workers are not acknowledged as officially employed, and this affects their status relative to government institutions.

Given these disadvantages of working in the informal economy and women's concentration worldwide within these jobs, it's safe to say that many women are at a distinct disadvantage when it comes to the work they do. On top of that, add the fact that no country keeps track of the "housework" that women do in their calculations of economic activity. When you hear figures about the gross domestic product (GDP) or gross national product (GNP) of a country—their gross domestic product or gross national product—this does not include the hours and hours of work necessary for caring for children and maintaining a household done by family members, unless that activity generates income for the family. Does changing dirty diapers generate income for a family? Not according to national statistics, but it is a task that has to be done to reproduce a new generation of children—and if there were not someone in the family doing this task for no pay, someone would have to be paid to do it. Almost all of the tasks that fall under the umbrella term of *housework* are tasks that can be, and often are, outsourced to someone outside of the family. Even what we think of as the highly emotional and intimate job of raising children has historically and sometimes continues to be done by employees who are not related to the children at all: nannies, babysitters, and au pairs. A person who does all these jobs for free within the family is saving the household a great deal of money and so contributing to the family's income. In the United States, the annual cost of child care for a 4-year-old ranges from $3,016 to $9,628, while for infants, the cost can be as high as $13,840 annually (Matthews, 2006). For many poor, single working mothers, more than half of their cash income goes toward the costs of child care. Why are these activities, then, treated as if they exist outside of the real economy when they are done by family members?

What would happen if women (and other people) around the world demanded to be paid for the labor they currently do for free? What would a world in which whoever does the child care and housework in your family had to be paid for that labor look like? How would the ways in which we think about work be different?

Theory Alert!
Institutional

The simplest answer is that work has become a deeply gendered institution. Just as the world of the family is presumed to be a feminine domain, work is considered to be the territory of men. This ideology is so strong that it makes the work that women do around the world largely invisible. Their work is not counted in official statistics gathered by international and national organizations (Bonvilliain, 2007). The role of women in determining the distribution of economic resources within families is often ignored and subsumed under the larger unit of the household, without analyzing the complex gender dynamics that take place within any given household. As we've discussed with women like Haruko, women themselves do not consider what they do work but merely as helping their husband or being a housewife and mother. Given all these realities, it is important at the beginning of this chapter to point to the ways in which much of the world's work currently is and always has been done by women. The story we will explore later in this chapter about the increasing labor force participation among women in many Anglo-European cultures is just one particular narrative about women's relationship to work. This story suggests that women working is a relatively new and novel phenomenon, but to tell the story in that way ignores the many other kinds of work women have always done and continue to do around the world.

A MAN'S JOB: MASCULINITY AND WORK

Although much of the work that women do is ignored or made invisible, work becomes for men an unspoken expectation that is often perceived as defining the very core of a man's identity. In almost any industrialized society, adult men are expected to have jobs, and any deviation from this expected course of events is viewed as not just a personal failing on the part of the man but also as a failure of his masculinity. A man who is successful in the world of work is perceived as successful at the all important task of being a man. Thus, in places like the United States, we follow the dictum, the "bigger the paycheck," the "bigger the man" (Gould, 1974, p. 99). This lines up with the dictum of contemporary manhood we discussed in the last chapter: that men should "Be a Big Wheel." To demonstrate the particular power that work has in men's lives, we need only look at what happens to men who do not work or men whose work does not provide the kind of status derived from having a big paycheck. A wealth of sociological research demonstrates what happens to men with work lives characterized by various kinds of failure, whether that means actual unemployment or working-class jobs that do not provide the kind of status associated with high-paying, high-status careers (Liebow, 1967; Rubin, 1976; Rubin, 1994). These failures in their work lives affect other relationships, souring friendships, placing a strain on relations between husbands and wives, and altering patterns of fatherhood. As one unemployed machinist in Lillian Rubin's (1994) study of working-class families described it:

> When you get laid off, it's like you lose a part of yourself. It's terrible; something goes out of you. Then, on top of that, by staying home and not going to work and associating with people of your own level, you begin to lose the sharpness you developed at work. Everything gets slower; you move slower; your mind works slower. (p. 110)

As Rubin (1994) pointed out in her study, when you ask most men to tell you about their identities, they are most likely to first describe their job or what they do for a living. Some women, and probably increasing numbers of them, will also mention their jobs, but Rubin argued that a woman's work identity is still situated within a more complex set of competing interests, including her family obligations. For men, work is connected to the very core of their selves, and it's therefore not surprising that men who lose their jobs feel like they have lost a part of who they are.

In her study, Rubin (1994) detailed the sense of loss and helplessness experienced by men who have lost their jobs largely as a result of layoffs. Some of these men are afraid to sit still and stop doing things around the house for fear that they'll never get up again or feel that they have not earned the "right" to sit still anymore since they are no longer employed. Although they are generally part of mass layoffs, and so are the victims of larger economic trends in society, they still often blame themselves and try to figure out what they could have done differently to keep their jobs. Men facing unemployment describe having lost not just a job but also a whole way of life because many of their friendships were grounded in work experiences. These men often become violent or turn to alcohol to deal with what are feelings of loss and depression as a result of their unemployment. Children describe their fathers as suddenly always mad about something and therefore unapproachable. Even the sexual lives of these men are affected, as a result of the anxiety, fear, anger, and depression that begin to characterize the lives of unemployed men.

The importance of work to men's lives is also demonstrated in the ways in which social class identity and masculinity intersect, affecting the dynamics of family life. In her study of class-based masculinities, Karen D. Pyke (1996) argued that the ways in which masculinity is enacted differ greatly among working-class and middle-class men, in part because of the status they are able to gain from their work experiences. Lower-class husbands follow a form of **compensatory masculinity**, an exaggerated form of masculinity involving drugs, alcohol, and sexual carousing that is used to demonstrate defiance and independence from both the control of their wives and the establishment (higher status men) (Pyke, 1996). This form of masculinity helps working-class men to compensate for their experiences in the everyday work world, where they occupy a subordinated status relative to higher status men who are their bosses and employers.

Theory Alert!
Intersectionality

This does not necessarily mean that working-class men are more likely to use drugs, go out drinking with their friends, or engage in sexual infidelity; rather, Pyke (1996) and others argue, working-class men are less able and willing to disguise these activities than are middle-class men, for whom the same set of activities are perceived as markers of ambition. For example, the working-class man who goes out drinking with his coworkers is perceived by his wife as being "immature," "lazy," and "not ambitious." The upper-class man who engages in many leisure activities outside of the home, like playing golf, going to the gym or club, or having drinks with coworkers, is perceived by his wife and others as seeking to further his career. Compensatory masculinity not only allows working-class men to express their own masculinity and to provide justification for their activities, but also it allows upper-class men to define themselves in contradiction to the behaviors of working-class men. Because upper-class men do not supposedly engage in these types of behaviors, they are able to represent their own version of **egalitarian masculinity** as more civilized, refined, and closer to a situation of gender equity.

This relationship between compensatory masculinity among working-class men and egalitarian masculinity among upper-class men is a good example of the ways in which gender is important not just in making distinctions between women and men but also in creating hierarchy within groups of men or groups of women. The version of masculinity performed by working-class men is judged inferior to the masculinity of upper-class men, and it therefore leaves working-class men less privileged relative to their higher status counterparts. Pyke (1996) pointed to the way these inequalities become manifested in the division of household labor within these families. Although both working-class and upper-class husbands by and large tend to avoid household labor, the explanations for opting out of housework provided by upper-class men are granted more legitimacy than those provided by lower-class men. The career priorities and role as main provider give upper-class men more legitimate power to be excused from helping out around the house in ways that working-class men's status as workers does not. Because working-class men were not perceived as having careers and because their lower earnings provided them with less status as providers within the household, Pyke argued that working-class families rely more on a rigid gendered division of labor based on traditional gender ideology to divide household tasks. This lower status for working-class husbands was often also exacerbated by the need for wives to work to supplement family income. In this way, the work life of working-class men and their upper-class counterparts has real effects on the amount of power they can exert within their household. It's not much of a stretch to argue that the status of a man's job also has an effect on how he is perceived in the world outside of work as well.

CULTURAL ARTIFACT 1: MEET THE PARENTS

The importance of a man's job to his identity is demonstrated to comic affect in the film *Meet the Parents,* starring Ben Stiller and Robert De Niro. In this film, Ben Stiller's character, Gaylord "Greg" Focker, meets his former CIA father-in-law Jack Byrnes (Robert De Niro) and acts out many of the common anxieties experienced by heterosexual men meeting their future in-laws. Consistent with the doctrine of separate spheres and the good provider role, many of these anxieties center around the career and economic prospects of the future son-in-law. In *Meet the Parents,* these issues are crystallized in Greg's job: male nurse. The very first scene of the movie plays on the novelty of Greg's profession as we see him rehearsing his marriage proposal to his girlfriend with a patient in a hospital who mistakes him for the doctor. Later in the movie, Greg's future mother-in-law mistakes triage for a job that's "better than a nurse," while his future father-in-law, Jack, notes the absence of many men in Greg's profession. When Greg meets two friends of the family who are doctors, he's introduced as being "in medicine." When he clarifies that he's a nurse, both male doctors laugh, assuming he's making a joke. Later, his fiancé's ex-boyfriend assumes that Greg does male nursing as charity work, rather than as his paid job. The plot of the movie hinges on a claim Greg makes about his performance on his MCATs, the exam used to gain entrance to medical school. Greg says he did very well on the MCAT and decided not to become a doctor because he wanted more interaction with his patients. Jack, as former CIA, uses his connections to investigate Greg's claims and initially concludes that Greg is lying because no Greg Focker is registered as having taken the MCAT. It is eventually revealed that *Gaylord* Focker did do well on the MCATs, and so Greg and Jack come to a grudging acceptance of each other. Besides being funny, *Meet the Parents* reveals a great deal about the intersections between jobs, money, prestige, and masculinity. That he's a male nurse isn't the only reason Greg's father-in-law finds him suspect as a husband for his daughter, but throughout the movie, it lowers Greg's status in countless interactions and makes him an easy object of ridicule. What's more, the movie suggests the ways in which Greg's status as a male nurse is perceived as a sign of a potential failure (he wasn't smart enough to be a doctor), rather than as an intentional decision. Would a future daughter-in-law's status as a nurse be called into question in the same way? Would the premise of *Meet the Parents* work if the gender of the main character were changed? Are women judged on the basis of their careers or jobs in the same way that men are? And if not, what criteria would be used to judge the suitability of a future wife for one's son?

When Men Can't Work

Given this connection, depriving men of the ability to work at all or to work in the kind of job that provides a degree of status has important repercussions for their ability to perform masculinity successfully. In the United States, one group that faces some of the most serious and seemingly intractable barriers to employment is African American men. Interviews with employers in the Chicago area revealed a multitude of reasons used to explain why Black people from inner-city areas of the city do not make good employees, including their lack of education, lack of basic skills, lack of work ethic, and lack of socialization into good work habits. These beliefs extended to African Americans in general, but they were especially directed at Black men. One study conducted by the Urban Poverty and Family Life study revealed that 77.9% of all employers were more positive toward the employment of Black women over Black men (Wilson, 1996). The problems employers cited for Black men ranged from the high level of job turnover demonstrated in their work histories to their tardiness and absenteeism. Especially, young African American men from inner-city areas in Chicago were described as having problems taking directions from authority figures and being undesirable employees because of the public perception of Black men; employers acknowledged that most people are afraid of young Black men, and although this is largely beyond the control of any individual Black man, it is still used as a criterion that makes African American men less desirable as employees. Few of the employers being surveyed cited discrimination as a reason for the high levels of unemployment among inner-city Black men, although 179 of them mentioned engaging in discriminatory activities in their own decisions about whether to hire Black men.

Many African American young men are prevented from using work as a resource and way to demonstrate masculinity. In fact, some scholars who study the lives of inner-city youth argue that whereas past generations of Black men were esteemed for being hard workers and important role models for their race and community, younger generations of Black men do not consider work to be an important way to express their masculinity (Anderson, 1990, 2000). On a structural level, this may be partly explained by the different types of jobs that are now available to many African Americans living in inner-city neighborhoods. In previous generations, many factories were still located in cities, providing unionized and therefore higher paying manufacturing jobs for minorities living in the inner city. With these jobs, it was often possible for a man to support his family with a blue collar, but respectable, job.

With deindustrialization beginning in the 1970s, many of these factories left the cities, especially northern cities, and moved out of the United States or in search of cheaper labor in places like the American South and Southwest. The fewer jobs that are available to unskilled workers in today's cities tend to be service jobs, and these jobs are generally lower paid and bring with them fewer benefits such as health insurance. In addition, these jobs have a high turnover rate built in; employers know that if their employees only work for short periods of time, they will not be with the company long enough to earn benefits such as health insurance or vacation time. This high turnover rate is acceptable to the employees, but it creates an uneven work history for those forced to take these jobs (Wilson, 1996). Service jobs also require the ability to serve effectively and relate to the public, and it was precisely these skills that many of the employers in the survey emphasized that they believed inner-city Black men lack.

These barriers to employment for many young Black men in the inner city help to explain why this generation would be less invested in work as a potential source of masculinity. If a good job that allows you to support your family is increasingly hard to come by, it makes sense to turn to other sources for displaying masculinity. For some urban African American men, this means turning to illegal activities that exist outside of the formal economy, such as dealing drugs. But even for those Black men who do not turn to crime as a means to achieve financial success, the very orientation necessary for success in the white, middle-class world becomes stigmatized and labeled as not masculine.

These explanations of differing forms of masculinity echo our previous discussion of compensatory masculinity and egalitarian masculinity, and they draw on arguments about oppositional culture. **Oppositional culture** is "a coherent set of values, beliefs, and practices which mitigates the effects of oppression and reaffirms that which is distinct from the majority culture" (Mitchell & Feagin, 1995, p. 68). As applied to the specific situation of young, Black, urban men, oppositional culture explains that one response to the oppression experienced by this group in the form of blocked and reduced job opportunities is to develop their own beliefs and practices for defining what it means to be masculine. This oppositional way of defining masculinity both helps to deal with the lack of job opportunities and demonstrates clear distinctions with the hegemonic, or dominant, masculinity. In this particular version of masculinity, the main provider role that is defined as so important to masculinity in the doctrine of separate spheres becomes less important than the kind of toughness (Anderson, 2000) exemplified in music forms like rap music, which is itself a form of oppositional culture (Mitchell & Feagin, 1995). Some evidence suggests that oppositional culture is also important in explaining the lower academic achievements of African American youth relative to other groups, especially when gender is combined with the effects of race (Farkas, Lleras, & Maczuga, 2002). African American high-school students are more likely to experience stigma from their peers for their academic achievement because doing well in school is perceived as "acting white," but survey evidence suggests that although both Black women and men experience this stigma, it is greater for young Black men.

All of these examples demonstrate the ways in which work and masculinity are connected and the ways in which the relationship between gender and work are not uniform across different groups in any given society. Work can be an important source of status for men, but when these opportunities are blocked, the content of what it means to be masculine can change and shift to reflect the particular social context of that group. This reminds us again that part of our quest in understanding gender is not to identify what is the norm or standard in terms of how gender is lived but, rather, to identify which particular definition of gender has the most power in society. The particular relationship in which masculinity becomes an important source of status and identity may not be true of most men around the world, and even if it is, that is not to say that this is the normal relationship between work and masculinity.

What is true is that gaining status through work is a task that is most easily achieved by men in powerful positions, which in the United States generally means white, upper-class, heterosexual men. Although working-class men and African American men may resist this particular hegemonic version of masculinity, this dominant form of masculinity remains something that exists to be resisted, and it thereby exerts some pressure on their lives.

Drawing on theories of intersectionality, which we discussed in Chapter 2, this becomes a good example of the ways in which gender, race, and class overlap. In these situations, gender is used to enforce what are class and racial differences. The case of African American men and employment specifically demonstrates that race and gender do not add up in simplistic ways, so that being Black and female (and therefore occupying two oppressed identities) is more difficult than being Black and male (and occupying only one oppressed identity). Rather, gender and race interact in complicated, historically, and structurally contingent ways so that, at least in the case of inner-city employment, the gender of Black men can put them at a distinct disadvantage relative to their female counterparts.

In many inner-city African American communities, there are more female-headed households (households in which a woman is the main provider) than male-headed households. How might this fact also affect the way gender works in these communities?

Men in Predominantly Female Occupations

One last example of the ways in which the work a man does and the status he occupies can be connected comes from research on men in predominantly female occupations. These cases demonstrate the ways in which the status of any particular job or career is determined not just by how much the job pays or how much authority and prestige is gained from the job but also from what kind of people generally do the job. Jobs that are done primarily by women tend to be perceived as lower status, especially for men in these particular occupations. For example, the general set of daily tasks performed by a general practitioner doctor and a nurse practitioner are not that different, as the increasing prevalence of nurse practitioners in the United States demonstrates. Yet doctors, who are still predominantly male, have higher status (and higher pay) than do predominantly female nurse practitioners. Historically, as women enter a particular occupation or field in large numbers, the status associated with that occupation decreases, along with the compensation. How does working in a predominantly female occupation affect the status and overall identity of men?

Christine Williams (1991, 1992) studied both men in predominantly female occupations and women in predominantly male occupations. Focusing on men's experiences in predominantly female occupations such as librarian, elementary school teacher, nurse, and social worker, Williams found that men faced little discrimination within their jobs, either in their hiring or from their coworkers once they were on the job. For heterosexual men in occupations such as librarian or elementary school teacher, they often found a distinct advantage in being able to bond with other males in their work environment, and these other males were often in supervisory positions. Gay men in these occupations were sometimes forced to disguise their sexuality, in part by exaggerating acceptable "masculine" qualities. For example, one male physician stated that he preferred all male nurses in his operating room unless the male nurses were gay. Men in general in predominantly female occupations face little hostility from their female counterparts. Although they are sometimes forced to take care of the more "masculine" tasks associated with their jobs, like discipline cases in elementary schools, on the whole, the men Williams interviewed found their female coworkers to be supportive of their careers and eager to have men enter into their fields. Although men often felt uncomfortable in these settings, their female coworkers often went as far as to include the men in social gatherings of coworkers, including bridal showers and baby showers.

Within the actual workplace environment, then, men receive something of a status boost from being one of the few men in a predominantly female occupation. The stigmatization of a man working as a nurse or librarian comes largely from the world outside of work for these men—from patients, parents, and clients as well as from friends, family members, acquaintances, and strangers. From these sources, men receive messages that male nurses are gay, male librarians are wimpy and asexual, and male social workers are feminine and passive, while male elementary school teachers are suspected of being pedophiles. The principal of one male elementary school teacher shared the following concerns expressed by the parents of one student: "How can he love my child; he's a man?" and "Aren't you concerned about homosexuality?" (Williams, 1992, p. 261). Another male elementary school teacher described the reaction he gets from other men when he tells them what he does for a living: "If I tell men that I don't know, that I'm meeting for the first time, that that's what I do . . . sometimes there's a look on their faces that, you know, 'Oh, couldn't get a real job'" (Williams, 1992, p. 262). This male teacher's wife, who was also an elementary school teacher, was perceived by her husband as receiving an increase in status from the fact that she worked at all, while the specific nature of his job decreased his status in the eyes of other men. Most of the men Williams (1992) interviewed felt that to outsiders, their job represented a distinct step down in status, and this may be part of the reason why there are so few men in these occupations. Doing work that is labeled as women's work makes it more difficult for men to make the connection between the status derived from a job and masculinity.

Theory Alert! Doing Gender Theory

THE GLASS CEILING AND THE GLASS ESCALATOR

One of the most important concepts that came out of Williams's (1991, 1992) work on men in predominantly female occupations was the idea of the glass escalator, which is a very different phenomenon from the glass ceiling faced by women. The glass escalator identified by Williams describes the experiences of men when they enter occupations that are predominantly held by women. Do they face the same set of disadvantages described by the metaphor of the glass ceiling? As we already discussed, men in female occupations face little job discrimination when it comes to their hiring, and they often enjoy a preference based on their gender. Librarians described a preference for hiring males over females, while male elementary teachers talked about how marketable they are because of their gender. One pediatric nurse was told by the supervisor who hired him that it was nice to have a man around in such a female-dominated occupation. Once these men are hired for the job, they are often tracked into certain areas or specialties in large part because of gender expectations. Male social workers, nurses, elementary teachers, and nurses are all tracked toward administrative jobs more so than their female counterparts, and often regardless of whether they express an actual interest in doing administrative work. This is what Williams (1992) called the **glass escalator**, the invisible pressure that men in these occupations face to move upward in their professions. On the glass escalator, men in these specific professions have to work extra hard just to stay in the same place and to resist the pressure from their superiors, coworkers, and others to move up in their career.

The **glass ceiling**, on the other hand, refers to the fact that despite the progress women have made into many managerial positions in the business world, they are still far less likely than men to have jobs that involve exercising authority over people and resources (Wharton, 2005). The glass ceiling is often described as invisible and unbreakable, and it serves as a useful metaphor in describing the experiences of both women and people of color in attempting to obtain positions of power and authority both inside and outside of the business world. Hillary Clinton used the glass ceiling as a metaphor in her speech conceding the U.S. Democratic presidential nomination when she said, "Although we weren't able to shatter that highest, hardest glass ceiling this time, thanks to you, it's got 18 million cracks in it" (Milbank, 2008). Women's blocked opportunities to positions of authority often also translate into an inability to climb beyond a certain point in the corporate ladder and, therefore, into an inability to achieve a certain level of earnings or to penetrate certain occupations or types of jobs (Cotter, Hermsen, Ovadia, & Vanneman, 2001). Although the glass ceiling is sometimes also used to describe the situation of people of color in the corporate world, it is largely a term for women's experiences in business settings that are male dominated.

How does the concept of androcentrism connect to the idea of the glass ceiling or the glass escalator? What about the idea of gender polarization?

The example of one librarian in Williams's (1992) study demonstrates how the pressure of the glass escalator works on men in female-dominated occupations. This particular public librarian specialized in children's collections, a heavily female-dominated specialty within a career that is already populated primarily by women. In his evaluation for his very first job, his supervisors were happy with his work in storytelling and other activities related to the children's collection, but they critiqued him in his evaluation for "not shooting high enough" (Williams, 1992, p. 257). These negative reactions to his desire to remain where he was, as a librarian in the children's collection rather than as an administrator, followed him through his 10-year career. Some men appreciate being given a ride on the glass escalator, but others, like this children's librarian, would actually prefer not to do administrative work.

SEX SEGREGATION IN THE WORKPLACE

The phenomena of both the glass escalator and the glass ceiling raise a more fundamental question about the relationship between gender and work: How do occupations such as social worker, nurse, elementary school teacher, and librarian become predominantly female in the first place? And why are some other occupations and jobs decidedly male, like CEO of a large corporation or U.S. senator? We've already discussed the sexual division of labor, a very old and universal way of dividing all kinds of tasks in a society—a concept that covers all the kinds of work people do on a daily basis around the world, including taking care of children, caring for crops, and writing memos or fetching coffee. The household division of labor covers the way in which people in a family divide up the tasks that are necessary to keeping a household up and running and is something that makes sense only with the separation of spheres and a clear division between the work that takes place outside of the house and the work that takes place inside of a house.

When we shift our focus specifically to the world of paid work within the formal economy, a new gendered division emerges—occupational sex segregation. **Sex segregation** refers to the concentration of women and men into different jobs, occupations, and firms (Wharton, 2005, p. 167). Sex segregation in general is something that exists to varying degrees in many cultures around the world; in the United States, we use sex segregation for most public bathrooms and locker rooms (although interestingly, we do not sex segregate the private bathrooms in our homes). In some cultures, sex segregation is enforced within the family home, as women are secluded in certain rooms or certain parts of the house while men are confined to others. When we use sex segregation here, we are focusing specifically on the ways in which not just physical spaces, but whole career paths, become dominated by either women or men.

You might think that in places like the United States, with programs such as affirmative action in place to redress past discrimination against women and people of color, occupational segregation should largely be a thing of the past. Some of the questions with which we began this chapter had to do with the aspirations you might have had as a child and whether anyone encouraged or discouraged you based on your gender. Surely we live in a time where all little girls are told they can be whatever they want to be, and the same is true for little boys. Or is it? The rates of occupational sex segregation (as well as occupational segregation by race) are still considerably high in the United States, and they have become a part of the taken-for-granted landscape of our everyday lives. In the United States, fewer than 10% of workers have a coworker or colleague of a different gender who does the same job, for the same employer, in the same location, on the same shift.

This statistic begins to point to the complexity of sex segregation in the workplace and to the many levels at which sex segregation exists. Women and men can be segregated by the particular shifts they work (graveyard or day shift), the locations (branch office or headquarters) in which they are placed, the employer for whom they work (cashier at Wal-Mart as opposed to cashier for a locally owned health food market), as well as by the seemingly simplistic notion of the actual job they do. When you throw all these complex factors together, very few women and men work alongside people of a different gender who are doing exactly the same work. Many of us have a sense that there are large numbers of women in the paid workforce in the United States, but how are those women actually distributed across jobs, occupations, and firms? Of the 66 million women in the United States who do paid work, 30% of them work in just 10 out of the 503 occupations listed in the U.S. Census. Another way of saying this is that a third of all women in the United States who work are concentrated in just 10 out of the myriad occupations supposedly available to them. Sex segregation in middle-class jobs has decreased significantly, although the trend lessened its pace beginning in the 1990s. But the levels of sex segregation in working-class jobs is basically the same as it was in 1950 (England, 2010).

These figures may be hard to believe, although you can try them out for yourself by looking at Table 9.1, which lists rates of sex segregation by some selected occupations. Most of us don't spend a lot of time thinking about the prevalence of sex segregation in occupations until someone points it out to us in part because in the complex, bureaucratic world of today's society, it's sometimes hard to figure out exactly who does what. But to begin to put these numbers into context, you need only begin to think about all the different types of workplaces you encounter and to reconstruct what you know about the gender of the people in those settings and the kinds of jobs they're doing. If you're a regular at a particular restaurant or coffee shop, is there a gender pattern to who is more likely to be serving you your food and who is more likely to be preparing it?

Think about all the teachers you had throughout your school career, from kindergarten all the way up to your current classroom. How many male teachers did you have in elementary school, or throughout your whole educational career? What kinds of subjects did those male teachers tend to teach? Who worked in your school libraries, or who works in the public libraries you may go to? How many doctors have you ever encountered who were women, and what kind of doctors were they? What was the gender of the person who sold you your car, or your computer, or your shoes, or your clothes? Is a man or a woman more likely to be the person cashing your check at the bank, changing the oil in your car, doing your dry cleaning, preparing your taxes, renewing your driver's license, delivering your mail, or pulling you over for a speeding ticket? We'll discuss what the actual answers are for some of these occupations, jobs, careers, and specialties in more detail below. For now, some of them may seem obviously sex segregated; fewer of us have probably had a female police officer give us a ticket because only 20% of law enforcement officers in the United States in 2006 were women. In other settings, the degree of sex segregation might be more difficult to see from the outside; how do you know who's cooking back in the kitchen at a restaurant, and how can you tell whether the person cashing your check at the bank is a cashier or a manager filling in for the moment?

Make a list of all the jobs you've occupied. In those jobs, did you ever have someone of the same gender in the exact same job as you? Were there people of a different gender in that workplace, and were there trends in the types of jobs they tended to do? Can you think of examples of racial segregation in jobs you've worked?

Sex segregation is complicated and can exist at multiple levels within a single institution or company. Take a college campus. At first glance, a college campus might look like an institution with a pretty good balance of women and men working together. If you begin with the faculty, chances are still good that you have more male professors than female professors, and this will especially be true at the higher ranks of academia (associate and full professor, as compared with assistant professor). Nationwide in the United States, college

Table 9.1 Sex Segregation by Occupation and the Gender Wage Gap for Selected Occupations, 2009

Occupation	Percent Women	Gender Wage Gap
Management Occupations	37.4	.72
Chief executives	25.0	.72
Food service managers	45.7	.80
Financial managers	54.7	.67
Computer and information systems managers	29.0	.79
Legal Occupations	49.8	.57
Lawyers	32.4	.75
Paralegals and legal assistants	85.9	*
Healthcare Practitioners and Technical Occupations	74.6	.78
Dentists	30.2	*
Pharmacists	49.3	.75
Physicians and surgeons	32.2	.64
Physicians assistants	57.1	*
Registered nurses	92.0	.95
Veterinarians	61.2	*
Education, Training, and Library Occupations	74.3	.78
Preschool and kindergarten teachers	97.8	*
Elementary- and middle-school teachers	81.9	.86
Secondary-school teachers	54.9	.91
Life, Physical, and Social Science Occupations	46.8	.79
Community and Social Service Occupations	62.9	.83
Social workers	80.7	.94
Clergy	17.0	*
Arts, Design, Entertainment, Sports, Media Occupations	46.6	.81
Athletes, coaches, umpires, and related workers	40.1	*
Writers and authors	62.7	*

(Continued)

Table 9.1 (Continued)

Occupation	Percent Women	Gender Wage Gap
Healthcare Support Occupations	89.4	.85
Dental assistants	97.6	*
Protective Service Occupations	22.3	.75
Police and sheriff's patrol officers	15.5	.83
Firefighters	3.4	*
Food Preparation and Serving-related Occupations	55.7	.91
Chefs and head cooks	20.7	*
Cooks	41.5	.93
Bartenders	56.4	.75
Waiters and waitresses	71.5	.87
Construction and Extraction Occupations	2.6	.94
Construction laborer	2.7	*
Pipe layers, plumbers, pipefitters, and steamfitters	1.3	*
Brickmasons, blockmasons, stonemasons	.1	*
Office and Administrative Support Occupations	74.5	.92
Secretaries and administrative assistants	96.8	*

Source: Based on data from the Bureau of Labor Statistics, Table 39. Median weekly earnings of full-time and salary workers by detailed occupation and sex and Table 11. Employed persons by detailed occupation, sex, race, and Hispanic or Latino ethnicity.

*Data are not shown where base (number of people in that category) is less than 50,000. This means that for these particular occupations, there are not more than 50,000 women or more than 50,000 men in that occupation.

professors as a whole are about 41.2% female, which is a marked improvement from the past (American Association of University Professors, 2010). But what happens when you begin to look at how those positions are distributed across disciplines? How many female professors are there in the natural and physical sciences as compared with the English department? If your college follows a typical pattern, you'll probably have a great many more women in the English department than in the biology or physics departments. What happens when we look at the security personnel on the average college campus? The student life and resident life staff? Cafeteria workers? "Maintenance" at a college might include those who clean the buildings, those who do physical repairs, and those who maintain the physical grounds of the college (landscaping), but the gender makeup of each of those types of jobs is very different. Women are more likely to do indoor cleaning, while men are concentrated in grounds keeping and physical repairs. What about the administration at your college? What's the gender makeup of the president, vice presidents, deans, provosts, or department chairs? How many men and women sit on the board of trustees? How many men serve as secretaries or executive assistants on your campus?

To help make sense of this complexity in relation to sex segregation in the workplace, researchers look at sex segregation at two different levels, focusing on three different units of analysis at each level. The first level is **occupational sex segregation**, and this simply refers to the concentration of women and men in different occupations. Occupations are categories defined by the U.S. Census to describe the kind of work people do, and because the census seeks to collect information about large proportions of the U.S. population, their categories are fairly broad. The more people you're gathering information about, the harder it is to include a lot of detail about their lives. So examples of census occupations include college and university teacher, truck driver, cook, salaried manager, cashier, nurse's aide/orderly, secretary, guard, and bookkeeper.

Most of the research that has been done on sex segregation was done at the level of occupational sex segregation because the census conveniently provides a lot of information on this particular level of sex segregation. The examples seem straightforward enough until you start to unpack all the different types of people that might fit into some of these broad occupational categories. Who exactly is a salaried manager? The manager at a fast food restaurant is a salaried manager, as is the person who works as an office manager at a prestigious New York advertising firm, the plant manager at a slaughterhouse in Iowa, and the manager of a publishing house in San Francisco whose position is the second highest in the company as a whole. How similar are these jobs to each other? Looking at sex segregation at the level of occupation is easiest, but because it is the broadest category, it will also tend to capture less segregation than more finely tuned units of analysis. If women tend to be the managers in fast food restaurants rather than plant managers, those distinctions are lost in the broad outline of the census category of salaried manager.

Job-level sex segregation focuses on "the specific positions that workers hold within specific establishments" (Reskin, 1993, p. 243). This is a more finely tuned way of examining sex segregation in that it allows us to make distinctions between the McDonald's manager and the manager at the Iowa slaughterhouse. Information on job-level sex segregation is harder to come by than is occupational-level sex segregation because large data sources like the U.S. Census don't collect information about the exact type of position someone holds or the specific place he or she works. Collecting this kind of information is painstaking social research work, but when it has been done, the extent of job-level sex segregation is much higher than sex segregation by occupation. One study found that during the 1970s, the index of job-level sex segregation was 30 points higher than that for occupational-level segregation (Bielby & Baron, 1984). The same study, based in California, analyzed 645 occupations in 290 work organizations and found that more than three fourths of the men or women would have to be reclassified to degender the occupational categories (Bielby & Baron, 1986). In that same study, 96% of the 10,525 different job titles were gender segregated, and only 8% of the workers shared a job title with a member of a different gender.

Whichever particular level of sex segregation is used, the evidence indicates that workplaces in the United States are highly segregated along both gender and racial lines, and these identities often intersect and cross-cut in interesting ways. In his reporting on the social world inside a North Carolina slaughterhouse for *The New York Times,* Charlie LeDuff (2000) documented the complex ways in which jobs in the plant were distributed along gender and racial lines. Black women were assigned to the chitterlings room, where they scraped feces and worms out of intestines. White men were the supervisors and mechanics, except for the occasional white prisoner who might be assigned the dirty work along with other African Americans and Mexicans. Some Native Americans and Mexicans were given the job of assembling boxes. Outside of the kill floor, a mixture of Black women and Mexican men and women skinned the hog carcasses and butchered them into various parts. Outside of a rare Mexican man, all the people on the kill floor where the pigs were ushered into the plant and killed were Black men. Jobs on the kill floor set the pace for the rest of the factory, and they paid a higher wage than others in the factory—as high as $12 an hour. There were quotas to be met, and the competition for jobs on the kill floor was high. The reasons given by workers for why the kill floor consisted primarily of Black men rather than the Mexican, Native American, or white men included that Mexicans were too small or that they didn't like blood or heavy lifting. There were no reasons for why any of the women—Black, Mexican, or Native American—didn't work in the kill floor because it was assumed that this was work that women were not capable of doing. One of the final reasons for why Black men worked on the kill floor, as provided by one Black male worker, was that, "We built this country and we ain't going to hand them everything," where the "them" referred to Mexican immigrants.

Levels of both sex and racial occupational segregation are fairly high in the United States, and yet they seem to go largely unnoticed. Why do these forms of segregation seem to be taken for granted? What does this reveal about how we think about gender, race, and work?

A Historical Look at Gender and Work

The quote by the Black male worker begins to suggest the level of tension that exists between the various racial groups in that particular North Carolina slaughterhouse. Black workers felt that the entrance of Mexican immigrants into a factory or an occupation drove down the wages because the immigrants were willing to accept less money than their African American counterparts. That assessment accurately reflects what employers have in mind because one of the central dynamics driving both sex segregation and racial segregation in the workplace is precisely the desire of business owners to lower their costs and increase their profits by paying their workers less. If they can create a sense that there are two different groups competing for jobs, this helps along the process of driving down the wages for a particular occupation or in a particular firm. When places like the United States began to industrialize in the 19th century, whether it would be primarily women, men, children, or some combination of all three, working in the factories was up for grabs. As we mentioned briefly in Chapter 8, farming families initially sent their unmarried daughters and younger sons to work in factories because their labor was least needed on the farm (Lamphere, 1987). Young women in this system could be paid less because there was no expectation that they needed a wage that could support a family; manufacturers boarded and watched over the moral lives of these young women, while paying them a wage that allowed them to save for a trousseau (a dowry, or money or goods a groom received upon marrying a woman, sometimes in the form of a wedding dress, clothes, or other items necessary to getting married or setting up a first home), pay off a mortgage (for their family), or send a brother to college.

Thus, the work of these young women benefited their family as a whole back on the farm, including contributing to their potential upward mobility through wealth creation (owning their own home or property) and increasing levels of education (sending a son to college). In addition to not having to pay female factory workers as much, many owners preferred women because they also felt they were more docile than their male counterparts, despite early attempts by some women to unionize themselves. The men at the head of the families to which these young women belonged were initially happy to have them working for the extra income they brought into the family (Lorber, 1994). What changed in places like the United States, so that now our typical image of a factory worker is certainly not a young, unmarried farm daughter?

As industrialization progressed, farms disappeared while factory jobs increased. Jobs in manufacturing became a larger percentage of all paid work, and that women were working these jobs became a threat to men's interests. If women could earn their way in the factories far from home and support themselves while men could not, the power of men within families and in society in general to control women would be greatly reduced (Walby, 1986). At first, men began to work in factories as the supervisors of women and children workers, but as industrialization progressed, supervising did not provide enough jobs for all the men displaced from farms and in need of work. The initial solution to this problem of monopolizing some jobs as solely for men was to declare certain machines within any given factory as "women's machines" and others as "men's machines." Sonya Rose (1987) described how this process worked within the context of the hosiery industry in England in particular. At first, hosiery was organized as a **putting-out industry**, which meant that rather than workers traveling to one central building (the factory) where everyone assembled hosiery, the raw materials needed to assemble hosiery were distributed to individual households where individuals assembled them on their own schedules and within their own homes.

This method of production, also called **piece-meal work**, or cottage industry, is still used in many places around the world, including the United States. In the hosiery industry, families bought or rented frames, a kind of machinery used to make the hosiery, and usually the entire family participated in the job of producing the final product, for which they were paid on a piece-meal basis; rather than being paid for working so many hours in a factory, families received so much money for every finished product they made. Mothers and daughters were more likely to be engaged in the stitching and knitting part of production, but they also operated frames. The production that took place within the household was under the supervision of the husband or father, and the income received was a family income, rather than an individual wage.

Gradually, the work of making hosiery shifted from individual homes into the factory setting. These factories were gender segregated from the beginning, with women again doing much of the needlework. But as many women had operated the framing machines in their own homes, there were also small numbers of women who operated frames within these first factories. However, a woman working in a factory was no longer under the control of her husband or father; she was paid an individual wage directly to her rather than a family wage that was filtered through the male head of the household, and this introduced potential tension within the gendered power structure of families. The fierceness of the competition between women and men for these jobs was driven by the fact that employers had great incentive to want to hire women as their workers. In a system dictated completely by economic rationality in terms of minimizing labor costs, employers would always prefer female over male employees because they could consistently pay women less. The owners of hosiery factories in England were no exception, and the all-male unions of hosiery workers responded by striking when they began to be replaced by women who were perfectly capable of doing their jobs. The response of the factory owners was to threaten to take the factory somewhere else, and editorials in local papers urged the male workers to accept their female counterparts rather than risk losing the factories as a source of jobs for the community.

The solution to this problem was gradually worked out by the unions when they integrated and allowed women workers to join. Men in the union wanted to demand a set price for each piece of hose produced, but the women were afraid that such a demand would cost them their jobs. The women were therefore willing to accept a lower price for their labor than the men; even at this lower price, these factory jobs were far superior to the other jobs available to women during that time period. Obviously, if women were willing to make hosiery for a lower price than were men, factory owners were still motivated to replace men with women, which they continued to do. The best solution the union could settle on was to designate some framing machines arbitrarily as "women's machines" and some as "men's machines" and to develop different pay scales for each type of machine. The differences between these machines were fairly minimal, but the solution allowed men and women to work in the same factory while doing what could be defined as slightly different jobs and being paid differently.

This specific story contains all the basic building blocks of our current system of sex segregation. The differences in the jobs (i.e., working on different types of machines) came to be justified through the use of gender even though originally it was clear that both women and men could effectively operate the machines. Men's machines came to be defined as more expensive and complex, and women were perceived as lacking the technical skills necessary to either use or take care of these machines. Men and women became sex segregated within the same factory, and the justification for women's lower wages, should they ever question them, became built into the very structure of the workplace; women worked on inferior machines and, therefore, should be paid less. This method of associating a certain set of skills with a type of job or work and then defining those skills as masculine or feminine to monopolize access to those jobs took place in other industries, including among cotton spinning and English printing (Rose, 1987).

Gender and Dangerous Work: Protective Labor Laws

Allowing women to continue working within a factory but creating sex segregation within the factory was one solution to the competition between women and men for jobs in early industrialization. Another widely employed solution was the passage of protective labor legislation. Keep in mind that in the case of hosiery work and many other kinds of work, women and children had historically performed all kinds of labor within the household unit of production. **Protective labor legislation** argued that women could not work too many hours in factories or do certain kinds of labor because of the detrimental effects on women's health and their ability to perform their crucial household duties as wife and mother (Kessler-Harris, 1982). Even feminists of the period, such as Lucy Stone, argued that the special sensibilities developed by women in the private sphere as mother and wife (such as compassion, nurturance, and a better developed sense of morality) needed to be preserved in the competitive world of wage labor (Kessler-Harris, 1982, p. 185).

So women were gradually prevented from working in places that sold alcoholic beverages, as it might corrupt their morality, and then from grinding and polishing metal, as it would clog their lungs. Work in underground mines would coarsen women's gentler natures, while any kind of work that could bring women into contact with strangers was also perceived as putting them in danger; this made the jobs of messenger, elevator operator, meter reader, letter carrier, and taxi driver all off-limits for women. Other legislation focused on limiting not the types of jobs women could work but their working conditions. The most familiar of these was the length of the work day, and arguments to change to a 10-hour work day depicted women as physically weaker than men and desperately in need of protection from long work hours because of their "special physical organization" (Kessler-Harris, 1982, p. 187), including their childbearing and maternal functions. In hearings before the U.S. Supreme Court in 1908, physicians argued that women had a greater disposition to diseases and were in periodic "semi-pathological" states of health. Between 1909 and 1917, legislation that restricted the number of hours women could work in a day passed in 19 states.

Is protective labor legislation a bad thing? Should women and children specifically be protected from certain working conditions? Or should men be protected from these working conditions as well? What does the idea of protective legislation reveal about femininity and masculinity?

The reduction in the number of hours a woman could work benefited married women, and many more married women began to move into the labor force in the period after World War I. Married women did not have to support their families on their income, and a shorter working day gave them more time to do household chores and care for children. The major problem with protective labor legislation aimed at women was that it assumed all women were in fact mothers or wives and needed to be protected primarily as such. The work day of women was limited in a period well before the institution of any minimum wage, and for a woman trying to support herself or her family through her job, cutting short her working hours meant cutting short her total income. This effort to protect women, then, also had the convenient effect of reducing their overall earnings and providing another justification for occupational sex segregation by excluding women from some of the better paying jobs or defining their jobs differently and as less valuable within the same factory walls.

Thus, the early history of industrialization in places like the United States and England was a history characterized by competition between women and men for the same jobs. Many who study the intersections between work, gender, and race argue that this competition continues in contemporary labor markets, like those of the slaughterhouse described previously. Especially since the second wave of the women's movement in the 1960s and 1970s, women do enter into occupations, jobs, and firms that are predominantly inhabited by men. But the usual result when enough women do so is resegregation of that particular area of the workforce. During the 1970s, when women were believed to be making significant progress into many predominantly male occupations, only 33 of the 537 occupational categories actually saw an increase in the numbers of women of at least 9% (Reskin & Roos, 1990). For the smaller percentage of women who did actually enter predominantly male occupations, they most often found themselves witnesses to a process of gender turnover in that occupation. In 1970, women were 33% of all personnel, training, and labor relations specialists, but by 1980, they made up 47%; by 1988, they made up 59%; and by 2005, they made up 71%. Women made up 30% of insurance adjusters, examiners, and investigators in 1970, 60% in 1980, 72% in 1988, and 87% in 2005 (Bureau of Labor Statistics, 2005; Reskin & Roos, 1990). Almost all of the jobs we currently think of as predominantly female occupations were once predominantly male, including real estate agents, secretaries (who used to be called clerks), and elementary-school teachers. In these occupations, men exited as women entered in large numbers, resegregating the occupation.

The Anatomy of Sex Segregation

How exactly does this process happen for any given occupation? When the first woman is hired in a particular factory or office, do the men pick up and leave at once? Where do all those men go, and what predicts which jobs in particular are likely to resegregate? In the beginning, before either segregation or resegregation occurs, an employer would be smart to discriminate against men rather than against women because women are generally conscientious, hard workers, and can be paid less (Bielby & Bielby, 1988). Men can prevent women from being hired in a particular job or occupation by monopolizing the skills and training necessary to do the job. The simplest example for many years in the United States was to prevent women from attending colleges and other educational training programs. If women cannot acquire the skills or knowledge necessary to do a certain job or are prevented from doing it effectively or at all by protective labor legislation, then employers have no choice but to hire only men. If you're an employer and the particular job for which you want to hire workers seems to carry a certain set of skills or knowledge, you're probably going to have to pay your workers more to do it, and if you're hiring mostly men, you'll probably have to pay them more than women. This is in part because women are willing to accept less pay because of the fewer jobs that are available to them, but it is also because of the connection between the perceived knowledge and skills necessary to do a job and gender; as we've discussed, the jobs that men do are generally perceived as more skilled than the jobs women do.

The hypothetical employer who wants to lower labor costs has several options available. She or he can move her or his factory somewhere with a ready supply of cheaper labor. Usually, this means an area with weaker unions (as unions can negotiate for higher pay) and more competition among potential employees for jobs. For example, many factories have moved to the rural South in the United States both because there is little tradition of labor organization in these communities and because the decline of agriculture and the high levels of poverty mean that there are few jobs available to workers in those areas. If you're desperate for a job and there aren't a lot of other options around, as a worker you're willing to accept lower pay for your work than you would in a situation with more available options. Another option available to the business owner who wants to reduce his or her labor costs is to change the nature of the work being done by de-skilling or automating it, or making it part time or home based. If you can break what was a complicated and skilled job into a series of routine and easily learned steps, usually along with automating many of the steps, you can justify paying your workers less than if more training is required to perform the job.

Part-time workers cost less because you don't have to provide them with benefits like health insurance or vacation time, and you can maintain high rates of turnover. If none of your workers stay at the job very long, they're not around long enough to get a raise and therefore cost you less to employ. Moving labor into the homes of your workers saves you overhead costs you would have to pay to provide space to work in, as well as the cost of paying someone to supervise workers.

In their study of several occupations that had transitioned from predominantly male to female beginning in the 1970s, Barbara Reskin and Patricia Roos (1990) found that these jobs became available to women largely because they became less attractive to white men. Men left occupations such as personnel manager, computer operator, insurance agent, and real estate salesperson when the jobs became de-skilled, earnings declined, job autonomy was reduced, and working conditions deteriorated. Employers contributed to these processes by creating better jobs for men within their organizations or by making some positions part time (Lorber, 1994). Women took these jobs that had just been made less attractive because they were still better than many of the jobs the women had occupied before. This process of downgrading jobs as a way to reduce costs and then resegregating the jobs by gender creates a strong connection between gender and the type of work people do. Women are willing to accept lower salaries because their jobs are structured in ways that justify lower salaries. Hiring more women to do a job, paying less for the job, and reducing the quality of the job are all processes that go hand in hand.

Once a job has been converted from a man's job into a woman's job, gender ideology is employed to justify that occupation as inherently suited to women. This naturalizes some occupations as appropriate for women and others as appropriate for men, reinforcing the system of sex segregation in the workplace. For example, in the 1970s, residential real estate agent work began to transition from a predominantly male to a predominantly female occupation. In a process similar to many other occupations, men did not completely vacate the real estate business, but the real estate industry underwent the process of ghettoization by gender. **Ghettoization** happens when lower paid "women's" jobs are separated from the better paid "men's" jobs within an occupation, job, or firm through the use of informal gender typing (Lorber, 1994). During the economic recession of the 1970s, residential real estate became much less profitable than it had been in the past, and it therefore became a less desirable occupation for white men. Men moved into commercial real estate, which brought much higher commissions and therefore better earnings, leaving residential real estate open for an influx of women. When employers began to hire large numbers of women to sell residential real estate, they began to argue that women were uniquely suited to these jobs because of their experience running their own homes as wives and mothers and their knowledge of neighborhood schools and playgrounds; that women were good at residential real estate came to be perceived as natural, despite the fact that this had been a primarily male job until this time (Thomas & Reskin, 1990).

Resegregation happens in the workplace when an entire occupation transitions from one gender to another, usually from a predominantly male to a predominantly female occupation. This term covers the whole process of disproportionate numbers of women entering a male occupation and their eventual disproportionate representation in that occupation, which results in the work associated with that occupation being labeled as women's work (Roos & Reskin, 1992). During the 1970s, the occupations of typesetter and composer underwent a process of resegregation from a predominantly male to a predominantly female occupation, from 30% female in 1970 to 72% by 1989. In their study of various occupations that had experienced a large shift in the gender composition of workers, Roos and Reskin (1992) found no examples of occupations that truly represented integration of women and men into the same workplace and the same jobs. Ghettoization was the most common outcome, with men and women existing in the same occupation but doing slightly different jobs or having different specialties. Resegregation was the other path, but no occupations actually achieved a level of integration, meaning anything approaching equal numbers of women and men.

Today, electronics manufacturers prefer women to men because they argue that women have greater finger dexterity than do men, the same argument used to justify why women should do needlework in 19th-century hosiery factories. Other jobs are perceived as off-limits for women because of the amount of physical strength required, such as construction jobs. Teaching from the level of kindergarten through high school is now considered a job uniquely suited to women in the United States because of the flexibility it provides, especially for mothers. A teacher's schedule generally coincides with the schedule of children, making it easier for women to combine work and child care, and therefore, teaching becomes women's work. But in the 19th century in rural America, teaching was viewed as and designed to be most compatible with men's work. Farmers could use teaching as a source of income during the months of the year when their labor was not needed in the fields. Male ministers, politicians, shopkeepers, or lawyers could use teaching as a good avenue for increasing their public visibility within the community (Strober, 1984). When the number of children attending school increased and schools were broken down into individual grades with specific curriculums, more formal credentials were required and the school year lengthened. Rather than increasing the compensation for teachers because of the increase in the skill level required for the job, schools kept the pay low and began a concerted effort to recruit more women into the field (because they knew women would accept the lower pay).

Teaching became the ideal profession for single, educated, middle-class women, and for many years, female teachers were not allowed to continue to work when they married. This historical evolution of teaching from good work for men, to the perfect job for women—but only single, unmarried women—to teaching as well suited to married women with children demonstrates the ways in which the gender of different jobs and occupations is highly flexible and can change over time to justify whatever the gender status quo is in that occupation at the time. Ghettoization exists in the education field as well, as elementary teaching is now perceived as women's work in that it involves nurturing and is low paid, brings little autonomy, and has high rates of turnover. Men are overrepresented in the ranks of administrators and principals, where their work is stable, has high chances for promotion, and offers higher wages (Lorber, 1994).

Although much of the research on sex segregation in the workplace has been performed in the United States and other parts of the global North, these trends persist in the global South as well. Just as with farmers' daughters in the United States in the 19th century and in developing nations in Asia and South and Central America, daughters from farming families are sent to work in factories where their lives are largely controlled by their employers. Women between the ages of 16 and 24 make up 80% to 90% of the export-processing labor in Mexico, Southeast Asia, and other nations that engage in large-scale export production (Tiano, 1987). In Mexico's ***maquiladoras,*** assembly plants largely owned by American and other foreign companies that are located along the Mexico–U.S. border, to take advantage of the supply of cheap labor across the border, 85% to 90% of the workers are women (Lorber, 1994). Most of the people working in the export zones around the world are women, and this means that the access to cheap products in the developed world depends on the low wages and poor working conditions of these women in the global South. Sex and racial segregation in the workplace are crucial components of what make the global economy as we know it function.

Gender and Precarious Work

At a global level, women are also more likely to be employed in precarious work. **Precarious work** is an employment that is "uncertain, unpredictable and risky from the point of view of the worker" (Kalleberg, 2009, p. 2). Precarious jobs exist across a wide range of sectors, including domestic work, manufacturing, and agriculture. In the fruit industry in South Africa, women constitute 69% of the temporary and seasonal workers and only 26% of the long-term jobs. In Chile, only 5% of women in the fruit industry have long-term jobs. Countries in the global North like the United Kingdom have armies of home-based workers, 90% of whom are women (Kidder & Ratworth, 2004).

What's so bad about precarious work, though? Granted, it doesn't sound particularly good to call something precarious, especially when that something happens to be your job. But what are the costs of doing precarious work? The most obvious cost of precarious employment is the uncertainty; precarious workers can lose their source of income from one day to the next with no warning and no unemployment or pension to fall back on. This unpredictability comes with its own on-the-job stress. Precarious workers are also vulnerable to intimidation and coercion from their employers based on the insecurity of their positions.

Less obvious costs to precarious work include income and benefits foregone. For example, Morrocan garment workers put in 90 hours of overtime in 2003 to meet a buyer's tight deadline. But their employer didn't count these hours as overtime, and so the women lost 50–60% of their rightful pay. Precarious workers in the United Kingdom who do piece-rate work in their homes lost £25 per day compared with what they would earn as a long-term employee. In the United States, employers save $979 per month by using young Mexican men as precarious workers in farming. There are other costs to self-esteem and human development. Among female Moroccan garment workers, 80% take their daughters younger than the age of 14 out of school to care for younger siblings. In their jobs, workers are abused by their supervisors and sometimes ostracized by members of their community.

Why is precarious work now such a large part of the way people do work around the world? The gender stereotypes discussed earlier come into play with precarious work; employers may believe that women are more dexterous or compliant. But precarious work also subsidizes the lifestyle of thoese in the developed parts of the world. Precarious work is part of the explanation for why the t-shirt at Wal-Mart and the banana in the supermarket cost so little. Food and clothing retailers source the products they sell through very complex global supply chains, and they are increasingly powerful to demand lower costs and faster production. In the Sri Lankan garment industry, for example, production times have fallen from 90 to 45 days, while prices to suppliers have fallen 35% in an 18-month period. Powerful retailers must cut costs somewhere to make a profit by providing the cheapest products to their customers. They do so in part by using precarious workers.

Do you think most people think about the connections between the cost of the products they buy and the workers around the globe who make those products? What would be the consequences of reducing the amount of precarious workers?

The Wage Gap: Why Sex Segregation Matters

When we take a step back from the specific details of sex segregation, we're left with a familiar question. All around the world, women and men do different types of paid work, just as women and men do different kinds of work inside and outside the household (as in the previous chapter). Does this difference have to carry negative implications? Is difference necessarily bad? Some of our discussions of sex and racial segregation have already begun to suggest an answer in this specific case. Regardless of whether women are working in completely different occupations or doing different kinds of jobs within the same occupation, women are paid less than their male counterparts. Researchers call this rather stark fact the **gender wage gap**, and it is usually calculated as the ratio of women's average earnings in an occupation to men's average earnings in the same occupation. In the United

States for 2006 among full-time, full-year workers, the gender wage gap was 76.9 (Institute for Women's Policy Research, 2008). This ratio is often translated into real dollar amounts, meaning that in 2006, women on average in the United States made about 77 cents for every dollar earned by a man. The gender wage gap varies across countries, reaching a high of 89 in Sweden and a low of 71 in Greece and the Netherlands (Wharton, 2005). The gender wage gap exists even when women and men have comparable qualifications, and it is uniform across occupations, regardless of the degree of sex segregation within that particular occupation.

Historically, the gender wage gap has decreased in the United States since 1955, but in recent years, progress has largely stalled. Between 1980 and 1990, the gender wage gap gained 11.4 percentage points, but it gained only 5.4 percentage points over the next 15 years (Institute for Women's Policy Research, 2008). The current gender wage gap has been relatively stable since 2001, indicating no progress in closing the gap between what men and women earn for doing the same jobs and occupations. These trends are especially true for women with higher levels of education. In 2006, college-educated women between the ages of 36 and 45 earned 74.7 cents in their hourly wage for every dollar earned by men in the same group. In 1996, women in the same group were earning 75.7 cents for every man's dollar (Leonhardt, 2006). White women as well as women of color have lost ground relative to white men in recent years in regard to the gender wage gap.

Making Connections: Sex Segregation and the Gender Wage Gap

How are sex segregation in the workplace and the gender wage gap connected? This returns us to our recurring question about the relationship between difference and inequality. Clearly, sex segregation enforces a sense of difference; the fact that women are concentrated in certain types of jobs and occupations helps to create and reinforce gender ideologies about the different skills, abilities, and needs of men and women relative to work. This relationship between gender and work becomes a kind of endless loop, where because men do certain types of work, this work is perceived as better suited to men, and therefore primarily men will continue to do what is defined as men's work. This endemic system of sex segregation makes the idea of equal work for equal pay attractive in principal but difficult to reinforce in the real world. In a world where men and women so rarely do the exact same kind of work, whether it's the type of machines they're using in a hosiery factory, the exact task they perform in a slaughterhouse, or whether they're selling a house or a commercial building, how can we establish exactly what equal work is? The differences that are established by sex segregation make the gender wage gap easier to justify by establishing two very different work worlds for women and men. As William Bielby and James Baron (1987) described this situation, "Men's jobs are rewarded according to their standing within the hierarchy of men's work, and women's jobs are rewarded according to their standing within the hierarchy of women's work. The legitimacy of this system is easy to sustain in a segregated workplace" (p. 226).

The gender wage gap is therefore connected to sex segregation in two ways. First, in any occupation, job, sector, or segment of the workforce that is composed primarily of dominant men, which in the United States means white middle- or upper-class men, the wage scale will be higher than in those occupations, jobs, sectors, or segments composed primarily of either men of color or of women, regardless of their racial background. When women or people of color are concentrated in some area of the workforce, the compensation will be less (Tomaskovic-Devey, Kalleberg, & Marsden, 1996), and when women enter an occupation, the wages for that occupation fall (Treiman & Hartmann, 1981). This is true because of the processes we outlined previously; women's entry into an occupation or job coincides with a loss of wages in that occupation or job, and so when women enter new areas of the workforce in large numbers, the pay for workers declines. There are some men who work in predominantly female occupations and some women who work in predominantly male occupations. Men in areas of the economy dominated by women make less on average than their male colleagues in predominantly male occupations, but they still make more than their female counterparts in those same jobs; remember the case of male elementary-school teachers, nurses, social workers, and librarians riding the glass escalator. The women who work in predominantly male occupations also make more than their female counterparts in occupations that are dominated by women; but unlike men in predominantly female occupations, these women still make less than their male counterparts in the same occupation or job. This is largely because women in a predominantly male workplace are less likely to be promoted, and therefore, their earnings do not compete with those of men.

The second way in which the gender wage gap and sex segregation are related is in the tendency for higher paying jobs also to involve more authority or supervision than lower paying jobs. A plant manager within a factory is paid more than the worker on the floor, and a doctor is paid more than a nurse—in part because both of these positions are endowed with more authority and involve supervising the work of others. Jobs that require workers to interact with people in a nurturing or interpersonal way pay less than jobs that do not. For example, when the primary role of flight attendants was perceived as caring for airline passengers, keeping them happy, and creating a pleasant trip, flight attendants were paid less. When this job was upgraded to include safeguarding the safety of the passengers in the event of a crash or other emergency, the pay for this job increased to reflect the new authority associated with the position; literally, the authority of flight attendants became enforced in the United States by the federal government because it's against federal law not to follow the directions of a flight attendant. Obviously, authority and supervision versus nurturing and interpersonal interaction have a distinctly gendered component. Women are supposed to be better at interpersonal interactions and nurturing, while men are perceived as natural authority figures. The sorting of men and women into these different types of jobs helps to re-create the gender wage gap by placing men in higher paying positions with authority and women in lower paying jobs that draw on their skills as nurturers.

EXPLAINING SEX SEGREGATION AND THE GENDER WAGE GAP

Researchers have established that as much as 90% of the explanation for the gender wage gap lies with the prevalence of sex segregation (Petersen & Morgan, 1995; Tomaskovic-Devey, 1993b). The gender wage gap is important because it remains an important determinant of inequality between women and men. We live in a culture where money and income provide power and prestige. That women make less money on average than men gives them less power within family settings, where we've already discussed how income relates to decision making in the home. In female-headed households in the United States and around the world, women's income is the sole source of support for children and other family members, and the gender wage gap makes it more difficult for women to provide sufficiently for their families. On a global scale, the lower pay and stability inherent in the jobs into which women are segregated have implications for global levels of poverty; development organizations and workers have discovered that women's income can affect poor families in ways that men's income does not. Namely, women's income is more likely to be directed toward supporting a family and potentially providing upward mobility for a family than are men's wages. The gender wage gap matters at the lower economic levels of society as well as at some of the highest levels. The lack of women in positions of power and prestige in corporations, universities, and the government means that women have less power over the shape these various institutions take and, therefore, less of an impact on the important decisions made in these institutions in general.

In all these ways and more, the gender wage gap remains an important barrier to equality between women and men. How, then, do we explain why this gap exists, which means explaining why sex segregation persists across the workplace? We've discussed how an occupation ghettoizes or becomes resegregated historically, but how does segregation continue to persist across many different occupations today? There are several different theoretical answers to these questions that help explain the different forces at play in the process of sex segregation. Like the sociological theories of gender we explored in Chapter 2, these theories can be grouped into different levels: individual, interactional, and institutional. Here we will discuss individual and institutional approaches, but keep in mind that no single theory is sufficient to explain in full the existence of sex segregation; each perspective reveals a different aspect of how this complicated process works. As many researchers who study these theories concede, a social phenomenon that is as widespread and entrenched as sex segregation is probably kept in place simultaneously by many of the causal factors identified in these theories.

Socialization as an Explanation for Sex Segregation

We'll start with individual theories, which focus on the decisions workers make about the types of jobs they want to have and the ways in which the skills they have can influence the types of jobs they seek. The first individual-level theory is **socialization theory**, and you should not be surprised to find a discussion of socialization categorized as an individual-level theory of gender. Remember that individual theories include theories of socialization because these theories view gender as something that becomes internal to an individual. Once gender is internalized, it impacts the decisions, thoughts, and feelings of an individual in uniquely gendered ways. Like other individual theories, gender operates from the inside out according to this perspective. As applied to the question of sex segregation, socialization theory argues that our experiences with gender socialization lead men and women to prefer different types of jobs, and so socialization is important to explaining sex segregation through the mechanism of shaping worker preferences. The question posed in the beginning of this chapter that asked you to think back on what your career aspirations were when you were younger is important to a socialization argument about sex segregation.

Theory Alert!
Individual

Theory Alert!
Social Learning Theory

Socialization theory as an explanation for sex segregation receives some support from a study on the differences in job aspirations between older and younger women (Tomaskovic-Devey, 1993b). If socialization is an important component in determining the types of jobs women and men aspire to do, then shifting gender roles should alter patterns of socialization. In his study, Donald Tomaskovic-Devey (1993b) did find that younger women were more likely than older women to be employed in sex-integrated jobs. This could suggest that younger women responded to changes in gender roles by choosing jobs that were less typically defined as women's work. This pattern did not hold true for men, however, as there were no differences between the level of sex segregation in the jobs of younger and older men. This could indicate that the socialization experiences have changed for women in ways that they have not changed for men.

Socialization theory as an explanation for sex segregation works well as good common sense, and there's a chance it has a role to play in explaining the grouping of men and women into different types of jobs. We learn a great deal about what it means to be gendered and how to be gendered at a fairly early age, and these lessons affect the future lives we imagine for ourselves in a wide variety of ways. It makes sense that these gender lessons would also include predispositions to do certain kinds of work and not others. Empirically, there's little research beyond the study by Tomaskovic-Devey (1993b) that supports this particular approach, and even this study has potential flaws. Older women could be in more sex-segregated occupations because they chose those jobs based on their gender socialization, but it could also be the case that more sex-integrated jobs were simply not open or available to them when they began their work lives. Changing jobs becomes more difficult with age, and older women could simply be stuck in the sex-segregated occupations and jobs with which they began their careers, despite the fact that they'd like to be in more sex-integrated occupations or jobs.

Other research suggests that the aspirations we form when we're young, although they tend to be fairly gender typed, are usually not the jobs in which we eventually find ourselves (Marini & Shu, 1998; Stroeher, 1994). Just compare what you wanted to be when you were very young with what you envision currently as your future career or with the job in which you are currently engaged. Do they match? The answer to what you want to be when you grow up changes as children age, usually becoming less gender typed, but often bearing little relationship to the series of careers a person might have as an adult. Other empirical research reveals no real differences in the kinds of work-related values held by women and men. An extensive study using data from 12 national samples, and spanning a time period from 1973 to 1990, found that men and women consistently ranked the same set of values in the same order when asked what was most preferred by them in a job (Rowe & Snizek, 1995). The most important qualities in a job described by both men and women in order from most to least preferred were a feeling of accomplishment, high income, chance for advancement, job security, and short working hours.

Human Capital Theory

Theory Alert!
Individual

The next individual-level theory comes from an economic perspective. **Human capital theory** is influenced by economic thought in its emphasis on an individual's choice of occupation as resulting from her or his attempt to maximize the benefits and reduce the costs involved in any particular job. As its name suggests, a concept central to this theory is that of human capital. **Human capital** refers to the skills workers may acquire (through education, job training, and job tenure) that affect their ability to be productive in their job (Seccombe & Beeghley, 1992). A simple way to think about human capital is that it includes anything that increases a worker's productivity. In a manufacturing setting, productivity is simply the number of products produced on an hourly or weekly basis. Factory workers are generally faster when they have had some training and when they have been on the job longer, so training and job tenure increase their productivity. Education is an important source of human capital in that many jobs (like doctor, nurse, and lawyer) require workers to achieve a certain level of education before they can adequately perform the job. But education throughout one's career theoretically improves a worker's human capital, as in the case of those in business or education who go back to school for advanced degrees. In many cases, employers pay the costs for their employees to obtain these degrees and other kinds of education and training based on the belief that this education will improve the overall productivity of their employees.

Human capital theory is a general theory in the discipline of economics, but as it applies to the specific case of sex segregation, there are three important assumptions. The first is a basic assumption of economics as a discipline in general—that human beings are rationally motivated. Within sociology, this approach to the study of behavior is called rational choice, and it assumes that people act in ways to maximize the rewards and minimize the costs for any particular course of action. According to this perspective, for each decision that we make as individuals, we perform a careful calculus, like a list of pros and cons assembled in our head, and pursue the action with more pros than cons. The second assumption of human capital theory as it applies to sex segregation follows from the first: If we are rational actors who seek to maximize rewards and minimize costs, then it is rational for workers to invest in their human capital whenever possible. Increasing our human capital increases our productivity, and increasing our productivity means that we make more money at our job. Rational choice theory allows that we can be motivated by rewards that are not necessarily monetary, but certainly monetary rewards are among the more powerful rewards sought by people in capitalist societies like the United States.

Gender enters into the picture with human capital theory in the third assumption. According to human capital theory, women and men, on average and as groups, make different kinds of human capital investments. Everyone may be rational, but the unique position of women leads them to make decisions that take into account a different set of factors than do men when it comes to investing in their human capital. Namely, this difference is women's ability to bear children. According to human capital theory, not all women will bear children, but most do and many more plan to do so at some point in their lives. Women's experiences of both marriage and childbearing pose problems to their future careers that are not faced by men, and therefore, when women think about their future career trajectories, they calculate how to maximize their human capital investments differently than do men. Women anticipate that the demands of being a wife and mother will have an effect on their jobs and careers in ways that being a husband and father will not. Based on these assumptions, women have two options to pursue relative to their human capital. First, they might choose simply to invest less in their job-related human capital

(Chang, 2004). If their job or career will always be in competition with their role as mother and housewife, one option is to choose rationally to invest less in that particular area of their life. The second option open to women is to choose jobs or occupations that will allow them to maximize their human capital while still meeting the demands placed on them by their roles within the family.

This second option open to women becomes the focus of much of the research on human capital theory because it implies concrete predictions about what kinds of jobs most women will gravitate toward. If you're a woman who wants to maximize your human capital while retaining room for the demands that will be made on you in your role as mother and wife, what kinds of jobs would you be more likely to choose? Human capital theory predicts that women will choose jobs and occupations that are more family or child friendly, or jobs that include flexible hours and lesser penalties for reentry into the job after a period of absence. Flexible hours allow women to fit the responsibilities of child care or elder care into their work schedules. Low penalties for reentry into a job or occupation make it easier for women to take time off for giving birth and caring for children or other family members. Teaching, as we discussed previously, is perceived as an ideal occupation in which women can maximize their human capital. The work hours often correspond to the hours during which children are in school, making child care easier for women with children. In addition, there are few penalties for teachers for leaving the workforce for a period and then returning. The converse is that jobs or occupations that require full-time or beyond full-time hours and continuous labor force participation in order to be successful will have fewer women. High-powered managerial positions often require more than a full-time work week and provide little in the way of flexibility. In addition, leaving these jobs for a year or two to take care of family members would have negative effects on an employee's chances for upward mobility. According to human capital theory, a rational woman seeking to maximize her human capital should avoid these types of jobs and occupations (Chang, 2004).

How would you rank the qualities that are most important to you in a job or career? Do you think you would find differences between your ranking and the rankings of those of a different gender?

Human capital theory is another example of an individual approach to explaining sex segregation because women and men as individuals are choosing different occupations based on a rational calculus that seeks to maximize returns on their human capital. When many individual women and individual men make these same sorts of decisions, the end result is the systematic sex segregation we observe in societies like the United States. The evidence in support of human capital theory is mixed: Some studies demonstrate potential support for this way of thinking about sex segregation, while others find weak relationships between human capital variables and sex segregation. Certainly, when we look at a profession such as teaching, which at the level of elementary- and middle-school education was

82% female in 2005, this seems to provide some support for human capital theory. Among chief executives, on the other hand, only 24% were women in 2005. But you may have already thought of some potential flaws in the basic argument proposed by human capital theory. Men and women are different because women bear children, but what about women who have no desire to bear children? What predictions would human capital theory make about them? Or what about men who might want to take on full responsibility for child care or caring for other family members? Would human capital theory predict that these men would behave more like women in their choice of occupation and their methods for maximizing their human capital?

The research testing human capital theory explanations of sex segregation has little to say about the case of men who "look like women" in their desire to assume responsibility for care in the family, but there is research on the women who "look like men" in the sense of women who are not married and do not have children. According to the predictions of human capital theory, single and childless women should be less likely to work in predominantly female occupations; women without these family demands should rationally choose to maximize their human capital in other occupations and jobs. Research indicates, though, that single and childless women are just as likely to be employed in predominantly female jobs as are married women and women with children (Beller, 1982; Tomaskovic-Devey, 1993a). Another prediction of human capital theory is that any job that is predominantly female should also be more family friendly. After all, the family-friendly nature of the job is the reason that so many women choose to work in that particular area. Again, research suggests that jobs that are predominantly female are not any easier to reenter after leaving the labor force than are predominantly male jobs and are no more flexible in regard to the working hours (England, 1982, 1984; England & Farkas, 1986; Glass, 1990).

CULTURAL ARTIFACT 2: TRANSGENDER IN THE WORKPLACE

Discussions of gender discrimination in the workplace don't even begin to touch on the discrimination faced by those who live outside our convenient gender and sex categories. All the different kinds of discrimination discussed in this section are illegal in the United States when directed against women or men. An often costly and lengthy court case is required to prove when discrimination has taken place, but sex category is one of the categories legally protected against workplace discrimination (along with race, ethnicity, and religious background). Gender identity and sexual identity are not statuses that are protected uniformly across the United States. Although various cities, states, and individual employers have passed laws and policies prohibiting discrimination on the basis of gender identity or sexual identity, there is no federal law that prevents gays, lesbians, bisexuals, or transgender individuals from being discriminated against in the workplace. Many transgender individuals face a unique set of concerns in the workplace as they transition from living as one gender to living as another. The number of individuals to whom these concerns apply is difficult to measure,

(Continued)

(Continued)

but one survey conducted by the financial firm JPMorgan Chase estimated that roughly 0.2% of their employees self-identify as transgender. The number of businesses with policies directed toward the lesbian, gay, bisexual, and transgender (LGBT) community has increased dramatically in recent years. In 2003, only 27 of the Fortune 500 companies had policies in place prohibiting discrimination on the basis of gender identity. By 2008, that number had increased to 153 of the Fortune 500 companies (Luther, 2008). Four hundred of the 500 companies attend conferences like the Out and Equal Workplace Summit, while companies like General Motors, New York Life, and IBM recruit at the Transgender Career Expo (Belkin, 2008). Much of this progress is a result of the tireless work of the Human Rights Campaign, the largest LGBT civil rights organization in the United States. In addition, corporations are motivated to adapt their policies to be more inclusive because diversity has become a powerful recruiting tool for the innovative employees necessary to compete in today's marketplace. Personal stories bear witness to this change, as in the case of Jillian T. Weiss, who transitioned from a male lawyer in the '90s but saw keeping her job in a New York firm after her surgery as simply outside the realm of possibility. Instead, in an interesting sex segregation twist, she got a job as a legal secretary after her transition, where her paper trail was harder to follow. Weiss's gender transition literally led her to follow the contours of sex segregation by occupation, going from an occupation (lawyers) that was 29.2% female in 2002, to an occupation (legal secretary) that was 82.2% female. Ten years later, Breanna L. Speed found the workplace component of her transition from living as a man to living as a woman to be the "simplest" part (Belkin, 2008). She stayed in the same job as a data administrator and was assisted by the human resources department in her company with her gender transition. After laying the groundwork for several months, Wendell eventually left work on Friday to be replaced by Breanna on Monday. Although not all transgender individuals are as lucky as Ms. Speed, not all transgender individuals are necessarily forced to leave their jobs altogether as did Ms. Weiss. How do transgender individuals fit into this discussion of sex segregation in the workplace and the gender wage gap? In the past, women have sometimes passed as men to gain access to fields that were denied to them based on their gender. This is certainly not the motivation of transgender individuals today, but how will their presence potentially reshape the way gender operates in the workplace? Do male-to-female transsexuals experience the same kind of downward mobility in the workplace as Ms. Weiss? What about female-to-male transsexuals? What would a genderless workplace look like, or a workplace in which switching gender from Friday afternoon to Monday morning was not perceived as such an unusual thing?

Gendered Organizations

Individual-level explanations of sex segregation focus on the decisions made by individuals and how those decisions result in a workforce that is segregated along gender lines. Other potential explanations focus on discrimination that might occur interactionally in the workplace and on how the structure of work itself has gendered implications. This institutional-level approach examines how gender gets built into the very functioning of organizations. It makes sense to call this approach **gendered organizations** or institutionalized barriers. Gendered organizations theory should be familiar from our discussion of this theory in Chapter 2. Joan Acker (1990), one of the main proponents of gendered organizations, argued that organizations like businesses are gendered in fundamental ways. Sex segregation is maintained not just because of an aggregation of individual choices or because of tendencies toward discrimination in interactional processes. Rather, any workplace is an inherently gendered institution from the beginning. Organizational routines, jobs, and behavioral assumptions are all fundamentally gendered, and the fact that employers and employees are operating within a context that is fundamentally gendered is also a fact that is by and large taken for granted (Tomaskovic-Devey, 1993a). This explains the power and persistence of sex segregation because sex segregation is not just one particular outcome of the workplace; it is a fundamental part of how businesses work and therefore an integral part of their functioning.

Theory Alert! Organizations Theory

The gendered nature of organizations in the workplace becomes taken for granted, in part, through the supposedly neutral ways of defining basic concepts like jobs and wages, which actual carry highly gendered meanings. Acker (1990) pointed to the way in which a *job* is supposedly a gender-neutral term, a position that can be filled by any abstract human being. Yet for many employees, the ideal human being to fill that job is a male worker whose family responsibilities are being taken care of by a wife at home. In her historical examination of wages, Alice Kessler-Harris (1991) pointed to a similar dynamic in regard to how men and women are paid. *Wages* at first appearance is a gender-neutral term that simply describes the money that a worker receives in compensation for her or his labor. But throughout the history of business and government debates about what constitutes an acceptable wage, gender has always been front and center in discussions about how much money women and men should make.

Traditionally, debates about appropriate wages reflected a long tradition concerning the **family wage**, or the sum necessary to sustain all family members. Regardless of what the reality on the ground is or has been, the ideal family wage has often been conceived of as a male wage, reflecting the idealization of the sole male breadwinner. If a family wage is supposed to be a male wage, where does that leave women's wages? A woman's wage was often conceived of as supplementing those of other family members, rather than as solely providing for the sustenance of all family members. This obviously provided justification for paying women less than men; a whole family was depending on a man's wage, while women's income was merely a welcome addition to the main source of income. Of course, this ignores the many women historically who were the sole breadwinner for their families and the men who were not, in fact, supporting a family but still experiencing the benefit of receiving a family wage. Maintaining this way of thinking about wages was a way of simultaneously preserving a social order in which women were dependent on men and were kept out of the labor force. The idea of a wage, then, is not a gender-neutral term but one that carries a heavy history of ideas about how wages could be used to maintain women's dependency.

In a more contemporary context, gendered organizations theory points to the way in which other seemingly gender-neutral practices have very gendered implications. **Internal labor markets** describe the ways in which larger firms frequently provide structured opportunities for advancement to those who are currently employed within the firm (Wharton, 2005). Essentially, many large organizations promote from within. This practice in and of itself seems to have few gendered implications, until we consider the ways in which gender segregation works within many workplaces and the complex system through which internal labor markets function. Promotions are more likely to be given to individuals in certain types of jobs and areas within a firm than to others. Research demonstrates that women tend to be segregated into occupations and jobs that have fewer positions of authority than male occupations and positions (Huffman & Cohen, 2004). Women segregated within a firm in an area with few opportunities to demonstrate their authority are less likely to be promoted within the firm. A secretary or administrative assistant has little job authority and is much less likely to be promoted to management level than other jobs within a firm that are less female dominated. Existing systems of sex segregation and ghettoization of women make the apparently gender-neutral practice of using internal labor markets into a system that favors men's internal promotion over that of women.

Another example of the ways in which work organizations have gender built into their basic functioning is in the tools and techniques used in different types of jobs and occupations. In occupations or positions that have traditionally been performed by men, the tools and techniques used to do those jobs come to reflect the needs and interests of men. An example of how the very design of tools can have gendered implications comes from a study by Barbara Reskin and Heidi Hartmann (1986) of women working outdoor jobs for the telephone company, AT&T. Injury rates for women working these jobs were higher than the accident rates for men until lighter and more mobile equipment was introduced. The choice of the previous heavier and less mobile equipment was probably not intentionally chosen to make it more difficult for women to perform the job; rather, they were chosen for the men who had in the past been the ones primarily doing the job. This neutral decision about what kind of equipment to use in a job setting can have exclusionary practices, making it more difficult and dangerous for women to do certain types of jobs.

TRANSMEN AT WORK

These three theories of sex segregation suggest very different mechanisms through which the sorting of men and women into different jobs happens. In her study of female-to-male (FTM) transsexuals in the workplace, Kristen Schilt (2006) suggested that the case of FTMs can serve as a unique case for testing these different theories of sex segregation and discrimination in the workplace. Female-to-male transsexuals have a potentially unique perspective on gender discrimination in the workplace as a result of their "outsider-within" position. Female-to-male transsexuals were born as women, and in her study, many had experienced being a woman in the workplace. Although they became categorized as *within* the category of men, their prior lives as women also provided them with an *outsider's* perspective on masculinity at work. What happens to these individuals when many transition into being a man, some within the same workplace environment? Will their gender socialization as women trump their new adult transition to living as men? Will the same individual with the same education, skills, and abilities (human capital) be treated differently

Theory Alert!
Status Characteristics Theory

when her gender status changes to that of a man? How does the gendered organization of their workplace settings affect those who are moving out of one sex category and into another? From a status characteristics perspective, what will the performance expectations be for a transman in the workplace?

Although there are important variations in the experiences of the female-to-male transsexuals in Schilt's (2006) study, which we will discuss later, two thirds of the 29 respondents she interviewed reported receiving some kind of advantage at work after their transition to being a man. Schilt categorized these advantages into four main types: gaining authority and competency, gaining respect and recognition for hard work, gaining "body privilege," and gaining economic opportunities (p. 476). Several interviewees, like Henry, reported experiencing firsthand the authority gap that researchers have identified between women and men in the workplace (Elliott & Smith, 2004); of his interactions in the workplace after having transitioned to being a man, Henry said, "I'm right a lot more now. . . . Even with folks I am out to [as a transsexual], there is a sense that I know what I am talking about" (Schilt, 2006, p. 476). Another respondent noticed the way customers bypassed his female boss and came directly to him with questions, reflecting an assumption that he would know more as a man. Paul, who openly transitioned in the field of secondary education, noticed how he was called on to speak at meetings more often as a man than he had been as a woman. In one especially interesting anecdote revealing the increased competency attributed to these individuals post-transition, Thomas told a story about a fellow attorney who congratulated his boss for firing "Susan" (a pseudonym for Thomas's name before his transition) because she was incompetent, while noting that the new guy (Thomas) was "just delightful" (Schilt, 2006, p. 477). Thomas and Susan, in this anecdote, are the exact same person, but while Thomas was perceived to be delightful, Susan was perceived to be incompetent.

In the area of respect and recognition for hard work, transmen reported often having to work less hard than they had as women while receiving more respect and recognition as men. Preston transitioned openly while remaining at his blue collar job as a forklift operator. As a woman, Preston received little support as a female crew supervisor, often finding herself short of staff yet still expected to carry out the job competently. These problems disappeared when he transitioned to being a man, and in his last three performance reviews on the job, he received the highest performance ratings of his career. These high ratings were despite Preston's admission that he was doing nothing differently than he had done as a woman and that actually he was working less in his transition to part time. Another transman, Crispin, reported losing jobs as a female construction worker despite his extra effort because employers gave preferential treatment to male employees based on the assumption that men had families to support (Schilt, 2006).

Gaining body privilege was another kind of reward that men received after their transition. *Body privilege,* here, means the freedom from unwanted sexual advances or undue attention to their sexuality. One man reported being relieved to be free from "having my boobs grabbed and being called 'honey' and 'babe'" as he was in a job waiting tables before his transition. Another FTM reported his relief at being free from the sexual harassment of his boss, while a transman in a blue collar job noted that "obvious dykes" (Schilt, 2006, p. 480) in these workplaces were still subject to sexualized comments and unwanted inquiries into their sexuality; as a man, these concerns largely disappeared.

Finally, becoming a man brought many of the men in Schilt's (2006) study increased economic opportunities in a variety of forms. Carl, who started and now owns his own business, believed this career path would have been impossible for him had he remained a woman. Another respondent, Henry, attributed his success as a lawyer to his gender transition, especially because he would not have been able to succeed in law as an obvious dyke and he had no interest in conforming to the gender norms that would have been necessary for him to succeed in law otherwise; although Henry admitted that he might have been successful as a female lawyer, it would have involved "dressing the part" in skirts and other attire worn by female lawyers, and to him, this felt like a kind of drag. Another transman was openly discriminated against in his job before his transition because of his gender ambiguity. His boss told him he could not appear in the front, public area of the restaurant at which he worked because his appearance would scare people. This discrimination is something he no longer feels he has to face after his transition. Although none of the men in Schilt's study magically found themselves making more money after their transitions to life as men, they did encounter a decided lack of barriers to employment, promotion, and entrepreneurship.

Based on these findings, Schilt (2006) concluded that the experiences of the female-to-male transsexuals in her study provided little support for gender socialization or human capital theory as explanations for workplace discrimination and the gender wage gap. The unique case of female-to-male transsexuals who openly transitioned in the same work setting provided strong evidence against human capital theory. In the case of the lawyer, Thomas, his exact same set of skills, capabilities, and education were judged completely differently (as incompetent) when he was Susan than they were after he became Thomas (as delightful). Female-to-male transsexuals were socialized as women, and gender socialization theory predicts this should condition the jobs they choose for themselves, as well as how they approach those jobs and interact within them; because FTMs have already been socialized as women, they should retain these internalized disadvantages even with their transition to men. Yet this does not seem to be the case for the men in Schilt's study; although they were raised as women, they were still able to succeed as men after their transitions.

The theory we have discussed that the case of female-to-male transsexuals seems to most clearly support is that of gendered organizations. The experiences of transmen are important because the advantages that biomen (men who were born biologically male rather than becoming male through surgical intervention) receive in the workplace are often invisible to them. Those within the gendered organizations perspective argue that this is true because of the subtle ways in which gender is built into the functioning of workplaces as organizations. Take the case of Crispin, the construction worker discussed previously. Crispin knew after her transition that women were often let go from construction jobs because of an assumption that jobs were more valuable to men with families than they were to a single woman. She knew this because she had actually been told this by foremen in her life as a woman. But to the biomen who do construction work, it is difficult to understand that keeping your job is the result of gender bias built into the functioning of an organization; you may not even be aware that anyone else has lost his or her job while you kept yours, and you are more likely to assume that you kept your job because you are competent, rather than because you are a man. The outsider-within perspective of transmen allows them to see the hidden dimensions of gender that are hard at work in the world of their jobs.

Masculinity benefits some, but not all, of the transmen in Schilt's (2006) study. Pointing us again to the importance of considering overlapping identities, Schilt found that the *patriarchal dividend* that accrues to female-to-male transsexuals is most powerful for tall, white men who are farther along in their hormonal treatments. The **patriarchal dividend** is "the advantages men in general gain from the subordination of women" (Schilt, 2006, p. 466) (Connell, 1995). But this gain is not distributed equally to all men in Schilt's study. A Black transman summarized his experiences in the following way: "I went from being an obnoxious Black woman to a scary Black man" (Schilt, 2006, p. 486). This African American female-to-male transsexual as well as several others interviewed by Schilt felt they had to manage their identities as Black men carefully and that this sometimes negatively impacted their work experiences. After their transition, Black men were labeled as threatening or aggressive. An Asian American transman felt that he was denied any advantages at work because of his height (being short) and the stereotype of Asian men as passive and, therefore, less masculine.

Height points us to other ways in which the patriarchal dividend was sometimes denied to female-to-male transsexuals. Some FTMs who did not use hormonal treatments felt there was little change in their work experiences perhaps because this group was still perceived to be women by their coworkers who often slipped into the use of feminine pronouns to refer to them. In addition, some FTMs who were in the early stages of their transition using hormones also felt less of an advantage because of their status as men. Early in this transitional process, FTMs may not be taking hormones or may have just begun hormone therapy, and therefore, they may not have facial hair. This lack of facial hair makes many female-to-male transsexuals appear like younger men. One FTM commented that he went from looking like 30 (as a woman) to looking like 13 (as a man). FTMs in this category also reported not experiencing a clear advantage after their transition, indicating the important ways in which gender and age intersect to provide privilege. This also seemed true for height. Hormones can make biological women grow facial hair, but they cannot change their height. FTMs who were shorter (between 5'1" and 5'5") felt that their height also placed them at a disadvantage as men and reduced the amount of authority they had in workplace settings.

Considerations of age, height, appearance, and race all demonstrate the important ways in which the patriarchal dividend is not equally accessible to female-to-male transsexuals, but also to men in general. This is a good example of the importance of an intersectional approach to gender, focusing our attention on the ways in which race, age, and other identities condition how we experience gender. It is also consistent with what we know about hegemonic masculinity as a theory. Remember that hegemonic masculinity teaches us that in any specific setting, certain masculinities are privileged over others. For many of the transmen in Schilt's (2006) study, being white and tall and looking older brought the greatest advantages post-transition.

COMPARABLE WORTH

If the research of Schilt (2006) at least suggests some evidence in support of gendered organizations as an explanation for sex segregation and the wage gap, what are some possible solutions based on this theoretical approach? Comparable worth is a method proposed by many in the second wave of the feminist movement for redressing the gender wage gap, and it can be aligned with a gendered organizations approach. **Comparable worth** as a policy seeks to raise the wages of the low-paying jobs occupied predominantly by women by demonstrating the gendered ways in which jobs are socially constructed. Using job evaluation, comparable worth seeks to unpack the complex relationship between the gender composition of a particular job or occupation, the relative evaluation that job or occupation receives, and the compensation for that job or occupation. Under the Equal Pay Act of 1963, men and women are supposed to be paid the same amount in jobs that are "equal" (Giele & Stebbins, 2003). But how exactly do you determine how jobs are equal when no two jobs are ever really the same? **Job evaluation** is one way of answering this question by determining how pay is assigned to jobs and by evaluating those pay rates as fair or unfair (Wharton, 2005). In comparable worth cases, job evaluation is used to generate rankings of jobs within an organization or firm, and these rankings consistently demonstrate that women receive lower average wages in jobs that are comparable with those occupied by men. In cases where one consistent job evaluation plan is used to set pay throughout a firm or organization, women nearly always receive higher wages relative to men than is true in firms or organizations that do not use job evaluation (England, 1992). In this way, comparable worth corrects for this particular way in which gender is built into the structure of organizations by standardizing the methods for distributing pay for different jobs regardless of the particular gender composition of that job.

Gendered organizations as a theory explaining sex segregation and the gender wage gap draw our attention to the structure and rules of institutions like corporations and other workplaces. A strength of this theory is that it helps to explain the persistence of these practices because they are deeply embedded in the working of organizations. Undoing sex segregation from a gendered organization perspective involves changing the ways in which organizations work in fairly fundamental ways. The ways in which the normal practices of any business can have unintended gender implications must be revealed and then redressed, and this is an extensive project to say the least. One potential weakness in gendered organization theory emerges from this connection between the normal functioning of organizations and the gendered consequences. If all organizations are gendered in these fundamental ways, how do we imagine what a genderless organization would look like (Britton, 2000)? It is also difficult within a gendered organizations perspective to account for variations in the extent to which sex segregation occurs in different businesses or across different occupations. Why are some firms more or less gendered, exemplifying higher or lower levels of sex segregation? Gendered organizations draw our attention to the subtle ways in which gender can be built into the basic structures of business and the economy, but it leaves other questions unanswered.

BIG QUESTIONS

- In this chapter, we discussed how much of the work performed by women throughout the world goes largely unmeasured by official labor statistics and, therefore, remains invisible. What questions does this raise about what we consider to be work and what we don't? How is gender built into the ways in which we think about what work is and what it means to work?

- In Chapter 8, we talked about the sexual division of labor and that some gender theorists argue this division is the key to understanding gender inequality. How does what you've learned in this chapter on the gender of work connect to this perspective on the sexual division of labor? Does the information in this chapter seem to support the view that the sexual division of labor is central to gender inequality, and how so?
- Throughout this book, we've asked questions about the relationship between gender difference and gender inequality. In this chapter, we discussed sex segregation and that women and men generally work different kinds of occupations and jobs. This leads to the gender wage gap, but can you imagine a world in which men and women worked different kinds of jobs and occupations but this did not lead to inequality in the form of a gender wage gap or other kinds of inequality?
- How important is it that women earn less than men on average? In many parts of the developing world, this affects overall poverty rates. What are other implications of the gender wage gap? How important do you think this gender inequality is relative to other examples of gender inequality we have discussed so far in this book?
- The implications for how to correct sex segregation in the workplace are very different depending on whether you take an individual-, interactional-, or institutional-level approach. What are the implications of an individual-level approach? An interactional-level approach? An institutional-level approach? How do those implications influence the way you think about these different approaches to explaining sex segregation?
- Several of the theories and approaches discussed in this chapter suggest that gender inequality is built into the very structure of economic systems and society. Do you agree with this perspective? What evidence might you assemble in support of this argument? What evidence might you assemble to contradict this argument? What are the implications of this perspective for reducing gender inequality in relation to work?

GENDER EXERCISES

1. Do your own investigation on the gender wage gap and sex segregation. Information for the United States is available at the following website: www.bls.gov/cps/cpsaat11.pdf. Pick some occupations you're interested in and look up the percentage of females in that occupation, and then calculate the gender wage gap. What's surprising about this information? Can you find an occupation in which women make more than men? What theories might you develop about which occupations tend to be more segregated than others and which have larger gender wage gaps?
2. How have attitudes about gender and work changed across time? Interview people from different generations—older and younger women and men—about their experiences with work and how they feel about work. You might ask older women and men questions about what kind of jobs they had, what kind of jobs were appropriate for different genders during their time, what the attitudes were then toward women and men working, and so on. Interview younger individuals and compare how the gender of work has changed over time.

3. Pick a firm or organization with which you are familiar, and conduct an audit of its level of sex segregation. Make a list of all the jobs within that firm or organization and the gender of the people occupying those jobs. Pay attention to details and to the subtle differences between jobs. For example, which jobs are supervisory (have authority over other workers), and is there a gender trend in who occupies those jobs? What are potentially subtle differences in what seem like relatively similar jobs? For example, in a law firm, who does more pro bono (free) legal work? Do different lawyers tend to take different kinds of cases?

4. Test some of the assumptions of the theories we discussed in this chapter explaining sex segregation in the workplace. Interview people who are already in the workplace or those who are preparing for future jobs. Ask them how they decided on their particular career or occupation. As predicted by human capital theory, do women choose careers with child-care and family considerations in mind while men do not? How does socialization seem to impact their career choice? If they are already working, have they faced any discrimination or advantages in the workplace as a result of their gender? What is their sense of the gender breakdown of their particular job, occupation, or firm?

5. Go to the library and find books or articles on the work lives of women or men in several different societies. Find a mix of industrialized societies and other kinds of societies. What are the important tasks to be accomplished in each of these societies? Among these tasks, which seem to be considered work and which are not? Is this related to gender (are the things men do more likely to be considered work than the things women do, or vice versa)? What kinds of tasks seem to be considered men's work or women's work, and how does that vary across the different societies?

6. Do a content analysis of some popular culture medium. What does this particular media outlet say about the gendering of work? For example, you might compare men's and women's magazines. How does the amount of content dedicated to thinking about the workplace in these magazines compare? How are the messages being sent about what work means to men and women different? Do men's magazines reinforce the idea that men should "Be a Big Wheel" or the ideals of Marketplace Manhood?

TERMS

formal paid labor

informal labor

own-account workers

contributing family workers

vulnerable employment

compensatory masculinity

egalitarian masculinity

oppositional culture

glass escalator

glass ceiling

sex segregation

occupational sex segregation

job-level sex segregation

putting-out industry

piece-meal work
protective labor legislation
ghettoization
resegregation
maquiladoras
precarious work
gender wage gap
socialization theory
human capital theory
human capital
gendered organizations
family wage
internal labor markets
patriarchal dividend
comparable worth
job evaluation

SUGGESTED READINGS

Sex segregation and the gender wage gap

Anker, R. (1998). *Gender and jobs: Sex segregation of occupations in the world.* Geneva, Switzerland: International Labor Office.

Cockburn, C. (1991). *In the way of women: Men's resistance to sex equality in organizations.* Ithaca, NY: ILR Press.

Eisenberg, S. (1998). *We'll call you if we need you: Experiences of women in working construction.* Ithaca, NY: Cornell University Press.

England, P., & Farkas, G. (1986). *Households, employment and gender: A social, economic and demographic view.* New York, NY: Aldine de Gruyter.

Giele, J. Z., & Stebbins, L. F. (2003). *Women and equality in the workplace.* Santa Barbara, CA: ABC-CLIO.

Glass, J. (1990). The impact of occupational segregation on working conditions. *Social Forces, 68,* 779–796.

Reskin, B. (1984). *Sex segregation in the workplace: Trends, explanations, remedies.* Washington, DC: National Academy Press.

Reskin, B. (1988). Bringing the men back in: Sex differentiation and the devaluation of women's work. *Gender & Society, 2,* 58–81.

Reskin, B. (1993). Sex segregation in the workplace. *Annual Review of Sociology, 19,* 241–270.

Reskin, B. F., & Hartmann, H. I. (1986). *Women's work, men's work: Sex segregation on the job.* Washington, DC: National Academy Press.

Reskin, B., & Padavic, I. (1988). Supervisors as gatekeepers: Male supervisors' response to women's integration in plant jobs. *Social Problems, 35* (5), 536–550.

Tomaskovic-Devey, D. (1993). *Gender and racial inequality at work.* Ithaca, NY: ILR Press.

Treiman, D. J., & Hartmann, H. I. (1981). *Women, work and wages: Equal pay for jobs of equal value.* Washington, DC: National Academy Press.

Walby, S. (1986). *Patriarchy at work: Patriarchal and capitalist relations in employment.* Minneapolis: University of Minnesota Press.

Theories of sex segregation and the gender wage gap

Acker, J. (1990). Hierarchies, jobs, bodies: A theory of gendered organization. *Gender & Society, 4* (2), 139–158.

England, P. (1992). *Comparable worth: Theories and evidence.* New York, NY: Aldine de Gruyter.

Kanter, R. M. (1977). *Men and women of the corporation.* New York, NY: Basic Books.

Marini, M. M., & Shu, X. (1998). Gender-related change in the occupational aspirations of youth. *Sociology of Education, 71* (1), 43–67.

Reskin, B., & Roos, P. A. (1990). *Job queues, gender queues: Explaining women's inroads into male occupations.* Philadelphia, PA: Temple University Press.

Rowe, R., & Snizek, W. E. (1995). Gender differences in work values. *Work and Occupations, 22,* 215–229.

Weeden, K. A. (2002). Why do some occupations pay more than others? Social closure and earnings inequality in the United States. *American Journal of Sociology, 108* (1), 55–101.

Cross-cultural perspectives

Chang, M. L. (2004). Growing pains: Cross-national variation in sex segregation in sixteen developing countries. *American Sociological Review, 69,* 114–137.

Clark, R., Ramsbey, T. W., & Adler, E. S. (1991). Culture, gender and labor-force participation: A cross-national study. *Gender & Society, 5,* 47–66.

International Labor Office. (2008). *Global employment trends for women, March 2008.* Geneva, Switzerland: Author.

Massiah, J. (1993). *Women in developing economies: Making visible the invisible.* Providence, RI: Berg.

Tiano, S. (1987). Gender, work, and world capitalism: Third world women's role in development. In B. B. Hess & M. M. Ferree (Eds.), *Analyzing gender* (pp. 216–243). Newbury Park, CA: Sage.

Historical perspectives

Kessler-Harris, A. (1982). *Out to work: A history of wage-earning women in the United States.* New York, NY: Oxford University Press.

Kessler-Harris, A. (1991). *A woman's wage: Historical meanings and social consequences.* Lexington: University of Kentucky Press.

Lamphere, L. (1987). *From working daughters to working mothers: Immigrant women in a New England industrial community.* Ithaca, NY: Cornell University Press.

Milkman, R. (1987). *Gender at work.* Urbana: University of Illinois Press.

Rose, S. O. (1987). Gender segregation in the transition to the factory: The English hosiery industry. *Feminist Studies, 13* (1), 163–184.

Masculinity, the glass ceiling, and the glass escalator

Cotter, D. A., Hermsen, J. M., Ovadia, S., & Vanneman, R. (2001). The glass ceiling effect. *Social Forces, 80* (2), 655–681.

Gould, R. (1974). Measuring masculinity by the size of a paycheck. In J. Pleck & J. Sawyer (Eds.), *Men and masculinity* (pp. 96–100). Englewood Cliffs, NJ: Prentice-Hall.

Pyke, K. D. (1996). Class-based masculinities: The interdependence of gender, class, and interpersonal power. *Gender & Society, 10,* 527–549.

Rubin, L. (1976). *Worlds of pain: Life in the working class family.* New York, NY: Basic Books.

Williams, C. L. (1991). *Gender differences at work: Women and men in non-traditional occupations.* Berkeley: University of California Press.

Williams, C. L. (1992). The glass escalator: Hidden advantages for men in the "female" professions. *Social Problems, 39,* 253–267.

WORKS CITED

Acker, J. (1990). Hierarchies, jobs, bodies: A theory of gendered organization. *Gender & Society, 4* (2), 139–158.

American Association of University Professors. (2010, March-April). No refuge: The annual report on the economic status of the professor. *Academe,* pp. 4–33.

Anderson, E. (1990). *Streetwise: Race, class and change in an urban community.* Chicago, IL: University of Chicago Press.

Anderson, E. (2000). *Code of the streets: Decency, violence and the moral life of the inner city.* New York, NY: W.W. Norton.

Belkin, L. (2008, September 3). Smoother transitions. *The New York Times.* Retrieved December 12, 2008, from http://www.nytimes.com/2008/09/04/fashion/04WORK.html?_r = 1&scp = 1&sq = smoother%20 transitions&st = cse

Beller, A. H. (1982). Occupational segregation by sex: Determinants and changes. *Journal of Human Resources, 17* (3), 371–392.

Bernstein, G. L. (1993). Haruko's work. In C. Brettell & C. Sargent (Eds.), *Gender in cross-cultural perspective* (pp. 225–234). Englewood Cliffs, NJ: Prentice-Hall.

Bielby, W., & Baron, J. (1984). A woman's place is with other women: Sex segregation within organizations. In B. Reskin (Ed.), *Sex segregation in the workplace: Trends, explanations, remedies* (pp. 27–55). Washington, DC: National Academy Press.

Bielby, W., & Baron, J. (1986). Men and women at work: Sex segregation and statistical discrimination. *American Journal of Sociology, 91,* 759–799.

Bielby, W., & Baron, J. (1987). Undoing discrimination: Job integration and comparable worth. In C. E. Bose & G. Spitze (Eds.), *Ingredients for women's employment policy.* Albany: State University of New York Press.

Bielby, D., & Bielby, W. T. (1988). She works hard for the money: Household responsibilities and the allocation of work effort. *American Journal of Sociology, 93* (5), 1031–1059.

Bonvilliain, N. (2007). *Women and men: Cultural constructs of gender.* Upper Saddle River, NJ: Prentice-Hall.

Britton, D. M. (2000). The epistemology of gendered organizations. *Gender & Society, 14* (3), 418–434.

Bureau of Labor Statistics. (2005). *Table 11: Employed persons by detailed occupation and sex, 2005 annual averages.* Washington, DC: Author. Retrieved July 31, 2008, from http://www.bls.gov/cps/wlf-table11-2006.pdf

Chang, M. L. (2004). Growing pains: Cross-national variation in sex segregation in sixteen developing countries. *American Sociological Review, 69,* 114–137.

Connell, R. (1995). *Masculinities.* Berkeley: University of California Press.

Conway-Turner, K., & Cherrin, S. (1998). *Women, families, and feminist politics.* New York, NY: Haworth Press.

Cotter, D. A., Hermsen, J. M., Ovadia, S., & Vanneman, R. (2001). The glass ceiling effect. *Social Forces, 80* (2), 655–681.

El Amouri Institute. (1993). Women's role in the informal sector in Tunisia. In J. Masiah (Ed.), *Women in developing economies: Making visible the invisible* (pp. 135–194). Providence, RI: Berg.

Elliott, J. R., & Smith, R. A. (2004). Race, gender, and workplace power. *American Sociological Review, 69,* 365–386.

England, P. (1982). The failure of human capital theory to explain occupational sex segregation. *Journal of Human Resources, 13* (3), 358–370.

England, P. (1984). Wage appreciation and depreciation: A test of neoclassical economic explanations of occupational sex segregation. *Social Forces, 62* (3), 726–749.

England, P. (1992). *Comparable worth: Theories and evidence.* New York, NY: Aldine de Gruyter.

England, P. (2010). The gender revolution. *Gender & Society, 24,* 149–166.

England, P., & Farkas, G. (1986). *Households, employment and gender: A social, economic and demographic view.* New York, NY: Aldine de Gruyter.

Farkas, G., Lleras, C., & Maczuga, S. (2002). Does oppositional culture exist in minority and poverty peer groups. *American Sociological Review, 67* (1), 148–155.

Giele, J. Z., & Stebbins, L. F. (2003). *Women and equality in the workplace.* Santa Barbara, CA: ABC-CLIO.

Glass, J. (1990). The impact of occupational segregation on working conditions. *Social Forces, 68,* 779–796.

Gould, R. (1974). Measuring masculinity by the size of a paycheck. In J. Pleck & J. Sawyer (Eds.), *Men and masculinity* (pp. 96–100). Englewood Cliffs, NJ: Prentice-Hall.

Huffman, M. L., & Cohen, P. N. (2004). Occupational segregation and the gender gap in workplace authority: National versus local labor markets. *Sociological Forum, 19* (1), 121–147.

Institute for Women's Policy Research. (2008, February). *The gender wage gap, 2009.* Washington, DC: Author. Retrieved July 31, 2008, from http://www.iwpr.org/pdf/C350.pdf

International Labor Office. (2008). *Global employment trends for women, March 2008.* Geneva, Switzerland: Author.

Kalleberg, A. L. (2009). Precarious work, insecure workers: Employment relations in transition. *American Sociological Review, 74,* 1–22.

Kessler-Harris, A. (1982). *Out to work: A history of wage-earning women in the United States.* New York, NY: Oxford University Press.

Kessler-Harris, A. (1991). *A woman's wage: Historical meanings and social consequences.* Lexington: University of Kentucky Press.

Kidder, T., & Ratworth, K. (2004). "Good jobs" and hidden costs: Women workers documenting the price of precarious employment. *Gender and Development, 12,* 12–21.

Lamphere, L. (1987). *From working daughters to working mothers: Immigrant women in a New England industrial community.* Ithaca, NY: Cornell University Press.

LeDuff, C. (2000, June 16). At a slaughterhouse, some things never die. *The New York Times.* Retrieved December 11, 2008, from http://partners.nytimes.com/library/national/race/0616001eduff-meat.html?scp = 1&sq = at%20a%20slaughterhouse&st = cse

Leonhardt, D. (2006, December 24). Gender pay gap, once narrowing, is stuck in place. *The New York Times.* Retrieved December 18, 2008, from http://www.nytimes.com/2006/12/24/business/24gap.html?scp = 1&sq = gender%20pay%20gap%200nce&st = cse

Liebow, E. (1967). *Tally's corner.* Boston, MA: Little, Brown.

Lorber, J. (1994). *Paradoxes of gender.* New Haven, CT: Yale University Press.

Luther, S. (2008, April 22). *Transgender inclusion in the workplace* (2nd ed.). Washington, DC: Human Rights Campaign Foundation. Retrieved May 25, 2009, from http://www.hrc.org/issues/transgender/1561.htm

Marini, M. M., & Shu, X. (1998). Gender-related change in the occupational aspirations of youth. *Sociology of Education, 71* (1), 43–67.

Matthews, H. (2006, April 3). *Childcare assistance helps families work: A review of the effects of subsidy receipt on employment.* Washington, DC: Center for Law and Public Policy. Retrieved May 6, 2009, from http://www.clasp.org/publications/ccassistance_employment.pdf

Milbank, D. (2008, June 8). A thank-you for 18 million cracks in the glass ceiling. *Washington Post.* Retrieved December 12, 2008, from http://www.washingtonpost.com/wp-dyn/content/article/2008/06/07/AR2008060701879.html

Mitchell, B. L., & Feagin, J. R. (1995). America's racial-ethnic cultures: Opposition within a mythical melting pot. In B. Bowser, T. Jones, & G. Auletta (Eds.), *Toward the multicultural university* (pp. 65–86). Westport, CT: Praeger.

Okine, V. (1993). The survival strategies of poor families in Ghana and the role of women therein. In J. Massiah (Ed.), *Women in developing economies: Making visible the invisible* (pp. 167–194). Providence, RI: Berg.

Petersen, T., & Morgan, L. (1995). Separate and unequal: Occupation-establishment segregation and the gender wage gap. *American Journal of Sociology, 101* (2), 329–365.

Pyke, K. D. (1996). Class-based masculinities: The interdependence of gender, class, and interpersonal power. *Gender & Society, 10,* 527–549.

Reskin, B. (1993). Sex segregation in the workplace. *Annual Review of Sociology, 19,* 241–270.

Reskin, B. F., & Hartmann, H. I. (1986). *Women's work, men's work: Sex segregation on the job.* Washington DC: National Academy Press.

Reskin, B., & Roos, P. A. (1990). *Job queues, gender queues: Explaining women's inroads into male occupations.* Philadelphia, PA: Temple University Press.

Roos, P. A., & Reskin, B. (1992). Occupational desegregation in the 1970s: Integration and economic equity? *Sociological Perspectives, 35* (1), 69–91.

Rose, S. O. (1987). Gender segregation in the transition to the factory: The English hosiery industry. *Feminist Studies, 13* (1), 163–184.

Rowe, R., & Snizek, W. E. (1995). Gender differences in work values. *Work and Occupations, 22,* 215–229.

Rubin, L. (1976). *Worlds of pain: Life in the working class family.* New York, NY: Basic Books.

Rubin, L. (1994). *Families on the fault line.* New York, NY: HarperCollins.

Schilt, K. (2006). Just one of the guys? How transmen make gender visible at work. *Gender & Society, 20* (4), 465–490.

Seccombe, K., & Beeghley, L. (1992). Gender and medical insurance: A test of human capital theory. *Gender & Society, 6,* 283–300.

Strober, M. H. (1984). Toward a general theory of occupational sex segregation. In B. Reskin (Ed.), *Sex segregation in the workplace: Trends, explanations, remedies* (pp. 144–156). Washington, DC: National Academy Press.

Stroeher, S. (1994). Sixteen kindergarteners' gender-related views of careers. *The Elementary School Journal, 95* (1), 95–103.

Thomas, B. J., & Reskin, B. F. (1990). A woman's place is selling homes: Occupational change and the feminization of real estate sales. In B. Reskin & P. A. Roos (Eds.), *Job queues, gender queues: Explaining women's inroads into male occupations* (pp. 183–204). Philadelphia, PA: Temple University Press.

Tiano, S. (1987). Gender, work, and world capitalism: Third world women's role in development. In B. B. Hess & M. M. Ferree (Eds.), *Analyzing gender* (pp. 216–243). Newbury Park, CA: Sage.

Tomaskovic-Devey, D. (1993a). *Gender and racial inequality at work.* Ithaca, NY: ILR Press.

Tomaskovic-Devey, D. (1993b). The gender and race composition of jobs and the male/female, white/black pay gaps. *Social Forces, 72,* 45–76.

Tomaskovic-Devey, D., Kalleberg, A. L., & Marsden, P. V. (1996). Organizational patterns of gender segregation. In A. L. Kalleberg, D. Knoke, P. V. Marsden, & J. L. Spaeth (Eds.), *Organization in America: Analyzing their structures and human resource practices* (pp. 276–301). Thousand Oaks, CA: Sage.

Treiman, D. J., & Hartmann, H. I. (1981). *Women, work and wages: Equal pay for jobs of equal value.* Washington, DC: National Academy Press.

United Nations Development Fund for Women. (2005). *Progress of the world's women, 2005: Women, work & poverty.* New York, NY: Author.

Walby, S. (1986). *Patriarchy at work: Patriarchal and capitalist relations in employment.* Minneapolis: University of Minnesota Press.

Wharton, A. S. (2005). *The sociology of gender: An introduction to theory and research.* Malden, MA: Blackwell.

Williams, C. L. (1991). *Gender differences at work: Women and men in non-traditional occupations.* Berkeley: University of California Press.

Williams, C. L. (1992). The glass escalator: Hidden advantages for men in the "female" professions. *Social Problems, 39,* 253–267.

Wilson, W. J. (1996). *When work disappears.* New York, NY: Random House.

CHAPTER 10

How Does Gender Affect What You Watch, What You Read, and What You Play?

The Gender of Media and Popular Culture

What kinds of things do you do when you're not working, at school, studying, or hanging out with family or friends? In other words, what do you do with your leisure time? How much time do you spend watching television, going to movies, checking your profile on Facebook or MySpace, watching clips on YouTube, listening to your iPod or browsing the Internet on your iPhone, playing with a Wii or an Xbox, reading magazines, or watching sports? How different do you think your answers would be if you were of a different gender? Does gender influence what you watch on television or how much or how often you watch? Are there such things as "chick flicks," and if so, what would be the male equivalent? Why do boys seem to spend so much more time playing video games than girls, and why do certain games appeal more to both women and men in ways that others don't? Do video games like *Grand Theft Auto, Halo, Doom,* or *Battlefield* encourage boys and men to connect violence with masculinity? How do advertisements use gender and sex to sell us things, and how does that impact the way we think about what it means to be men and women? What do all these media outlets tell us about our sexuality and about how sexuality is connected to gender? How many of our assumptions about what people who are different from us are like are formed by what we see in the media and popular culture? How do the media and popular culture influence the way we understand racial and class identities and the ways in which these identities intersect with gender? And who's in charge of dictating all these images we see, read, and absorb in our daily lives? How does gender matter behind the scenes of the media, as an institution, in dictating who gets to be on television, in the movies, on iTunes, or in a magazine? What subtle messages about gender are being sent by this complicated world of popular culture?

THE MEDIA: AN INTERESTING INSTITUTION

These are the questions we'll be exploring in this chapter, and it's a wide and varied bag of topics related to gender. Before we dive in, we need to think about the media as an institution and about exactly what it means to think about the media as an institution. We've already discussed several examples of institutions and explored the gendered nature of these institutions. Institutions are a basic building block of society and of sociology as a discipline because they help us explain how society works as a whole. Institutions can be made up of a lot of different things, but all institutions in the end are composed of people. The institution of marriage is best described as a set of behaviors, but there are still large groups of people engaging in those sets of behaviors. If there weren't large numbers of people engaging in the institution of marriage, marriage would gradually cease to be an institution. As part of the process of change in society, institutions do sometimes disappear or lose their power.

In Chapter 6, we discussed the practice of courting, a system for finding a marriage partner that existed before the system we now call dating. Courting was an institution in the sense that it was a set of behaviors with very specific rules and guidelines that everyone was expected to follow. Girls asked boys to call on them, and boys could not call on a girl without having first been asked. Courting took place under the supervision of an adult, usually in a family parlor or on the front porch of the girl's home. Like marriage, courting was an institution that involved a set of behaviors, but when people stopped engaging in those behaviors, the institution of courting disappeared.

The media is a slightly different kind of institution than marriage, courting, or dating. The media is certainly also a set of behaviors. Few people in places like the United States, or increasingly around the world, would set up a household without purchasing a television. Ninety-nine percent of people in the United States have a television in their home, while globally, people watch an average of 22 hours of television each week (Graham, 2001). Purchasing a television and turning the television on are both behaviors that are important to the media as an institution. Deciding who gets to control the remote, and therefore how to watch television, are also behaviors that are part of the institution of the media—and behaviors that have gendered implications, which we will explore later. But the media as an institution is also made up of many formal organizations.

A **formal organization** is a group of people who interact on a regular basis and have a set of explicit, written rules. Courting was a set of behaviors, but there were no formal organizations that regulated how people courted because there were no written set of rules that people were obliged to follow in order to engage properly in courting behavior. There are formal organizations that are involved in marriage as an institution, including churches and government bodies that make decisions about the legal definitions of marriage. But unlike courting or marriage, formal organizations are a central component of the media as an institution. The media is supported by the individual behaviors of millions of people buying and turning on their televisions, as well as going to the movies, browsing the Internet, and playing video games, but without all the organizations that produce the content that comes out of the television, film projector, Internet, or video game, those behaviors would not take place.

Because of the importance of these formal organizations to a consideration of the media in general, there are two different ways of approaching our own investigation of gender and the media. The first is to examine the structure of the formal organizations that are involved in the creation and production of all the "stuff" we call the media. This approach means taking a behind-the-scenes look at what goes on at television and cable networks, movie studios, publishing firms, advertising firms, and Internet content providers. Most of the formal organizations involved in the media are also businesses and corporations in the United States, although in some instances, media organizations are government owned or controlled. Media corporations are concerned with making a profit, while media organizations that are owned or controlled by the government may be invested in educating the public or serving the interests of the government itself. Regardless of the purpose of the particular media organization, revealing how gender is a part of these behind-the-scenes activities is an important part of understanding the gendered content that these organizations produce. For example, some research demonstrates a significant gap between women and men in the management ranks for the media and advertising industries in the United States. In advertising agencies, only 28% of women held upper management positions compared with 54% of men, while in broadcasting, 37% of women were in management compared with 54% of men (Elliott, 1993).

The second way of approaching the study of gender and the media focuses both on the actual media content that gets produced and on what exactly people do with that content. In this approach, we shift from behind the scenes to the actual stuff that's on the television, the movie screen, the magazine page, or the computer screen. There are many theories from within sociology, feminism, and cultural studies that help to unpack (or interpret) the gendered messages contained in television, the movies, advertising, and magazines. This content-focused perspective is obviously connected to the first approach, but it also assumes that once media content is created, it takes on a life of its own. Exactly what the people who wrote, produced, directed, and acted in a particular television commercial intended to say about gender is less important once the commercial hits the airwaves. The meaning they originally intended may or may not be successfully conveyed, and this is why it is also important to ask what people do with media content. Exactly how do people make sense of the gendered messages that are being produced? Do people blindly and unthinkingly absorb the gender messages broadcast to them through the media? Do they largely ignore those messages? Or do they use those messages in creative ways, decoding media content in a way that conforms to what they already thought and believed to be true about gender and the way it works in the world?

As we'll see later in our examination of different theories about the media and its relationship to gender, these two approaches are necessarily connected. Any assumptions you make about how gender operates behind the scenes in media organizations are going to have an effect on the assumptions you make about the media content produced by those organizations. On the other side of the equation, the ways in which you imagine people use and make sense of media content will also influence how you identify the particular messages that are being sent by that particular media form. For simplicity, we'll separate these different sides of the dialogue between media organizations, media content, and media audiences, but keep the more complicated relationships between all three in mind.

Behind the Scenes: The Gender of Media Organizations

Theory Alert! Gendered Organizations

We'll begin our examination of the relationship between gender and the media with a look at what goes on inside media organizations. Gender shapes the structure of these organizations as well as the interactions that take place inside of them, and therefore, gender is an important part of the picture before any television show ever airs, magazine gets published, or CD gets recorded. Among the top national advertising and public relations agencies in the United States, only a handful of women occupy the top management positions, despite the fact that women make up 85% of all consumer purchases in the United States (Phillips, 2005). These are the agencies responsible for producing much of the advertising content viewed by consumers in the United States across a broad range of media outlets, and those who work within this industry acknowledge that gender often plays an important role in determining who works with what account. Despite the fact that many advertising executives claim to believe that there are no gender differences in the way male and female creative workers (as those working within the advertising industry are called) think, women are more likely to be assigned to working on perfume and tampon accounts, while their male counterparts get the beer and car assignments (Phillips, 2005).

Although some advertising executives try to assign their employees against gendered assumptions, putting their male workers on makeup accounts, the desire for a match between the gender of the advertising team and the gender of the product is driven in part by client expectations, their clients here being the companies who hire them to do their advertising. The perfume company executive expects to have women working on their advertising team, while the beer company expects mostly men. This is a good example of consumer-driven discrimination, as we discussed in Chapter 9, where advertisers allow the gender stereotypes of their clients to drive their own form of sex segregation by product. Of course, this has an effect on the careers of women and men within advertising and public relations because the account with the car manufacturer probably carries more prestige than the tampon or makeup account. But it also has an effect on the gender content of the media being produced by these agencies, as the ad you see for a car or beer in a magazine, on television, or on the Internet is more likely to incorporate the perspectives of men than it is to be designed to appeal to women.

Women Making Movies

This example and others demonstrate the ways in which media organizations, like almost all business organizations, face a set of gender problems related to sex segregation in the workplace, the gender wage gap, and the glass ceiling. In the world of advertising and public relations, women face a glass ceiling that prevents them from rising to higher levels of management and a kind of segregation that confines them to work on products that are considered to be more feminine. In their research on gender among television writers and screenwriters, Denise D. Bielby and William T. Bielby (1992, 1996) suggested that some of the features of these popular culture industries might intensify the ways in which gender stereotypes are employed, making them perhaps even worse than in other occupations and industries. One of the more obvious of these features is that both industries—television and film—are dominated by men, especially at the most powerful levels and positions. Of the

250 top grossing feature films released in 2009, only 7% (about 18 movies) were directed by women, a decrease of 2 percentage points from 2008 (Lauzen, 2010b). On television, of the top 40 television shows of the 2009–2010 season, only 16% were directed by women, representing a recent historical high (Lauzen, 2010a). In 2007, three of the four women who held top jobs at Hollywood's major studios left or were forced out of their positions, and all of them were replaced by men, leaving Amy Pascal (co-chairwoman of Sony Pictures Entertainment) as the only remaining woman (Waxman, 2007).

One study suggests that the number of women working as directors, writers, producers, and editors in Hollywood is declining (from 16% in 2005 to 15% in 2006) while the number of total female studio executives stays at about 20% (Waxman, 2007). Another feature that contributes to the role gender plays in these areas is the high-risk nature of both industries. Of the thousands of television pilots that are written and shows that are proposed, very few ever make it into a network executive's office, fewer ever make it onto the air, and only a miniscule number become successful shows that run for more than one season. The same funneling process occurs in the film industry, where the 36,000 scripts or script treatments that are registered with the Writers Guild each year compare to only 300 feature films that are actually released in the United States (Bielby & Bielby, 1996). Given these numbers, studios are averse to taking any risks, such as producing the first action movie written by a woman screenwriter.

CULTURAL ARTIFACT 1: GOING TO THE MOVIES TOGETHER—WHAT DO MEN AND WOMEN WANT?

Are the movies gendered in ways similar to the structure of television? Television has its daytime programming for women and prime time for a mixed-gender crowd. Are there women's movies and men's movies? One of the reasons cited for the declining numbers of women in top positions in Hollywood studios in recent years is the decline of the romantic comedy. In the 1980s and 1990s, top female Hollywood stars like Julia Roberts, Meg Ryan, and Sandra Bullock were consistent box office draws. With no major female stars to replace them, the romantic comedy geared toward female audiences has become something of an afterthought for major studios. Or perhaps the romantic comedy has just changed shape. Some followers of popular culture argue that the new genre is the "bromance." The idea of the bromance, defined primarily as a "man-crush," or close, nonsexual relationship between two men, has gained such popularity as to warrant its own MTV reality TV show, in which men compete to become best buddies with Brody Jenner and to share in his Beverly Hills lifestyle. In the recent Hollywood film, *I Love You, Man*, Paul Rudd plays a "wuss who mans up by befriending a guy's guy (Jason Segel) whose masculinity is so secure he wears Ugg boots and shorts to walk his wee dog" (Dargis, 2009). In the film, Paul Rudd's character, Peter, is about to get married and has no "dude to call his own" as his best man. Within the logic of the bromance, this makes Peter less than a man (something of a "semi-man" or "femi-man"), and his fiancée and family encourage

(Continued)

(Continued)

him to seek out his male soul mate. Comic moments of the film derive from Peter's mishap-prone search for a soul brother, including a misconstrued man-date. He finally meets "guy's guy" Sydney at an auction where they bond over their mutual love of the band Rush.

Other popular culture manifestations of the bromance include the HBO show *Entourage* and the *Ocean's Eleven* series of movies. In this sense, romantic comedies, or perhaps romances in general, have written women out of the picture altogether. What are women watching these days? Some evidence suggests that the movie tastes of women and men have converged in recent years. The horror movie is a genre that many might label as geared toward a more masculine audience. Many feminist analysts of popular culture have pointed to the misogynistic tendencies in horror films, which usually include a young, nubile, and scantily clad woman being gruesomely slaughtered. But recent horror films such as *Bug* and *Hostel* draw a large female audience. Fifty percent of the audience for the shriek-fest *Hostel* were women (Waxman, 2007). What do these changes to the gendering of Hollywood movies suggest, and what does the potential emergence of new genres of film imply about the status of gender? What does the bromance reveal about changing views of masculinity and of relationships between men? Is this really the new form of the romantic comedy? How are horror movies able to appeal to male and female audiences, and how might these changes affect the structure of media organizations like Hollywood studios?

All of these features of the television and film industries translate into a situation that makes gender an important component of decision-making processes. Callie Khouri, who won an Academy Award for her screenplay for the feminist-styled buddy movie, *Thelma and Louise,* explained the ways in which gender can affect a screenwriter's career in the following way:

> There is a certain stigma, I think that there is a set of expectations that women write a certain type of picture, so you don't look for an action movie that's written by a woman. You don't look for a thriller. There are certain types of movies that you don't expect to be written by a woman. People still call things "women's pictures." If it has a female audience then there is always a somewhat derogatory connotation to a so-called woman's picture. (Bielby & Bielby, 1996, p. 249)

Theory Alert! Status Characteristics Theory

Another female screenwriter described the ways in which studio executives and producers are simply not comfortable with women pitching scripts for action movies, as action movies are perceived as more masculine. From a status characteristics perspective, you might argue that executives had certain performance expectations about women and men. Confining women to a select genre of movies (romantic comedies or other "women's pictures") reduces the range of possible movie scripts they can write. It also makes female screenwriters more vulnerable to the ever-shifting trends in television and film; if the blockbuster films of that particular period tend to be action movies or thrillers, women confined to writing romantic comedies will find it even more difficult to get their movies produced.

Is believing that women can't write or produce a good action movie and men can't write or produce a good romantic comedy an example of essentialist thinking about gender? Do any of the theories we've explored suggest that men should be better at writing action movies or women should be better at romantic comedies? Do you believe these assumptions are true?

Gender and the History of Screenwriting

The specific case of screenwriters is especially interesting in regard to gender because it is one of the few professional occupations that has experienced a transition from predominantly female to predominantly male. In the early days of the silent film industry in Hollywood, many of the most successful *scenarists,* as screenwriters were called during that period, were women (Bielby & Bielby, 1996). Hard statistics for the period are difficult to come by, but some estimates suggest that during the silent film era (from the 1900s to 1927), women made up somewhere between 50% and 90% of all scenarists. The highest paid writer of the 1920s was Frances Marion, whose career extended into the era of sound movies, or *talkies.* Because of their predominance, these women scenarists were responsible for establishing many of the narrative forms and scenarios that are essential to the structure of films and, therefore, influenced the very basic idea of what makes a movie a movie today.

Women were displaced by men in the film industry as the innovation of sound in movies increased the importance of storytelling, thereby raising the prestige and desirability of screenwriting jobs. Under the growing studio system, the conception, writing, production, and distribution of movies were centralized under the control of a few large studios. The job of screenwriting was rationalized along with the process of production in general because story departments in major studios became responsible for finding literary material that could be turned into movies, as well as for writing the screenplays for those studios. Some women in this new era, such as Frances Marion, were able to hold onto their jobs, and Marion was a founding member and the first vice president of the Screen Writers Guild (now the Writers Guild of America). But the transition from women to men was nearly complete by the 1930s, with women making up only 15% of working screenwriters. In this new landscape, women became responsible not for writing scripts but for helping to write "women's films," writing dialogue for female actors, or generally adding a "woman's angle" to films. Today, the percentage of female screenwriters is only slightly higher than in the 1930s, with women making up 19% of all screenwriters.

Women make up 27% of television writers perhaps in part because the risks associated with producing a television show are slightly lower than those for movie production, as producing a television show costs considerably less than producing a film. But gender becomes a part of the structure of television in different ways than it does within the movie industry. For example, the very way in which television networks have traditionally structured the scheduling and content of their programs is based on gender assumptions, assumptions that may be increasingly inaccurate reflections of the new gender reality. Television networks divide their programming into daytime, prime time, and late night, reflecting their beliefs about who is at home and watching television during those periods.

Daytime television includes genres of shows—soap operas and game shows—that are assumed to be directed toward women, following the assumption that women are more likely to be home and watching television during the day. Prime time reflects the assumption that men come home and spend their evenings watching television, presumably along with their wives and children (Watkins & Emerson, 2000). This gendering of television is reflected in the structure of media organizations, where networks have separate development divisions for drama, comedy, daytime, and news (Bielby & Bielby, 1996). Female executives in television are more likely to be confined to "feminine" divisions, supervising female-typed genres such as television movies and miniseries, children's programming, and daytime programming. This gendered structure of television networks as media organizations continues for the major broadcast networks despite the fact that fewer and fewer women but increasing numbers of men are at home taking care of children during the daytime.

Limiting women's opportunities to have their screenplays made into movies or confining them to working with certain types of television shows has obvious effects on the gender content of the media that these organizations produce. Some Hollywood insiders have suggested that the decline in female top executives at major studios affects the types of movies that get produced and released (Waxman, 2007). Namely, the few female producers have to fight to get movies that might appeal to a predominantly female audience made in what is described by one female producer as a "boys' era" in movies. Lynda Obst, the producer of films such as *Hope Floats* and *How to Lose a Guy in 10 Days,* described having to fight hard to keep Disney from canceling her production for a remake of *Adventures in Babysitting,* another movie geared toward female audiences.

Sharry Lansing, the former chairwoman at Paramount, suggested that many women leave the top studio jobs because of both the difficulty of wedding these careers with love and family and the subtle shift in focus from quality content in films to the box office bottom line. Realizing that they probably cannot make the kind of character-driven movies with social and political content, on top of the 6 A.M. to 11 P.M., six-days-a-week schedule, leads many women to choose having a life over their high-powered careers, according to Lansing. As she described, "Women want to be in love. A huge percentage want children. They want friends. They want life" (as quoted in Waxman, 2007). Other women in the movie business suggest that women at lower levels of the studio system suffer from a lack of support, preventing them from ever moving up in the ranks to the top positions. Regardless of the underlying reasons, many believe the situation for women in movie studios may be getting worse, not better, and that this will continue to affect the kind of movies that get produced in Hollywood.

An example of the more subtle ways in which gender can influence media content comes from the world of news reporting. Carolyn M. Byerly (2004) shared her experiences with sexual harassment early in her career as a newspaper reporter in the 1960s and 1970s when one Pulitzer Prize–winning male journalist she interviewed asked her to spend the night with him after she completed the interview. In the early 1990s, Byerly worked in a newsroom in which all but one of the major gatekeepers were male and the newsroom was full of men's commentary about sports, women, and war. Working for the newspaper during the first Persian Gulf War, Byerly believed the predominantly male reporters were more interested in covering the war itself than the growing antiwar responses that were erupting in Seattle and around the world. In addition, coverage of "women's issues" at the newspaper rarely incorporated the perspectives of feminists or members of other women's groups who might be positioned to possess some expertise or a different perspective. Another female reporter working with Byerly would not call staff at rape crisis centers for background on stories about sexual assault because she saw these sources as biased. One female reporter for the *Los Angeles Times* who was known for her aggressive style was given a jockstrap as a going away present by her colleagues, with a note that said, "Sniff for luck."

How do experiences like the ones described discourage more women from working in media organizations? What kind of workplace discrimination discussed in Chapter 9 does this sound like?

Gender, Advertising, and the Commodification of Gender

All of these examples from the world of television, movies, and news media demonstrate the ways in which the gender composition of a media organization can have an effect on the gender content that gets produced. In addition, gender gets built into the very ways in which some media organizations function, as in the case of the division of television into daytime and prime time. Advertising is one part of this structure that is growing in importance across many different types of media outlets and has important gender implications. As technologies like the Internet and TiVo, or other digital recording devices, make it more difficult for advertisers to reach their viewers, or to compete for attention among the barrage of advertisements most people face on a daily basis, targeting advertising toward a specific group becomes increasingly important. Gender is one type of identity that marketers use in **market segmentation**, a sales approach in which corporations divide a large and diverse market for a product into smaller segments based on characteristics like gender, social class, and race, thereby creating many small, homogeneous markets. Creating advertising that is specifically geared toward one gender or the other is an example of market segmentation. For example, advertisers and marketers have traditionally ignored men as potential buyers of household goods, including cleaning products and groceries (Newman, 2008). In recent years, as increasing numbers of men have begun to clean and to go to the supermarket, advertisers and marketers have begun to reconsider their decisions to aim their commercials and print ads for groceries and cleaning products exclusively at women.

Advertising and marketing are an especially powerful component of the structure of many media organizations because of the way advertising works. In our consumer-driven world, advertising is no longer, as it once was, about convincing audiences that one product is better than another or cheaper than another. Advertising and marketing create a sense of need in consumers by convincing them that they can attain some desirable state of being by buying their product because most of the products and services people in the developed world purchase are nonessential goods. Think back on all the things you've bought in the last few months. How many of those products or services were absolutely necessary to your survival? You may very well feel that the soy, double-shot cappuccino from Starbucks or the song you downloaded from iTunes is necessary to your survival, but you won't starve or die of thirst or exposure without the latte or song, and this is true of most of the things we purchase on a daily basis.

No one (or at least no one that I can think of) actually needs an iPhone to survive. Advertisers and marketers work at convincing us not just that we should buy an iPhone because it works better than other cell phones or mp3 players, or that we really actually need it, but that owning an iPhone will make us cooler and happier. Beginning in the 1920s, advertisers increasingly began to create connections between social values and **commodities**, or the products they were selling. Prior to the now famous DeBeers advertising campaign reminding us that "diamonds are forever," diamonds were considered primarily to be financial investments, not symbols of romantic love and long-term marital commitment. The advertising campaign, which began in 1947, forever changed the meaning of diamonds, so that in the United States, many couples could not conceive of becoming engaged to be married without sealing the deal with a diamond (Jhally, 1995). These advertisers created an association such that buying a diamond is associated with romance, happiness, and sustained marital bliss in a very specifically gendered way (so far, men do not seem to need diamonds to achieve marital bliss). Some who study popular culture and media identify this process as **commodification**, which means turning any object, idea, or behavior into something that can be bought and sold. With diamonds, the ideas of love, commitment, and marital bliss were commodified.

Gender becomes an important tool of advertisers and marketers when it is used to attract attention and persuade consumers, largely by drawing on assumptions about the ways in which we believe men and women are supposed to behave. Using gender in advertising is especially powerful for two reasons. The world of advertising is a highly gendered world because it reflects these often stereotypical assumptions and projections, rather than portrayals of women and men that are grounded in the experiences of real people. These gender displays are intimately familiar to consumers because socialization ensures that we are all well acquainted with a detailed sense of how to enact gender norms. The second added advantage of using gender in advertising is that it allows advertisers to reach consumers through the medium of a core individual identity and one of the most fundamental ways we have of thinking about ourselves—as a woman or a man. Being a man or a woman is an important identity for a great many people, and therefore, advertisements that allow us to feel more masculine or feminine through buying a certain product can have a very powerful appeal. Gender in advertising, then, is especially effective because it's very easy for consumers to recognize and highly salient to their own identities and their own lives (Jhally, 1995).

There's nothing necessarily wrong with using gender in advertising, but the particular ways in which many advertisers use gender begins to create a distorted picture of the world and the way gender operates in it. The complex ways in which we experience and live gender in our everyday lives, and the countless variations in how gender is expressed, are largely lost in advertising depictions. For example, advertisements still rarely depict working-class women and men or men and women of color, and when they do, these depictions are often negative. In her analysis of advertisements, Jean Kilbourne (Jhally & Kilbourne, 2000) identified the ways in which women of color are often depicted as animals, animalistic, or primitive. Later in this chapter, we'll discuss portrayals of working-class men as buffoons.

Even more rarely do advertisements portray gender norms that do not conform to the most traditional assumptions, regardless of whether the men and women involved are white or middle class. How many times have you seen a commercial in which a man is cooking dinner for his family (leaving out those in which he's barbecuing, which is largely considered an appropriately masculine cooking activity) compared with the number of commercials in which women cook dinner for their family? How many commercials have you seen that depict women fixing cars, using power tools, or driving large trucks as compared with the commercials in which you see men doing these things? These distortions are amplified by the connections between gender and sexuality that are increasingly used in the United States to sell products that have little apparent connection to either gender or sexuality.

In one example from Kilbourne's video, the copy for a magazine advertisement for rice read, "Whatever you're giving him tonight, he'll enjoy it more with rice," and featured a picture of a woman smiling seductively (as quoted in Jhally & Kilbourne, 2000). In this particular advertisement, a satisfying sex life was commodified and connected to the unlikely food product, rice. As the constant bombardment of advertising increases, finding some way to grab the attention of consumers amidst the overall noise becomes increasingly difficult. Sexuality, and use of sexuality that continually pushes the limits, is one way to win the contest for consumer attention. Although this has been especially true of women's sexuality, it is also increasingly true that men's sexuality is being used to sell products from clothes to cleaning products.

Can you think of any examples of advertisements you've seen that stick out specifically because they challenge the gender status quo in some way? How many of such advertisements can you think of? Why do you think these particular advertisements use nontraditional depictions of gender, and does this make the advertisement effective?

The primary problem that emerges in the use of gender in advertising is that these words and images begin to create an increasingly distorted sense of what is really true about gender and sexuality. One of the most powerful examples of the way in which advertising creates this distorted view of gender is through its effects on body image. This has been especially true of women, who are bombarded with images of ideal beauty as primarily white, young, and thin. But it's also increasingly true of men. In one study of the images in *Men's Health* magazine, Susan M. Alexander (2003) found a strong emphasis on images of white, well-toned but not overly muscled bodies represented on the magazine's cover. The articles inside *Men's Health* sent a message similar to that seen in women's magazines like *Cosmopolitan,* with their focus on fat, weight, diet, and food. The overall message of the magazine was that men should work to build and maintain a hard body, which may also be a healthy body, but it is important for its ability to "Pull Her Sex Trigger." The word *sex* permeates the magazine, suggesting that obtaining a hard body is important because it brings the added perk of being able to engage in dynamic sex (presumably with a woman).

Despite these emphases on obtaining a hard body, the actual advertising content of the magazine focuses primarily on clothes and automobiles. Alexander (2003) argued that this combination of advertising and magazine content in *Men's Health* sends the clear message that automobiles are men's domain and that the correct brands of cars and clothes are important status symbols for successful masculinity. The end effect of this combination of images of hard bodies and advertisements is to turn men's bodies into "walking billboards for brand-name products" (p. 550) in a manner that has been true of women and advertising for many years. Men's bodies become sexualized because of the need for men to be able to perform sexually, an important component of dominant ideals of masculinity. As we discussed in Chapter 5, men's bodies are sexualized in ways different from the sexualization of female bodies. Men are sexualized as sexual subjects, with expectations to perform, while women are sexualized as sexual objects. *Men's Health* reinforces these messages about masculine sexuality.

The central role of advertising in driving the structure and functioning of media organizations is an important part of the behind-the-scenes component of the relationship between gender and the media. The need to attract attention through advertising drives media organizations to use gender in ways that create a distorted view of gender. Shifting our attention to the content of these advertisements as well as the media that goes along with them moves us from our consideration of media organizations themselves and toward the second approach to gender and the media—the content of the messages produced by the media and the question of what consumers and viewers do with those messages. Exploring this particular approach to gender and the media raises questions about exactly how we make sense of what television, movies, the Internet, popular music, and magazines are telling us about gender, and here it's useful to explore several different theoretical perspectives on those questions.

Most media organizations are businesses, and their bottom line is to make money. Should media organizations have a responsibility to portray gender in a certain way? Do media organizations create these gendered images, or are they merely reflecting the stereotypes that already exist?

MEDIA POWER THEORY: WE'RE ALL SHEEP

There are two contrasting theories about the social impact of the media that come from the disciplines of sociology and cultural studies. Because there are important connections between the structure of media organizations and the media content that they produce, these theories also have something to say about the media organizations themselves, reflecting the complex dialogue between the structure of organizations and the content they produce. The first theory focuses on media power and was developed primarily by Theodor Adorno, a sociologist who was part of a group of German, Jewish intellectuals who fled to Los Angeles and New York when the Nazis rose to power in Germany in the 1930s (Gauntlett, 2002). Adorno, like other intellectuals in this group that came to be known as the Frankfurt school, was deeply affected by witnessing firsthand the ways in which the Nazis used the media to spread propaganda in Germany. These experiences influenced the **media power theory** Adorno developed, in which he described the "culture industry" as a "well-oiled machine producing entertainment products in order to make a profit" (Gauntlett, 2002, p. 20).

To describe media organizations as industries doesn't seem very shocking to us, but Adorno and his counterparts in the Frankfurt school were deeply disappointed that the making of culture had become subject to the same rules of mass production used to produce cars and other manufactured goods. Under this system, culture ceases to be something that is made by "the people"; rather, it becomes something that is imposed from above and churned out by industrial entities like the studio system in Hollywood or television networks. Under this system, culture can no longer be called art—it becomes another commodity, exactly like a car, a bar of soap, or a refrigerator; culture becomes something that exists merely to be bought and sold. Given this view of how the culture industry works, it's not surprising that Adorno had a pretty dismal view of the quality of the media produced by this **culture industry**.

According to those in the Frankfurt school, all the media products of the culture industry are essentially "exactly the same" in the sense that they all embody the particular values of the established system (Gauntlett, 2002, pp. 20–21). Any potentially transgressive modes of expression are inevitably absorbed by the culture industry so that what might have been a rebellious act becomes just another way of contributing to the status quo. Many people found the rapper Eminem's lyrics offensive toward gay people and women, but his rebellious disregard for societal norms and the fears he supposedly generated in his song, "White America" in no way hindered his marketability, as his CDs sold millions of copies. According to the Frankfurt school perspective, consumers may have believed they were being rebellious by buying an Eminem CD, but in reality they were merely giving their full support to the status quo of the consumer society in which they live; rebellion through the consumption of mass-produced art is simply not real rebellion for Adorno and the other members of the Frankfurt school.

CULTURAL ARTIFACT 2: *30 ROCK* AND MEDIA POWER

Metafiction is a literary term used to describe a type of story that self-consciously addresses its existence as a story. In examples of metafiction, characters step outside of their roles to poke fun at the story itself. It sounds like a fancy term, but if you're a fan of the NBC television show *30 Rock*, you already know what metafiction is. *30 Rock* is a comedy show on NBC about a fictional comedy show on NBC. It stars

(Continued)

(Continued)

former *Saturday Night Live* stars such as Tina Fey and Tracy Morgan, as well as the actor Alec Baldwin as NBC executive Jack Donaghy. Tina Fey is an executive producer and writer of the real show and plays Liz Lemon, head writer for the fictional show, TGS. Tracy Morgan plays an outrageous comedian named Tracy Jordan. Examples of metafiction in *30 Rock* include an episode in which, after Liz Lemon expresses her reluctance to do product placement in her show, TGS, numerous, gratuitous product placements appear throughout the episode of *30 Rock*. This use of metafiction in *30 Rock* is interesting to our discussion of media power theory because the show pokes fun at the idea that the "creative" world of television is really always about the bottom line, as in trying to work product placements into a script to generate advertising revenue. The NBC executive played by Alec Baldwin is famous within the larger parent corporation of NBC for developing toasters, and he has no experience with television before he is assigned to become the "Head of East Coast Television and Microwave Oven Programming." In this and other ways, the show references the fact that NBC is owned by the larger corporation, General Electric, but in a comedic twist that pokes fun at the corporate conglomerations behind most media organizations, GE itself on *30 Rock* is owned by the fictional Sheinhardt Wig Company. In these and other ways, *30 Rock* as a show lends some support to the media power theory. Especially in early seasons, the "artist" Liz Lemon battles against the executive Jack Donaghy and often loses out to Jack's bottom-line, industrial approach to entertainment. In this sense, *30 Rock* is a fictional portrayal of Adorno's theory that making television shows is really not much different in the end from making toasters or microwave ovens. On the other hand, you could read *30 Rock* in a very different light. Doesn't the fact that NBC and its parent company, GE, produce a show that largely makes fun of the network itself and the business of television entertainment in general demonstrate the lack of power big corporations have? Or is the critique presented by *30 Rock* harmless in the end? Are *30 Rock* and other examples of metafiction on television and in the movies really able to be critical about media power, or are they still largely serving the interests of the Sheinhardt Wig Company and their kind?

The wide variety of styles and forms of entertainment that are available in today's consumer society may look different, but they merely ensure that no one truly escapes from the grip of the culture industry. Television and cable networks, movie studios, magazine publishers, and websites all provide us with a set of choices, but the range of those choices is always controlled by the media organizations themselves. And from the viewpoint of the culture industry, television shows, movies, and magazine articles consist largely of interchangeable elements that are reused over and over again. This was true in the 1940s when the Frankfurt school was developing its theory of popular culture, and it is still true today in the era of endless movie sequels and television programs that are revised versions of older programs or are borrowed from other sources (like the BBC in Great Britain, which has given the United States shows such as *Friends, The Office, Big Brother,* and *American Idol*). Adorno would be skeptical as to whether the culture industry really ever produces anything that can legitimately be called new, let alone anything that can be called art (Gauntlett, 2002).

Given their perspective on how the culture industry works and the quality of the media that these organizations produce, Adorno's particular view on the impact of the media should come as no surprise. From the perspective of Adorno and others in the Frankfurt school, the main effect of the media on its audience is to cultivate a kind of overall passivity. The content of popular culture forms such as television, movies, and increasingly the Internet are less important than the amount of time and energy that all these media outlets consume in our everyday lives. Our favorite TV show or Internet site may very well bring us enjoyment and, from time to time, may spark an interesting conversation with fellow viewers. But on the whole, our participation in media culture teaches us to conform to the status quo without even questioning what that status quo is or why it exists. According to Adorno, the culture industry creates a world in which "conformity has replaced consciousness" (Adorno, 1991, p. 90). Adorno would have agreed with your parents' admonishment that watching too much TV or playing too many video games kills brain cells, although, more specifically, Adorno would argue that our participation in these media destroys our ability to be critical thinkers. From the perspective of the Frankfurt school, then, the culture industry produces banal media content that has the effect of turning its audience into passive, conformist sheep.

So do you feel like a sheep? Most of us would probably say no, although Adorno (1991) would argue that because of false consciousness, sheep are never really aware that they are, in fact, sheep. **False consciousness** is an idea that comes from Karl Marx's theory of class relations but has been used in other contexts to explain gender and race relations. The basic idea of false consciousness is that some institutions (in this case, the culture industry) can exert enough power over the way people think to mislead them about the true state of affairs, and usually about relations of power in a society. False consciousness, in a very *Matrix* kind of way, leads you to believe you know things that are not, in fact, actually true. In the case of Adorno (1991) and the culture industry, false consciousness means that the media, in addition to turning us into passive, conformist sheep, accomplishes the neat trick of convincing us that the media really has no influence on us whatsoever. You don't feel like a sheep, but you wouldn't notice it anyway because the culture industry has done such a good job creating a sense of false consciousness and, therefore, of convincing you that you're not a sheep at all. This concept has been applied to gender, as well. Women who claim that they experience no oppression in male-dominated society are victims of false consciousness who have had the wool pulled over their eyes by the institutions of a patriarchal society. But there is also the possibility that we may not be sheep, after all, and John Fiske's (1989) theory of audience power gives us one perspective from which to make that argument.

Theory Alert! Feminist Theory

If you believe in the idea of false consciousness as it relates to gender, it follows that any men or women who claim that gender does not affect their lives at all are misguided victims of false consciousness. What are problems with using this concept? How would you feel about being told you were a victim of false consciousness? If you buy this concept, how do you help people to understand gender "reality" as it really is? And how do you ever know for sure that you're not a victim of false consciousness?

Audience Power Theory: Power to the People

John Fiske comes from a background in cultural studies and communication rather than from sociology, and in his book, *Understanding Popular Culture* (1989), he begins by rather emphatically rejecting many of the ideas of Adorno and the Frankfurt school. On the subject of media organizations and their relationship to their audiences, Fiske (1989) argued that, "Popular culture is made by the people, not produced by the culture industry" (p. 24). Fiske made this argument by shifting his emphasis from the structure or motives of the culture industry to the ways in which real people use the texts and media that are produced by the culture industry. The television show produced by some network or cable station cannot be considered without also examining how people make sense of that show, its characters, and the messages it sends. This theory is called **audience power theory** because Fiske puts the power to determine the popularity of any given media content squarely in the hands of the people who are making informed decisions about what they like and what they don't like. According to the theory of audience power, we are not sheep but consumers capable of decoding and interpreting the media we see and hear to suit our own unique needs and lives and then making decisions about whether that particular media is something in which we'd like to participate.

Fiske's emphasis on the ability of the audience to *decode* media content in different and unpredictable ways is borrowed from the work of Stuart Hall (1973/1980), another influential figure in the world of cultural studies. That audiences interpret and make sense of the messages in media content rather than unthinkingly absorbing the messages created by the culture industry makes good sense, and it is this freedom that Fiske celebrates in his theory. We may very well live in a capitalist society where much of the media content that gets produced has the aim of making a profit; but according to Fiske (1989), it is also important to remember that culture is something that cannot be imposed from outside or above but only created from within by a group of people. The culture industry, in the end, is a group of people who, whether Adorno approves or not, are still members of the larger culture.

Fiske (1989), like Adorno, acknowledges that power has a role to play in examining the workings of the media. The production of media content does take place within a "determining framework of power relations" (p. 58), but in some ways Fiske argues that the "meaning" of any media product is always escaping from whatever the preferred or intended meaning of its creators may have been. For example, the advertisers, writers, directors, and actors who all cooperated to create a certain television commercial may have one **preferred meaning** for that commercial while they are engaged in the process of its creation. Perhaps they wanted their beer commercial to send the message that real men don't drink beer with fruit in it (a message that was sent humorously in a Miller Lite commercial featuring the men of the "square table" developing their own set of "Man-Laws"). But even media content as seemingly simple as a beer commercial contains what Fiske (1989) calls an **overspill of meanings** and, therefore, the possibility for viewers to create their own alternative or resistant meanings. As a viewer of the beer commercial, you may read its message as saying that only a certain *kind* of man (white, middle-class men or famous men) think drinking fruit in their beer is not masculine. Or you may decide that the message being sent by this commercial is that masculinity is an inherently unstable identity that is *so* ever-changing in today's societies that actual rules need to be negotiated. Or you might just decide that the commercial is making fun of men in general. There are countless other meanings any number of viewers might glean from any particular commercial, television show, movie, magazine, song, or Internet site, but according to Fiske, most of those will not match up perfectly with the particular meaning that the creators of the media content originally intended for the audience to receive. In this sense, the power to determine what media content means is not solely in the hands of the media organizations that create the media content (Gauntlett, 2002).

These two theories have very different perspectives on both the nature of media organizations and the relationship between media organizations and their audiences. To understand how these two perspectives would make sense of a specific media phenomenon, we can take the case of *Chapelle's Show,* a half-hour television show that aired for two seasons on Comedy Central beginning in 2003 and included sketch comedy written by African American comedian Dave Chapelle and his white writing partner, Neal Brennan. Comedy sketches on *Chapelle's Show* covered a range of topics, from stories about 1980s pop star Rick James and Prince, to a mock documentary on a blind, Black Ku Klux Klan member. Many of the sketches, such as a made-up game show called "I Know Black People" and a mock racial draft, focused on issues related to race in a controversial manner. Many sketches also touched on gender issues, as in one spoof on the movie *It's a Wonderful Life,* in which a woman's wish to no longer have large breasts is granted and she sees what life as a woman without large breasts would be like. Many of the sketches on *Chapelle's Show* generated controversy for their offensiveness to racial groups, women, and gays and lesbians, and in general, the show pushed the boundaries of acceptable dialogue on many of these issues.

How would the two perspectives on the media—media power and audience power—approach *Chapelle's Show?* Adorno and the Frankfurt school would say that the comedy in *Chapelle's Show* may seem transgressive or risky—but in reality, the messages contained in *Chapelle's Show* do not truly threaten the racial and gender status quo; that they are produced by the culture industry is proof that they are harmless enough not to rock the boat of existing racial and gender norms. The widespread success of *Chapelle's Show,* measured through its ratings and advertising revenue, is proof of the way the culture industry works. *Chapelle's Show* does not and cannot help people to think critically about racial and gender norms in our society; rather, the audience watches a sketch, finds it funny, and then moves on with their conformist lives. The culture industry can mass-produce this television show so that people, rather than thinking for themselves about racial and gender issues, merely settle for the prepackaged version of this conversation that *Chapelle's Show* represents. Far from being edgy or rebellious by choosing to watch *Chapelle's Show,* viewers are merely confirming the workings of the culture industry and thereby contributing to the existing status quo. This particular television show is amusing and nothing more.

From the perspective of audience power, *Chapelle's Show* at least has the potential to be much more than just an amusing television show. Regardless of the particular meanings Dave Chapelle and his writers intended to send with their sketches, once the content is on the air, the meanings people attach to them are up for negotiation. Like all popular culture texts, according to Fiske, *Chapelle's Show* is full of potential contradictions. When the game show sketch testing people's knowledge of Black people trades on stereotypes about African Americans, such as a predisposition to smoke menthol cigarettes and listen to rap music, audiences may find this comedy funny precisely because they believe in the stereotypes themselves. "Yes," they may say to themselves, "that's the way Black people are and therefore that's funny." But other people watching the same sketch may feel the show questions and plays with those stereotypes, suggesting that the idea of "knowing Black people" is a ridiculous idea and that the point of the sketch is precisely to demonstrate that ridiculousness. According to audience power, any given sketch from *Chapelle's Show* becomes a site of semiotic struggle, where groups compete to substantiate their own meaning for the television show. Audiences do not have *all* the power to win this struggle, but they do have *some* power, which is more than is provided for them by Adorno and the Frankfurt school.

What makes *Chapelle's Show* an especially interesting case to demonstrate these two theories is that its creator, Dave Chapelle, struggled himself with exactly these issues as they related to his own show. Chapelle became afraid that the meaning he intended for much of his comedy—to serve as a critical interrogation of contemporary racial assumptions—was not at all the meaning to which audiences were responding. In this sense, Chapelle subscribed to the audience power theory of media and popular culture; stories suggested that Chapelle decided to end *Chapelle's Show* because he could not truly control the ways in which audiences encoded and decoded his comedy, and he didn't want to be responsible for creating a television show that further confirmed many of the racial stereotypes that many viewers already held. Or perhaps, Adorno would argue, Dave Chapelle could not remain a part of the culture industry because these media organizations cannot tolerate the kind of actual dissent that *Chapelle's Show* represented.

As with most theoretical perspectives, there's probably some truth in both ways of seeing the relationship between media and their audiences. Regardless of whether media power or audience power seems more plausible to you, both theories seek to make sense of media content and the effects of the media on the audience of people who watch, surf, and listen. These two theoretical perspectives were not developed specifically to explain how gender becomes a part of the media or how people translate those media messages in their own lives. But gender norms, in addition to racial norms, class norms, and norms about sexuality, are all part of the status quo that Adorno believes the culture industry supports. In Fiske's theory of audience power, gender, race, class, and sexuality are all important sites where the media messages contained in television, movies, magazines, and the Internet are all negotiated and contested by audiences. The central question posed for us about gender by these two theories might be phrased in the following way: Does the media support and confirm our preexisting assumptions about gender, thereby contributing to the maintenance of gender inequality, or does the media challenge what we think we know about gender and therefore serve as a potential source of resistance to the gender status quo? If the media is an increasingly important institution that contributes to our gender socialization, exactly what messages is it sending us and what do we do with those messages?

TRANSGENDER IN THE MEDIA

On one level, we might begin our exploration of the gender messages being created and interpreted in the media by looking at various depictions of women and men. What are the views about what it means to be masculine or feminine that are perpetuated through the media? But if we take a step further back, a better first question might examine what the media tells us about what gender is in the first place. We have learned that there are many different ways of thinking about gender as well as about how gender relates to sex category and sexuality across different times and places. The dominant notion in Anglo-European societies like the United States is that there is a fairly tight linkage between sex category, gender, and sexual identity. People with penises, XY chromosomes, and the right mix of testosterone act in masculine ways and sexually desire women. Do movies, television shows, Internet sites, popular music, newspapers, and magazines reflect this same belief system? Or can we find examples of places where this particular way of thinking about gender is challenged or subverted?

Sexual dimorphism is the belief that there are two discrete sexes (male and female) that are physically and genetically distinguishable. From the perspective of sexual dimorphism, you are either male or female, and there is nothing else in between. We have already discussed how sexual dimorphism does not apply across all cultures or all historical time periods, as some cultures have more than two sex categories or genders while some cultures think of sex or gender categories as continuous rather than as discrete. In a continuous formulation of sex category or gender, female and male are two ends of a spectrum, with many points existing somewhere between female and male. Most people in the United States subscribe to a belief in sexual dimorphism, and so it's not surprising that much of the content of our media reflects that belief. Generally, the characters and personalities we see on television and in the movies are clearly identifiable as one gender or the other. Instances of people who are not clearly identifiable as one gender or another are never seen as normal, but they are usually the subject of comedy, tragedy, or a kind of voyeurism. Gender ambiguous characters like Pat in a famous *Saturday Night Live* skit are objects of laughter, while the transvestite character played by Hillary Swank in *Boys Don't Cry* is sympathetic but ultimately comes to a tragic end. The freak show–like attraction of gender ambiguity dates as far back as the bearded lady attraction at circuses, and in television talk shows, people with ambiguous gender become objects of fascination because of their supposed freakishness.

Yet as transgender individuals have become an increasingly vocal and recognized group in society, depictions of transgender individuals in the media have become more sophisticated and less likely to emphasize the abnormality of these groups. The British comedian Eddie Izzard openly discusses his transvestism in his comedy routines, and he describes his decision to wear women's clothes and makeup as relatively unrelated to either his gender identity or his sexual identity, sometimes describing himself as a "male tomboy" or a "male lesbian" (Jordan & Izzard, 1999). Izzard says of his cross-dressing, "Women wear what they want and so do I" (Gordillo & Izzard, 1994), and in this he suggests that transvestism could potentially be perceived of as just one particular choice of self-expression, and not that different from choosing to wear a t-shirt or a suit. In this sense, Izzard cross-dresses not to convince anyone he is really a woman or really female, as even in women's clothes and makeup he clearly looks like man. But as a celebrity, he disturbs the assumed alignment between sex category, gender, and sexual identity, suggesting that it is possible to pick and choose among these categories. You can be a male who sexually desires women but demonstrates feminine qualities in the way you dress. Other cross-dressers have shown up on the television show *Friends,* where Kathleen Turner played Chandler's male transvestite father, and the *Drew Carey Show,* where Drew's brother, Steve, was a cross-dresser.

Transsexuals, or those individuals who wish to live as a different gender or sex, are also being depicted in the media with increasing frequency in ways that also portray the full humanity of these individuals. In the HBO movie *Normal,* a middle-aged, Midwestern factory worker with a wife and children struggles as he comes to term with his desire to be a woman, including pursuing sex reassignment surgery. In the movie, the main character's wife of several decades is forced to ask herself whether love is gendered or genderless. Can you still love your husband if he's no longer a man? *TransAmerica* is a humorous depiction of the journey across the United States of one preoperative male-to-female transsexual and the teenage son he fathered years ago as a man. Although the transsexual status of the main character is certainly a major part of the film, it is neither the sole source of humor nor the cause of any tragedy that occurs in the film. *TransAmerica* is largely a classic travel movie with a character who also happens to be transsexual, and in this sense, it is an important step in establishing transsexual individuals as people who have many problems that are not so different from those people who live (or believe they live) relatively content with their own sex category and gender.

Theory Alert!
Queer Theory

Movies like *TransAmerica* and *Normal* are important in exposing many people to the existence of transgender people and exposing something of their lives and the problems they face. But to some extent, by focusing on that one particular aspect of their lives, these movies still create a sense that transgender individuals are uniquely different from the rest of us. Being able to create a neat category for transgender people, queer theorists might say, only reinforces the sense that transgender people are different while everyone else is normal. Having a movie or television show about transgender people does nothing to disturb the notion that there are real differences between transgender people and other "normal" people, rather than suggesting that the differences are really only those of degree. All of us violate gender norms to varying extents, so why make distinctions between Eddie Izzard wearing makeup and a male athlete's refusal to tell degrading jokes about women in the locker room? Both people are saying no to some prescribed aspect of hegemonic masculinity. Are they really so different that they belong in completely different categories?

The Gender of Facebook

Adorno and the Frankfurt school would say, of course, that the media doesn't challenge our underlying notions about transgender people or the existence of these categories. These are the real limits of the culture industry, which is predisposed not to go too far in disturbing our taken-for-granted ideas about sex category, gender, and sexuality. Perhaps they are right, but perhaps something different can happen in relation to gender on the unique, ever-evolving medium of the Internet. The Internet has certainly changed the way many people consume and use traditional media like television, movies, magazines, newspapers, and books. Television shows can be watched on iTunes or Hulu. Movies can be downloaded instantly from online rental sites like Netflix. Some pop music albums are released first on the Internet or exclusively at websites like iTunes. Google is attempting to create an online library that will contain millions of books, easily accessed over the Internet. Even the smallest town newspapers, as well as major newspapers and magazines, now generally have online versions, while some magazines exist solely on the Internet. Blogs compete with traditional journalistic methods for breaking the latest news story and allow people from all over the world with access to a computer and the Internet to become their own publishers. YouTube, of course, allows anyone with a webcam and some creativity to attain their 15 (or so) minutes of fame through the spread of "viral" videos.

The Internet has transformed traditional media outlets, but it is also uniquely different from these other media formats in its interactivity. Surely one of the most important advantages of the Internet has been its ability to transform the ways in which people connect and interact with each other. Online dating, online gaming, chat rooms, instant messaging, and social networking sites have all become important new ways of communicating and interacting. The ability to interact with someone in a virtual world has the potential to change radically the way we think about identity in general and gender as a specific form of identity. Online, people can create their own identities, complete with virtual avatars, and there are few checks on whether those online identities correspond to real-life identities. This at least creates the potential for people to try on different identities within the anonymity of cyberspace. If you're chatting or playing online poker with someone you've never met, how do you know for sure that they actually are the 18-year-old white male they claim to be? And does it really matter (in the context of online poker) if they're not telling the truth about their gender, their age, or their race? In this sense, some forms of the Internet do seem to have the potential to change our ways of thinking about gender and other identities.

In your experience, are people generally honest about their online identities? Do people tell the truth on Facebook, MySpace, Twitter, or online dating sites like Match.com? Have you ever misrepresented yourself online? Can the Internet open up a space where people can play with their identities?

In the summer of 2008, Facebook, the popular social networking site, announced that it would begin to change the way in which gender was used on the website (McCarthy, 2008). Previously, Facebook did not require that users identify their gender on their profiles. This created grammar problems in the ways in which genderless users were described, leading to status updates such as, "Debbie changed their profile picture" or "Joe tagged a photo of themself." Grammatical challenges aside, the larger problems were created for Facebook in trying to translate this awkward English into other languages that lack any gender-neutral options. Gender is built into the English language in the form of gendered pronouns, but in many other languages, gender is even more pervasive, making being genderless on Facebook linguistically difficult. But forcing people to choose one gender or another on their profile raises another set of problems. A Facebook profile is an online expression of a person's identity, where in-person symbols of self-expression are translated into virtual methods of self-expression, including friends, photos, group memberships, lists of favorite movies and books, quotes, links to other websites, and various applications. Forcing users to identify with one gender or the other impinges on the ability of transgender individuals to express fully their own identity through their Facebook profiles.

Expressing their sensitivity to these groups, Facebook announced that they would still allow users to remove gender entirely from their profile if they so chose.

At first glance, the freedom to not have to place yourself into a category—male or female—that may not perfectly describe your own gender identity seems like a relatively small matter. But the case of Facebook and gender raises interesting questions about gender and the Internet more generally. As we create a virtual world online, will that new world be significantly different in regard to gender, sex category, and sexual identity, or will the Internet simply continue to re-create the gender system as it exists in the "real" world? The decision of Facebook to accommodate the needs and desires of transgender individuals is different from the real world, where almost all the required documentation of self includes information on a person's biological sex, from birth certificates to driver's licenses to medical records. Transgendered individuals are therefore forced to fit themselves into categories that may not necessarily reflect or describe the reality of their own lives. On Facebook, transgender individuals can create reflections of their identity that escape these categories by not being forced to be either female or male, feminine or masculine.

Other evidence points to the ways in which the Internet does *not* necessarily create a virtual world where users are able to escape traditional ways of thinking about gender. Rather, many of the assumptions about gender are carried over into the virtual world. MySpace is another popular social networking site. In one study, researchers analyzed the gender of MySpace profiles in the extent to which white male and white female users between the ages of 17 and 29 featured their boyfriends or girlfriends in their profiles (Magnuson & Dundes, 2008). This study revealed few gender differences or patterns among the friends listed by MySpace users; men and women were equally likely to have friends of another gender listed among their Top Eight friends on their profile. Gender differences did emerge in the focus placed on romantic relationships in MySpace profiles. They found that women were much more likely to mention their boyfriends in the "About Me" and "Interests" sections of their profiles than men were to mention their girlfriends. In the "About Me" section of their profiles on MySpace, 43% of males mentioned their girlfriends 0 times, compared with only 16% of females who mentioned their boyfriend 0 times.

Fourteen percent of males mentioned their girlfriend from 2 to 10 times in the "About Me" section, while 37% of females mentioned their boyfriends that often. In the "Interests" section, 67% of men did not mention their girlfriend at all, while 53% of women mentioned their boyfriends in this section between 1 and 5 times. Women were also more likely than their male counterparts to mention love in their personal quotes in their profiles (Magnuson & Dundes, 2008). This research suggests that in the online world, traditional notions of femininity and masculinity are reproduced; through their MySpace profiles, women re-create the notion that women's lives revolve around men, while men in relationships maintain their independence and individuated sense of self. Following the assumptions of psychoanalytic theory, women experience more of a sense of connection between themselves and others than men, who have more strongly developed ego boundaries. In projecting their identities in the virtual world of MySpace profiles, these gender assumptions are re-created online.

Theory Alert!
Psychoanalytic Theory

Have you noticed any differences in how people of different genders represent themselves online through their Facebook or MySpace profiles? Do men and women use social networking sites differently? Does one group tend to use them more frequently or to form different kinds of friend networks?

We can ask this same set of basic questions about many different types of media content and analyze how these questions relate to the two theoretical perspectives we discussed previously. Media power theory tells us that because the culture industry re-creates the status quo, most media content will inevitably reproduce the dominant gender ideology of any given society. The audience will go on uncritically absorbing these messages. But according to audience power, the meaning of any given media outlet is open for debate. There are the set of meanings that the producers or creators of the media intended, but once in the hands of the audience, all bets are off; audiences have the power to interpret media content in ways that fit their own particular view of the world. In the next part of this chapter, we will explore these two contradictory views of the relationship between the media and gender. We'll start with an examination of some of the basic messages that the media sends about gender, and then we'll explore some research on the ways in which the audience makes sense of the media in relation to gender.

THE STRUGGLE OVER IMAGES

As identified by Adorno and the Frankfurt school, much of the power of the media as an institution lies in the ability to create a particular vision of reality. When we are bombarded with certain images and stories about any particular group of people, the images and stories depicted begin to take on the force of reality. This is especially true when we are not familiar with members of that group in our everyday lives. If you've never actually met or talked with an Arab man, the only information you have about what that particular group is like may come from the images you see in movies, on the television screen, or on your computer screen. If you live in the United States, the combined effect of all those images is probably not a very flattering portrait of what Arab men are like. Namely, based on the images and stories viewed on television shows, movies, and video games, you'd probably conclude that Arab men are dangerous terrorists and Islamic extremists whose version of demonstrating their masculinity involves blowing up buildings, as well as kidnapping, torturing, and murdering people.

Harems and Terrorists: Depictions of Arabs in the Media

Of course, the terrorist depictions of Arab men seen repeatedly in media in the United States are no more an accurate representation of what Arab men are like than Homer

Simpson is an accurate representation of what white, working-class men are like. But as we will see, Homer-like images are also re-created across many media outlets. Given this particular role of the media, much of the focus of research on the media as it relates to gender is on identifying the particular visions of reality that are created by various media outlets. In other words, this body of research investigates how the media perpetuates stereotypes about particular groups—a project that is very consistent with a media power approach.

If, in the United States, the media creates a view of Arab men as extremist terrorists, what is the image of women in the Arab world? For much of the Anglo-European world, those living in the Middle East have been viewed as exotic others, as the other half of the division between the West and the Oriental world, which originally encompassed everything that was not European. Early depictions of women of the Middle Eastern world were based on Europeans' exposure to stories like *The Thousand and One Nights,* and therefore, they centered on depictions of women as exotic and mysterious objects of curiosity, locked away from Western eyes in luxurious harems (Salhi, 2004). Using the image of the harem as the primary lens through which to view Middle Eastern women led to an emphasis on the sensuality and submissiveness of Arab women (Khatib, 2004). A harem, after all, is associated with sexuality, and the concept of women being confined within the physical space of the harem implies women who are passive by nature. These images of Middle Eastern women as oversexualized and submissive are similar to the way in which those within Western culture have often depicted Asian women, particularly the geisha.

As in many cases, the view of Arab women from the outside differs considerably from the portrayals of Arab women from within the Arab world. In her study of the depictions of women in Egyptian national cinema, Lina Khatib (2004) explored the connections between the media portrayals of Egyptian women and ideas about Egyptian nationalism, as well as Arab nationalism more generally. Although the outside world may view Arab women in general as sexualized and passive, from the perspective of many of those individuals within Arab culture, women are perceived as the moral gauge for the rest of society, including Arab men. This is because of the connection in places like Egypt between women and nationalism, where women are considered to be symbols of the nation itself—whether that nation refers to Egypt specifically or to Arab nationalism more generally. As we discussed in Chapter 5 in relation to sexuality, when women become symbols of a nation, the representations of those women become important ways of transmitting messages about what it means to be a citizen of that particular nation. In Egyptian movies, women come to represent the values of the nation as a whole. A film about Egypt's former president depicts her as "a virtuous female who does not pose a threat to patriarchy" (Khatib, 2004, p. 73).

Given this connection between honorable, subdued femininity and Egypt as a nation, it makes sense that many Egyptians informally refer to their nation as *Umm al-Dunya* (Mother of the World). Many Egyptian movies also juxtapose the morals of Egyptian women with non-Egyptian others, thereby using gender to send messages about the superiority of their own nation. The sexual permissiveness of women from countries such as the United States and Israel are depicted in many Egyptian films to form a contrast between the values of Egyptian women and the moral depravity of countries that might be considered the enemy. These connections between national identity and gender are not unique to Egypt, as we saw in Chapter 5.

Beware of Black Men: Race, Gender, and the Local News

This struggle over media depictions of Arab women is just one example among many in media research that explores the misconceptions of various groups that are perpetuated in popular culture. In the United States, studies have demonstrated how the media plays an important role in convincing the public that Black men are a dangerous group that should be avoided at all costs (Anderson, 1990; Glassner, 1999; Hogobrooks, 1993; Orbe, 1998; Staples, 2009). While local television news programs show images of Black men robbing, raping, looting, and pillaging night after night, the truth is that being an African American man is truly most dangerous for African American men themselves.

The media do not provide stories about the high rates of disease among African American men, including that Black men are twice as likely to suffer from prostate cancer and heart disease as are white men. The local news does not tell us that teen suicide rates for African American men increased 146% between 1980 and 1995, compared with 22% for white males. We do not learn from the media that these figures combined with other factors result in an average life expectancy for Black men in the United States of 66, a full 7 years less than their white male counterparts (Galliano, 2003). Instead, the media shows us stories about Black gangster rappers like Fifty Cent, which are geared toward convincing viewers of the incredible power and influence of these figures. When two white boys opened fire on their classmates in Jonesboro, Arkansas, in 1998, assorted politicians, teachers, and so-called experts blamed their actions on the Black rap musicians to whom these boys had listened.

African American men are portrayed as dangerous, in contrast to Asian men, who are often depicted as passive and feminine, lacking the very masculine characteristics that might make them a threat to white male authority. This has not always been the case, as prior to World War II in the United States, the immigration of Chinese and Japanese men was often described as the "Yellow Peril." Asian men in California competed with white, working-class men for jobs, and there was a long history of racial animosity toward Asian immigrants as "unassimilable," and therefore incapable of joining other European immigrants in the great American melting pot. Not surprisingly, early media depictions of Asian men often cast them as the exotic villain, sometimes with sexual designs on white women. More recently, Asian men in film and television were generally depicted as action heroes but rarely as leading men with romantic interests and sex appeal. In the recent film adaptation of the Broadway musical *The King and I, Anna and the King,* Chinese action star Chow Yung-Fat was the romantic lead as the king of Siam, but he never actually kissed his white, female love interest, Jodie Foster. This lack of Asian men in romantic leads reflects both the current views of Asian men as asexual and effeminate and the lingering fears of interracial mixing between Asian men and white women.

Homer and Ralph: White, Working-class Men on TV

But men of color are not the only men who are stereotyped in many media representations. Social class in the United States also forms an important area of distinction, and there is a long history in television of depicting white, working-class men as buffoons. Working-class people of both genders rarely appear on television in general, and especially in the genre of situation comedies that tend to focus on wealthy or middle-class families (Butsch, 1995). When working-class men do appear on television, they are almost universally depicted as buffoons. The **white working-class male buffoon** as a type is shown on television as dumb, immature, irresponsible, and lacking in commonsense. The proverbial television buffoon is Homer Simpson, whose ineptitude is legendary and forms the critical plot point for most of his family's misadventures—including *The Simpsons Movie,* where Homer's incompetence single-handedly causes an environmental disaster for the entire town of Springfield.

But the buffoon has been around on television since its very earliest days, as seen in Ralph Kramden on *The Honeymooners.* Other versions of the buffoon include Fred from *The Flintstones,* Archie from *All in the Family,* Doug Heffernan from *The King of Queens,* and Peter Griffin from *Family Guy.* The buffoon in all these versions is typically well intentioned, and sometimes even lovable, but he's never someone to imitate or respect. We may laugh at Homer Simpson and Peter Griffin, but probably few men actually want to be Homer or Peter, both of whom would be hopelessly lost without their more mature and sensible wives to save them from disaster. In contrast, men, and especially father figures, in middle-class situation comedies are generally perceived as wise and capable, working cooperatively with their wives to raise their almost perfect children. Think of Bill Cosby from *The Cosby Show,* Mike Brady from *The Brady Bunch,* or Jason Seaver from *Growing Pains.* These television shows create a reality where working-class men are perceived as incompetent, while their professional and managerial male counterparts make fulfilling their role as main provider and father figure look like a breeze.

Theory Alert! Intersectionality

Examining these variations in how different groups are depicted in the media demonstrates some of the key insights of an intersectional approach to gender. To discuss the ways in which men as a group or masculinity in general are depicted in the media would be difficult given the stark differences in the ways in which different types of men are depicted. African American men are perceived as hypermasculine in their aggressiveness and violence, while Asian men are perceived as passive and feminized. Working-class white men are viewed as buffoons in a family dynamic where wives and mothers are infinitely more capable than are husbands and fathers.

These examples demonstrate that there is no one version of masculinity that is being depicted in the media but many specific and often contradictory versions of masculinity depending on other social identities. In addition, intersectionality as a perspective draws attention to the ways in which gender is employed to enforce other types of inequality. In the case of the working-class male buffoon, showing how these men are inferior to the women in their lives is a way to demonstrate that working-class men are inferior to their middle- and upper-class counterparts. Trying to show that the normal gender status is inverted within a particular group is a way of demonstrating the inferiority of that group; working-class men are inferior because they cannot maintain the necessary dominance over their wives and other women in their lives.

Do these various media stereotypes shape the ways in which you perceive particular groups? Is there any group with which you have little face-to-face interaction for whom you believe many of your assumptions about that group are based on media depictions? Does the media have the power to shape how we perceive certain groups of people? Do you assume that white, working-class men are like Homer Simpson or Peter Griffin?

SEXUALITY IN THE MEDIA

The relationship between sexuality and various forms of media has a very long history. The bawdy humor of William Shakespeare includes its fair share of sex jokes, and pornographic writings are as old as the printing press itself. Photography had not been invented for long before a thriving business in nude photographs developed. The media technology of moving pictures was used to film a strip tease as early as 1896. So it should come as no real surprise to anyone that the Internet, too, has become a forum for the expression of many different types of sexuality. The word *sex* is the most popular search term used on the Internet (Brown, 2002). Although many people feel that the ways in which today's popular media uses sexuality is different in its frequency or content from past eras, the truth may be that when it comes to the use of sexuality in the media, there's nothing new under the sun. Many historians of sexuality might point out that the relationship between various kinds of cultural expression and sexuality is very old. But this still leaves us with questions about what in particular is new or different about our own contemporary portrayals of sexuality.

In the brief history of media forms like television and the movies, there has been a considerable shift in the depictions of sexuality in mainstream outlets. The kind of sexual acts depicted in today's movie theaters and on television screens would have been unheard of in the early days of Hollywood. We have already suggested in Chapter 6 that recent years have seen an increasing sexualization of men and masculinity, similar to the ways in which female movie and television stars have always been perceived as potential sex objects. Another trend that gives many media watchers pause is the way in which this sexualization of women seems to begin at an increasingly younger age. This particular concern about sexuality and the media might be dated to the highly visible murder of JonBenét Ramsey in 1996 and the corresponding images of the six-year-old girl wearing full makeup and a ball gown for a beauty pageant that bombarded the media landscape. Images of JonBenét on the cover of magazines were visible representations of the ways in which little girls were growing up too quickly and trying to look and act like grown-up women, including making themselves into potential sex objects.

The PBS documentary *Merchants of Cool* (Goodman, 2001) focused on the ways in which the attempts of media organizations to market ideas like *cool* and *sexy* result in a co-optation of "real" youth cultures. In the course of this exploration, several popular media types are identified, and for females, this type is the midriff. The idea of the midriff might have begun with Madonna's very intentional manipulation of female sexuality in the 1990s, but it has since evolved with young, female pop culture icons like Brittney Spears, Christina Aguilera, and Paris Hilton. The **midriff**, as described in *Merchants of Cool,* is a young woman who intentionally uses her sexuality as a means of empowerment.

For some media scholars, the midriff represents one way of working through some of the emerging contradictions regarding gender and sexuality in the wake of the second wave of the women's movement in places like the United States. Previously, women could be perceived as exchanging sex and emotional support for financial support in the context of marriage (Wilkins, 2004). As women gained economic independence in the workplace, a new space was provided for experimenting with ideas about gender and sexuality, including the exploration of what women's sexuality means when it is disconnected from the context of marriage. One theme of the second wave of the feminist movement was reclaiming women's sexuality, and developing a sense of sexual agency therefore seems consistent with those goals. What happens in the case of the midriff, where that sexual agency is largely disconnected from a broader discourse about gender and it is young girls who are laying claim to their sexual subjectivity?

The midriff and other examples of the increasing sexuality of young women and girls are a source of alarm for many because of the assumed connection between these images and the behaviors of real girls and young women. Rising rates of teenage pregnancy and the spread of sexually transmitted diseases (STDs) are considered to be consequences of this youthful exploration of sexuality as empowerment (Wilkins, 2004). Note that this particular perspective is most closely in line with a media power perspective, where girls and young women uncritically absorb the messages about sexuality portrayed in the media and reflect these messages in their own behavior. This argument suggests that a media environment that depicts young women and girls as sexual agents creates added pressure on women and girls to engage in sexual activity before they may be ready and without the negotiating power to demand safe sexual practices (like the use of contraceptives). On the other hand, the idea that women need to be pressured into having sex assumes a kind of passive and reluctant sexuality for women in general.

Sexuality and Subculture

One study of gender in a particular subculture of young women suggested a more complicated answer to questions about the representation of women's sexuality in the media and the exact ways in which young women make sense of those messages. Goth is a subcultural youth movement that many date to an early 1980s blending of the punk music scene with that of glamour rock (Wilkins, 2004). Goth is a music-based subculture, and therefore, it has important ties to the media and popular culture, but Goth subculture as it is experienced by young people is about more than the music, implying an entire aesthetic that involves both ways of seeing and being seen. In her study of the Goth scene in one small Eastern community in the United States, Amy C. Wilkins (2004) found that Goth subculture is characterized by an atmosphere of open sexuality that many Goth women find liberating and empowering. The Goth women whom Wilkins interacted with found the subculture to be supportive of women's sexual power and, in this sense, reflective of a feminist sensibility. This sensibility was reflected in two specific norms of the Goth scene, both of which contain their own complications as evidence of Goth women's sexual empowerment.

The first norm of this particular Goth scene as described by Wilkins (2004) involved rules about spatial boundaries at Goth clubs. At Goth events in clubs, the preeminent social rule is for all individuals to respect the social space of others. Goth men or women who violate this rule by impinging on someone else's social space in a club are ostracized, and although the rule is theoretically gender neutral, many Goth women and men emphasize the specifically gendered implications of the rule. As one Goth woman described the gender consequences of this rule, "If a guy dances closely to you, people will come down on him with a vengeance. They don't say, 'Oh, you wore a corset, what did you expect'" (Wilkins, 2004, p. 336)? Thus, many Goth women in Wilkins's study enjoyed Goth events because they could dance and enjoy themselves without having to deal with the unwanted physical advances of men, whether that meant having their butt grabbed or having guys "grind" on them on the dance floor.

In this environment, both Goth men and women supported this rule, not because it enforced sexual purity, but because it shifted power in heterosexual interactions toward Goth women. As one Goth woman described, Goth women "are more comfortable initiating relationships" and, as another said, are "perfectly capable of letting people know I'm interested in them" (Wilkins, 2004, p. 336). Goth subculture in this particular setting, then, seems to encourage women's ability to be sexually active rather than sexually passive, with the added benefit for Goth men of reducing the amount of effort required to get sex on any given night in a Goth club. The Goth clubs Wilkins studied were not desexualized spaces but spaces where the particular rules governing how people hooked up had been changed.

The second norm related to gender and sexuality discussed by Wilkins (2004) related to Goth women's style of dress. The Goth women who showed up to club events in this particular scene wore outfits that fetishized images of the whore, including corsets, fishnet stockings, vinyl, and leather (Wilkins, 2004). All Goth women, including those with many different body types, wore fairly revealing, sexy clothing in the club scene. Goth women interpreted this not as an objectification of their own bodies but as taking advantage of the safe space of the Goth scene to take pleasure in dressing in ways that expressed their sexuality without sending the message that they were "there to get laid" (Wilkins, 2004, p. 337). Thus, these Goth women were engaged in a classic feminist dilemma, the tension between pleasure and oppression as it relates to sexuality. Goth women want to be able to take pleasure in their bodies by dressing in sexy and provocative ways without becoming the objects of sexual objectification. The Goth women in Wilkins's study were fully aware of these tensions, as in one online conversation about whether looking at women's breasts in the club context was ever appropriate. When Goth women wore clothes that were especially revealing of their breasts, were they doing so for completely internal reasons of their own pleasure in their bodies or to draw the attention of others to their bodies and therefore objectify themselves?

This basic dilemma faced by Goth women brings us back to the midriff and to the issues of women's sexuality in general in the media. Some feminist observers of the media argue that what disturbs many people who object to figures like Brittney Spears is not their youth but that they are young girls trying in various forms to claim the power of female adulthood for themselves (Projansky, 1998). The issue is specifically gendered in that very rarely do you hear concerns about young boys engaging in sexual behavior or making other claims to male adulthood at too early an age. The essential issue, then, regardless of whether we are discussing Goth women, midriffs, Madonna, or Lil' Kim, is the way in which our ideas about femininity always complicate how we read and understand images of women's sexuality as depicted in the media. Goth women and men in the social scene studied by Wilkins (2004) tried to use the aesthetic and rules of the Goth subculture to create a space where women could take pleasure in their own sexuality and therefore gain access to this power without suffering the negative consequences of objectification and oppression. How do other women and young girls make sense of images of the midriff, and how do they use this media image to make sense of gender and sexuality in their own lives?

What are other examples of women in the media facing similar dilemmas in terms of how to be sexual subjects without becoming sexual objects? Is it possible for a woman to express positive sexuality without potentially becoming a sexual object?

These questions bring us to an examination of our final question related to gender and the media. In John Fiske's (1989) audience power theory, the ways in which audiences, like the Goth men and women just discussed, make sense of the media is more important than trying to discover some definitive meaning intended by the creators of that particular media. In some disciplines, this focus is called *audience research* or *reception studies.* The central question shifts from the researcher trying to decode the particular meaning of a television show or song to researchers talking to real people to investigate how these people make sense of the content of a television show or song. So what do we know about how people make sense of the gendered messages contained in the media?

SOAP OPERAS, *TELENOVELAS*, AND FEMINISM

If you were asked to imagine the popular culture genre that is most stereotypically and unquestionably feminine, what would you say? For many people around the world, the unqualified answer would be soap operas in all their countless permutations on television or on the radio, in daytime or prime time, and in countless languages and lengths. Soap operas in a wide variety of formats are generally daytime programming, although occasionally soap operas like *Dallas* or *Dynasty* make it onto prime-time television—while others argue professional wrestling in its current incarnation is basically a soap opera for men. The positioning of soap operas in daytime partially explains the low-budget nature of most of these programs because the larger quantities of advertising time help to subsidize the more expensive and prestigious shows produced by networks for prime time (Dines & Humez, 1995). Soap operas began as 15-minute daily offerings in the 1930s on the radio, but like many other radio genres, they made the transition to television in the 1950s and 1960s (Williams, 1992).

It might be harder to imagine a popular culture format that, on the surface, better fits the view of popular culture presented by Adorno (1991) and the media power perspective. Soap operas are generally produced on a grueling daily schedule, with little time for rehearsals, writing, or production. They seem to be simple products of the culture industry that propagate conservative messages about home, family, and sexuality that are specially designed to entertain their assumed audience of housewives (Lopate, 1976). When media critics and scholars first began to study soap operas, their initial focus was on the ways in which soap operas reflected patriarchal ideology (Rogers, 1995). To name a few examples of this particular reading of soap operas, female characters who devote "too much" time to jobs at the expense of their families are usually punished in soap operas; having a baby is defined within the soap opera genre as "the single most important thing in a woman's life" (p. 326), while women without children are perceived as incomplete; and women are often treated *like* children by their male counterparts in the soap opera world (Rogers, 1995). This side of the argument views soap operas as uncritically reflecting a status quo ideology about women's subordinate place in society relative to men, and these messages are directed at a primarily female audience.

If you watch soap operas yourself, what are other ways in which soap operas represent a patriarchal point of view? What are ways in which they also present positive images of women or images of empowered women? Do you agree that soap operas are generally perceived as feminine?

Other media scholars, many of them influenced by feminism, have made a very different argument about soap operas—that they can serve as a critique of existing patriarchal structures. John Fiske (1995) focused his attention on the particular form that soap operas take in the United States. Although there are variations across different cultures, in the United States, soap operas are characterized by endless storylines. Plotlines in a film or in many other television genres eventually come to an end. Fiske (1995) pointed out that the structure of many traditional narratives, whether in novels, plays, films, or video games, begin with some disturbance in the state of equilibrium. The middle of the narrative is composed of the events that take place during this disturbed equilibrium, and the end happens when equilibrium is restored, although it may be a new and different kind of equilibrium. In *Romeo and Juliet,* the equilibrium is disturbed by the love of Romeo and Juliet, and the middle consists of their attempts to realize that love in the face of the obstacles put in place by their families.

The equilibrium that is restored in the end is certainly new and different because it involves both Romeo and Juliet being dead, but it signals an end of the initial disturbance of their love affair and perhaps a better future for the two warring families. But as Fiske pointed out, in the soap opera version of Romeo and Juliet, one or both of the two have a very good chance of coming back from the dead and eventually falling in love with someone else. Thus, although large and long-anticipated weddings are often important staples of daytime soap opera plots, marriage as a happy ending for women is destabilized by the viewer's knowledge that hardly any soap opera marriage actually lasts. Female soap opera fans often root for extramarital affairs and divorce, taking pleasure in the disruption of the patriarchal status quo that the world of soap operas represents.

Are You a Feminist If You Watch Soap Operas?

So what exactly is the truth about soap operas and their relationship to the gender status quo? Do they support traditional gender roles by emphasizing the importance of women's role in the family and the importance of having children? Or do they provide an outlet for women to cheer on the disruption of patriarchal power? In debating these questions, some media scholars concluded that perhaps the best way to find the answers was to go and talk to the actual women who were watching the soap operas. Research on soap operas began to include not just an analysis of the soap operas themselves by the researcher but also in-depth interviews and observations conducted with real soap opera fans. This ethnographic approach to the study of popular culture emphasizes a concern with how the audiences and consumers of media make sense of and interpret the words, images, and sounds they absorb. Some of these studies have suggested that female viewers do, in fact, take pleasure in the ways in which soap operas challenge the gender status quo, and especially in the ways in which the deviant behaviors of female soap opera characters "destroy the ideological nucleus of the text—the sacredness of the family" (Seiter, 1982, p. 38). A study of Korean women in the United States revealed the ways in which these women used soap operas as a vehicle to challenge a sexual double standard legitimized by Confucian tradition and to create a forum in which to "evaluate and criticize husband's behavior" (Lee & Cho, 1995, p. 358).

This last study of Korean immigrant women in the United States (Lee & Cho, 1995) suggested another important facet of soap operas: They are a global phenomenon. In fact, although soap operas differ in their particular content and format across the world (*telenovelas* in Latin America, "daily weepies" in Great Britain, and *Oshin* in Japan), they are so widespread that they potentially serve as an international language for the many viewers across the world. *Brookenya!* is just one interesting testament to the ability of soap operas to communicate across global borders because it is a soap opera that brought together 150 people in three cities and on three continents to create a "grassroots global soap opera" (Gardner, 2006). In India, soap operas like *Kyunki Saas Bhi Kabhi Bahu Thi* (*Because the Mother-in-Law Was Once the Daughter-in-Law*) reflect the importance in Indian families of relationships between mothers, sons, and daughters-in-law. In South Africa, the soap opera *Soul City* was started by health nongovernmental organizations (NGOs), and it therefore focuses on the looming health issues of South Africa, including the specter of AIDS/HIV, disability, substance abuse, and rape (Fulker, 2008). *Soul City* has become a media phenomenon in South Africa, where it reaches an audience of 34 million and has expanded into radio, information booklets, and a spin-off show for young children. This particularly successful case (HIV prevalence among South African women younger than 20 dropped from 16.1% in 2004 to 13.7% in 2006, a drop attributed primarily to the effects of *Soul City*) represents a trend for NGOs and African governments to use soap operas as a means to combat the dire threat of HIV/AIDS and other health problems across the continent.

In Latin American countries, the ***telenovela*** serves as the ubiquitous soap opera equivalent, sharing soap operas' emphasis on serial melodrama as well as the simultaneous popular success and critical disdain of soap operas. The *telenovela* differs from soap operas in the United States and other places around the world in that it has a finite number of episodes (usually 120–200), rather than running continuously for decades. *Telenovelas* can also be broadcast both in the afternoon and in prime time in many Latin American countries. Carolina Acosta-Alzuru (2003) studied the creation, representation, and ultimate consumption of feminist ideas in one particular *telenova de ruptura* produced in Venezuela and broadcast in Venezuela, Argentina, Peru, Puerto Rico, and the United States. A ***telenovela de ruptura*** is a genre of *telenovela* that incorporates social and cultural issues in Latin America, thus, breaking with the traditional ***telenovela rosa,*** which focus on "heart-rending, tragic suffering" (p. 271) and portrays a fairly one-dimensional set of characters who are clearly good or evil. In this sense, the specific *telenovela de ruptura* that Acosta-Alzuru studied, *El Pais de las Mujeres* (The Country of Women), showcases characters that are more "complex, ambiguous and unpredictable" (p. 271), and it combines personal and social problems in the storylines.

Acosta-Alzuru (2003) used textual analysis of the *telenovela* itself, as well as interviews with the shows' creators (writers and actors) and with audience members, to explore the "circuit of culture" surrounding *El Pais.* This approach involves examining all the moments involved in the cultural process (representation, identity, production, consumption, and regulation) as distinct (different) but not discrete (capable of being divided into completely separate entities). Using this approach, Acosta-Alzuru explored the representations of femininity, machismo, and feminism in the *telenovela.* One goal of the creators of *El Pais* that clearly comes across to audience members is to depict in a more realistic fashion the qualities, rights, and struggles of Venezuelan women. As one *El Pais* fan said, "They present how Venezuelan women *really* live," rather than relying on stereotypes (as quoted in Acosta-Alzuru, 2003).

One particular form that traditional ideas about femininity take in Latin American cultures is **marianismo**, a cultural tradition of female superiority and the ability to endure suffering and self-sacrifice (Galliano, 2003). This cultural tradition derives in part from the strong influence of Catholicism in Latin America and the ideal of spirituality and suffering portrayed by the image of the Virgin Mary. The character of Miranda in *El Pais* is one example of the departure from marianismo and other stereotypical portrayals of women because Miranda begins the show as an engineer at the Venezuelan oil rigs where she is the victim of sexual harassment. This experience, as well as her past experience of sexual abuse by a male relative, "turned" Miranda against men, and her choice of wardrobe (black clothes and leather) depicts Miranda as *poco femenino* (not very feminine). After leaving the oil rig, Miranda works as a bartender and begins a relationship with the male cook, representing a reversal of traditional roles where women cook and men tend bar. Although Miranda was a favorite character for many of the *El Pais* fans interviewed by Acosto-Alzuru, her popularity was strongest among those between the ages of 25 and 45, suggesting that other age groups found Miranda's version of femininity "a bit extreme and strange" (p. 281).

Miranda and other female characters represent what many viewers find to be a more realistic depiction of Venezuelan women, while male characters such as Jacobo represent a different version of Venezuelan masculinity. Like many Latin American societies, masculinity in Venezuelan society is associated with ideas of **machismo**, a cultural tradition of male dominance, particularly as it relates to matters of sexuality and family (Galliano, 2003). As a form of hegemonic masculinity in many Latin American societies, machismo perceives that men are entitled to authority and privilege. In Venezuela, machismo as a

Theory Alert! Hegemonic Masculinity

cultural ideal is realized in the incidence of male infidelity in marriages as well as in the sociocultural legitimization of this behavior. In addition, domestic violence is a widespread problem within Venezuelan families, across class and age categories. In *El Pais,* several archetypical male characters represent this machismo orientation, but the character of Jacobo contradicts these stereotypes by portraying *el hombre ideal* (the ideal man). Jacobo stands in contrast to his wife, Chiqui, who is explosive compared with Jacobo's more even temperament.

Although the couple eventually gets divorced, Jacobo is consistently portrayed as considerate toward the women in his life and as possessing an exceptional sense of empathy with their plight. When compared with the other more stereotypical, macho men who treat women as trophies to be won or food to be consumed, it's easy to understand why his character became a special favorite among *El Pais* viewers, even though some of the female writers for the show complained that Jacobo was *muy perfecto* (too perfect). The female actress who plays Jacobo's wife agreed, saying, "Someone like Jacobo doesn't exist" (Acosta-Alzuru, 2003, p. 282). But audience members interviewed by Acosto-Alzuru loved Jacobo, often listed him as their favorite character, and wished that there were more Jacobos in Venezuela—even though they felt that the other macho male characters on the show were more representative of most Venezuelan men.

Feminists as "Poisonous Serpents"

When it comes to feminism, though, even the sympathetic character of Jacobo expressed negative views. Although the storylines of *El Pais* frequently criticize women's oppression and provide examples of strong, empowered women, feminism is still perceived as a less-than-attractive ideology by the show's creators, characters, and audience. On the show itself, feminists are referred to as "irrational," "wild," and "*cuaimas* (poisonous serpents)," while stereotypes of feminists as superwomen, man haters, and aggressive are also present. Macho male characters dismiss female characters when they refer to or complain about the multiple roles juggled by women on a daily basis by calling this kind of talk *discurso feminista* (feminist discourse). These are the macho guys, and so their dismissal of feminism is somewhat to be expected, but the female characters on *El Pais* also often apologize for their emotional outbursts by explaining they "acted like a feminist" (Acosta-Alzuru, 2003, p. 284). When Miranda, the strong and independent female character discussed previously as an alternative example of femininity, describes men as "selfish beings interested in only one thing from us women, and it happens that I have the bad habit of considering myself as a whole, complete human being," her sister replies, "Miranda, please! There is nothing more fatuous than being a feminist" (Acosta-Alzuru, 2003, p. 284). The creator of *El Pais* affirmed that he did not want the show to be a feminist *telenovela* because feminism for him represents a rejection and condemnation of men.

The ways in which audience members make sense of the messages of *El Pais* and feminist ideology reflect the complicated ways in which feminism is understood in Venezuela. Women's organizations in Venezuela have changed their names to remove the word *feminist* because of their belief that the word has been largely discredited, and this reflects the belief that feminism seeks to subvert traditional gender roles. Because these traditional gender roles are perceived as the cornerstone of Venezuelan society, there is a widespread reluctance on the part of many Venezuelan women to identify themselves as feminists or to identify with feminism as an ideology. Not surprisingly, the viewers of *El Pais* expressed conflicting views about feminism in general and feminism specifically in relation to the *telenovela.* Some viewers were confused as to exactly what feminism was, with some conflating feminism with femininity. Other viewers considered feminism to be something that threatened femininity or "the essence of being a woman" (Acosta-Alzuru, 2003, p. 285). A few audience members did consider feminism to be something positive in that it was related to women's "struggle to be better" (p. 285), but many other audience members made a firm separation between the message of women's empowerment contained in *El Pais* and the ideology of feminism. As one woman explained, her friends and workmates may call her a feminist, but she responds with, "I'm not a feminist. I only defend women as human beings" (Acosta-Alzuru, 2003, p. 286).

Acosta-Alzuru (2003) argued that this close examination of the *telenovela El Pais* gives us some insight into the relationship between the cultural messages being sent by the media and the ways in which viewers make sense of those messages. Remember that the audience power perspective on media argues that regardless of the meaning originally intended by the creators of a cultural form, audience members have some freedom to interpret these messages as they choose. In the case of *El Pais,* there seemed to be a close match between the intentions of the creators and the ways in which audience members understood the show. For example, both the show's creator and its audience made a strong separation between the emphasis on women's independence and power in *El Pais* and feminist ideology. For the creator of the show, this was intentional because he felt a *telenovela* that was very obviously *feminista* would not do well with his Venezuelan audience. But other moments in the show's history suggest ways in which the original meaning of characters or situations did escape their original intentions. The character of Miranda, with her strong independence and aversion to men, began to be understood by audience members as a lesbian. The writers of *El Pais,* worried about the effect this perception would have on the show's popularity given the traditional homophobia of many Venezuelans, decided to correct this particular audience reading by writing dialogue for Miranda's character that addressed her sexual identity. This example lends support to the audience power perspective, in that it demonstrates the power of the audience to understand characters in this particular *telenovela* in ways not originally intended by their creators; audiences can, in fact, be more than mindless sheep.

But the case of *El Pais* can also serve as an example of some of the arguments made by the media power perspective. The content of the messages that the writers of *El Pais* could attempt to convey in their program was limited by the need to stay on the air. Why not continue to allow viewers to think Miranda was a lesbian? Why should this matter to the *El Pais* writers? Although the show later introduced gay characters, the creators were concerned that having a major gay character early in the show's history would affect its popularity and therefore its ability to survive on the air. Acosta-Alzuru (2003) suggested that although *El Pais* served as a challenge to some of the traditional gender assumptions in Venezuela, there were real limits to how far the show could go. Although it may be OK to show powerful women and to make fun of the machismo version of masculinity, doing so without also ridiculing feminism would be too much for a successful and popular *telenovela*. From the audience power perspective, this demonstrates the ways in which even media like *El Pais* are limited in their ability to challenge the status quo. The question raised by this case, though, is whether those limitations are the result of the television industry as a media institution or the particular demands of the audience. The writers corrected the perception of Miranda as a lesbian character because they feared audiences would stop watching the show if this perception continued. But it is impossible to know whether audiences really would have left the show had Miranda's sexual identity remained ambiguous or whether this was, as Adorno might argue, merely another example of the how the culture industry functions to maintain the gender status quo.

Can you think of other examples where writers or producers of television shows may have made adjustments to characters, plotlines, or other aspects of their programs to cater to audience demand? Especially in today's world of fan sites and blogging, do the creators of television shows respond to their audience's tastes and demands? How might this affect how gender is depicted on television?

MASCULINITY AND VIDEO GAMES: LEARNING THE THREE RS

Next to alarm about the sexualization of young girls in the media, probably the second most popular topic of alarm about the media's effects on gender is the spread of violence across many media outlets, and specifically within video games. Violent video games, violent song lyrics, and violent movies have been blamed for a wide variety of real-life acts of violence, from the school shootings like those at Columbine High School in Colorado to copycat crimes like the murders committed by Benjamin James Darrus and Sarah Edmondson in Mississippi in 1995, supposedly inspired by the Oliver Stone film *Natural Born Killers*. As we pointed out with the relationship between sexuality and the media, the use of violence as entertainment is nothing new to human societies. A historical perspective might draw our attention to the fact that other societies and cultures have used spectacles of real violence (gladiators in ancient Rome or lynchings in the American South) as forms of entertainment. From this point of view, video games like *Grand Theft Auto* don't look so bad.

Nonetheless, an abundance of research has been conducted that examines the connections between the violence in video games and other media outlets and the behaviors of the individuals who consume these types of media. These questions are distinctly gendered because the audience for the most violence-prone video games is predominantly young men. For game consoles like XBox and PlayStation, players are 75% male; for online games, 85% of players are male (Kimmel, 2008). The players of sports and adventure games, from *Madden NFL* to *Duke Nukem*, are 95% male, and the only genre of games into which women have made significant inroads is strategy games like *The Sims*. In his book on the culture of young men in the United States, *Guyland,* sociologist Michael Kimmel (2008) argued that many of these games are violent, misogynistic, heterosexist, and hostile toward racial and ethnic others. In one of the most notorious of this particular genre, *Grand Theft Auto,* your character, or avatar, is a criminal, and your goals are to sell drugs, build your criminal empire, and kill cops. Along the way, you can kill anyone else you want, and having sex with a prostitute in your car increases your health. You can get back the money you paid afterward by following the prostitute out of the car and killing her.

Grand Theft Auto is probably an extreme example of the misogyny in video games, but many adventure games in general reflect a world of hypermasculine male characters and ultrafeminine female characters. The male characters generally have bulging biceps and an upper body so out of proportion to their lower half that they would not be able to stand up as real, three-dimensional (3-D) entities. The women are usually blond, with disheveled, "bedroom" hair, large breasts, and tiny waistlines (Kimmel, 2008). Although a few women, like Lara Croft of *Tomb Raider,* are also the heroes, they are still highly sexualized and always heterosexual. People of color in video games are generally portrayed as part of the criminal underworld, as in *The Warriors,* which involves a multiracial street gang trying to work its way out of the Bronx.

In the world of guyland described by Kimmel (2008) in his book, young men on college campuses spend countless hours in their dorms or fraternity houses playing just these types of games or, the other growing Internet pastime, gambling online. The average kid in the United States between the age of 13 and 18 spends 2 hours a day playing video games and about 7 hours a day interacting with some form of electronic media (Kimmel, 2008, p. 145). But moving beyond the actual content of the video games themselves, the question becomes, "What exactly do these guys get out of these hours spent playing *Grand Theft Auto* or *Doom*"? Given that the primary audience for these games seems to be young and male, what is it about these games that seems to give them such a powerful masculine appeal? Kimmel argued that for many of the young, white, male college students he interviewed, playing video games did not actually contribute to any increase in real-life violence. Rather, these video games reinforced the three Rs of the parallel education received by these men through the predominant, guyland media. The first R, as described by Kimmel, is **Relaxation**, or an escape from the "weight of adult demands and of the rules of social decorum (also now known as political correctness)" (Kimmel, 2008, pp. 153–154). This is why strategy games like *The Sims* are not as popular among young males; *The Sims* involves domestic situations that are fairly close to those of the real, adult world. One sociology professor who studies video games and teaches a college course on them explained that games like *The Sims* are a little "too realistic" for the residents of guyland in that the game involves the adult world of getting a job, getting married, and having kids, which is precisely what these young men are trying to escape (Lugo, as quoted in Kimmel, 2008, p. 155).

The second R described by Kimmel (2008) is **Revenge**, "against those who have usurped what you thought was yours" (p. 154). Regardless of whether their perceptions are correct, many of these white, male college students feel they are victims of a world that places the needs and interests of women and people of color above their own. As this particular rule applies to video games, this revenge is accomplished through the fantasy world that is created within video games, where as one 24-year-old male described to Kimmel, "I can just relax, be myself and not worry about offending anyone. I can offend everyone" (Kimmel, 2008, p. 150)! The revenge experienced through video games takes the form of stereotypical portrayals of women who are helpless and want to be saved or minorities who are incompetent or criminals. In the extreme example of *Grand Theft Auto,* this revenge can take the actual form of killing these groups, but in general, the revenge provided in video games is merely a return to a world in which these young men are in complete control—even if that control is obtained through violent means.

Kimmel argued that these sentiments are encouraged among the residents of guyland by the messages sent by "Guy Radio" personalities like Howard Stern and Rush Limbaugh. One senior, white male college student at Vanderbilt University who had just been accepted to law school explained to Kimmel his feeling that both women and minorities with lower grades and abilities prevented him from getting into universities like Duke and the University of Virginia. He concludes by saying, "It's not fair. My family didn't own slaves. We're from Pennsylvania, for Chrissakes. I'm not racist; I don't care what color you are. But I shouldn't be penalized because of my race, my color, right? I mean, that's just not fair" (Kimmel, 2008, p. 161). Given this sense of victimhood among many white, college-age men, one appeal of video games is to exist for a time in a world where they are clearly not the victims, but they are able to take some revenge on those they perceive to be their real-life oppressors. In this sense, video games also provide the final R, **Restoration**. Part of the appeal of video games is the sense of control they give their players—the ability to decide what another person does, even if that person isn't necessarily real. Video games allow the men of guyland to feel they are restored to a state of power and privilege that they feel no longer exists for them outside the world of *Grand Theft Auto*.

If we are interested in how audience members, in this case the white, college-age residents of guyland, do with the media produced by media organizations, the answer here is that video games are used largely to escape the real world as this particular group of young men see and understand it. Although this particular group of men certainly does face a world in which women and minorities are closer to a level playing field than they have ever been in U.S. history, it is important to note that there are still considerable privileges that come with being a white, heterosexual, middle-class male. Research on affirmative action policies does not support the contention that white men are the victims of discrimination because of preference policies, despite the strength of this belief among many people. A 1994 survey revealed that 70% to 80% of whites believed that affirmative action sometimes discriminated against whites, while other studies demonstrated that "men are more likely to believe that a woman will get a job or promotion over an equally or more qualified man than they are to believe that a man will get a promotion over an equally or more qualified woman" (Steeh & Krysan, 1996).

This survey compares with data on the number of cases of reverse discrimination that are filed with the U.S. Department of Labor. **Reverse discrimination** is the perception that affirmative action policies result in discrimination against dominant groups such as whites and men. According to Department of Labor statistics, out of more than 3,000 cases, less than 2% involve charges of reverse discrimination against whites or men (U.S. Department of Labor, 1994). So although there are few actual reports of reverse discrimination against whites or men, the belief that this type of discrimination exists and is widespread is fairly common. Regardless of the chances that the young Vanderbilt student mentioned previously actually *was* beat out of schools like Duke and the University of Virginia because of competition from women and minority students, his perception is that this was the case. As sociologists, we understand that those perceptions can be important and, in this instance, important enough to make the fantasy world of video games seem like a safe and relaxing escape from the world.

What Kimmel (2008) argued that video games by and large do *not* do for the particular groups of young men he studied is make them more prone to violence or necessarily numb to real-life violence. This is the argument commonly reflected in media discussions of violence in video games and other media formats, and it reflects a belief that observing and participating in the fantasy violence of video games actually leads players to become more violent individuals. There is little research that actually supports this causal connection between exposure to violent interactive games and real-life violence or crime (Kimmel, 2008, p. 152). At the most, some research has suggested that children become agitated and aggressive after playing video games, and this may apply to the young men playing *Halo* in college dorms.

But this research does not offer any clue as to whether the aggression and agitation experienced immediately after playing video games is sustained over long periods of time. There is no evidence to suggest that the avid video game player maintains a high and prolonged level of agitation and aggression. This research is consistent with the audience power approach to thinking about how people relate to the media. The violence, misogyny, and racism contained in many video games are not simply and uncritically absorbed by the many people who play these games. Rather, their appeal is grounded in the particular historical and social context of these young men's lives. The residents of guyland have the ability to choose to use video games as a means of relaxation and escape rather than being hopelessly brainwashed into becoming violent zombies.

Do you agree with Kimmel's research and assessment of college-age men playing video games? Does he accurately describe the appeal of these games? Are these games largely just about entertainment and escape? Are they harmless?

THE BATTLE OF THE SEXES AND THE BATTLE FOR THE REMOTE CONTROL

Turning our attention toward the behaviors of people who consume the products of media institutions raises some interesting questions about what happens in the actual settings where people play video games, surf the Internet, or watch television. Some researchers have suggested that technologies like the television, and especially the remote control, have changed the daily lives of families as well as the gender dynamics within families. To understand how this might be true, you need only ask yourself a series of questions: Who in your family usually gets to operate the remote control? If you're one of the few families with only one television, who generally gets to decide what everyone watches? And how many arguments are likely to be spawned by all the difficulties that can come with watching television together in today's world of infinite options and channel surfing as an art form?

If your family is like most, when family members are watching television together, the remote control is probably controlled by dad or another male figure. Studies of television-watching behavior have revealed that the remote control is generally considered the "symbolic possession" of the father, or of the son if dad's not around (Thoman & Silver, 1995; Walker, 1996). Because the person in possession of the remote control generally gets to decide what and how everyone else watches, this means that fathers and husbands obtain a kind of power through their possession of the remote control. This is certainly not a fact that has gone ignored in popular culture in general, where comedy routines and sitcoms joke about the battle for the remote control, and in the Adam Sandler movie, *Click,* the magical powers of a remote control provided by the mysterious character played by Christopher Walken give the main character real power to alter his world outside of television. Studies suggest that this power doesn't fall to fathers, husbands, or sons simply because mothers, wives, and daughters are not interested or invested in television viewing. In one study, three fifths of men and two thirds of women reported that there were things about their joint television watching that were frustrating to them. But the things that women and men found frustrating about their joint television watching tended to be different, as you might guess based on your own experiences in front of the tube.

Men express frustration with the quality of television programming in general—the classic, "There's nothing good on." But when their complaints focus on the interactions with their fellow audience members, they focus on the way in which their wives or significant others might pout if they don't turn the channel to the program they would like to watch. Even more commonly, a clear gender difference emerges in the way men and women watch television, and this itself becomes a source of tension. Men in general are less likely to combine their television watching with other kinds of activities, and they are definitely less likely to combine their television watching with the specific kinds of activities that women are likely to engage in while watching TV. Men express a clear preference for attentive television viewing, watching without interruption "in order not to miss anything" (Thoman & Silver, 1995, p. 364). It therefore makes sense that the kinds of activities men are more likely to combine with their television viewing include pleasurable activities such as doing nothing, eating, drinking, or playing computer games. These activities (aside from the computer games) detract minimally from placing full attention on the television.

Women, on the other hand, often treat television watching as a social event. Women are more likely to carry on a conversation (or attempt to carry on a conversation) while they watch television or to engage in some other domestic activity. And research suggests that, in general, women are probably more likely to combine family work activities (child care, laundry, cooking, etc.) with their television viewing, especially in families that include children (Walker, 1996). Studies suggest that for many women, the idea of doing nothing else but watching television can seem like a difficult-to-defend waste of time, and this is probably in part a result of the added sense of domestic obligations that women still face. Given the typical distribution of household labor we discussed in Chapter 8, in which women always do more housework and child care than their male counterparts, it should come as no surprise that sitting down and watching television can become a luxury many women feel they cannot afford. Some studies suggest that women feel most comfortable watching television when they have the house to themselves, usually during the day, and are therefore free from the domestic obligations that come with a house full of children and husbands (Thoman & Silver, 1995).

These differences in general orientation toward television viewing explain the dynamics of complaints that many women and men have about each other as partners in TV watching. Men complain about the fact that women attempt to have conversations while they're watching the television together because this interferes with their ability to focus fully on the program. It's probably not surprising that in Adam Sandler's movie, *Click,* one of the first uses to which Sandler puts his magic remote is to mute his wife while she's complaining to him about his behavior. But for women, doing nothing but watching television can seem like a waste of time. Men are frustrated by the lack of programming on television and at being insulted by commercials, and they therefore tend to channel surf a great deal more than women. This includes flipping through the channels rather than settling on one program, flipping through the channels during commercials, or trying to watch two shows at once by flipping back and forth between channels. But this is one of the common sources of frustration for women watching television with male partners, as described by one woman: "I would say that the only thing that's frustrating for me is when we first turn on the TV and he just flips through the channels. It drives me crazy because you can't tell what's on, because he just goes through and goes through and goes through" (Walker, 1996, p. 817).

Perhaps these descriptions sound similar to your own experiences with television viewing. Or perhaps, given that the average U.S. household now has 2.24 televisions, you no longer have to debate about who gets to watch what or who gets to hold the remote control. In her study of a wide range of couples and their television watching behaviors, Alexis J. Walker (1996) suggested that the battle over the remote control (or the lack of a battle, if no one argues with dad about what to watch or when to change the channel) represents an important example of how couples do gender, and therefore power, on an everyday basis. That the possession of the remote control is connected to issues of power is demonstrated by the main exception to the male dominance over this technological device; in families where the husband is unemployed while the wife is working, it is slightly more common for the man to let other family members decide what to watch together (Thoman & Silver, 1995; Walker, 1996). The expectation in these families is that the unemployed husband or father has a more flexible schedule, allowing him to tape shows he might want to view and to watch them later at night or during the following day. In addition, Walker (1996) included gay male and lesbian couples in her study of joint television watching and found some evidence for a more egalitarian approach to viewing than among heterosexual couples.

Theory Alert! Doing Gender Theory

Among gay and lesbian couples, one partner was still more likely to use the remote control than the other, and partners often engaged in a sometimes pitched battle over what to watch. But Walker also found an example of a conscious attempt among one lesbian couple at negotiating the conflicting styles of television watching. One partner, Mary, was aware of her tendency to channel surf, and so would give the remote control to her partner, Becky, if they were watching a show in which Becky was especially interested. For this couple, Mary was not letting Becky have the remote, implying that the remote was still primarily under her control, but rather was allowing Becky to help her curb her own surfing behavior. In this instance, the couple was aware of the power dynamics and tension that resulted from different orientations toward television watching and had intentionally attempted to develop a solution.

In the end, how important is who gets the power that goes along with the remote control? For researchers like Walker (1996), the answer is that power over the remote control is actually fairly important because it is one example of the way in which inequality between genders is acted out on a daily level. In this sense, it is a concrete example of how gender inequality works. In heterosexual relationships, men have more power over what they watch on television, without having to take into consideration their partners' wishes. They have more power over when they watch, what they watch, and how they watch. The women in Walker's study expressed frustration over these patterns as well as a sense of resignation to the remote control status quo; they generally predicted a negative reaction from their male partner to any possible change in remote control behavior. Some couples solved their television conflicts through the use of a second television or a VCR (and probably more recently, a digital video recording [DVR] system), but this solution reduced the amount of leisure time spent together as a couple and often meant women had to schedule their television watching around their husband's schedule.

THE GENDER OF LEISURE

It's easy to dismiss these findings because they deal with gender patterns in our leisure activities. You might call it the gender of what we do for fun, and much of this chapter could fall under this heading. Media in all its various forms has come to occupy a large segment of what many of us around the world do with our leisure time. Is there a gender inequality in the amount of leisure time available to men and women, and if so, how important is that inequity? **Leisure** is generally perceived by economists as the opposite of paid work, or what we do with our time when we're not working jobs in the formal economy (Bittman & Wajcman, 2000). You might remember from Chapter 9, on gender and work, that defining exactly what work is can get rather complicated. Many women and men may be working in their homes or other locations even when they're not officially doing paid work, and in the age of cell phones, personal digital assistants (PDAs), and laptop computers, the line between work and home is becoming increasingly blurred.

Most of us noneconomists generally think of leisure as *free time,* or time that is completely at our own disposal to spend as we choose. In a study of leisure in 10 different countries, Michael Bittman and Judy Wajcman (2000) found that men and women living in Denmark and the Netherlands have by far the most amount of leisure time, with an average amount of weekly free time that exceeds 40 hours per week. In the United States, men average only 30 hours of free time per week, but there is still a gap between free time for all women and men in the United States. For all men and women, men have 1 hour and 22 minutes more free time while for married men and women employed full time, the gap is 39 minutes. In their analyses, Bittman and Wajcman (2000) found varying levels of inequality in the amount of free time available to women and men across countries (Italy had the greatest gender disparity, with Italian men having 6.5 more hours of free time than Italian women), but these differences were still fairly small overall.

The more important gender differences in free time emerged when the researchers took into account not just the amount of free time available to men and women but also the quality and character of that leisure time. Bittman and Wajcman (2000) were able to explore this characteristic of leisure because of their use of time diaries as data. With time diaries, respondents are asked to record what kind of activities they are engaging in at various intervals throughout their day. The analysis of leisure activity includes a broad range of different kinds of underlying activities that all qualify as leisure. Some of these consist of **pure leisure**, where the respondent records engaging in only leisure activities at that moment, with no secondary activities taking place. But in today's world, we know that we rarely engage in just one activity at a time. Some respondents combined one leisure activity with another (watching television combined with conversation), while some respondents combined leisure activities with unpaid work activities, such as housework and child care.

The wife or mother described previously who watched television while she did laundry would be a good example of what the researchers call **contaminated leisure**. When these characteristics of leisure are considered, important gender differences begin to emerge. For men, more than 61% of their leisure is pure leisure, with no other accompanying activity, while little more than half of women's leisure is pure leisure. This means that men spend more than 24 hours on average per week in pure leisure activities, compared with women's 21 hours per week, a statistically significant difference. In addition, women experience a larger proportion of contaminated leisure, or leisure time that is combined with the unpaid work of child care and other household labor.

Examining other characteristics of leisure also reveals important gender differences. Leisure that occurs in long, uninterrupted blocks of time is qualitatively different from leisure that is widely distributed in short increments. Bittman and Wajcman (2000) described the latter type of leisure as **harried leisure**, indicating that it must be taken in short segments as opposed to the more relaxed and comfortable extended block of leisure time. Again, they found that women's leisure was much more likely to be interrupted, indicated by the larger number of actual leisure episodes in women's schedules and the shorter duration of their longest episode of leisure. In other words, women are more likely to squeeze leisure into their busy schedules throughout the day, and their longest, uninterrupted period of leisure is shorter on average than those for men.

One final characteristic of leisure considered by the researchers was leisure spent purely in the company of adults—**adult leisure**—as opposed to leisure spent with children—**family leisure**. Here, as one might imagine, there are significant differences between parents with young children and everyone else, as both fathers and mothers of small children spend more of their time overall in the company of children. But in examining those parents with small children, men still spend more time engaged in adult leisure than do women. Perhaps even more important, Bittman and Wajcman (2000) revealed that the kind of activities fathers engaged in with their children were very different from the kind of activities mothers were likely to be doing. More than half of the time mothers spend with their young children is spent doing physical care such as feeding, clothing, bathing, changing, and tending to injuries and ailments. For men, almost a third of the eight hours per week they spend in child care is dedicated to playing with their children. Thus, even when men are engaged in unpaid work like child care, their work is more like fun and less like work than for women who spend a smaller proportion of their time engaged in playing with their young children.

This research demonstrates that there are important differences in the quality of leisure time available to women and men—even if the actual amount of leisure time is not significantly different. Moreover, these differences can be traced back to the structure of family life and the doctrine of separate spheres. Although large numbers of women do paid work, full time outside of the home, the doctrine of separate spheres stating that a woman's place is in the home continues to place an extra burden on their time. Walker (1996) suggested that part of the reason many women allow their husbands control over the television is the belief that his role as main provider means that his leisure time should be prioritized over her own. This is true even for couples who both work full time outside of the house. In developed countries like the United States, we profess to place importance on the amount and quality of leisure time that we have. Huge industries, including the media, have emerged with the express purpose of providing us with things to do in our free time. If leisure contributes to our overall sense of physical, mental, and emotional well-being, women seem to be at a decided disadvantage relative to their male counterparts.

Questions about who gets to change the channel, what college men like so much about video games, what soap operas say about gender, and how gender is built into the structure of Facebook may seem like relatively trivial matters compared with the issues of family, work, and bodies that we've discussed in other chapters. But media institutions are an increasingly powerful influence in societies around the world. Media outlets like the Internet, music, television, and films are important ways in which we experience the world. They have the potential to teach us both what is normal and right (media power) and to show us what is possible and different (audience power). Gender is built into their structure as organizations, and it in turn affects the gender messages of the media products that are produced. Gender also affects the way we interact with those media products, in terms of what we watch and consume and how we watch and consume. As we discovered in the end, gender affects the amount of time we have available to engage in leisure activities, including consuming various media products in general. If the media is the background noise against which we live our lives, its buzz is becoming increasingly louder. It is therefore important that we continue to investigate what these sounds tell us about gender.

BIG QUESTIONS

- Some social scientists argue that the increasing power of the media is a result of the decreasing importance of face-to-face communities in places like the United States. As people become less connected to their family, friends, and neighbors, what they buy, watch, or listen to becomes increasingly important. Do you agree with this theory about the rising importance of the media? Does this theory have gendered implications? What other explanations might you provide for why media is becoming increasingly important, or do you believe it is?

- The two theories we discussed in this chapter—media power and audience power—have different perspectives on the relative power of the media to influence our attitudes and behaviors. How powerful do you believe the media is? Are we sheep who are being influenced by the media without knowing it? Or do we have the power to shape and alter the messages the media sends? What evidence would you use in support of either theory?

- Advertising as a media institution helps distort our ideas about gender by playing on stereotypes and focusing on narrow definitions of what it means to be masculine or feminine. Do advertisers have some responsibility to reflect more accurately the reality of the gendered world? How might they do this and still be effective as advertisers? Does the main goal of advertising—to sell you something—inevitably lead to exaggerations of gender stereotypes?

- In Chapter 4, we read about agents of socialization that teach us about how to be gendered members of our society. How is the media an example of an agent of socialization that helps to teach its members how to be men and women, masculine and feminine? What are examples of the shape this socialization would take based on what you've read in this chapter or on you own experiences with the media?

- Does the gender makeup of the formal organizations that produce media content impact the gendered nature of the media content they produce? Can a man make a movie that appeals to women, and can a woman make a movie that appeals to men? Does an emphasis on the gender makeup of advertising firms or television or film studios exaggerate the importance of sex differences, assuming that only women can properly represent femininity and vice versa?

- In this chapter, we discussed how the ways in which people decide who gets to use the remote control can be perceived as an example of the doing gender perspective. What are other media-related examples that might be explained using the doing gender perspective? Is the media an example of a resource that people use to do gender, and what does that mean?

- What would a genderless world of media and popular culture look like? How would movies, television shows, music, magazines, and the Internet be different in such a world? How would the structure of media institutions change in a world without gender? Would this world be better or worse than the currently gendered media?

- How important are the gender inequalities we discussed in relation to leisure? Does it matter that women and men have different kinds of leisure? Is different in this case necessarily unequal? What might be the larger implications of having different kinds of leisure time? How might this contribute to other issues of gender inequality?

GENDER EXERCISES

1. In today's increasingly globalized world, it's relatively easy to gain access to media and popular culture from many other societies and other cultures. Use your library, rental services, or the Internet to explore movies, television shows, or magazines from a different society. You might pick one particular type of media genre and compare across two different societies: for example, women's magazines in the United States, France, and China. How are the messages being conveyed about gender different across these different societies?

2. The media and popular culture produced in different time periods can tell us a great deal about the behaviors and attitudes of the people of that time period—and, specifically, about their ideas about gender. Pick some media genre—magazines, comic books, popular songs, television shows, or movies—and compare examples across several different time periods. For example, you might look at Superman comic books from the 1940s to the 1950s, the 1960s to the 1970s, the 1980s to the 1990s, and 2000 and beyond, paying attention to the various messages being sent about gender in these different time periods.

3. Explore the ways in which the gender of the artist or creator of various media genres does or does not impact the ways in which gender is depicted in their creations. You might pick music produced by female and male musicians and then analyze the gender content of their music and lyrics. Do women and men seem to sing about different things, or is their gender unimportant to their music?

4. Research by Michael Kimmel (2008) that was discussed in this chapter suggested that many college-age men play video games as a means of Relaxation, Revenge, and Restoration. Find some people who play video games and interview them about their experiences with playing video games. You might ask them what kinds of games they like, what they enjoy about playing, and what they think the effects of playing video games might be, as well as what gender messages they believe are contained in the video games they play. How do their responses compare with Kimmel's findings?

5. Try an experiment with the power of the remote control. Set up a room with one television and one remote control, and arrange for a group of mixed-gender individuals to come into the room, instructing them that they are just to watch television together for a certain period of time. Observe how the group negotiates the use of the remote control. You can retry the experiment with different combinations (same gender, smaller groups, homosexual couples, heterosexual couples, older people, younger people, etc.). What does the experiment reveal about norms considering the use of the remote control? Does gender matter?

6. Conduct your own research into the amount and type of leisure time available to men and women. Pick some sample of people, trying to include a range of different genders, age groups, and other backgrounds (for example, a married woman with young children compared with a young, male, high-school student). Ask them to keep detailed time diaries, writing down what they're doing in a given day every 30 minutes. Are there differences in the amount of leisure? In the type of leisure (contaminated, harried, pure, adult, or family)? How do your findings line up with those discussed in this chapter?

TERMS

formal organization
market segmentation
commodities
commodification
media power theory
culture industry
false consciousness
audience power theory
preferred meaning
overspill of meaning
white, working-class male buffoon
midriff
telenovela
telenovela de ruptura
televovela rosa
marianismo
machismo
relaxation
revenge
restoration
reverse discrimination
leisure
pure leisure
contaminated leisure
harried leisure
adult leisure
family leisure

SUGGESTED READINGS

Gender and the media in general

Bielby, D. D., & Bielby, W. T. (1996). Women and men in film: Gender inequality among writers in a culture industry. *Gender & Society, 10,* 248–270.

Brown, J. D. (2002). Mass media influences on sexuality. *The Journal of Sex Research, 39,* 42–45.

Dines, G., & Humez, J. M. (1995). *Gender, race, and class in media.* Thousand Oaks, CA: Sage.

Gauntlett, D. (2002). *Media, gender, and identity: An introduction.* New York, NY: Routledge.

Jhally, S. (Director), & Kilbourne, J. (Writer/Director). (2000). *Killing us softly, III* [Motion Picture]. United States: Media Education Foundation.

Ross, K., & Byerly, C. M. (2004). *Women and media: International perspectives.* Malden, MA: Blackwell.

Watkins, S. C., & Emerson, R. A. (2000, September). Feminist media criticism and feminist media practices. *The ANNALS of the American Academy of Political and Social Science, 571* (1), 151–166.

Wilkins, A. C. (2004, June). "So full of myself as a chick": Goth women, sexual independence and gender egalitarianism. *Gender & Society, 18* (3), 328–349.

Soap operas

Acosta-Alzuru, C. (2003, September). "I'm not a feminist . . . I only defend women as human beings": The production, representation, and consumption of feminism in a telenovela. *Critical Studies in Media Communication, 20* (3), 269–294.

Williams, C. T. (1992). *It's time for my story: Soap opera sources, structure, and response.* Westport, CT: Praeger.

Media and popular culture theory

Adorno, T. W. (1991). The culture industry: Selected essays on mass culture. London, England: Routledge.
Fiske, J. (1989). Understanding popular culture. London, England: Unwin Hyman.

Media, popular culture, and gender in cross-cultural perspective

Khatib, L. (2004). The Orient and its others: Women as tools of nationalism in Egyptian political cinema. In N. Sakr (Ed.), *Women and media in the Middle East: Power through self-expression* (pp. 72–88). London, England: I.B. Tauris.

Salhi, Z. S. (2004). Maghrebi women film-makers and the challenge of modernity: Breaking women's silence. In N. Sakr (Ed.), *Women and media in the Middle East* (pp. 53–71). London, England: I.B. Taurus.

Masculinity, the Internet, and remote controls

Alexander, S. M. (2003). Stylish hard bodies: Branded masculinity in "Men's Health" magazine. *Sociological Perspectives, 46,* 535–554.

Bittman, M., & Wajcman, J. (2000). The rush hour: The character of leisure time and gender equity. *Social Forces, 79* (1), 165–189.

Magnuson, M. J., & Dundes, L. (2008). Gender differences in "social portraits" reflected in MySpace profiles. *CyberPsychology and Behavior, 11* (2), 239–241.

Walker, A. J. (1996). Couples watching television: Gender, power and the remote control. *Journal of Marriage and the Family, 58,* 813–823.

WORKS CITED

Acosta-Alzuru, C. (2003, September). "I'm not a feminist. . . I only defend women as human beings": The production, representation, and consumption of feminism in a telenovela. *Critical Studies in Media Communication, 20* (3), 269–294.

Adorno, T. W. (1991). *The culture industry: Selected essays on mass culture.* London, England: Routledge.

Alexander, S. M. (2003). Stylish hard bodies: Branded masculinity in "Men's Health" magazine. *Sociological Perspectives, 46,* 535–554.

Anderson, E. (1990). *Streetwise: Race, class and change in an urban community.* Chicago, IL: University of Chicago Press.

Bielby, D. D., & Bielby, W. T. (1992). The Hollywood "graylist"? Audience demographics and age stratification among television writers. In M. G. Cantor & C. Zollars (Eds.), *Current research on occupations and professions (Creators of culture)* (pp. 141–172). Greenwich, CT: JAI.

Bielby, D. D., & Bielby, W. T. (1996). Women and men in Film: Gender inequality among writers in a culture industry. *Gender & Society, 10,* 248–270.

Bittman, M., & Wajcman, J. (2000). The rush hour: The character of leisure time and gender equity. *Social Forces, 79* (1), 165–189.

Brown, J. D. (2002). Mass media influences on sexuality. *The Journal of Sex Research, 39,* 42–45.

Butsch, R. (1995). Ralph, Fred, Archie, and Homer: Why television keeps recreating the white male working-class buffoon. In G. Dines & J. M. Humez (Eds.), *Gender, race, and class in media: A text-reader* (pp. 403–412). Thousand Oaks, CA: Sage.

Byerly, C. M. (2004). Feminist interventions in newsrooms. In K. Ross & C. M. Byerly (Eds.), *Women and media: International perspectives* (pp. 109–131). Malden, MA: Blackwell.

Dargis, M. (2009, March 20). Best man wanted. Must be Rush fan. *The New York Times.* Retrieved May 15, 2009, from http://movies.nytimes.com/2009/03/20/movies/2010ve.html?scp = 1&sq = best%20man%20wanted%20must%20be%20rush%20fan&st = cse

Dines, G., & Humez, J. M. (1995). *Gender, race, and class in media.* Thousand Oaks, CA: Sage.

Elliott, S. (1993, May 10). Survey details job gender gap. *The New York Times.* Retrieved May 12, 2009, from http://www.nytimes.com

Fiske, J. (1989). *Understanding popular culture.* London, England: Unwin Hyman.

Fiske, J. (1995). Gendered television: Femininity. In G. Dines & J. M. Humes (Eds.), *Gender, race, and class in media* (pp. 340–347). Thousand Oaks, CA: Sage.

Fulker, J. (2008, July 13). Body and soul. *The Observer.* Retrieved December 19, 2008, from http://www.guardian.co.uk/lifeandstyle/2008/jul/13/observerhealth.observerhealth1

Galliano, G. (2003). *Gender: Crossing boundaries.* Belmont, CA: Thomson-Wadsworth.

Gardner, K. (2006, March 23). Soap opera meets community empowerment: Brookenya! *MediaRights.* Retrieved December 19, 2008, from http://www.mediarights.org/news/2006/03/23/soap_opera_meets_community_empowerment_brookenya

Gauntlett, D. (2002). *Media, gender, and identity: An introduction.* New York, NY: Routledge.

Glassner, B. (1999). *The culture of fear.* New York, NY: Basic Books.

Goodman, B. (Director/Producer). (2001). *Frontline: Merchants of Cool* [Motion Picture]. United States: PBS.

Gordillo, J. (Director), & Izzard, E. (Writer). (1994). *Unrepeatable* [Motion Picture]. United States: Universal Pictures.

Graham, I. (2001, April 1). Television viewing (most recent) by country. *NationMaster.com.* Retrieved May 10, 2009, from http://www.nationmaster.com/red/graph/med_tel_vie-media-television-viewing&b_desc = 1

Hall, S. (1973/1980). Encoding/decoding. In S. Hall, D. Hobson, A. Lowe, & P. Willis (Eds.), *Culture, media, language* (pp. 107–116). London, England: Hutchinson.

Hogobrooks, H. (1993). Prime time crime: The role of television in the denigration and dehumanization of the African-American male. In J. Ward (Ed.), *African-American communications: An anthology in traditional and contemporary studies* (pp. 165–172). Dubuque, IA: Kendall/Hunt.

Jhally, S. (1995). Image-based culture: Advertising and popular culture. In G. Dines & J. M. Hume (Ed.), *Gender, race, and class in media: A text-reader* (pp. 77–87). Thousand Oaks, CA: Sage.

Jhally, S. (Director), & Kilbourne, J. (Writer/Director). (2000). *Killing us softly, III* [Motion Picture]. United States: Media Education Foundation.

Jordan, L. (Director), & Izzard, E. (Writer). (1999). *Dress to kill* [Motion Picture]. United States: WEA.

Khatib, L. (2004). The Orient and its others: Women as tools of nationalism in Egyptian political cinema. In N. Sakr (Ed.), *Women and media in the Middle East: Power through self-expression* (pp. 72–88). London, England: I.B. Tauris.

Kimmel, M. (2008). Guyland: The perilous world where boys become men. New York, NY: HarperCollins.

Lauzen, M. (2010a). Boxed in: Employment of behind-the-scenes women in the 2009–2010 prime-time television season. *Center for the Study of Women in Television and Film.* Retrieved November 3, 2020, from http://womenintvfilm.sdsu.edu/files/2009-10_Boxed_In_Sum.pdf

Lauzen, M. (2010b). The celluloid ceiling: Behind the scenes employment of women of the top 250 films of 2009. Retrieved November, 3, 2020, from http://womenintvfilm.sdsu.edu/files/2009_Celluloid_Ceiling.pdf

Lee, M., & Cho, C. H. (1995). Women watching together: An ethnographic study of Korean soap opera fans in the United States. In G. Dines & J. M. Humez (Eds.), *Gender, race, and class in media* (pp. 355–361). Thousand Oaks, CA: Sage.

Lopate, C. (1976). Day-time television: You'll never want to leave home. *Feminist Studies,* 3, 70–82.

Magnuson, M. J., & Dundes, L. (2008). Gender differences in "social portraits" reflected in MySpace profiles. *CyberPsychology and Behavior, 11* (2), 239–241.

McCarthy, C. (2008). Facebook 'gender policy' has grammar in mind. *cnet news.* Retrieved November 3, 2010, from http://news.cnet.com/8301-13577_3-9978875-36.html

Newman, A. A. (2008, August 11). The man of the house. *Adweek*. Retrieved May 12, 2009, from http://www.adweek.com/

Orbe, M. (1998). Constructions of reality on MTV's The Real World: An analysis of the restrictive coding of black masculinity. *Southern Communication Journal, 64,* 32–47.

Phillips, E. (2005, November 28). Does gender matter? Ad execs weigh in on whether male and female creatives think differently. *AdWeek*. Retrieved May 12, 2009, from http://www.adweek.com/aw/esearch/article_display.jsp?vnu_content_id = 1001570859

Projansky, S. (1998). Girls who act like women who fly: Jessica Dubroff as cultural troublemaker. *Signs, 23* (3), 771–807.

Rogers, D. D. (1995). Daze of our lives: The soap opera as feminine text. In G. Dines & J. H. Humez (Eds.), *Gender, race, and class in media* (pp. 325–331). Thousand Oaks, CA: Sage.

Salhi, Z. S. (2004). Maghrebi women film-makers and the challenge of modernity: Breaking women's silence. In N. Sakr (Ed.), *Women and media in the Middle East* (pp. 53–71). London, England: I.B. Tauris.

Seiter, E. (1982). Eco's TV guide—the soaps. *Tabloid, 5,* 35–43.

Staples, B. (2009). Just walk on by: A black man ponders his power to alter public space. In E. Disch (Ed.), *Reconstructing gender: A multicultural anthology* (pp. 204–207). New York, NY: McGraw-Hill.

Steeh, C., & Krysan, M. (1996). The polls—trends: Affirmative action and the public, 1970–1995. *Public Opinion Quarterly, 60,* 128–158.

Thoman, E., & Silver, R. (1995). Home, home on the remote: Does fascination with TV technology create male-dominated entertainment. In G. Dines & J. M. Humez (Eds.), *Gender, race, and class in media* (pp. 362–366). Thousand Oaks, CA: Sage.

U.S. Department of Labor. (1994). *The rhetoric and the reality about federal affirmative action programs.* Washington, DC: Author.

Walker, A. J. (1996). Couples watching television: Gender, power and the remote control. *Journal of Marriage and the Family, 58,* 813–823.

Watkins, S. C., & Emerson, R. A. (2000, September). Feminist media criticism and feminist media practices. *The ANNALs of the American Academy of Political and Social Science, 571* (1), 151–166.

Waxman, A. (2007, April 26). Hollywood's shortage of female power. *The New York Times.* Retrieved May 12, 2009, from http://www.nytimes.com/2007/04/26/movies/26wome.html?_r = 1&scp = 1&sq = hollywood's%20shortage%200f%20female%20power&st = cse

Wilkins, A. C. (2004, June). "So full of myself as a chick": Goth women, sexual independence and gender egalitarianism. *Gender & Society, 18* (3), 328–349.

Williams, C. T. (1992). *It's time for my story: Soap opera sources, structure, and response.* Westport, CT: Praeger.

CHAPTER 11

How Does Gender Help Determine Who Has Power and Who Doesn't?

The Gender of Politics and Power

What is power and how does it work? What kind of power do you have because of your gender, and how do you know you have that power? When you think about all the people in your life who have some power over you, how does that list break down in terms of gender? Is there such a thing as masculine power and feminine power, and how would those two be different? How does the balance of power between people of different genders shift over time? Do women have more power relative to men than they did in the past, or do they just have different kinds of power? How is the power to inflict physical violence on a person different from the power that comes from nurturing and caring for someone? How is the power exercised by individuals different from the power of institutions and organizations? Where and how do people acquire power, and are there different methods open to women and men? How are issues like sexual assault and sexual harassment manifestations of power relations? How do power and gender permeate the various institutions in our lives, and how does power in one institution translate into power in another institution? Are certain types of governments more conducive to gender equality than others? What are the barriers that exist to women obtaining power in democratic societies? What would a society in which the balance of power between men and women were more even look like, or what would it look like if the balance were radically shifted in one direction or another? Do you believe, like many feminists, that we live in a world ruled largely by men, and what would life be like if you lived in a world ruled by women?

A BRIEF WARNING

These questions help to get us started thinking about some of the issues centered on the connections between gender and power. It is impossible for sociologists (or perhaps anyone) to discuss gender without also discussing power. So in every chapter so far, we have dealt with issues of gender and power. Power exists at a micro level in the form of the everyday decisions made by individuals and groups, as well as the ways in which they are influenced by the dominant institutions in their society. In Chapter 10, we discussed the power differences involved in who gets to hold the remote control and change channels when men and women watch television together. In Chapter 9, we examined how men in predominantly female occupations such as nursing and social work have more power relative to their female counterparts even though they are the numerical minority in these organizations. In Chapter 6, we looked at the ways in which different historical modes of courtship and dating shifted the power between men and women. These are all examples of the ways in which power is enacted through gender at an everyday level. In this chapter, we will focus on the relationship between gender and power in a more specific group of settings and institutions. These include large institutions, such as governments and religious institutions, as well as how power is manifested in specific issues such as sexual assault, sexual harassment, and modern-day slavery or human trafficking.

These are far heavier topics than who gets to work the remote control or what it means to go out on a date; this chapter has the potential to be upsetting for several reasons that may or may not be true of other chapters. Many feminists and gender scholars would argue that the reason it is impossible to have a conversation about gender without also discussing power is because power is, in the end, what gender is all about. From this perspective, gender categories exist solely to distribute power in certain ways, ways that generally benefit men over women. Within this perspective, the answer to the question, "Why does gender exist?" is fairly simple. Gender exists as a category that enforces differences that create and preserve power for one group while depriving another group of access to that power. The purpose of gender is to ensure that men maintain power over women. The same is true of other categories of difference, such as race, ethnicity, and social class. From this perspective, the differences between women and men that gender categories presume cannot be separated from a larger system of inequality. Every difference that is attributed to gender is also and always a form of inequality. If we go with this particular perspective, it's fairly difficult to argue that there's much of anything that's good or positive about the existence of gender, and our most logical course of action as people who are invested in reducing inequality would be to get rid of gender as a category altogether.

But for some of us, that might be a rather difficult pill to swallow. Being a woman or a man, feminine or masculine, is an aspect of our identity that can be fairly important to our sense of ourselves. Many women find great comfort in engaging in feminine activities, including putting on lipstick, control-top pantyhose, headscarves, or hijab, and participating in female circumcision, a practice that we will discuss later in this chapter. If we subscribe to the perspective described previously, these women are taking comfort and pleasure in activities that are designed for the sole purpose of depriving them of power relative to men. Although many women consider their commonalities through femininity to be a potential source of community, cohesion, and solidarity, how do we make sense of this if that femininity is merely a means for enforcing power? From within this perspective, women become mindless dupes in their own domination (a good example of false consciousness), while men become the dominators—the bad guys. If gender is a system for distributing power, men are the main beneficiaries, even though different men may benefit more or less based on their position within other structures of power. This perspective on gender, then, can be perceived as not particularly flattering to either gender. Those individuals who seek to escape the boundaries of this rigid gender system in a variety of ways—**gender outlaws**, as you might call them—are heroes who seek to dismantle these categories of power by demonstrating their contradictions and flaws.

What would a gender outlaw look like? How would they behave? Have we read any examples of people who might be considered gender outlaws? Do you know anyone yourself whom you think would qualify as a gender outlaw?

This preamble is certainly not included to dismiss this particular perspective on the relationship between gender and power. Perhaps gender is, in the end, like the emperor's new clothes in the fairy tale. We believe we look good in the gender suits we wear around, but in reality, we're hurting ourselves as well as others. Perhaps the only true way to achieve a leveling of power in relation to gender is to do away with the category altogether, an important question that we have dealt with throughout this text. The point is that when we take questions of power in relation to gender head on, as we will do in this chapter, many of the issues that have been floating around the periphery of our conversation become much harder to avoid. Power has real consequences in people's lives, and many of them are ugly, violent, and unpleasant. Some of us may be so lucky as not to have been directly or indirectly affected by the violence that can be a part of gender and power relations, but others may not have been so fortunate. As we proceed, it's important to keep in mind that we should be sensitive to the experiences all of us bring to these issues and the ways in which those experiences shape our particular point of view.

POWER: GOOD AND BAD

How can we understand power, outside of the vague sense that it's the thing supervillains are after and that superheroes reluctantly accept? **Power** is often defined by sociologists as that ability of some actors to influence the behavior of others, whether through the use of persuasion, authority, or coercion. In the world of supervillains and superheroes, power is already defined as something that is ruthlessly craved by the bad guys and perceived largely as an annoying burden by the good guys. This is the popular culture version of the maxim, "Absolute power corrupts absolutely," which is probably drawn from a quotation by Lord Acton in 1887: "Power tends to corrupt, and absolute power corrupts absolutely.

Great men are almost always bad men" (Hill, 2000). Hence, the classic superhero struggle not to become the supervillain, leaving us with the deep philosophical question: Is power a bad thing?

To answer that question, it's important to think about all the different kinds of power that exist. The power sought by supervillains is usually coercive power. **Coercive power** is the ability to impose one's will by force, threats, or deceit, and that certainly sounds very Lex Luther-ish, to invoke just one supervillain. But there are many people in society who have power but are generally not considered to be supervillains. If you think of the hypothetical list of all the people who have some power over you, probably very few of them are supervillains, and many are probably people whom you love and trust. Your parents have power over you, a power that varies in length and intensity depending on the particular family and culture in which you were raised. Your teachers have some power over you, as does the flight attendant when you're on a plane and the official in a soccer game you may be playing. When he or she gives you a red card, which means you're being ejected from the game, you'll probably leave the field (although you may not be happy about it).

Coercive power is probably the kind of power that first comes to mind when we begin to think about what power is, but for many (but not all) of us who live in the developed world, coercive power is probably not the kind of power we encounter most frequently. The power of parents, teachers, flight attendants, and referees is institutional power, or authority. **Authority** is power that comes from a position in an organization or institution that is widely regarded as legitimate. The referee in the soccer game doesn't make you leave the game by threatening you, forcing you, or deceiving you into doing so—unless you play a very interesting version of soccer. But the referee's position in an organization, the particular league to which you belong, provides some degree of authority. It's a limited kind of authority because there aren't a lot of other things you would probably do just because a soccer referee told you to. But it's a kind of power all the same, a power that's necessary to the smooth functioning of our society. If no one listened to sports referees, most games would become fairly chaotic; the same is true of society in general. Authority is also sometimes called **institutionalized power** because it's power that derives from the strength of an institution. Take the person out of the institution, the referee out of the game and out of his or her particular league or organization, and the power's largely gone.

The power that parents have over their children, teachers over their students, and flight attendants over their passengers isn't bad power. But just because authority as a form of power comes from institutions doesn't mean it's always power that's being used in a positive way, or that there aren't potentials for inequality in the use of that power. In this chapter, we'll first explore coercive power as it relates to gender, and then we'll look at the gender consequences of institutionalized forms of power, or authority. But first it's important to dive head first into some of the more contentious questions about gender and power more generally.

Masculinity and Power

You probably noticed in Lord Acton's quote that power corrupts "great *men*" who inevitably become (or already are, depending on your point of view) "bad *men*." Lord Acton was writing about power in 1887, and although the suffrage movement had begun in England more than 20 years earlier with a petition to Parliament requesting the right to vote for women and although Queen Victoria had been on the throne for 50 years, we can probably assume that Lord Acton literally meant *only* men, rather than implying the universal inclusion of *men* as humanity. Even though some of England's most powerful rulers included women like Elizabeth I and Victoria, the assumption in 1887 would have been that matters of real power in the political and economic realm were male matters, exclusively. For the Victorians, masculinity was equated with the possession of power. How much has that connection between masculinity and power changed in today's world?

Make a list of adjectives or characteristics you might use to describe someone who is powerful. How many of these words or characteristics would also describe what it means to be masculine? How many would also describe what it means to be feminine? What does this demonstrate about the connection between masculinity and power?

Another way to think about this question is to examine whether having power is an integral part of what it means to be masculine. This brings us back to our questions about the relationship between gender and power, and whether gender as a social system is fundamentally a way for distributing power between groups. Does being a man mean having access to power that being a woman does not include? In many of the tangible ways in which we would think to measure power, the answer seems to be yes, in a very broad, general sense. As we'll explore in our discussion of institutionalized power, the amount of power available to women fluctuates across time and cultures, but in even the most egalitarian societies about which we have information, men still seem to be in possession of more power than women.

We experience this in the way men make and enforce laws, rule governments, start and wage wars, speak to and interpret for others the word of God, and accumulate wealth. In some times and places, women have also had some power to do these things, but they have never done so either consistently across time or in as great a number as their male counterparts. Even today on a global scale, men have more control in courtrooms, the boardrooms of corporations, computer labs, medical centers, and university classrooms (Bonvilliain, 2007). Decisions made by certain kinds of men (usually white, upper or middle class, and heterosexual), whether as CEOs of large corporations, elected officials, or high-ranking religious leaders, still have the power to affect many of the basic aspects of our lives—from where we live, to how we work, to how we spend our free time.

That's an awful lot of power in the hands of men: supervillain-type power, you might say. But in reality, you probably don't actually know a lot of men who seem like supervillains either in the amount of power they have or in how they use it. And even if there are some men whom you suspect of having supervillain-type power, neither they nor their family and friends probably consider them to be supervillains or even as particularly powerful. Many men profess to feel pretty powerless and not at all like a supervillain. Is this all merely a complicated and devious supervillain ploy? The answer is probably no, and even if it *is* yes, it's really unclear exactly who the supervillain is in this scenario. Still, we're left with the fact that, although across many different societies men seem to be quite powerful, many individual men feel fairly powerless. How do we explain this?

Who Really Has the Power? Hegemonic Masculinity

The answer has to do with the way masculinity works both as a social system and as an identity, taking us back to our earlier discussion of hegemonic masculinity. Hegemonic masculinity is a standard against which all men are judged, even though few if any men actually conform to its ideals. Hegemonic masculinity is largely white, middle class, younger, and heterosexual—pointing already to the ways in which power is not distributed equally to all different types of men. Some men—men of color, working-class men, as well as gay and transgender men—do not have equal access to the power of masculinity because of the other disadvantaged statuses that they occupy. Nonetheless, the **hegemonic definition of manhood**, or what it means to be masculine, is "a man in power, a man with power, and a man of power" (Kimmel, 1994). Michael Kimmel (1994, p. 122) argued that this power is demonstrated in the endless quest by men to demonstrate this particular definition of manhood. Men must exert power over actual women but also over any feminine characteristics in themselves because hegemonic masculinity is defined in opposition to anything perceived as feminine. Although the primary audience toward which displays of manhood are directed is other men, men must also avoid at all costs the slightest indication of affection, love, or sexual interest in other men because this, too, calls into question their masculinity; this is because masculinity is inevitably equated with heterosexuality, and so to be homosexual is to make achieving real manhood impossible. Homophobia, then, is not just incidental to the question of manhood, it is essential to it because the quest to prove oneself masculine is always plagued by the fear of betraying any hint of homosexuality. As Kimmel pointed out, there is probably no worse accusation for many young boys and men in the United States than to be called a "sissy," with the implications it contains about one's sexuality and masculinity.

Theory Alert! Hegemonic Masculinity

Because manhood as an identity is structured in this way, Kimmel (1994) argued that most men spend their lives trying to avoid any form of **emasculation** (taking away or reducing masculinity), whether by women or other men. Women, men of color, working-class men, and gay men are the groups against which men act out their definitions of manhood—the other, "nonmen" against whom their masculinity is defined. Given that men are engaged in a constant game of defending themselves against emasculation, it's no surprise that violence and the willingness or desire to fight are perceived as key markers of manhood. In his account of a Bronx neighborhood rite of passage for boys, Geoffrey Canada (1995) recounted how the willingness to fight was a necessary component of membership in the street life of Union Avenue. Who won the fights arranged by older boys between younger boys on the block was unimportant; what mattered was that the younger boys proved that they were willing to fight, and this rule was enforced on Canada even by his older brother. The willingness and ability to use violence becomes, for many men, an important way in which to prove and defend their manhood.

This perspective on masculinity as an identity begins to suggest why individual men might not actually feel very powerful, especially if they are engaged in a perpetual battle to prove their manhood. When Canada (1995) was forced to fight by the older boys in his neighborhood, he had little power to refuse. Canada had already witnessed what happened to boys who didn't fight, and they basically received a far worse beating at the hands of the older boys than any of them received in their fights with other younger boys. This partly explains why many men do not feel powerful; they are often being coerced themselves by other men. But this contradiction is also connected to a sociological dynamic we discussed in Chapter 2: the difference between private troubles and public issues, or between our experiences as individuals and our knowledge of social structure. Remember that one of the revolutionary aspects of the second wave of the women's movement was the way in which feminists began to make connections between their personal lives and larger social structures—between their private troubles and public issues. For women, there is a more easily observed symmetry between their private experiences and public realities. Feminists have highlighted how women as a group do not have power structurally in society; as individuals, many personally experience this powerlessness in their own lives. For example, as we will explore, sexual assault is a crime experienced disproportionately by women, and the fear of becoming the victim of rape is a fear that many individual women experience, sometimes on a daily basis. Structurally, many institutions in our society reinforce this fear through the ways in which law enforcement officers deal with cases of sexual assault; the structure of the courtroom, which makes the rate of convictions for sexual assault so low; and the punishments that are mandated for the perpetrators of sexual assault. In this instance, there is a clear match between the lack of power experienced individually by women through a fear of sexual assault and the lack of power women as a group have in society to dictate procedures, laws, and punishments for sexual assault.

This relatively straightforward connection between feeling power in our private lives and power in the public sphere is not as true for men. Although evidence suggests that men as a group do have more power at a societal level, many individual men do not feel particularly powerful. One telling metaphor for this feeling of powerlessness comes from the mythopoetic men's movement, a social movement centered around figures like Robert Bly and characterized by an emphasis on reclaiming the "warrior within" men or exploring the inner power of "deep manhood," often through the use of developing and adapting myths and archetypes of masculinity from fairy tales or poetry. Some men within this movement use the metaphor of a chauffeur to demonstrate this inconsistency from their perspective on the position of modern men. From the outside, the chauffeur seems to be in command; he has on a uniform, he's driving the car, and he knows where the car's going. But we all know that in reality the chauffeur is taking directions from someone else and is, therefore, not in charge at all. In this metaphor, although feminists from the outside think men have all the power, the reality is that they're really just taking directions from someone else (Kimmel, 1994).

Who exactly is giving the chauffeur his directions in this metaphor? This metaphor as presented by those in the mythopoetic men's movement doesn't give us an answer, which exemplifies their emphasis on the individual experiences of men. Individually, men may feel bossed around by their wives, their children, their bosses, and the government, just to provide a small possible list. But Kimmel (1994) pointed out that in reality, the person giving the chauffeur his orders in this metaphor is really most likely to be another man. The powerlessness experienced by men is real in that it is something truly felt by men, but it is the product of relations between other men, rather than of any power being transferred away from men as a group in the wider social structure. Some men have more or less power than other men, and some men have more or less access to the power available to men as a group. For example, a working-class white man will find he has less power than many of the middle-class men he encounters in his life, such as his boss, doctors, landlords, bankers, teachers, and law enforcement officials. But most of the people who possess more power than a working-class white man will still primarily be other men, so that the primary power struggles are still primarily among groups of men. To conclude that the solution to the feelings of powerlessness of individual men is to give men more power is an imperfect understanding of the way in which power relations are arranged.

In the end, Kimmel (1994) concluded that most men feel powerless because of the way masculinity is constructed as an identity. If the ideals of the hegemonic definition of manhood are so difficult to achieve and so narrowly defined, very few men ever actually succeed in feeling completely and securely like a man and, therefore, powerful. So men who are not white, middle class, young, and heterosexual are likely to feel less masculine and therefore less powerful. But if Kimmel is correct in concluding that, at least in the United States, "masculinity has become a relentless test by which we prove to other men, to women, and ultimately to ourselves that we have successfully mastered the part" (p. 129), then how many men ever really feel they have passed the manhood test once and for all? Men may have more power as a whole than do women, but in the end, it doesn't seem like such a great gig—which brings us back to our superheroes and supervillains. In reality, men may be closer to the superheroes, and especially to our modern versions, for whom being a superhero never seems to end up as everything it's cracked up to be. In modern-day superhero stories like Spiderman, you never get the girl, you might get chased by the cops or perceived as the bad guy, and you have to hide the fact that you are, in fact, a superhero. Like superheroes, the power that men possess comes with a great burden, and this is an important point to consider as we move forward.

Is this way in which men often feel powerless in society an example of false consciousness, as we discussed in Chapter 10? What does it mean if this is an example of false consciousness?

COERCIVE POWER

As we discussed previously, coercive power is the ability to influence the behavior of others through the use of force, threats, or deceit. Much of coercive power in gender relations takes the form of violence or the threat of violence. Some feminist analysts argue that coercive power in this form is the building block on which all other forms of gender power are built. Men are more likely to inflict acts of violence on women (as well as on other men), and this essential fact affects all relations between women and men, permeating many different institutions throughout our society. In many societies, including the United States, the willingness and ability to be violent is a sure means to gain power, and according to this perspective, men's proven violence is therefore the primary source of their power.

It is verifiably true in the United States that men are responsible for the bulk of violence in a wide range of forms. Although their percentages of these crimes have been declining over the past 10 years, men still make up 99% of those arrested for rape, 79% of those arrested for aggravated assault, 88% of those arrested for robbery, 75% of those arrested for all other assaults, 75% of those arrested for all family violence, and 74% of those arrested for disorderly conduct (U.S. Department of Justice, 2007). Nearly 88% of murder victims are killed by men (U.S. Department of Justice, 2005). Men are more likely to commit violent crimes in general, and beginning in the second wave feminist movement, a whole host of crimes that were specifically gendered were for the first time identified and perceived as crimes. Feminist analyses point out that, until very recently, much of the coercive power exercised by men against women was not perceived as particularly wrong, let alone criminal. As recently as the 1950s in the United States, women who were not virgins could not legally accuse a man of raping them, and rape occurring within a marriage (**marital rape**) became illegal in the United States only as recently as 1976.

Until 1976, rape laws included a marital exclusion because marriage gave husbands the right to have sex with their wives whenever they chose, regardless of their wives' consent. Before the advent of the second wave feminist movement in many places around the world, there was no such thing as domestic violence, wife battering, or sexual harassment. Husbands beating their wives was not treated seriously, and certainly not as a crime, until efforts by the women's movement in the 1960s and 1970s. The murder of wives by their abusive husbands was often described by headlines such as "Husband Goes Berserk and Shoots Estranged Wife," ignoring the longstanding patterns of abuse that resulted in these deaths (Del, 1976). Head and master laws across the United States up until the 1960s and 1970s enforced the almost complete control of a husband over his wife, and so assaults by husbands against wives, as well as parents against children, were not within the purview of the legal system. **Head and master laws** dictated that husbands had final say regarding all household decisions and joint property in a marriage and that a husband could make decisions about these matters without his wife's knowledge or consent. A 1954 report from Scotland Yard demonstrates some of the prevailing attitudes toward family violence, reading as follows: "There are only about 20 murders a year in London and not all are serious—some are just husbands killing their wives" (Coontz, 2005). All these examples demonstrate that because of the fewer number of rights granted to women historically, many acts of violence against them were not perceived as criminal.

As we discussed in Chapter 9, women have participated in the paid workplace since the very beginnings of industrialization. Sexual harassment did not first emerge during the 1970s when activists brought it to the attention of the public, businesses, and governments. Rather, it was only during this period that behaviors that had been perceived as normal and expected aspects of women's lives in the workplace came to be perceived as immoral, unfair, and, eventually, illegal. Up until the 1950s, some laws in the United States made it perfectly legal for employers to deny employment to women in general, and especially to married women because of the assumption that married women didn't need jobs to support themselves and were therefore taking jobs away from married, male breadwinners. With this often base-level hostility toward women in the workplace, it's not surprising that treating women as sexual objects and expecting sexual favors from female coworkers was an acceptable aspect of work life in the not-so-distant past.

CULTURAL ARTIFACT 1: SEXUAL HARASSMENT AND *MAD MEN*

An inevitable part of the discussion of power and gender is uncovering the ways in which our sense of what's right and wrong is often historically and culturally contingent. Few people in contemporary Anglo-European societies would condone men hitting their wives or children. Most of us understand why it's probably not appropriate to expect sexual favors from the people who work for you. But as recently as 50 years ago, these behaviors were not known as domestic violence and sexual harassment, and they were often not considered to be wrong. A startling example of how quickly ideas about right and wrong can change is demonstrated by the AMC television show *Mad Men.* There's some debate about whether this show, which is based on the lives of male advertising executives and their female secretaries and wives, perfectly reflects what it would have been like to work in advertising in the 1960s (Witchel, 2008). But excessive smoking, drinking, womanizing, and anti-Semitic attitudes suggest a time before any of these things were considered wrong within certain communities. In the WASPish (White, Anglo-Saxon, Protestant) world of this television show, secretaries, who are exclusively female, get ahead by dressing more provocatively and are expected to serve as second wives to their bosses in the workplace. Male executives stand around the halls of their advertising firm leering at the secretarial pool, and sex in the workplace is not just OK, it is expected. In the first episode, one male executive advises a meeker colleague, "You got to let them know what kind of guy you are. Then they'll know what kind of girl to be" (Witchel, 2008). The divorced woman who moves into the neighborhood of the main character, Don Draper, is viewed as scandalous by the other housewives. Meanwhile, Don's wife Betty, a college graduate, languishes in their suburban home and eventually begins to see a therapist—who reports on all their sessions to her husband, Don. One enterprising secretary, Peggy Olson, is promoted to copy writer at the end of the first season because of her ability to provide insight into "what women want." In other episodes, pregnant women and gynecologists smoke constantly, while Don's boss treats his stomach ulcer with butter and milk, eventually suffering from a heart attack (no surprise to modern-day viewers). The scale of history is often hard for us to process, but *Mad Men* is set in a time period in the not-so-distant past and yet still seems like a different world. What does considering the historical and cross-cultural nature of right and wrong suggest about some of our current ideas about what is and isn't appropriate behavior? Which behaviors that we currently view as normal might be considered deeply immoral in the future?

In the 1990s, feminism came under some criticism and experienced internal divisions about the emphasis on these coercive aspects of gender inequality. Some women within and outside of the feminist movement felt that feminist analyses placed too much emphasis on issues such as sexual assault, domestic violence, and sexual harassment. Essentially, these critics were suggesting that violence is *not* a critical aspect of gender inequality, either because times have changed (marital rape is now illegal and women who are not virgins can now charge someone with sexual assault) or because, in their analyses, violence was never the central feature of gender inequality. According to feminist analyses that emphasize the importance of coercive male power, the effect of violence against women is not felt solely by women who are themselves the direct victims of violence. Even if you have never personally been the victim of sexual assault, domestic violence, or sexual harassment, the threat of becoming a victim of any of these types of violence remains a real part of your life as a woman.

This real threat influences the opportunities that are available to women and the decisions they make. It is now illegal to prevent a woman from taking a job in a male-dominated occupation such as law enforcement, but the sexual harassment that women are likely to face in the police force may drive many women out of this occupation as well as prevent many more from seriously considering the option. Researchers have shown that the perpetrators of domestic violence use the threat of abuse to create a wider pattern of control and coercion. Victims of domestic violence, then, come to exist in a world that is very much circumscribed by limited opportunities and choices as a result of the constant threat of violence. Within this situation, the reluctance of some women to leave their abusers, even at the risk of their own lives and those of their children, becomes a particularly strong example of the way in which the looming threat of violence can affect a person's perceived amount of freedom. Some feminists who place violence at the center of their analyses of power and gender argue that every male–female interaction is shaped by a bottom line of potential violence—in a physical fight, the man will win (Kaye/Kantrowitz, 2009). Expressing this perspective, radical feminist Ellen Willis said, "Men don't take us seriously because they're not physically afraid of us" (as quoted in Kaye/Kantrowitz, 2009). In this sense, women's status as primarily victims rather than as perpetrators of violence is at the core of gender inequality.

Do you agree with Ellen Willis's statement that men don't take women seriously because they're not afraid of them? What does this imply about the centrality of violence to power? If you agree with this statement, what should women do to get men to take them seriously?

The Geography of Fear

The possibility of becoming a victim of sexual assault that many women live with has a wide range of possible implications according to feminist analyses. Some argue that this threat leads to the creation of a **geography of fear** among women, or the assessments of vulnerability and fear in certain spaces that women are forced to make as a result of the threat of potential violence, assault, or harassment (Valentine, 1989). Despite the fact that most violence against women occurs within the private realm (inside the home) (Dobash & Dobash, 1992; Hollander, 2001), women are taught to avoid the sphere of public space (Bynum, 1992; Duncan, 1996; Valentine, 1992). A woman in a public setting may feel vulnerable to a wide range of encroachments on her privacy and safety, from objectification through lewd comments to violent crime in the form of physical assault (Wesely & Gaarder, 2004). Although women may actually be aware that the private realm is far more dangerous than the public, many constantly question and assess their ability to gauge whether spaces are or are not safe. In this way, the fear of becoming a victim of various types of violence in public spaces enforces the gendered nature of the private–public split, giving women incentive to cede more of the public realm to men. Fear of violence in public spaces literally becomes a means for keeping women "in their place," that place being the home and the private sphere.

In their study of women's use of an outdoor, urban park, Jennifer K. Wesely and Emily Gaarder (2004) examined the strategies and negotiations women use to balance a genuine enjoyment of outdoor recreation against the fears of possible violence and harassment that might take place in those spaces. Using both surveys and in-depth interviews with women who used South Mountain Park (a pseudonym), Wesely and Gaarder (2004) explored how women enjoy some of the special advantages of outdoor recreation. Women described their workouts in the park with words like "spiritual," "freeing," and "peaceful" (p. 653), which is consistent with an emphasis on the therapeutic value of outdoor recreation. Women also reported enjoying the advantages of getting a physical workout outside the confines of a gym, with its negative social aspects. Women described feeling that gyms, as opposed to the park, were places where people go to find dates and where they feel judged and observed. Getting physical exercise in South Mountain Park allowed them to avoid these downsides of the gym.

These advantages of outdoor recreation were balanced against the real fears that many women expressed about being at places like South Mountain Park. In their surveys, about 40% of women had experienced some kind of harassing behaviors while engaged in outdoor recreation in general, while 2.4% had reported being physically assaulted. Although the incidence of harassment and assault were lower for South Mountain Park specifically, women still reported that there were many trails and areas where they still felt unsafe. As one woman described, "I feel more unsafe on the less traveled trails—fear of being raped, murdered and left where no one will find me" (Wesely & Gaarder, 2004, p. 654). At the same time, many women also found these isolated trails the most enjoyable because of the opportunities they provided to experience the quiet of nature and avoid the crowds of bikers or other park users.

The fear of certain trails and areas of the park expressed by some women was often grounded in specific experiences described in Wesely and Gaarder's (2004) interviews and surveys. One park user, Mona, jogged alone at night in the park and described an incident in which a young man grabbed her breast and then ran away; after this assault, she was obviously more concerned about where and when she exercised outdoors.

Other women's sense of privacy and control were violated by being made to feel like objects during their experiences in South Mountain Park. Katie said, "I'll occasionally get a catcall or second glance. I'd like to be able to hike without feeling like someone is looking right through my clothes" (Wesely & Gaarder, 2004, p. 655). Another woman described being generally suspicious of men on trails, hoping they wouldn't turn around and look at her or look not "quite right"; she suspected that few men on the trails looked at or thought about women they encountered in the park in the same way. Through these incidences, women developed their own geography of fear, gradually beginning to associate the public space of South Mountain Park with male violence and violation. For some women, this was a result of direct experiences of being assaulted in the park, but for many others, it was the inability to "choose with whom they interact and communicate" (Wesely & Gaarder, 2004, p. 655) that affected their sense of their own safety.

How did women negotiate these fears of public space in South Mountain Park? The most common strategy was to avoid recreating in the park alone, which meant using the park with other women, men, or sometimes bringing a dog. Women also avoided certain isolated trails and tried not to use the park at dusk or at night. Some carried items for safety, most commonly a cell phone—but in at least one case, a woman carried mace with her. Other women had made the decision to alter their own patterns of behavior in the park. Lisa faced a situation in which the same man showed up at the park every time she was there, talking to her constantly and following her on her hikes. Although she tried going to the park at different times, the same man still showed up, so Lisa eventually stopped going altogether because she didn't want the man to get the "wrong idea" (Wesely & Gaarder, 2004, p. 656). In this sense, the responsibility for preventing Lisa from potentially being attacked is squarely with Lisa, who has to keep this man from getting the wrong idea by not going to the park at all.

This sense of responsibility for their own safety was one of the important dilemmas of the women at South Mountain Park as identified by Wesely and Gaarder (2004). No matter how hypervigilant the women were or what specific measures they took to protect themselves, there was no real way available to them to remove the threat of violence completely while they were recreating in the park. Much of this burden of making themselves safe fell on the women themselves, as opposed to on the men whose behaviors often made them feel unsafe. But even the possibilities available to these women that shifted the burden of ensuring their safety to others brought a new set of problems. Some women discussed how the park service itself could improve their safety, through the use of lighting, emergency phones, or call boxes, and the presence of park rangers on the trails. But these solutions present a situation in which women trade limits on their use of space and free range of movement because of the fear of assault, rape, and harassment for new limits as a result of surveillance and limited mobility (Wesely & Gaarder, 2004). Women trade the social control that results from the threat of violence and harassment for a new kind of social control designed to ensure their safety, but both are forms of social control nonetheless. In addition, the presence of lighting, call boxes, and park rangers interferes with many of the exact qualities of being in nature that these women and other users of the park are seeking.

Can you think of other examples of places where women might encounter the geography of fear? How do women deal with this geography of fear in these places? Does the geography of fear affect all women? Does it ever affect men? Why or why not?

This study demonstrates in one concrete setting how the fear of being assaulted, raped, or harassed has very real effects on basic behaviors for women, like deciding whether they can afford to take the perceived risk of hiking or biking in an urban park. This suggests that there is some truth to the arguments that coercive power is an important tool in maintaining gender inequality. The fear of what might happen to them can effectively work to keep women out of public spaces, and as we discussed in Chapter 9, the public sphere is where the important decisions about the world are made. In this case, the consequences seem relatively minor. But if we imagine other examples of the public sphere where coercive power might deter women from entering, the implications can seem more important. Police officers spend much of their time operating in the public sphere, and if some women choose not to enter this occupation because of these kinds of fears, the important job of ensuring public safety and investigating crime lays primarily in the hands of men. The competitive floor of the stock market and commodity exchange markets are also part of the public sphere, and when sexual harassment as a form of coercive power prevents women from entering these occupations, this considerable power over our economy is controlled largely by men.

The case of women and their use of South Mountain Park also highlighted the ways in which the problem of coercive power used against women is perceived as primarily a woman's problem. Well, of course it's a woman's problem because women are the ones primarily being victimized. But this perspective leaves aside the other side of the equation, that there are people who are responsible for assaulting, harassing, and raping women. As Wesely and Gaarder (2004) described, the women took on the burden of ensuring their own safety, rather than challenging the entire system that normalizes the harassment of women in public places. Viewing coercive power used against women as a problem for women leaves unexamined the ways in which men and the larger system of gender are also part of this dynamic. If women are solely responsible for ensuring their own safety, then we implicitly assume that nothing can really be done about the tendencies of some men to use coercive power against women. The connection between masculinity and violence is naturalized and assumed to be unchangeable. Besides placing the burden for addressing problems of coercive violence on women, this perspective is also hardly complementary toward men—especially the large numbers of men who do not harass, assault, or rape women.

Rape-Prone and Rape-Free Cultures

Many feminists describe this perspective on the relationship between coercive power and gender as representing a rape culture in our society. A **rape culture** is characterized by a certain set of ideas about the nature of women, men, sexuality, and consent, or **rape myths**, that create an environment conducive to rape (Armstrong, Hamilton, & Sweeney, 2006). An example of a rape myth is the belief that female victims of rape and sexual assault "ask for it" through their style of dress or their choice to put themselves in certain situations or locations, a conception that shifts the blame for these crimes from the perpetrator to the victim. Another rape myth is the belief we discussed previously that sexually aggressive behavior on the part of men is natural and that behaviors like harassment are merely an extension of these underlying male dispositions.

The rape culture perspective highlights the ways in which rape myths are damaging to both women and men while prioritizing the role of culture (rather than biology or individual psychology) in creating these myths. The rape culture perspective does not, however, point to the specific circumstances under which these cultural beliefs are more likely to be enacted. Rape myths may dominate our culture, but as we already pointed out, not all men are perpetrators of sexual assault or rape and not all men harass women. Some research suggests that the degree to which masculinity is connected to the use of coercive power against women varies from culture to culture, even sometimes within the small setting of one individual college campus. Many gender scholars argue that if we want a real solution to the problem of violence against women, we must examine both sides of the coin—the role of both women *and* men in this particular dynamic. For these reasons, many researchers interested in investigating and eventually proposing solutions to the problem of violence against women pay careful attention to the specific contexts that give rise to more or less violence against women in a culture or society.

In her research, anthropologist Peggy Reeves Sanday (1981, 1990) sought to identify those factors that affect the prevalence of rape, or the differences between rape-prone and rape-free cultures. In a **rape-prone** culture, the incidence of rape is reported by observers to be high, rape is excused as a ceremonial expression of masculinity, or rape is an act that men use to punish or threaten women (Sanday, 2004). By contrast, a **rape-free** culture is one in which the act of rape is either infrequent or does not occur, meaning that not all rape-free cultures are actually completely free of any instances of rape. Rather, in a rape-free culture, sexual aggression is socially disapproved of and severely punished (Sanday, 2004). In her comparison of 95 band and tribal societies, Sanday (2004) found that 47% were rape free while 18% were rape prone. What kind of factors explained the difference between these two types of cultures? In a rape-prone culture, rape is a part of a larger cultural system that includes high rates of interpersonal violence, male dominance, and sexual separation.

Sanday argued that rape-prone tribal societies are more likely to be environmentally insecure and that women are turned into objects to be controlled by men as part of men's struggle to gain control of the environment. Men and women live largely separate lives, and men are more concerned with proving their manhood than are the men in rape-free cultures. In contrast, in rape-free cultures, women have a great deal of ceremonial importance and respect is given to the unique contributions made by women to the coherence of the larger community. This ceremonial importance and respect means men and women are in relatively balanced spheres of power and are, therefore, characterized by a belief in **equality through complementarity**. This means women and men may not do exactly the same things in the culture, but the contributions of both genders are equally valued, and "each is indispensable to the activities of the other" (Sanday, 2004, p. 65). From this perspective, larger cultural beliefs, some of which are not specifically related to gender, affect the prevalence of rape in various cultures. There is nothing inevitable about this particular form of male violence; rather, the behavior is conditioned by the particular cultural context.

How does the idea that women and men are complementary affect the tension between difference and inequality? In this situation, are men and women perceived as different but still equal? Can you think of examples of complementarity in your own culture?

Sanday (1990) and others applied this perspective to a setting more familiar to many in Anglo-European societies and in which sexual assault is often a pressing issue: college campuses. One National Institute of Justice study estimated that between one fifth and one quarter of women are the victims of completed or attempted rape while in college (Fisher, Cullen, & Turner, 2000). Women on college campuses are therefore at greater risk of becoming victims of sexual assault or rape than women in the general population of a comparable age group. College campuses in general seem to be rape-prone cultures, as defined by Sanday, but why?

Several researchers have focused on the cultural context of fraternities on many college campuses and how some of the particular features of Greek life may contribute to an increased likelihood of sexual assaults. For example, fraternities often emphasize competition with and superiority over other fraternities as well as within individual fraternities, and this competition can sometimes take the form of sexual conquests. Many fraternities use alcohol as a weapon to obtain sex from women, and the structure of some fraternities commodifies women as bait (to lure new members), servers (in the form of little sister organizations), or sexual prey (Martin & Hummer, 1989). When these particular structures and norms are in place, fraternities can easily become rape-prone environments.

But this does not mean that all fraternities are rape-prone environments because a great deal of variation can exist within the culture of individual fraternities. For example, in her research, Sanday (2004) described one fraternity that was much more in line with the cultural context of a rape-free culture. In this fraternity, which she gave the pseudonym QRS, binge drinking was rare, and at parties, certain brothers took on the responsibility of patrolling to ensure that no intoxicated women were taken advantage of. Women were not perceived to be sexual objects in this fraternity, and there was no emphasis on attracting "party girls." Women's studies students, along with outspoken feminist activists, were regular guests to the fraternity, while one of the male members of the fraternity started an organization for men to talk about gender issues. Sanday also noted that the QRS fraternity was unusual in its openness to homosexuality and bisexuality. The fraternity was often labeled the "gay frat" or "faggot house," but the men in QRS were largely unaffected by those labels (Sanday, 2004, p. 68). This research suggests that although it may be true that the cultural context created by some fraternities contributes to an increased prevalence of sexual assault against women, there is nothing inherent in the structure of a fraternity that does so. Even the subtle variations in the culture of individual fraternities matter a great deal.

Sexual Assault on Campus

Theory Alert! Integrative Theory

One study that demonstrated the importance of multiple levels of context and structure in creating rape-prone cultures examined how a combination of individual, cultural, and structural factors contributed to the prevalence of "party rape" on college campuses. This research by Elizabeth Armstrong, Laura Hamilton, and Brian Sweeney (2006) is a good example of the integrative approach to gender that we discussed in Chapter 2. Integrative theory attempts to explain some gender phenomena using all three levels of analysis—the individual, the interactional, and the institutional. In addition, the approach of these researchers also examined how processes and characteristics that are not initially gendered can still have gendered implications.

They began with the party scene that dominates many college campuses and the ways in which both female and male college students are motivated to participate in the party scene. The researchers argued that long before they ever arrive on the college campus, many students have already been socialized into certain expectations about college life, which include the party scene as a central element. Going out, drinking, and having fun are considered things you're supposed to do in college and an important way to feel like part of the larger college community life. Many of the women interviewed in the study by Armstrong et al. (2006) had been popular in high school and saw partying in college as the new route to popularity and generally fitting in on the college level. This party motivation had no specifically gendered aspects, but when combined with other factors, it had important gendered implications.

Theory Alert! Social Learning Theory

These gendered implications were realized in part because, although the motivation to party in college was gender neutral, the party scene itself was not. In the particular large, Midwestern university setting of this study by Armstrong et al. (2006), partying was the primary way for heterosexual women to meet men. Although the dorm the researchers focused on was itself coed, men coming onto the women's floor had to be escorted, and the floor was locked to any nonresidents. In the large lecture classrooms, women found it difficult to meet or talk to men, and many women complained that they lacked casual, friendly contact with men in the college setting, especially compared with the mixed-gender groups many of them had become accustomed to in high school.

The party scene, then, became the primary venue available to these women for meeting men, as well as for obtaining the status and self-esteem that came with being perceived as desirable to men. The women in this study enjoyed dancing and kissing at parties in part because the attention they received from men required a "skillful deployment of physical and cultural assets" (Armstrong et al., 2006, p. 488). These skills involved wearing the right kind of outfit—"hot" but not "slutty"—as well as working to present the image of the "ideal" college girl with "white, even features, thin but busty, tan, long straight hair, skillfully made-up, and well-dressed in the latest youth styles" (Armstrong et al., 2006, p. 488). The psychological benefits of being admired in the party scene were such that women with boyfriends sometimes mourned their own inability to garner such attention without upsetting their boyfriends by making them jealous.

The motivation to party and the specifically gendered ways in which women experienced partying are important aspects of understanding party rape at the level of individuals. At the institutional and organizational levels, the rules of universities and Greek organizations become important. For example, colleges and universities have some discretion about how strictly and in what settings they enforce state drinking laws. Enforcement of drinking laws is usually especially rigorous in residence halls, where the researchers (Armstrong et al., 2006) observed both resident assistants (RAs) and even police officers patrolling residence halls for alcohol violations. The sanctions for these violations were often severe, including a $300 fine, an 8-hour alcohol class, and probation for a year; not surprisingly, students engaged in minimal amounts of alcohol consumption in their dorms. This was in comparison with the free flow of alcohol that took place in the fraternities. Public spaces for social activities that did not involve drinking were also hard to come by in the residence halls, and the nondrinking social options (a midnight trip to Wal-Mart) were perceived as uncool.

Theory Alert! Gendered Organizations

Meanwhile, although the Greek system requires the consent of the larger university for some of its activities, the university lacks full authority over what takes place in Greek houses. Greek houses are privately owned and governed in part by the rules of their national organizations as well as by the Interfraternity Council (IFC). Given this situation, according to Armstrong et al. (2006), fraternities at this university had almost complete control over what happened at their parties. When deciding party themes, fraternities often developed parties that required women to wear scant, sexy clothing and put them in subordinate positions. Examples of the parties the researchers observed during their study included "Pimps and Hos," "Victoria's Secret," "Playboy Mansion," and "CEO/Secretary Ho." Some fraternities required that their pledges transport first-year students, and particularly first-year women, to the parties, although transportation home for these women was not provided.

Fraternities controlled who came to their parties, and they showed a pronounced preference for first-year women, while often turning away unaffiliated men. In fact, the ability to attract women, and specifically attractive women, to their parties, was important to the ability of the fraternities to recruit new members. In these ways, the rules of both the larger university and the structure of the fraternities created an atmosphere that contributed to the incidence of party rape by funneling most of the unsupervised drinking into the Greek houses.

These individual and institutional factors come together to shape the actual interactions that take place during parties at many fraternity houses and create an atmosphere conducive to party rape. Ironically, it is the interactional production of fun that is a key component of this dynamic. In their research, Armstrong et al. (2006) identified the interaction routines that were important to the party scene at this particular university. **Interaction routines** are patterns or norms of speech or action that are followed with regularity to accomplish some task in an interaction, and to imply that they exist for the university party scene is another way of saying that there are taken-for-granted rules that dictate how to have fun at a party. One example of an interaction routine is when passing an acquaintance on the street, you might ask them, "How are you?" and understand that this is not so much a question (you don't actually expect them to explain to you how they are in any detail) as it is a greeting; the proper response according to this interaction routine is a simple answer, like "Fine." The idea that there are rules dictating how to have fun seems like a contradiction until you think about how you might react to someone who tried to tell you intimate details about a recent and traumatic breakup or illness in the middle of a party. A gender-neutral interactional routine for the party scene is that a participant "throw[s] him or herself into the event, drinks, displays an upbeat mood, and evokes revelry in others" (Armstrong et al., 2006, p. 490). The individual who very obviously displays unhappiness or creates tension is not participating in the interactional routines appropriate to the party scene.

Theory Alert!
Interactional

Can you think of other examples of interaction routines? Can you think of interaction routines that are gendered or interaction routines that are different for men and women?

Other interaction routines of the party scene at universities are gender specific. Armstrong et al. (2006) suggested that women at fraternity parties were expected to wear revealing outfits; cede control of turf, transportation, and liquor to the male fraternity members; and to be grateful to their male hosts as well as generally "nice" in ways that men were not necessarily expected to be. This overall "niceness" expected from women at parties is another way of describing an expectation of **deference**, or that women accept their status as a group subordinate to men and do not challenge men's control over them. Deference in other settings can take the form of women smiling even when they are actually angry or unhappy to disguise their emotions or choosing to be silent when they are threatened, challenged, or insulted in order not to cause a disturbance. In the context of university fraternity parties in this study, deference meant that women often did not call attention to inappropriate behaviors on the part of men because they did not want to make a scene. In one example, Amanda was drinking at a bar with an older guy named Mike.

Although she and Mike made out at the bar, Amanda made it clear that this did not mean she wanted to go home with him. When she found herself stranded at the bar with no ride home, Mike promised that a sober friend of his would take her home. Instead, the sober friend refused to take Amanda home and instead dropped her at Mike's house, asking her in the meantime whether she would be interested in a ménage à trois. Amanda stayed awake all night out of fear and woke Mike up early in the morning to take her home. Despite this ordeal, she described Mike as a "really nice guy" and exchanged telephone numbers with him. Despite being afraid and staying awake all night, in this story Amanda took no action that would suggest any of Mike's or his friend's behaviors were wrong, deferring her own fears to the need not to be considered a troublemaker.

Can you think of other examples of women demonstrating deference? Do you believe this is something that still happens frequently? How are deference and chivalry connected? Do other subordinate groups demonstrate deference? Servants? Minority groups? Children?

The party rapes that women reported in the Armstrong et al. (2006) study were generally not accomplished through the use of physical force in the form of knives, guns, or fists. Rather, sexual assault happened through the combination of several forms of low-level coercion. Perhaps most important among those forms of coercion was the pressure to consume large amounts of alcohol. As recounted by one fraternity member, his brothers encouraged women to drink alcohol partly because it was, after all, a party, but they were also aware of the ways in which drinking lowers inhibitions. Women often struggled to remember sexual encounters that occurred when they were intoxicated, suggesting that they were certainly not sober enough to give consent to sexual activity. In addition to alcohol, college men used persuasion and the manipulation of their environment, which was often as simple as locking a door, physically blocking a door, or using body weight to make it difficult for a woman to get up. In these situations, deference could prevent women from expressing their own desires in any form, whether that meant saying no, running away, knocking someone down to get out a door, beating on a locked door, fighting back physically, or screaming.

Armstrong et al. (2006) encountered at least one instance of sexual assault in every focus group that included heterosexual women. Most of the women complained about the men's efforts to control their movements within the party scene and the pressure to drink they experienced. Two women in the hall the researchers studied had been sexually assaulted during the first week of school, and later in the semester, another woman was raped by a friend. A fourth woman suspected she had been drugged at a fraternity party. Given this small sampling of the experiences women had in the party scene, why did they continue to participate in it, as many of them did? The answer is that even with these negative experiences, the party scene can also be fun for university women. Partying allows women to meet new people, experience and display belonging, and enhance their social position (Armstrong et al., 2006).

Many women in the study were invested in the party system, and this made it difficult for them to find fault with something they also enjoyed. Many women therefore responded to the reality of women who were victimized in the party scene by blaming the women themselves, rather than blaming the context of the party scene that they so valued. The sexual assault prevention campaigns waged on many campuses play into this tendency to blame the victim; many campaigns emphasize the precautions that women should take to prevent themselves from being sexually assaulted, and women who then do fall victim can be chastised for not following the prevention strategies laid out for them. Even a group of feminist women on the campus studied by Armstrong et al. (2006) argued that a woman who had been raped "made every single mistake and almost all of them had to do with alcohol." Blaming female victims of sexual assault is also part of the way in which campus women create a sense of status hierarchy between themselves and other women. Even though the party scene emphasizes the importance of women being perceived as "hot," many women convince themselves that they will never be victimized because men perceive them to be worthy of their sexual respect. Women who are viewed as sexually promiscuous are then considered to be responsible for their own assault because they did not properly earn the sexual respect of men.

Given this reality as outlined by Armstrong and her fellow researchers (2006), what options are available to women for protecting themselves? Some female students chose to opt out of the party scene altogether, but this made them social misfits in a residence hall dominated by the party scene. For the women who chose not to opt out, they continued to face the ongoing danger that full participation in the party scene, including drinking, having fun, and feeling attractive to male college students, carried with it. Women will continue to run a risk in the party scene at universities like the one studied by Armstrong et al. (2006) as long as the problem of sexual assault is perceived solely as either resulting from the decisions of individuals or resulting from different cultural contexts, rather than as a combination of factors at many different levels. This includes approaches that continue to treat violence against women as a woman's problem. This study provides another example of how the threat of violence makes women's participation in another public sphere—here, the party scene at college—complicated in ways that men's participation is not.

Violent Intersections: The Gender of Human Trafficking

In recent years, an old form of coercive power that many people believed had long been banished has reappeared and taken on a particularly gendered form. Most people assume that slavery is a thing of the past, largely abolished from the world when the United States officially ended the enslavement of African Americans in 1865. But **slavery** as an economic and social relationship between two people characterized at its core by violence never really disappeared (Bales, 2004). The increase in world population coupled with rapid worldwide social and economic changes and corrupt governments have all contributed to a booming business in what organizations like the United Nations call *human trafficking* (Bales, 2004; United Nations Population Fund, 2003). **Human trafficking** takes many different forms, but at its core, it means using coercive force to sell a person for exploitative uses. This is accomplished in many parts of India through debt bondage, in which a person pledges her or his labor against a loan of money, leaving unclear how long the person will work to pay off the loan or exactly what kind of work he or she will do. In addition, the labor the person performs in debt bondage does not actually diminish the amount of the original debt. In these situations, the lender does not legally own the person, but complete physical control is exercised over the debtor, qualifying as coercive force used to keep the person and his or her family in slavery for indefinite periods of time (Bales, 2004).

Human trafficking is a gendered issue because many of the forces that contributed to the rise of modern slavery had an uneven effect on women and men. The rapid social and economic changes that have taken place since roughly the middle of the 20th century have greatly exacerbated the inequalities between the global North and the global South. Within countries of the global South, women and children often bear more of this increasing burden of inequality, and therefore, they become more vulnerable to human trafficking. This increased vulnerability is in part a result of a phenomenon identified as the **feminization of poverty**, or the economic disadvantage experienced by female-headed households relative to male-headed households or couple-headed households.

Research on poverty both within the developed world and in the global South has demonstrated that when women are the main providers for a family, the family is more likely to struggle economically and therefore find themselves in poverty. Obviously, the feminization of poverty and the increased economic vulnerability of women are also caused by outright gender discrimination and the lack of educational and professional opportunities available to women around the world. The increased numbers of women experiencing poverty worldwide makes them more vulnerable to being caught up in the web of human trafficking. Although exact estimates of the volume and scale of human trafficking of women are difficult to come by given the clandestine nature of this activity, rough estimates from the United Nations suggest that between 700,000 and 2 million women are trafficked across international borders every year (United Nations Population Fund, 2003). If we included the number of women who are also trafficked domestically within the borders of a particular country, the number would probably climb to 4 million every year.

Women are more likely than men to become victims of modern slavery, and the type of labor performed by women and men also differs. Women who become caught up in the global system of enslavement are more likely to end up as domestic servants or to be engaged in various forms of prostitution, and in both situations, they are likely to become victims of sexual violence. The Central Intelligence Agency (CIA) in the United States estimates that 50,000 women are brought to the United States each year under false pretenses and eventually end up working as prostitutes, abused laborers, or servants (Goodey, 2004; United Nations Population Fund, 2003). The European Commission recently estimated that perhaps 500,000 women from Eastern Europe have been forced into commercial sex work. In other parts of the world, perhaps 200,000 Bangladeshi women have been trafficked from their home country to Pakistan, with many more ending up in the Middle East (Paul & Hasnath, 2000).

Accounts of women who have survived and escaped their experiences in the trafficking system reveal the varied backgrounds of those who become victims of this dark side of the global economy. In 2003, a multistate prostitution ring was discovered when a 17-year-old girl in a suburban Detroit strip mall burst into a store and asked the security guard for help (Bales, 2004). The girl had been abducted while she waited at a bus stop in downtown Cleveland, Ohio. Her captors drove her to Detroit, where she was kept in a house with other women and forced to have sex with male visitors. The women were escorted around the house at all times, and younger women were kept in line by older women who threatened and sometimes beat them.

The girls worked as prostitutes, but they were sometimes also sent to malls in metro Detroit to sell jewelry and trinkets, while others were forced to dance and strip for private parties. The captors, including Henry Davis, who the police discovered had been kidnapping girls for his prostitution rings in the Midwestern United States since 1995, controlled the girls through a system of rewards and punishments. If the girls obeyed orders, they were rewarded by being taken out to get their nails and hair done, and this was how the girl came to escape and bring the ring to the attention of law enforcement. If they disobeyed, they were beaten, with older women often acting as enforcers.

In a very different case, Hilda Dos Santos had worked as a domestic servant for her employees, the Bonnettis, in Brazil for many years before they asked her to move to the United States with them in 1979. Once in the United States, the Bonnettis stopped paying Hilda and made her into a slave in their suburban, Washington, D.C., home (Bales, 2004). For 20 years, Hilda cleaned the house, did the yard work, cooked the meals, cared for the pets, and shoveled snow without gloves, boots, or a coat. She slept on a mattress in the basement and was not allowed to use showers or bathtubs in the house. She was fed leftovers and scraps for food, and she was beaten when she made mistakes. When a cut on Hilda's leg became infected, the Bonnettis refused to provide any medical care. Hilda only escaped when a stomach tumor grew to the size of a soccer ball and a neighbor took her to the hospital because the Bonnettis refused to do so. It was only then that social workers were alerted and the law intervened to save Hilda.

Shanti Devi was 39 years old when she left her village in Bangladesh to look for work in town in order to support her ailing husband and three children, exemplifying the feminization of poverty as a female-headed household. A man named Monkhar, from a neighboring village, convinced Shanti Devi that her long lost brother had sent for her with a job offer in India. When she went with Monkhar, he took her to a slum in Delhi, India, and sold her to a pimp for $60 (2,000 rupees). When Shanti Devi realized what was happening to her, she was fortunately able to escape and take shelter in a Bangladeshi slum, where an Indian nongovernmental organization (NGO) was able to help her return to Bangladesh (Paul & Hasnath, 2000). Shanti Devi's case demonstrates a stark and frightening difference between historical slavery and new slavery: the affordability of slaves. In 1850, an average field laborer in the United States cost between $1,000 and $1,200. Adjusted for today's values, that would be about $40,000. But Shanti Devi was sold for only $60, and in some parts of the world, slaves can be bought for as little as $10.

Although some women, like the teenage girl in the United States, are abducted and forced into slavery, many others, like Shanti Devi and Hilda Dos Santos, are coerced into the clutches of human trafficking by the desperation of their economic circumstances. The problem of human trafficking, then, is a problem that reflects wider problems of gender inequality and inequality between the developing and the developed world. Through the booming business of sex tourism, those with the wealth to do so can increasingly travel to parts of the global South to participate in the system of modern slavery, but regardless of the particular route, the human trafficking system is driven by the desires of those with wealth, whether that be to satisfy themselves sexually or to have someone to clean their house for them. The issue of human trafficking is therefore a problem of both the very wealthy and the very poor. It is an issue that has received increasing amounts of attention in recent years, including an antitrafficking law passed by the U.S. Congress in 2000 and the appointment of a Special Rapporteur (Reporter) on human trafficking by the United Nations in 2004 (Bales, 2004). But the trafficking of women is also a new manifestation of a very old problem—the systematic use of coercive power against women. It is a new manifestation of the existing system of gender inequalities in global context.

GENDER RIGHTS AND HUMAN RIGHTS?

Human trafficking as a moral issue seems fairly cut and dry; it's difficult to imagine someone articulating a moral defense of some of the cases we've discussed. Forcing women to work in brothels or prostitution rings and to marry against their will seem to be clear-cut violations of basic human rights. But just as Sanday (2004) pointed us to the importance of cultural context in establishing the level of societal acceptance for rape, there are cultural and ideological variations in the extent to which certain behaviors are perceived as morally wrong. Prostitution itself is but one example. The women we discussed had no choice at all about being forced into lives of prostitution, but for some women, the extent to which the decision to become a sex worker is a choice becomes less clear. On one side of this debate lies the argument that all prostitution is a form of slavery and that prostitution in all its forms should be eliminated (Barry, 1995). But the other side argues that it's only forced or nonconsensual prostitution that should be made illegal and therefore perceived as immoral. From this perspective, it is not prostitution itself that demeans women but the outcast status that sex workers are given by the outside world. Setting aside this outcast status, prostitution could in some circumstances serve as a means for women to gain economic independence from men (Davidson, 2002; Kilvington, Day, & Ward, 2001; Songue, 1996). Only by removing that outcast status can the human rights of sex workers be fully protected.

This particular argument hinges on deciding exactly what we mean by *choice*. Those who argue that prostitution is wrong in every form it takes argue that the decision on the part of women to sell sex is never really a choice; women are *forced* to become prostitutes through violence but also because of economic pressures. The woman who becomes a prostitute to support her family financially, send her children to college, or buy a house is not really making a choice; she is being forced into prostitution by the lack of other economic avenues open to her as a woman. In a world where economic opportunities are equally provided to men and women, no woman would choose to sell her body, and prostitution would therefore not exist.

These debates become especially important when they help dictate the approach that governments and other organizations take to dealing with prostitution. Most of the prosecution for prostitution that takes place in countries where it illegally targets the women who provide sex rather than the men who take advantage of their services. This supports the argument that it is the outcast status of prostitution that makes the institution morally wrong; were sex work legal, the women themselves would not have to pay the price by being arrested. Different countries have developed varied ways of dealing with prostitution based on these opposing views. In Sweden, the purchase of sex is illegal, but the sale of sex is not (Bales, 2004), based on the assumption that the women providing sex are the vulnerable population who need to be protected, rather than the men exploiting their vulnerability. Germany and Holland, on the other hand, have legalized prostitution and instituted programs to inspect brothels, ensuring healthy and safe working conditions for prostitutes.

Hijab and Ethnocentrism

Prostitution is just one example of a larger debate about what exactly we mean by human rights and how we should go about defining what those are. Should women have the right to sell sex for a living, or should they have the right never even to face the possibility of selling sex for a living? The idea of human rights exists as a direct counter to the kind of coercive power we've been discussing thus far in this chapter. Many of the basic human rights that have been established by governments and the United Nations focus on the right to be free from coercive forms of power. But conversations about human rights can lead us down the hazardous path of ethnocentrism and into a tangle of decisions about who gets to decide what is right and wrong.

Ethnocentrism, as we discussed in Chapter 5, is the tendency to use the lens of your own culture to judge someone else's culture. When people in the United States began to turn their attention to Afghanistan and the Taliban in the wake of the terrorist attacks of September 11th, 2001, many people were shocked by the sight of Afghan women wearing **burkas**, a tent-like version of hijab that covers the woman's entire body, leaving a tiny transparent bit of cloth for women to see out of and negotiate the world. Insisting that women wearing burkas in public was strange, cruel, and wrong from our perspective, when the United States invaded Afghanistan, freeing many women from their burkas, it was considered a sign of clear moral victory. We liberated these women from this particular form of oppression.

Theory Alert!
Intersectional

But this position can be perceived as a good example of an ethnocentric view, and it is a manifestation of a larger struggle over the issue of hijab in general. As we discussed **hijab** briefly in Chapter 5, this belief originates from the Prophet Muhammad's directions toward Muslim women to "lower their gaze and be modest, and to display of their adornment only that which is appropriate, and to draw their veils over their bosoms" (Brooks, 1995, p. 21). Different Islamic cultures have interpreted this line from the Koran in different ways. Some assume this merely means that women should dress within the contemporary limits of modesty. Many Algerian women in France interpret this to mean that women must keep their head covered in public, and they therefore wear headscarves. In Saudi Arabia, women wear black cloaks that completely cover their bodies, while in Palestine and Egypt, some women wear floor-length coats with white headscarves. Other Palestinians and Egyptians wear calf-length skirts and head scarves. Muslim women in the United States sometimes wear headscarves, but they also often wear Western clothing that they consider appropriately modest. It's safe to say that the ways in which hijab is interpreted and enforced in the Islamic world varies a great deal.

From the perspective of many outside the Islamic world, forcing women to cover themselves completely, whether in the blue burkas of Afghanistan or the black hijab of Saudi Arabia, is a clear form of oppression. Yet many women in Islamic countries defend these practices and argue that depriving them of the ability to cover themselves is, in fact, the clear example of oppression. Their reasons for defending the practice of hijab vary, from a defense of their deeply held beliefs about gender differences to perceiving hijab to be a means of critiquing the colonialist domination of the Middle East by the West (Read & Bartokowski, 2005). From this latter perspective, hijab is a way to challenge the imperialism of Western women's dress and the ways in which Western fashion turns women's beauty into "a product of capitalism to be bought and sold" (Brooks, 1995, p. 25). Examined in this way, women in places like Iran who wear hijab are kin to militant feminists in Anglo-European society like Andrea Dworkin, who wore denim overalls to protest the exact same tendencies of Western capitalism. On a personal level, many women in the Islamic world appreciate the way hijab frees them from the exact kinds of harassment and threats of violence that were experienced by women in South Mountain Park in the United States.

Can you think of examples of practices related to gender in your own culture that might seem oppressive from an outside perspective? How would you go about explaining or defending these practices to someone from outside your own culture? Or would you defend these practices?

The dangers of ethnocentrism become clear when we look at the case of hijab. When we judge Islamic cultures by our own standards, being forced to cover your entire body looks weird and definitely wrong. But many women in the Islamic world mirror back their own ethnocentric perspective, in which what is weird and definitely wrong is the way in which Anglo-European women are objectified and made into moving targets in public space. Which perspective is the "right" one?

Ethnocentrism, although it is often used as such, is not necessarily identifying an intentionally malicious way of thinking; rather, an ethnocentric viewpoint is the natural product of successful socialization. When you learn how to be a member of your own culture, it's something of a given that you'll feel that the way things are done in your own society is the way things should be done. This doesn't mean that we can't be aware of our ethnocentric tendencies and begin to move in another direction. **Cultural relativism** is a perspective that encourages us to view other cultures through their own lens rather than through ours. Rather than asking just what *we* think about the practice of hijab, we should also be sure to examine what the people within that culture think about the practice. This doesn't necessarily give us a way to say who's right and who's wrong, but it certainly gives us a better and more complete wealth of information from which to draw conclusions.

The dilemma of ethnocentrism is one that's especially important to a discussion of gender equality and power. Feminism and other discourses about women's rights are largely products of the Western world and Western modes of thinking, and they are therefore prone to imposing their own views on what gender equality means. In Chapter 2, we discussed the ways in which feminism itself is sometimes perceived as a version of Western imperialism being imposed on people of the global South. The power dynamic that exists between the developing and the developed world is one that involves coercive power at times, as when countries like the United States invade other countries to bring about democratic change. At other times, the United States and international organizations speak from a position of authority granted them by the global economic system, using a form of institutionalized power. Many people argue that the prosperity and democratic institutions of countries in the Western world give them a kind of moral authority to dictate what should happen in the developing world. But power that comes from authority is still a kind of power, and it can still be resented by the people whose behaviors are the targets of change.

INSTITUTIONAL POWER: NATIONS AND GENDER

In the modern world, an important source of institutional power is the **nation-state**, a concept that links the nation as a sense of identity and community with the state as a political and geographic entity. The **state** is what we usually think of as government, the set of institutions that set laws and govern life within a set of geographical boundaries. The **nation** is akin to ethnicity in that it defines the cultural ties that unite (or attempt to unite) the group of people who may or may not reside within the borders of the state. In the United States, we live in a nation-state because we generally feel some sense of what it means to be American as a group of people and there is a political entity that governs that specific group of people called Americans. Many Palestinians have a sense of nationalism—that they constitute a nation of people—but they have no state that organizes those people. Jewish people saw their sense of nationality become a state with the founding of Israel as a country. This distinction between nations and states is important to make because both have gendered implications on the lives of people and exemplify how power works through institutions.

Earlier in this chapter, we explored how, for many gender scholars, the very construction of masculinity as a gender identity contributes to the equation of violence and masculinity. In this way, we examined the connection between masculinity and violence at an individual, psychological level. But gender scholars also point to the ways in which the militarism of nation-states also has important connections to ideas about masculinity. For much of the history of many countries, the ability to belong to the nation was a male privilege; the full rights of citizenship were denied to women in most nation-states up until the beginning of the 20th century (Hartstock, 1984). If for much of their history nation-states were composed primarily of men, it makes sense to argue that nation-states are by nature masculine entities.

My Missile's Bigger Than Yours

For evidence of this assertion that nations are inherently masculine, many scholars point out that much of what we call *politics* involves a great deal of militarism. President George W. Bush was the latest manifestation in a long line of U.S. presidents who combined the language and imagery of masculinity with that of statesmanship (Hartstock, 1984; Kimmel, 2006). Using his Texas background and Crawford ranch to cast himself as a new era cowboy (his wife described him as a "windshield cowboy" because he surveys his ranch largely from behind the windshield of a truck rather than on a horse), Bush is a clear example of how politicians use masculinity to construct an identity. In the wake of September 11th, Bush evoked the language of the Old West, promising of the terrorists to "smoke 'em out of their holes" and answering questions about whether he wanted to see Osama Bin Laden dead with references to Old West wanted posters reading, "Wanted: Dead or Alive" (Coe, Domke, Bagley, Cunningham, & Van Leuven, 2007).

Bush later became the first president in the history of the United States to don military attire (during his presidency) in his famous appearance on an aircraft carrier behind the banner, "Mission Accomplished." Some critics called Bush's appearance in military flight fatigues, like a scene out of the movie *Top Gun,* "a masculine drag performance," and critics included those who had served in the military themselves. General Wesley Clark described Bush as "prancing on the deck of an aircraft carrier" (Kimmel, 2006, p. 250). The military interventions of the Bush administration were clearly posed in masculine terms when one politician said, "Anyone can go to Baghdad. Real men go to Tehran" (Kimmel, 2006, p. 250); this reflected the belief that the resistance encountered in invading Iran (which was discussed as a military possibility) would be much greater than that encountered in the invasion of Iraq.

But there is nothing unique about President Bush or his particular administration in regard to their language of masculinity and statesmanship. Militarism and masculinity have a long history. In 1952, Edward Teller sent a telegram to his friends after testing a nuclear bomb with 1,000 times the power of the bomb that had destroyed Hiroshima. The telegram read, "It's a boy" (Kimmel, 2008, p. 323). Advisers close to President Lyndon Johnson argued that Johnson was driven to escalate the war in Vietnam in part by a desire not to be pushed around by the Vietnamese and to gain the respect of "real men," whom he imagined to be those men most hawkish about the war. President Johnson responded to those who criticized the war effort with attacks on their masculinity, and he said of one member of his administration who was becoming dovish (in favor of ending the war) on Vietnam, "Hell, he has to squat to piss" (Kimmel, 2008). President Jimmy Carter was perceived as less masculine by one academic security affairs specialist who was quoted as saying that, "under Jimmy Carter the United States is spreading its legs for the Soviet Union" (Cohn, 2004). Even at the level of international organizations, the Third Committee of the United Nations General Assembly—which deals with social, humanitarian, and cultural issues—is referred to as the "ladies' committee" (Cohn & Enloe, 2003, p. 1189) in comparison with the more masculine U.N. Security Council.

Theory Alert! Gendered Organizations Theory

Carol Cohn immersed herself in the world of North American nuclear defense intellectuals and security affairs analysts, a community that was composed almost entirely of white men and was also characterized by deeply masculinized styles of interaction and discourse (Cohn, 2004). In her research, Cohn identified the specific gender discourse that is used within this community—a community that essentially helps to underwrite American national security policy and therefore has important influence on decisions made by the U.S. government. Many feminists have pointed to the "missile envy" that seems to be evident in the language with which military personnel and government officials discuss nuclear and other weapons, and how this dynamic dominated much of cold war discourse, with its competition over who had the most and bigger bombs.

Cohn was surprised at the extent to which defense intellectuals used the language of **missile envy,** using the language of sex and anatomy to describe the weapons and their effects. The dependence of the United States on nuclear weapons was described as irresistible because you "get more bang for the buck" (Cohn, 1987, p. 693). Lectures by defense intellectuals were peppered with discussions of vertical erector launchers, thrust-to-weight ratios, soft lay downs, deep penetrations, and language that described "releasing 70 to 80 percent of our megatonnage in one orgasmic whump" (Cohn, 1987, p. 693). Cohn argued that this particular language was used because it helped these men defuse and downplay the ultimate effects of the weapons that they were discussing. The defense intellectuals often talked about being able to "pat" the weapons, meaning literally to touch a nuclear warhead, bomb, or military aircraft. In this way, these dangerously lethal devices were reduced to the same level as other objects that are patted—babies and pets. In describing the arms race as a boys' pissing contest, these groups were able to distance themselves from the life-and-death consequences of their conversations and their implications.

In a similar vein, Cohn (2004) found that gender discourse was also important to these conversations in the things that were left out and *not* said. Certain styles of communication and discussion were perceived as femininely gendered and, therefore, against the norm. She gave the example of one physicist discussing the number of immediate fatalities from a counterforce attack. A remodeled attack the group was working on had reduced the number of projected fatalities from 36 million to 30 million, and everyone was commenting on how great the reduction to 30 million was. The physicist telling the story described suddenly hearing what they were saying and blurting out, "Wait, I've just heard how we're talking—Only thirty million! Only thirty million human beings killed instantly?" As he described, "Silence fell upon the room. Nobody said a word. They didn't even look at me. It was awful. I felt like a woman" (Cohn, 2004, pp. 397–398).

In this story, the physicist violated the community norms by bringing attention to the bloody consequences of their work. Although these defense intellectuals may raise these questions outside of their professional lives, it is understood that they should not be discussed within professional settings; taking them into consideration on a daily basis would make their jobs impossible. But the physicist in the story violated a gender discourse in that he blurted out his statement, demonstrating impulsiveness, emotion, and a concrete attention to human life. He was not only behaving unprofessionally, but in a *feminine* manner, as the physicist himself identified when he said he "felt like a woman." In this way, Cohn (2004) argued, the gender discourse of these settings was about more than the gender composition or dispositions of the people in the room. The gender discourse dictated the types of things that could and could not be said and the degree to which things that were said would and would not be heard. When the physicist used a brand of communication deemed as feminine, his statements were discounted and, so, were effectively not heard. In this way, the community of defense intellectuals exhibits masculinity not merely in the fact that they are constituted primarily of white males, but also in the ways in which certain forms of expression and topics are deemed masculine (appropriate) or feminine (inappropriate).

Given this gender dynamic among defense intellectuals, what would be the implications for a woman working in these settings? What obstacles might she face, and how might she deal with them? Would a woman necessarily bring a different perspective to this setting?

Gender and Political Institutions

Cohn's (2004) research on gender discourses among defense intellectuals begins to suggest what it means to examine power that is institutionalized. Because the military is often perceived to be a masculine institution, it makes sense that those dealing with military matters would assume a masculine approach to those issues. The exercise of coercive power is forceful and difficult to deny; few people would argue that sexual assault, rape, harassment, or trafficking in women are morally defensible acts, and the stories of women who fall victim to these injustices are powerful. Some may argue with the ways in which the threat of violence or harassment serves as a means to control women and therefore to maintain male dominance, but it is hard to deny the negative effects on victims of coercive power. Institutionalized power is different in that it is a more subtle form of power, and it is therefore sometimes more difficult to identify in its workings, as well as its effects. Institutionalized power can result in its own forms of violence, although that violence may look very different from that resulting from coercive power.

An examination of institutionalized power leads us to an interesting series of questions about the relationship between gender and political institutions. Many gender scholars argue that the increasing numbers of women in the military have so far failed to transform the predominantly masculine culture and style of that institution. In other words, women have largely failed as yet to degender the military, assuming we have some sense of what a degendered military would look like. What would the military be like as an institution if it were gender neutral rather than masculine? What would a feminine army look like, and how would a feminine war be waged, or is the very idea of a feminine war a contradiction in terms? Would a military that represented feminine values simply cease to exist, along with war and combat in general? These are the kinds of questions that have often been posed by feminists and gender scholars under the general category, "If women ruled the world . . . ," or at the very least, if women had a more equal share in ruling the world than they currently do. When women enter into politics and other positions of power, do they necessarily alter the dynamics of those institutions, or are they merely forced to conform to the existing status quo?

Men and Women in Office

Much of the U.S. history that is taught in high schools and colleges is the history of American politics, what historians call *political history*. Political history focuses on who was elected to office, the decisions they made, the wars they fought, the peace they made, and the general happenings in the world of what political scientists sometimes call the political elite. The **political elite** are "the small group of people who hold high-level positions of power and responsibility in a government" (Conway, Steuernagel, & Ahern, 2005, p. 117). For most of the history of the United States, the political elite were all male, and so much of the history that we learn is also about men. For these reasons, when high-school students in the United States are asked to name 10 famous American women, most can do so only by adding "Mrs." to the last names of all the Presidents they know (i.e., "Mrs. Jefferson, Mrs. Washington, Mrs. Adams"); on average, students can generate 11 male names but only 3 female names (without using the "Mrs." strategy) (Sadker & Sadker, 2009). A history of the United States that focuses solely on the political elite is, therefore, a history of men—which is still not to say gender isn't an important part of that history. Nonetheless, the gender implications of political history as a history of masculinity go largely unexplored.

When the definition of political activity is expanded beyond membership in the small group of political elite, there is more evidence of women's important contributions to the history of the United States. In recent years, scholars have turned their attention to women's roles in social movements and civic organizations, groups that can also be important political actors. In this arena, women were much more active than in the capacity of elected officials. For example, 40,000 women joined to form the Women's Peace Party between 1914 and 1915, while thousands of women were involved in the suffrage (the movement for women's right to vote), settlement house (a movement focused on urban poverty), and temperance movements (movements to make alcohol illegal) of the 19th century (Githens, 2003). In the 20th century, women worked in the public arena to secure reproductive freedom and equal rights, as well as being active in peace movements and movements for the expansion of social welfare programs. Other women became involved with the International Workers of the World (IWW), or the wobblies, organizing for workers' rights. In other parts of the world, women were involved with the overthrow of the aristocracy in France, the bread riots that helped spark the Czarist revolution in Russia, and the Algerian war for independence (Githens, 2003).

These examples demonstrate that women have had an important part to play in politics historically. But social movements are social organizations that often work outside of the established world of official politics, and especially so in their early stages. Social movements happen when actors and groups in a society find that their interests are not served by the workings of established politics. The African American civil rights movement in the United States beginning in the 1950s came about because the government, at both the federal and local levels, was not taking any action to ensure the rights of Black citizens. As Martin Luther King, Jr. articulated, drawing on the philosophy of Mahatma Gandhi, when a government becomes unresponsive to the needs of its population, civil disobedience is a valid means of working toward social movement goals. In other words, when the laws of the land are unjust, the only strategy available to a group may be to disobey those laws. Thus, women's involvement in social movements, rather than in existing political structures, is evidence for the fact that women's access to these structures has historically been restricted, a situation that is only slowly beginning to change around the world.

As of 2009, there were 15 women serving as heads of state or heads of government, in countries as diverse as Switzerland, Mozambique, the Phillipines, Ireland, Liberia, Chile, Haiti, Bangladesh, and Germany. However, these 15 countries represent only 8% of the roughly 195 countries that exist worldwide, putting the percentage of women serving in the highest reaches of government far below a number that represents their total population. At a slightly lower level of government, women hold slightly less than 18% of all parliamentary seats worldwide (Inter-Parliamentary Union, 2010a). This average disguises a great deal of variation that doesn't always conform to standard expectations. Some established democracies, such as the United Kingdom and Canada, that might be expected to have higher percentages of women in parliament, actually have 20% and 22%, respectively (Inter-Parliamentary Union, 2010b). This compares to Sweden, South Africa, and Cuba with 46%, 45%, and 43% female legislators, respectively. In some countries, the amount of time that lapses between when women are given the right to vote and become eligible to run for election and their actual election or appointment to an office is measured in decades. In Australia, women earned the right to vote in 1902 and were eligible to be elected to office in the same year, but the first woman was not elected until 1943. Whereas in Ireland, women received the right to vote, became eligible for office, and the first woman was elected all in the same year: 1918. Rwanda is the only country at present in which the percentage of women in parliament exceeds 50%, at 56% (Inter-Parliamentary Union, 2010b).

Women have made significant gains in their representation among the political elite, although this progress has been unsteady at times, as we will discuss. Looking at women and men in political office raises two interesting questions about gender and power. First, how do we explain the fact that in many countries around the world, women's right to vote and be elected to office does not translate into equal representation in their national governments? What are the barriers that women face when they seek elected office? The second question is one that gender scholars could only speculate about until fairly recently: How would it matter to have more women in positions of political power?

Political scientists call this *substantive representation*, as opposed to *descriptive or numerical representation*. **Descriptive or numerical representation** is what we have been discussing—namely, the number of members of a particular group (here women) who hold office. **Substantive representation** is the representation of group interests, or the extent to which the members of the particular group represent the interests of that group when they are elected (Carroll, 2001). If descriptive or numerical representation is not connected to substantive representation, then there's not much point in keeping track of how many women are elected to government positions. But if electing women to positions of power does make a difference, what kind of differences does it make? We'll start with the first question about why women are still underrepresented in the halls of power, and then we'll turn to the question of how women's presence in those halls has the possibility to change them.

The Smoke-Filled Room: Descriptive Representation

Legend has it in the history of U.S. politics that a group of powerful Republican senators gathered in a room at the Blackstone Hotel in Chicago in 1920 and emerged with a cloud of smoke and the consensus that Warren G. Harding would be the next Republican nominee for president (Thale, 2005). Regardless of what *really* happened in that room, the Associated Press reported that Harding had been chosen as the nominee in a "smoke-filled room" (Thale, 2005), and since then, the phrase has come to stand as shorthand for a behind-the-scenes location where political party bosses secretly conspire to choose candidates. The image of the smoke-filled room captures the ways in which entrance into the world of politics is about much more than electability or the ability to do the job successfully. The smoke-filled room resonates with images of all-male spaces, like social clubs and bars, and it suggests the ways in which politics is also much like a social club. As the late New Jersey state legislator, Millicent Fenwick, put it, "Women are on the outside when the door to the smoke-filled room is closed" (Conway et al., 2005, p. 106). There is a whole series of gatekeepers who patrol access to the political realm, long before voters ever enter into the mix. Examining how gender affects who gets past these particular gates is one part of answering our first question about the descriptive or numerical representation of women.

In Chapter 9, we explored the historical division of the world into the public and private spheres in Anglo-European society that began roughly with industrialization in the 19th century. According to the doctrine of separate spheres, women belong in the home, taking care of children and providing a nurturing refuge for their family from the rational, dog-eat-dog world of the public sphere. The public sphere is the realm of men, and although it may be dog-eat-dog, it is also where the real and important actions of the world take place. The world of politics as it has traditionally been conceived of is clearly located in the public sphere. By this argument, politics is already by its very nature masculine and, therefore, better left to the male half of the population. The doctrine of separate spheres was based, as we discussed in Chapter 9, on a division that never applied to the majority of the population in the Anglo-European world and that certainly doesn't apply to the lives of most people around the world today. Nonetheless, the division of the world into a public and private sphere remains an important ideology shaping the way in which people around the world think about gender and politics.

Can you think of other ways in which politics as an institution is inherently masculine? How would politics be different if this were not true? What would a more feminine politics look like? A genderless politics? Are such things possible?

For example, in Peru, surveys demonstrate a strong level of support for women's participation in politics. This support includes quota measures for Peruvian political parties, which would dictate that a certain percentage (30% in 2000) of spots on "party lists" be set aside for female candidates. This system would be roughly equivalent to the Republican and Democratic parties in the United States being required to nominate women for a certain percentage of offices at the state and national level. In a 1998 survey, 75.6% of the Peruvian population supported the gender quotas, while in 2000, the percentage increased to 80.9% (Rousseau, 2005). These statistics demonstrate both that many Peruvians feel the participation of women in the political process is valuable and that they support state intervention (in the form of political party quotas) to ensure that participation. As further evidence for this support, in 2001, a Peruvian woman (Lourdes Flores Nano) came very close to winning the national presidential election, and she would have been the first woman to have done so in Peru without the benefit of family connections to a famous male politician.

This strong support for women's roles in politics in Peru is especially interesting because these values do not challenge the basic ideology of the public and private sphere. In the Peruvian political context, politicians are generally considered to be corrupt, authoritarian, lazy, and self-centered; because those politicians have also tended to be men, many Peruvians have come to believe that women are a positive alternative. One survey asked Peruvians to compare male and female mayors, and respondents noted that although male mayors work more and are more efficient, women as mayors were more honest, loyal, just, and sensitive to social and human issues (Rousseau, 2005). In this sense, women's mothering and caring characteristics are perceived as valuable and necessary to clean up the political sphere and provide leadership that is more trustworthy than that provided by men in the past. The divisions between the qualities of the feminine, private sphere are maintained, but they are considered to be valuable contributions to counterbalance some of the negative aspects of the public sphere. This particular dynamic resonates with arguments made by suffragists in the United States during the early part of the 20th century. Many women working to obtain the right to vote argued that women's unique moral superiority, as a result of their location in the caring and nurturing environment of the private sphere, would humanize politics in much the same way Peruvians today expect women to bring their own feminine perspective to the public sphere of politics.

In the Peruvian case, the barriers to women's political participation represented by the smoke-filled room are overcome partially by the use of government-mandated quotas, and by Peruvians' strong belief in the unique contributions women can bring to the political system. Other scholars who focus on understanding women's low rates of descriptive or numerical representation focus on the characteristics of women themselves. In essence, for a society to have a pool of potential women to be elected to political office, there must be a group of women who are willing to run and are capable of running for these offices. In this sense, all of the broader gender inequalities or gender ideologies of a society are crystallized at the level of political participation. If women in that society have lower levels of access to education, there will be fewer women with the means to run for and perform in an elected office. If women in a society have relatively little power in their families and marriages, there will be very few women who have the freedom within their families to decide to run for political office. These are some factors that are important to explaining political participation in many of the countries of sub-Saharan Africa.

In Anglo-European narratives of modernization and development, the introduction of democracy and capitalism is assumed to bring along a broad-scale liberalization of society, including improved gender equality. This was not true for the introduction of colonialism to many nations in the global South, and it is often no more true for the advent of democracy.

In many precolonial societies of sub-Saharan Africa, women exercised political and economic rights that did not translate into full equality with men but that did sometimes allow them to hold high political positions (Yoon, 2005). Women in precolonial societies often had control over economic resources, including access to land and the ability to control household budgets and to engage in trade. Under colonialism, women lost their political positions when colonial powers recognized only men as chiefs or intermediaries for administrating their organizational needs. When colonial administrators chose native Africans to train for low-level administrative positions or to receive education in European languages and other skills, they imposed their own European gender systems by choosing boys and men, rather than girls and women.

Women had little power in the colonial administrative system that was controlled largely by men, and laws imposed by colonial powers further eroded the economic and political powers women had once possessed. For example, colonial powers did away with communal land ownership in many African societies, and the individualization of property rights usually resulted in men receiving the titles. Customary laws imposed by colonial powers also treated women in sub-Saharan Africa by and large as minors under the authority of males. Although in some countries these laws are changing, there are still countries in this part of Africa where women are not entitled to own, inherit, or dispose of land, as well as being unable to receive financial credit legally (Yoon, 2001). Women cannot open a bank account without their husband's permission in the Democratic Republic of Congo, and in Botswana, a woman cannot have legal guardianship of her own children. In many of these countries, women's rights to property ownership may be protected by official laws, but they are not recognized by local customs (Benschop, 2004).

In many ways, then, the exposure to European gender ideology through colonization increased the levels of gender inequality in sub-Saharan Africa and left women legally deprived of many basic rights. Women were involved in the struggles against colonial powers in many countries, and after decolonization, women in most parts of sub-Saharan Africa were granted the right to vote and hold office at the same time that their nation gained its independence from colonial powers. However, until the 1990s, many of the nations of sub-Saharan Africa were characterized by single-party, authoritarian rule. In places like the Democratic Republic of Congo (formerly Zaire), strongmen like Joseph Mobutu ruled as dictators, outlawing the existence of any political parties besides their own and turning many African nations into effective tyrannies. Beginning in the 1990s, more than 40 African countries began the transition to a multiparty democratic system and began holding presidential and parliamentary elections with candidates from more than a single party. The process of democratization, not without its complications and setbacks, began for many African countries in earnest during this period. The question this transition raises for many gender scholars is whether the introduction of democracy increases women's numerical and descriptive representation in those nations and, ultimately, whether that translates into substantive representation.

On the one hand, scholars argue that democratization creates a more favorable environment for women to become more active in the public and, therefore, political sphere, thus, potentially increasing descriptive representation. Democracy should also encourage political parties to include more female candidates and to give more weight to women's issues.

This argument takes the basic position that democracy creates more space for women's political involvement and for the articulation of concerns related to gender inequality than exists in authoritarian regimes. From this perspective, women are just simply better off under a democracy. With the breakup of the Soviet Union and the collapse of many of the authoritarian regimes that existed under communist control in Latin American and Eastern and Central European nations, scholars had a chance to test these assumptions empirically.

Research often revealed that women's representation in governing bodies actually decreased in the transition from authoritarianism to democracy. In many of these countries, the numerical representation of women in institutional politics has decreased relative to their numbers under authoritarian regimes, as in the case of Soviet legislatures that witnessed the exit of women from elected office under the process of democratization associated with *perestroika* (Yoon, 2001). In the former communist nation of Hungary, women's representation in parliament before the first democratic elections in 1990 was as high as 30%. Candidates under the communist system were nominated by the state party, and quotas were allocated to both certain regions and demographic groups, such as women. In the first democratic elections in Hungary, the percentage of women in parliament dropped to 7% and has averaged around 10% ever since (Kardos-Kaponyi, 2005).

In light of these outcomes, some researchers began to revise their theories about the relationship between democracy and gender equality in the political realm. Gender scholars argue that although democracy may provide more opportunities and freedom for women to pursue political power through elected office, the process of democratization does not necessarily change the underlying structures of society that might make women's entry into politics difficult. Namely, the introduction of democracy does not change the social and economic inequalities that already existed between women and men, nor does it magically remove the gender ideologies that are resistant to women holding positions of power (Yoon, 2001). These three factors—social inequality, economic inequality, and culture in the form of gender ideologies—are also important factors in determining the numerical representation of women in elected office, in addition to the particular structure various governments take in different nations. Where democracy leaves these areas of social life untouched, democratization is less likely to lead to increasing numbers of women gaining access to political power.

Social and economic inequality are two important barriers to women's political participation, but gender ideologies can also serve as critical roadblocks. The doctrine of separate spheres is one part of this gender ideology that exists in sub-Saharan Africa and around the world, where women's primary roles are as mothers and wives. These cultural barriers mean that women are less likely to view themselves as potential politicians, while voters are less likely to vote for women who are stepping outside of their place within the private sphere. One study in the United States found that women are less than half as likely as men to have considered running for political office (8% of women compared with 18% of men) (Duerst-Lahti, 1998). In the U.S. study, women were also significantly less confident about their chances of winning than men—and more likely to think of themselves as unqualified and incapable. These internalized notions on the part of potential women candidates themselves are in part the result of socialization and, specifically, the political socialization that occurs in different countries. **Political socialization** is "the process by which people learn what is expected of them in their particular political system" (Conway et al., 2005, p. 22). At this point, it's not surprising to realize that the political socialization experienced by women is very different from that experienced by men in many parts of the world. Women are socialized to be more passive than men and to direct their concerns toward things that are not usually thought of as political (e.g., marriage, having children, doing housework, etc.). In many societies, historically and still today, women are not socialized into a sense of **political efficacy**, or the sense that "what a person does really matters" (Conway et al., 2005, p. 23), and women are not made to believe that they can make a real difference in the political realm.

These aspects of culture prevent many women from running for office, but those women who do run for office often face very real manifestations of these cultural beliefs that women are not fit to be political leaders throughout their campaigns, in the form of violence, intimidation, ostracism, and verbal abuse (Yoon, 2005). One woman involved in Zambian politics in the 1950s described the ways in which party officials warned her husband that allowing his wife to be involved in politics would cause others in the community to label him as "no man" because of his inability to control his wife's activities. The female politician herself was "despised" by other women who accused her of making adultery easier by mixing so much with men and of breaking up her own home (Geisler, 1995, p. 549). These kinds of hostilities on the part of women and men are in part because the doctrine of separate spheres means that women who run for political office are more likely to be perceived as private individuals, rather than as public figures (Conway et al., 2005). If women's natural place is in the home caring for husbands and children, those women who do seek political power are perceived as unnatural, and their means for gaining power are considered to be dubious at best (Dixon, 1992). In the United States, Congresswoman Nancy Pelosi, in her role as Speaker of the House of Representatives, has been questioned by the media about her ability to lead as a woman. An anchor at MSNBC questioned whether Pelosi's "personal feelings were getting in the way of effective leadership," suggesting that this was not a problem that arises with "male-run leadership posts" and that men were "more capable of taking personality clashes" than women (as quoted in Boehlert & Foser, 2006).

Other news figures ignored Pelosi's leadership style altogether, focusing instead on her appearance and attire, quipping that Pelosi "dresses a lot better" (as quoted in Boehlert & Foser, 2006) than Senator Hillary Rodham Clinton. In her own campaign for New York senator, Clinton was held by the press to narrow definitions of gender roles, and she received negative press coverage when she failed to conform to those prescribed gender roles (Scharrer, 2002). Other studies of the media and women candidates have suggested that the media portray women as less viable candidates and focus their coverage more on "feminine" issues and traits than on general political issues (Kahn & Goldberg, 1991). Research on voter preferences for female presidential candidates suggests that, in experimental designs, respondents who are presented with two presidential candidates with equal qualifications still give higher ratings to male candidates in terms of "effectiveness" (Rosenwasser & Dean, 1989). This suggests that, in many places, the competence of women as presidential candidates is called into question more than it is for men and that women are therefore held to higher standards in their candidacies for president.

CULTURAL ARTIFACT 2: CLINTON AND PALIN—GENDER, POLITICS, AND THE MEDIA

The 2008 presidential campaign in the United States was groundbreaking in many ways. It saw a woman make a viable run for Democratic nominee for the president, a Republican woman in the vice presidential spot, and the first African American to be elected to the highest office in the land. Both the Democratic primary and the presidential campaign itself made gender and race the center of a national conversation in ways they had probably not been since the civil rights era. How did Senator Hillary Clinton and Governor Sarah Palin fare in regard to sexism? What do their campaigns reveal about the current status of women in politics? In the wake of the election, media observers noted that both Clinton and Palin faced a great deal of sexism in the media's treatment of their candidacy. T-shirts reading, "Iron my shirt," showed up at Clinton's rallies, Clinton nutcrackers were sold at airports, one media pundit said on air that he crossed his legs every time he saw Clinton (Diffendal, 2008), and cameras caught and aired footage of a Senator John McCain supporter asking Senator McCain, "How do we beat the bitch?" Senator McCain's response was to call the question "excellent," to point to a poll that showed him beating Clinton in a head-to-head matchup, and then to note his respect for Senator Clinton (Diffendal, 2008). These are just a partial list of some of the clear instances of sexism encountered by Clinton, which the media by and large ignored. Only one media organization documented the exchange between McCain and his supporter. Journalists instead participated in the sexist coverage themselves, rather than choosing to make the sexism a story (Diffendal, 2008). Pundits focused on Clinton's "cackle" and suggested that her emotional breakdown on a campaign stop was calculated to gain voter sympathy. They suggested that people would vote for Clinton only because they felt sorry for her, and at other times, they portrayed her as a "nagging wife" or a "scolding mother" (Cutlip, 2009). Governor Palin encountered a different kind of sexism in media attempts to sexualize her and question her ability to be a mother and vice president. Commentators referred to her as a MILF (Mother I'd Like to F***), and she was sexualized especially on the Internet, where her face was photoshopped onto the bodies of bikini models. Palin faced questions about how she would mother and serve as vice president, which were not put to Senator Barack Obama, who also has two small children. Many media observers suggested that if similarly racist comments had been made about Senator Obama, the media would have made these into headlines and lead stories. In the aftermath of the campaigns, the only media figure who spoke out about the continuing legacy of sexism was CBS News anchor Katie Couric. On the evening news, Couric noted that the real lesson to be gained from Clinton's failed candidacy was to draw attention to the "continued and accepted role of sexism in American life, particularly in the media" (Couric, 2008). Like other media observers, Couric noted that had Senator Obama faced similar instances of racism, rather than the sexism faced by Senator Clinton and Governor Palin, "the outrage would not be a footnote. It would be front page news" (Couric, 2008). Was outrage over the sexism in media coverage of these two female candidates largely a footnote? What does this say about the current status of female candidates? What does it say about how people today, including those in the media, think about sexism? Do many of us assume sexism is a thing of the past? Do people in the United States have a blind spot to gender that they don't have concerning racial issues? And what would intersectional theory say about these comparisons of responses to sexism versus racism?

Strangers in the Halls of Power: Substantive Representation

Women have often found themselves on the outside of the smoke-filled room and, therefore, barred initial entry through election into the halls of power. Our second question addresses what happens to women once they gain entry into these halls and what further doors remain open or closed. In 1991, the U.S. Senate was debating whether to pursue extensive questioning of Clarence Thomas, a nominee for the Supreme Court who had been accused of sexual harassment by a former colleague, Anita Hill. Two days after the accusations were made, it remained unclear whether the House and Senate would explore the charges in Thomas's confirmation hearings. Seven Democratic women from the House and Senate, accompanied by reporters, walked to the committee room in the Senate where the Democratic caucus was discussing how to proceed and knocked on the door. Initially, a senior congressional aide told the contingent that they would not be allowed in, but with some pressure from Barbara Boxer, a senator from California, and the possibility of the story showing up in the press, the group of women was able to meet with the senate majority leader to express their concerns. Another senator later explained to Boxer that the women were refused entry because strangers were not allowed in that room, *strangers* here meaning anyone who is not a senator. Boxer responded by noting that, "the truth is that women have been strangers in the Senate . . . strangers in the highest, most powerful legislative body in the world" (Conway et al., 2005, p. 107). The accusation of sexual harassment against Thomas was discussed as part of his confirmation hearings, but only after female politicians were forced to, as described by Boxer, "pound on a closed door, to have to beg to be heard on a crucial issue that couldn't really wait for niceties" (as quoted in Conway et al., 2005, p. 107). What does this story begin to suggest about the degree to which numerical representation for women translates into substantive representation in the halls of power around the globe?

The issues of numerical or descriptive representation and substantive representation are closely related. In the previous example, seven women from the Senate and House of Representatives formed a large and influential enough group to have their voices heard in a way that would have been more difficult for one or two female legislators. But the kind of political positions to which women gain access also matters. In Senegal, female legislators were able to push for the successful passage of legislation that outlawed the practice of female circumcision in 1998. But other pieces of legislation related to women's issues were held up by the minority status of women in the legislature and by women's poor representation in the parliamentary executive. Not surprisingly, the ability of women to pass legislation is aided by women occupying powerful positions within the governing body, whether that is vice presidents and elected secretaries in Senegal or Speaker of the House or chair of committees in the United States.

In South Africa, when both the Speaker and Deputy Speaker of the legislature were women, as were chairs of 10 parliamentary committees, a wide range of legislation catering to women's issues was made into law. This included the legalization of abortion, the criminalization of domestic violence, granting wives equal legal status with their husbands in marriage, and banning discrimination in the workplace—as well as instituting affirmative action plans (Yoon, 2005). In the United States, female representatives working on issues of women's health were greatly aided in the 103rd Congress by the presence of women members on the Appropriations' Labor, Health, and Human Services subcommittee. As one lobbyist involved in the legislation recalled, the "women on the Appropriations Committee . . . really went to bat for us. . . . It's very easy to get members of Congress to say they're for more money for breast cancer research. . . . But it's not so easy to find members of Congress who are willing to actually pick up the banner and go into battle on the issue" (Dodson, 1998, p. 139). Women in positions of political power find it easier to translate their interests into real results when certain doors of power are open to them.

Interviews with women in the House of Representatives in the United States seem to suggest that women who are elected to political office feel some responsibility to represent the interests of other women, even in the face of initial reluctance. Representative Marge Roukema, a Republican congresswoman from New Jersey, described that when she first came to Congress, she wanted to avoid being stereotyped as a "woman legislator" and hoped to work on things like banking and finance. But she "learned very quickly that if women like me in Congress were not going to attend to some of these family concerns . . . then they weren't going to be attended to" (Dodson, 1998, p. 133). Representative Patsy Mink from Hawaii also felt initially that her role was primarily to represent all of her constituents, women and men, and her home state. But with time, she realized her far greater role was also to "speak for all the women in America" (Dodson, 1998, p. 133). Roukema, Mink, and the testimonies of other congresswomen demonstrate their sense of the special role they play as women in political office. Although they may, like Roukema, initially wish to distance themselves from "women's" issues, they discover that if they as women do not focus on these issues, no one else will. These women acknowledge a unique perspective and unique responsibility they have as women.

In fact, research demonstrates that women in elected and appointed offices in the United States have distinctive policy priorities compared with their male colleagues. Women have focused far more than men on issues related to women, children, and the family (Dodson, 1991; Palley, 2001; Swers, 1998; Thomas, 1994, 2003; Thomas & Welch, 1991). Although these are not the only issues women have focused on, they are the issues around which women in office have tended to have the most success, as women have achieved passage of legislation in these areas at a faster rate than men (Thomas, 2003). Women may also bring a different kind of process to political decision making in general. Research in the United States suggests that women tend to view public policy problems more broadly, leading them to approach solutions to those problems in a different manner. For example, whereas male politicians are more likely to view criminal behavior as a simple matter of an individual breaking a law, women are more likely to look for the societal factors that might lead to criminal behavior (Thomas, 2003). Legislation crafted by women is therefore more likely to address the roots of problems than is that written by men.

These established gender differences lead many scholars to wonder whether leadership styles themselves are gendered. Debates about exactly how to define leadership rage across many disciplines, but many gender scholars have pointed out that many of the social settings that serve as exemplars of leadership are settings that are traditionally dominated by men (Eagly & Johnson, 1990; Gilligan, 1982; Rosener, 1990). Carol Gilligan (1982) famously used psychoanalytic theory, as discussed in Chapter 4, to argue that women as leaders tend to emphasize cooperation over conflict, inclusion over exclusion, and consensus over hierarchy. Remember psychoanalytic theory is a theory of gender socialization and argues that because of women's primary role in mothering, women develop a gender identity that is based on strong connection and empathy with others. In Freudian language, women develop weak ego boundaries, and this predisposes them to emphasize connection over competition and conflict. Gilligan applied these theories specifically to explain differences in the moral reasoning of women and men. Men tend to be more concerned about people interfering with each other's rights in moral dilemmas, and this masculine approach is commonly called an **ethic of justice**. Women, according to Gilligan, are more worried about not helping others when they could be helped, or with the possibility of omission; this feminine approach is referred to as an **ethic of care** (Kathlene, 1998). A great deal of research in political science has focused on proving or disproving Gilligan's theories about these two different ethics, and the differences in approaches to public policy problems as discussed here fit into this line of research. Women's ethic of care leads them to form a more contextual view of the world, emphasizing the complex web of interrelationships that connect people together and make them deeply interdependent. With this worldview, it makes sense that the policy solutions often pursued by women reflect this broader perspective on how people's lives are involved in causing social problems and how this complex web might be affected by any proposed solutions.

Some research on women in positions of political leadership confirms this theory of feminine leadership styles. Women chairs of legislative committees place more emphasis on getting the job done and accomplishing their tasks in a team-oriented way, rather than using unidirectional power, than do their male colleagues (Bell & Rosenthal, 2003; Rosenthal, 1998). Another study shows that in state legislatures, women tend to exhibit consensus styles of leadership, rather than the command-and-control styles more likely to be used by male legislatures (Whicker & Jewell, 1998). This research suggests that if the number of women in positions of political power increases over time, their presence does have the potential to transform the operations of institutional power in significant ways. But associating women with certain styles of leadership can also serve as a means to exclude them from certain areas of decision making.

In the United States, foreign policy is an area that has traditionally been considered off-limits for women, with the exceptions of Madeleine Albright, Condoleezza Rice, and more recently, Hillary Clinton (McGlen & Sarkees, 2001). Women are seldom interviewed on television, quoted in newspapers, or appointed to positions of authority in foreign policy areas, and this may be attributed in part to the image of women as peace makers. Using a consensus leadership style may be useful in legislative bodies, but both men and women may view the feminine ethic of care as incompatible with the job of protecting national interests. Perceptions of women as exercising this more feminine leadership style may inhibit their access to these crucial global leadership roles. In addition, the debate about whether women truly do bring an ethic of care with them into the political world is ongoing and hardly resolved. Studies have found that women use both styles of moral reasoning—an ethic of care and an ethic of justice (Plutzer, 2000)—while others have pointed to the importance of context in determining leadership styles. For example, legislative and executive leadership are generally viewed differently, and in a distinctly gendered light (Tolleson-Rinehart, 2001). Legislative leadership is perceived as feminine, while executive leadership is perceived as masculine, and women who serve in executive positions (as governors, judges, cabinet secretaries, and in the military) certainly experience pressure to conform to the leadership style associated with that particular institution and position.

Many women who gain positions of political power certainly do feel an obligation to represent the voices of women, and although they face many obstacles in pursuing these issues, women's access to positions of institutional power within governments does seem to make a difference. As we have discussed, the effectiveness of women in politics is often limited by a variety of factors, sometimes including disagreement among women themselves about exactly what issues and agendas they should be pursuing. At this point, we should know to mistrust any quick and easy way in which to formulate a set of programs that covers the issues faced by many different types of women. In the United States, women in Congress find it easier to reach agreement on issues related to women's health, but they are sometimes less likely to form a coherent front on issues of abortion or health-care reform in general because party lines can dictate different positions on these issues (Dodson, 1998). One study revealed that women in Congress in the United States are less likely to vote together on foreign aid, budgetary, and defense issues (Clark, 1998).

In Senegal, women members of parliament experience divisions between those who are from polygamous families and those who are not (Yoon, 2005). Research on African American women in the Congress demonstrates that compared with their white female counterparts, Black female representatives stated stronger support for policies geared toward women's interests, such as state-mandated comparable worth policies, laws forbidding landlords from discriminating against families with children, and publicly funded parental leave (Barrett, 2001). This evidence reminds us that any two women could take completely opposite viewpoints on an issue or policy and both firmly believe that they are representing women's interests. Research also seems to suggest that as women's numerical representation increases, their substantive representation increases as well. When women find their way into political positions of power, they can have a real effect on gender inequality.

Recently, some gender scholars have begun to wonder whether focusing attention solely on elected positions in an examination of political power provides the fullest picture of women's roles in politics. **State feminism** refers to all the activities of governments that are officially charged with furthering women's status and rights (Stetson & Mazur, 1995). These may or may not involve elected officials and can include many different types of activities that still have important consequences for the lives of women in these countries. Gender scholars who study state feminism argue that the formation of structures within governments whose main purpose is to better women's social status is an equally important consequence of women's movements and can have important effects on gender equality in those nations (Stetson & Mazur, 1995). From this perspective, there are other ways to affect gender policies than merely electing women to political office.

SUMMING UP

The election of Barrack Obama as the first African American president in 2008 was a history-making event, as was the campaign for Democratic presidential nominee of Hillary Clinton and Sarah Palin's nomination as vice presidential candidate with John McCain.

A less well-known landmark happened in the small town of Silverton, Oregon, which elected the first ever openly transgender mayor (Murphy, 2008). Stu Rasmussen had previously served as mayor of Silverton in 1990 and 1998, but that was before he had breast implants and began wearing skirts, lipstick, and high heels, calling himself Carla Fong. Rasmussen won the vote in Silverton by a margin of 52% to 39% against an incumbent mayor and won residents' votes with his promise to halt the rapid development that was threatening Silverton's small town charm. That Stu also happens to have breasts and a longtime girlfriend was largely irrelevant to the voters in Silverton. Outside of Silverton, some groups like the Westboro Baptist Church of Topeka, Kansas, have been less sympathetic to Stu's situation, as church members planned to picket downtown Silverton to protest the "60-year-old pervert" (Murphy, 2008). But Rasmussen's election to mayor may be a sign of things to come and a logical outcome of the lesbian, gay, bisexual, and transgender (LGBT) movements. Democratic Congressman Barney Frank became the first openly gay representative in the United States when he came out in 1987 and continues to be a powerful political figure, while the recent film biography *Milk* brought attention to the career of Harvey Milk, the first openly gay man to be elected to public office in California in 1977. Outside of elected office, LGBT social movement organizations have worked at the state level to pass laws that would make discrimination against gays, lesbians, bisexuals and transgender individuals illegal.

When we focus our conversations on simple divisions between women and men, we ignore the complex ways in which masculinity and femininity are expressed. Where is the space in our political systems for those who do not fit into our assumed categories of sexual dimorphism? What happens when we move beyond the simple boundaries of women and men? How can we redistribute power in our society so that all forms of gender inequality are taken into consideration, and not just the power differential between those people we call women and those whom we call men? Remember we began this chapter with a consideration of whether the distribution of power is an essential and unavoidable aspect of any gender system. If this is the case, then perhaps those individuals who seek to escape the bounds of gender as a category can help lead us on the path to understanding how to undo gender as a system of inequality, or at least how to lessen its connection to power. Perhaps this move to positions that exist outside the traditional division of sex categories and gender can give us a clearer vantage point from which to understand the workings of power and gender.

BIG QUESTIONS

- How would you answer the questions raised in this chapter about the connections between gender and power? Do you agree that one of the main purposes of gender as a system is to distribute power to men? What evidence would you provide for your particular position? What are the implications if gender is inextricably linked to power? What does this imply for the connections we've discussed between inequality and difference?

- Which particular theories that we've discussed in this book fit in well with the perspective that one of the main purposes of gender is to distribute power? Think about the ways in which all these theories define what gender is and which seem to include a concern with gender and power. Which theories seem less compatible with this perspective?

- In this chapter, we discussed the characteristics of rape-prone and rape-free cultures. Do you live in a rape-free or a rape-prone society, based on the descriptions provided by Sanday (2004)? Which set of characteristics better describes the culture in which you live? Does this fit with Sanday's predictions in terms of the prevalence of rape in your particular culture?
- In this chapter, we raised questions about ethnocentrism and human rights and about the dangers of using our own cultural lens to judge the practices of other cultures. Is it possible to overcome this problem by developing a universal system of morality, or way of defining what is right and wrong? How would you go about doing that? What have you learned about what the potential dangers might be of developing one universal definition of morality?
- Some women in elected office experience a pressure to represent women and women's issues in their positions. Do you think men experience similar kinds of pressures? What are the potential effects of these pressures on women in elected office, and how might it limit the power of women in elected office? What are other situations in which people from a certain social group might experience pressure to represent that group and their interests?
- What would the world look like if the power differential between women and men were reversed, and women occupied most of the most powerful positions in society? Would the world be different? Is there something essentially different about women that would make a world in which they had more power different, or would having power make women act differently? In other words, is it about the power or about gender? How would your own life be different in this new world?
- What would the world look like if gender itself disappeared? How would power be distributed in a genderless world? Would some kinds of power be more or less important (coercive power versus authority)? How much impact would getting rid of gender have on the systems of power?

GENDER EXERCISES

1. Test your friends' knowledge of women's history and women's role in history. Ask them to name 10 famous historical men and 10 famous historical women in the United States. Entertainers (actresses and musicians) don't count. How many women on average can people name? You might then do some library research on famous women in history, starting with women's involvements with some of the social movements listed in this chapter.
2. Do some research and reading on your own about some of the issues surrounding prostitution. Find sources that represent both of the perspectives discussed in this chapter—that prostitution should always be illegal and that prostitution can be legal if the rights of sex workers are protected. You might start with the laws regarding prostitution in different countries. Write an essay in which you use the evidence you've gathered to explain your own position on this issue.

3. In this chapter, we discussed potential gender differences in political participation and political efficacy. Conduct some interviews or do a survey on these topics, comparing women and men. You might ask questions that measure how knowledgeable women and men are about political issues and whether they have ever run for an office or considered running for a political office. You might also include questions about political efficacy, or how powerful your respondents feel to make a meaningful difference politically. Do gender differences seem to emerge in your interviews or surveys? What other characteristics seem to matter?

4. Pick a political race that involves a woman candidate and do a content analysis of the coverage of that candidate. You might compare the coverage of the female candidate with a male counterpart. What, if any, gender differences do you notice in how the press talks about the two candidates? Do stories about the female candidate seem to place more emphasis on her personal life or family (private sphere), on feminine qualities (ability to compromise, empathize, etc.), on her lack of experience, or on her appearance? Is this significantly different than the coverage of male candidates?

5. Investigate the idea of the geography of fear as it connects to gender. Interview men and women about whether they have ever felt unsafe in a public space. Ask them about specific places in your community or neighborhood and how safe they assume those places to be. Ask them for stories about specific times when they felt unsafe of were harassed in a public space. If they had these experiences, how did they deal with them? Do gender differences emerge in your interviews? What kinds of things lead women and men to feel unsafe in public spaces? Do women seem to have a geography of fear?

6. If you have access to a college or university campus, do some observations of the party scene to compare to the descriptions of Armstrong et al. (2006) you read in this chapter. If you have access, you might observe in a Greek house as well as in other locations such as dorms or non-Greek parties. What are the gender dynamics of these social events? Do women and men seem to be following the interaction routines as described by Armstrong et al.? How prevalent is alcohol use? How does the social scene vary across different types of social groups on campus?

TERMS

gender outlaws

power

coercive power

authority

institutionalized power

hegemonic definition of manhood

emasculation

marital rape

head and master laws

geography of fear

rape culture

rape myths

rape-prone

rape-free

equality through complementarity

interaction routines

deference
slavery
human trafficking
feminization of poverty
burkas
hijab
cultural relativism
nation-state
state
nation
missile envy
political elite
descriptive or numerical representation
substantive representation
political socialization
political efficacy
ethic of justice
ethic of care
state feminism

SUGGESTED READINGS

Gender and violence

Armstrong, E. A., Hamilton, L., & Sweeney, B. (2006). Sexual assault on campus: A multilevel, integrative approach to party rape. *Social Problems, 53* (4), 483–499.

Bales, K. (2004). *New slavery: A reference handbook.* Santa Barbara, CA: ABC-CLIO.

Del, M. (1976). *Battered wives.* New York, NY: Pocket Books.

Dobash, R., & Dobash, R. (1992). *Women, violence and social change.* New York, NY: Routledge.

Fisher, B., Cullen, F., & Turner, M. (2000). *The sexual victimization of college women.* Washington, DC: National Institute of Justice and the Bureau of Justice Statistics.

Hollander, J. (2001). Vulnerability and dangerousness: The construction of gender through conversations about violence. *Gender & Society, 15,* 83–109.

Martin, P. Y., & Hummer, R. A. (1989). Fraternities and rape on campus. *Gender & Society, 3,* 457–473.

Sanday, P. R. (1981). The socio-cultural context of rape: A cross-cultural study. *Journal of Social Issues, 37,* 5–27.

Sanday, P. R. (1990). *Fraternity gang rape.* New York, NY: New York University Press.

Sanday, P. R. (2004). Rape-prone versus rape-free campus cultures. In M. S. Kimmel (Ed.), *The gendered society reader* (pp. 58–70). Oxford, England: Oxford University Press.

Geography of fear

Valentine, G. (1989). The geography of women's fear. *Area, 21,* 385–390.

Wesely, J. K., & Gaarder, E. (2004). The gendered "nature" of the urban outdoors: Women negotiating fear of violence. *Gender & Society, 18,* 645–663.

Prostitution

Barry, K. (1995). *The prostitution of sexuality.* New York, NY: New York University Press.

Paul, B. K., & Hasnath, S. A. (2000). Trafficking in Bangladeshi women and girls. *Geographical Review, 90,* 268–276.

Songue, P. B. (1996). Prostitution, a petit-metier during economic crisis: A road to women's liberation? The case of Cameroon. In K. Sheldon (Ed.), *Courtyards, markets, city streets: Urban women in Africa* (pp. 241–255). New York, NY: Westview Press.

Gender and politics

Bynum, V. (1992). *Unruly women.* Chapel Hill: University of North Carolina Press.

Carroll, S. J. (2001). *The impact of women in public office.* Bloomington: Indiana University Press.

Carroll, S. J. (2003). *Women and American politics.* Oxford, England: Oxford University Press.

Conway, M. M., Steuernagel, G. A., & Ahern, D. W. (2005). *Women and political participation.* Washington, DC: CQ Press.

Dodson, D. L. (1991). *Gender and policymaking: Studies of women in office.* New Brunswick, NJ: Center for the American Woman and Politics.

Galligan, Y., & Tremblay, M. (2005). *Sharing power: Women, parliament, democracy.* Burlington, VT: Ashgate.

Garlick, B., Dixon, S., & Allen, P. (1992). *Stereotypes of women in power: Historical perspectives and revisionist views.* New York, NY: Greenwood Press.

Inglehart, R., & Norris, P. (2003). *Rising tide: Gender equality and cultural change around the world.* Cambridge, England: Cambridge University Press.

Kahn, K. F., & Goldberg, E. N. (1991). Women candidates in the news: An examination of gender differences in U.S. Senate campaign coverage. *Public Opinion Quarterly, 55* (2), 180–199.

Lawless, J. L. (2004). Women, war and winning elections: Gender stereotyping in the post-September 11th era. *Political Research Quarterly, 53* (3), 479–490.

Rosenthal, C. S. (1998). When women lead: Integrative leadership in state legislatures. New York, NY: Oxford University Press.

Thomas, S., & Wilcox, C. (1998). Women and elective office: Past, present and future. New York, NY: Oxford University Press.

Tolleson-Rinehart, S., & Josephson, J. (2001). Gender and American politics. Armonk, NY: M. E. Sharpe.

Waylen, G. (2007). Engendering transitions: Women's mobilization, institutions and gender outcomes. Oxford, England: Oxford University Press.

Yoon, M. Y. (2001). Democratization and women's legislative representation in sub-Saharan Africa. Democratization, *8* (2), 169–190.

Gender and the military

Cohn, C. (1987). Sex and death in the rational world of defense intellectuals. *Signs, 12,* 687–718.

Cohn, C. (2004). Wars, wimps, and women: Talking gender and thinking war. In M. S. Kimmel (Ed.), *The gendered society reader* (pp. 397–409). New York, NY: Oxford University Press.

Enloe, C. (1988). *Does khaki become you? The militarization of women's lives.* London, England: Pandora.

Enloe, C. (2000). *Maneuvers: The international politics of militarizing women's lives.* Berkeley: University of California Press.

Enloe, C. (2004). Wielding masculinity inside Abu Ghraib: Making feminist sense of an American military scandal. *Asian Journal of Women's Studies, 10* (3), 89–102.

Hartstock, N. C. (1984). Masculinity, citizenship and the making of war. *PS, 17* (2), 198–202.

Levy, C. J. (1992). ARVN as faggots: Inverted warfare in Vietnam. In M. S. Kimmel & M. A. Messner (Eds.), *Men's lives* (pp. 183–197). New York, NY: Macmillan.

Peach, L. (1996). Gender ideology in the ethics of women in combat. In J. Striehm (Ed.), *It's our military too!* (pp. 156–194). Philadelphia, PA: Temple University Press.

WORKS CITED

Armstrong, E. A., Hamilton, L., & Sweeney, B. (2006). Sexual assault on campus: A multilevel, integrative approach to party rape. *Social Problems, 53* (4), 483–499.

Bales, K. (2004). *New slavery: A reference handbook*. Santa Barbara, CA: ABC-CLIO.

Barrett, E. J. (2001). Black women in state legislatures: The relationship of race and gender in legislative experience. In S. J. Carroll (Ed.), *The impact of women in public office* (pp. 185–204). Bloomington: Indiana University Press.

Barry, K. (1995). *The prostitution of sexuality.* New York, NY: New York University Press.

Bell, L. C., & Rosenthal, C. S. (2003). From passive to active representation: The case of women congressional staff. *Journal of Public Administration Research and Theory: J-PART,* 65–81.

Benschop, M. (2004, April 22). Women's rights to land and property. *UN-Habitat.* Retrieved May 18, 2010, from http://www.unhabitat.org/downloads/docs/1556_72513_CSDWomen.pdf

Boehlert, E., & Foser, J. (2006, November 20). Gender stereotypes and discussions of Armani suits dominate media's coverage of Speaker-Elect Pelosi. *Media Matters for America.* Retrieved February 23, 2009, from http://mediamatters.org/items/200611210002

Bonvilliain, N. (2007). *Women and men: Cultural constructs of gender.* Upper Saddle River, NJ: Prentice-Hall.

Brooks, G. (1995). *Nine parts of desire: The hidden world of Islamic women.* New York, NY: Anchor Books.

Bynum, V. (1992). *Unruly women.* Chapel Hill: University of North Carolina Press.

Canada, G. (1995). *Fist stick knife gun: A personal history of violence in America.* Boston, MA: Beacon Press.

Carroll, S. J. (2001). Introduction. In S. J. Carroll (Ed.), *The impact of women in public office* (pp. xi–xxvi). Bloomington: Indiana University Press.

Clark, J. (1998). Women at the national level: An update on roll call voting behavior. In S. Thomas & C. Wilcox (Eds.), *Women and elective office: Past, present and future* (pp. 118–129). New York, NY: Oxford University Press.

Coe, K., Domke, D., Bagley, M. M., Cunningham, S., & Van Leuven, N. (2007). Masculinity as political strategy: George W. Bush, the "war on terrorism," and an echoing press. *Journal of Women, Politics, and Policy, 29,* 31–55.

Cohn, C. (1987). Sex and death in the rational world of defense intellectuals. *Signs, 12,* 687–718.

Cohn, C. (2004). Wars, wimps, and women: Talking gender and thinking war. In M. S. Kimmel (Ed.), *The gendered society reader* (pp. 397–409). New York, NY: Oxford University Press.

Cohn, C., & Enloe, C. (2003). A conversation with Cynthia Enloe: Feminists look at masculinity and the men who wage war. *Signs, 28* (4), 1187–1207.

Conway, M. M., Steuernagel, G. A., & Ahern, D. W. (2005). *Women and political participation.* Washington, DC: CQ Press.

Coontz, S. (2005). *Marriage, a history: From obedience to intimacy, or how love conquered marriage.* New York, NY: Viking Press.

Couric, K. (2008). Sexist media hurt Hillary Clinton's chances [video]. *CBS Evening News.* Retrieved from http://www.youtube.com/watch?v = VyjEGZSM83Y

Cutlip, E. (2009, February 5). Author: Clinton, Palin coverage tainted by gender stereotyping. *MediaMouse.* Retrieved May 26, 2009, from http://www.mediamouse.org/news/2009/02/jamieson- women-politics-media.php

Davidson, J. O. (2002). The rights and wrongs of prostitution. *Hypatia, 17,* 84–98.

Del, M. (1976). *Battered wives.* New York, NY: Pocket Books.

Diffendal, A. (2008, October 15). Palin, Clinton have one thing in common: Both candidates faced media sexism—but chose to respond in different ways. *Orbis.* Retrieved May 26, 2009, from http://www.vanderbiltorbis.com/2.12253/palin-clinton-have-one-thing-in-common-1.1612025

Dixon, S. (1992). Conclusion—The enduring theme: Domineering dowagers and scheming concubines. In B. Garlick, S. Dixon, & P. Allen (Eds.), *Stereotypes of women in power: Historical perspectives and revisionist views* (pp. 207–225). New York, NY: Greenwood Press.

Dobash, R., & Dobash, R. (1992). *Women, violence and social change.* New York, NY: Routledge.

Dodson, D. L. (1991). *Gender and policymaking: Studies of women in office.* New Brunswick, NJ: Center for the American Woman and Politics.

Dodson, D. L. (1998). Representing women's interests in the U.S. House of Representatives. In S. Thomas & C. Wilcox (Eds.), *Women and elective office: Past, present and future* (pp. 130–149). New York, NY: Oxford University Press.

Duerst-Lahti, G. (1998). The bottleneck: Women becoming candidates. In S. Thomas & C. Wilcox (Eds.), *Women and elective office: Past, present and future* (pp. 15–25). New York, NY: Oxford University Press.

Duncan, N. (1996). Renegotiating gender and sexuality in public and private spaces. In N. Duncan (Ed.), *Bodyspace: Destabilizing geographies of gender and sexuality* (pp. 127–144). New York, NY: Routledge.

Eagly, A. H., & Johnson, B. T. (1990). Gender and leadership style: A meta-analysis. *Psychological Bulletin, 108,* 233–256.

Fisher, B., Cullen, F., & Turner, M. (2000). *The sexual victimization of college women.* Washington, DC: National Institute of Justice and the Bureau of Justice Statistics.

Geisler, G. (1995). Troubled sisterhood: Women and politics in southern Africa. *African Affairs, 94* (377), 545–578.

Gilligan, C. (1982). *In a different voice: Psychological theory and women's development.* Cambridge, MA: Harvard University Press.

Githens, M. (2003). Accounting for women's political involvement: The perennial problem of recruitment. In S. J. Carroll (Ed.), *Women and American politics* (pp. 33–52). Oxford, UK: Oxford University Press.

Goodey, J. (2004). Sex trafficking in women from Central and Eastern European countries: Promoting a "victim-centred" and "woman-centred" approach to criminal justice intervention. *Feminist Review, 76,* 26–45.

Hartstock, N. C. (1984). Masculinity, citizenship and the making of war. *PS, 17* (2), 198–202.

Hill, R. (2000). *Lord Acton.* New Haven, CT: Yale University Press.

Hollander, J. (2001). Vulnerability and dangerousness: The construction of gender through conversations about violence. *Gender & Society, 15,* 83–109.

Inter-Parliamentary Union. (2010a, March 31). *Women in national parliaments: World average.* Retrieved May 18, 2010, from http://www.ipu.org/wmn-e/world.htm

Inter-Parliamentary Union. (2010b, March 31). *Women in national parliaments: World classification.* Retrieved May 18, 2010, from http://www.ipu.org/wmn-e/classif.htm

Kahn, K. F., & Goldberg, E. N. (1991). Women candidates in the news: An examination of gender differences in U.S. Senate campaign coverage. *Public Opinion Quarterly, 55* (2), 180–199.

Kardos-Kaponyi, E. (2005). Hungary. In Y. Galligan & M. Tremblay (Eds.), *Sharing power: Women, parliament, democracy* (pp. 25–36). Burlington, VT: Ashgate.

Kathlene, L. (1998). In a different voice: Women and the policy process. In S. Thomas & C. Wilcox (Eds.), *Women and elective office: Past, present and future* (pp. 188–202). New York, NY: Oxford University Press.

Kaye/Kantrowitz, M. (2009). Women, violence and resistance. In E. Disch (Ed.), *Reconstructing gender: A multicultural anthology* (pp. 504–516). Boston, MA: McGraw-Hill.

Kilvington, J., Day, S., & Ward, H. (2001). Prostitution policy in Europe: A time of change? *Feminist Review, 67* (1), 78–93.

Kimmel, M. (1994). Masculinity as homophobia. In H. Brod (Ed.), *Theorizing masculinities* (pp. 119–141). Thousand Oaks, CA: Sage.

Kimmel, M. S. (2006). *Manhood in America: A cultural history.* New York, NY: Oxford University Press.

Kimmel, M. S. (2008). *The gendered society.* New York, NY: Oxford University Press.

Martin, P. Y., & Hummer, R. A. (1989). Fraternities and rape on campus. *Gender & Society, 3,* 457–473.

McGlen, N. E., & Sarkees, M. R. (2001). Foreign policy decision makers: The impact of gender. In S. J. Carroll (Ed.), *The impact of women in public office* (pp. 117–148). Bloomington: Indiana University Press.

Murphy, K. (2008, November 20). The mayor-elect's new clothes: Silverton, Oregon elects a transgender leader. *The Los Angeles Times.* Retrieved May 13, 2010, from http://articles.latimes.com/2008/nov/20/nation/na-transgender20

Palley, M. L. (2001). Women's policy leadership in the United States. *PS: Political Science and Politics, 34,* 247–250.

Paul, B. K., & Hasnath, S. A. (2000). Trafficking in Bangladeshi women and girls. *Geographical Review, 90,* 268–276.

Plutzer, E. (2000). Are moral voices gendered? Care, rights and autonomy in reproductive decision making. In S. Tolleson-Rinehart & J. Josephson (Eds.), *Gender and American Politics* (pp. 82–101). Armonk, NY: M. E. Sharpe.

Read, J. G., & Bartokowski, J. P. (2005). To veil or not to veil? A case study of identity negotiation among Muslim women in Austin, Texas. In M. B. Zinn, P. Hondagneu-Sotelo, & M. A. Messner (Eds.), *Gender through the prism of difference* (pp. 94–107). New York, NY: Oxford University Press.

Rosener, J. B. (1990). Ways women lead. *Harvard Business Review, 68* (6), 119–125.

Rosenthal, C. S. (1998). *When women lead: Integrative leadership in state legislatures.* New York, NY: Oxford University Press.

Rosenwasser, S., & Dean, M. (1989). Gender role and political office. *Psychology of Women Quarterly, 13* (1), 77–85.

Rousseau, S. (2005). Peru. In Y. Galligan & M. Tremblay (Eds.), *Sharing power: Women, parliament, democracy* (pp. 91–105). Burlington, VT: Ashgate.

Sadker, M., & Sadker, D. (2009). Missing in interaction. In T. E. Ore (Ed.), *The social construction of difference and inequality: Race, class, gender and sexuality* (pp. 331–343). New York, NY: McGraw-Hill.

Sanday, P. R. (1981). The socio-cultural context of rape: A cross-cultural study. *Journal of Social Issues, 37,* 5–27.

Sanday, P. R. (1990). *Fraternity gang rape.* New York, NY: New York University Press.

Sanday, P. R. (2004). Rape-prone versus rape-free campus cultures. In M. S. Kimmel (Ed.), *The gendered society reader* (pp. 58–70). Oxford, England: Oxford University Press.

Scharrer, E. (2002). An "improbable leap": A content analysis of newspaper coverage of Hillary Clinton's transformation from First Lady to senate candidate. *Journalism Studies, 3,* 393–406.

Songue, P. B. (1996). Prostitution, a petit-metier during economic crisis: A road to women's liberation? The case of Cameroon. In K. Sheldon (Ed.), *Courtyards, markets, city streets: Urban women in Africa* (pp. 241–255). New York, NY: Westview Press.

Stetson, D. M., & Mazur, A. G. (1995). Introduction. In D. M. Stetson & A. G. Mazur (Eds.), *Comparative state feminism* (pp. 1–21). Thousand Oaks, CA: Sage.

Swers, M. (1998). Are women more likely to vote for women's issues bills than their male colleagues? *Legislative Studies Quarterly, 23* (3), 435–448.

Thale, C. (2005). Smoke-filled room. *Encyclopedia of Chicago.* Retrieved February 22, 2009, from http://www.encyclopedia.chicagohistory.org/pages/3217.html

Thomas, S. (1994). *How women legislate.* New York, NY: Oxford University Press.

Thomas, S. (2003). The impact of women in political leadership positions. In S. J. Carroll (Ed.), *Women and American politics* (pp. 89–110). New York, NY: Oxford University Press.

Thomas, S., & Welch, S. (1991). The impact of gender on activities and priorities of state legislators. *Western Political Quarterly, 44,* 445–456.

Tolleson-Rinehart, S. (2001). Do women leaders make a difference? Substance, style and perceptions. In S. J. Carroll (Ed.), *The impact of women in public office* (pp. 149–165). Bloomington: Indiana University Press.

United Nations Population Fund. (2003). *Gender equality: Trafficking in human misery.* New York: Author. Retrieved January 24, 2009, from http://www.unfpa.org/gender/violence1.htm

U.S. Department of Justice. (2005). *Homicide trends in the U.S.* Washington, DC: Author. Retrieved August 24, 2009, from http://www.ojp.usdoj.gov/bjs/homicide/gender.htm

U.S. Department of Justice. (2007). *Uniform crime reports, 2007.* Washington, DC: Author.

Valentine, G. (1989). The geography of women's fear. *Area, 21,* 385–390.

Valentine, G. (1992). Images of danger: Women's source of information about spatial distribution of male violence. *Area, 24,* 22–29.

Wesely, J. K., & Gaarder, E. (2004). The gendered "nature" of the urban outdoors: Women negotiating fear of violence. *Gender & Society, 18,* 645–663.

Whicker, M., & Jewell, M. (1998). The feminization of leadership in state legislatures. In S. Thomas & C. Wilcox (Eds.), *Women and elective office: Past, present and future* (pp. 163–187). New York, NY: Oxford University Press.

Witchel, A. (2008, June 22). 'Mad Men' has its moment. *The New York Times Magazine.* Retrieved May 30, 2010, from http://www.nytimes.com/2008/06/22/magazine/22madmen-t.html?_r = 1& ref = television

Yoon, M. Y. (2001). Democratization and women's legislative representation in sub-Saharan Africa. *Democratization, 8* (2), 169–190.

Yoon, M. Y. (2005). Sub-Saharan Africa. In Y. Galligan & M. Tremblay (Eds.), *Sharing power: Women, parliament, democracy* (pp. 79–90). Burlington, VT: Ashgate.

Glossary

Part I

Chapter 2: What's the "Sociology" in the Sociology of Gender?

Accountability: the ways in which we gear our actions with attention to our specific circumstances so that others will correctly recognize our actions for what they are.

Accounts: the descriptions we engage in as social actors to explain to each other the state of affairs, or what we think is going on.

Allocation: in doing gender theory, the way decisions get made about who does what, who gets what and who does not, who gets to make plans, and who gets to give orders or take them.

Alters: in social network theory, the people related to ego (the focal person) in the network.

Breach: a social experimental method used by ethnomethodologists that disrupts normal social interactional rules to reveal the taken-for-granted norms of our everyday lives.

Confirmation bias: the tendency to look for information that confirms our preexisting beliefs while ignoring information that contradicts those beliefs.

Consciousness-raising: activities that seek to help women see the connection between their personal experience with gender exploitation and a larger sense of the politics and structure of society.

Density: in social network theory, a measure of how interconnected the alters are in an ego network.

Diversity: in social network theory, the number of contacts with people in multiple spheres of activity.

Ego: in social network theory, the focal person.

Ego network: in social network theory, the network consisting of an ego and her alters.

Ethnomethodology: a particular subfield in sociology concerned with the taken-for-granted assumptions of social interaction.

Expressive: a role that is oriented toward interactions with other people.

Fundamental attribution error: the tendency to explain behavior by invoking personal dispositions while ignoring the roles of social structure and context.

Gender status belief: superiority/inferiority of one gender to the other.

Gendered organization: a social aggregate in which advantage and disadvantage, exploitation and control, action and emotion, and meaning and identity are patterned through and in terms of the distinction between male and female, masculine and feminine.

Homophilous: in network theory, the tendency for like to attract like in a social network.

Individual approach to gender: a perspective that locates gender inside individuals and assumes gender works from the inside out.

Instrumental: task-oriented role.

Job: supposedly gender-neutral term for a position within an organization that can be filled by any human being

Liberal feminism: a version of feminist thought that posits that inequality between men and women is rooted in the way existing institutions such as the government treat men and women.

Master frame of equal rights: an ideology that assumes that diverse groups of individuals in society, such as African Americans, women, and gays and lesbians, are entitled to the same rights as everyone else in society because we are all fundamentally the same.

Master status: a social position that cuts across all other identities and situations and that is the most important status in dictating how people respond to us.

Matrix of domination: in intersectional theory, the way in which the social structures of race, class, gender, and sexual orientation work with and through each other so that any individual experiences each of these categories differently depending on his or her unique social location.

Men as a proxy: sociological studies that include only men as research subjects, assuming that the experiences of men are universal.

Multiple consciousnesses: the way of thinking that develops from a person's position at the center of an intersecting and mutuality reliant system of oppression.

Organizational logic: the taken-for-granted assumptions and practices that underlie an organization and that often have gendered implications.

Performance expectation: a guess about how useful your own contributions will be to accomplishing the goal of the interaction, as well as about how useful the other people in the group will be.

Private troubles: those problems we face that have to do with ourselves and our immediate surroundings.

Privilege: a set of mostly unearned rewards and benefits that comes with a given status position in society.

Public issues: issues beyond the individual and are located within the larger structures of our societies.

Radical feminism: version of feminist thought that posits that gender is a fundamental aspect of the way society functions and serves as an integral tool for distributing power and resources among people and groups.

Self-fulfilling prophecy: a prediction or assumption that becomes true solely because the prediction or assumption was made.

Sex assignment: putting someone into one or the other sex category, usually at birth.

Sex categorization: the way we use cues of culturally presumed appearance and behavior to represent physical sex differences that we generally cannot see.

Sex role: set of expectations that are attached to your particular sex category.

Size: in social network theory, the number of others to whom someone is linked in a network.

Social aggregates: building blocks of society that are composed of individuals but become more than the sum of the individuals within them.

Social movement abeyance: a way to keep the basic ideas of a movement alive during a period of decreased activism, often a result of increased resistance and hostility to the movement or to a shift in the opportunities that make movements more or less successful.

Social movement cycle: a period of increasing frequency and intensity in social movement activities that spreads throughout various parts of society and globally across countries.

Social role: a set of expectations that are attached to a particular status or position in society.

Sociological imagination: the ability to see the connection between your own life and larger social structures.

Status characteristic: some kind of difference that exists between people in society and to which a sense of lesser or greater value and esteem is given.

Chapter 3: How Do Disciplines Outside of Sociology Study Gender?

Andocentric: theories that assume that what men do is more important than what women do.

Biological determinism: attempts to reduce some type of behavior to unchangeable biological roots as the sole explanation.

Castration complex: in Freudian theory, the fear of boys that their fathers will castrate them if they act on their sexual desire toward their mother.

Commodity crops: agricultural products that can be exported and sold on the world market.

Development project: post–World War II, the division of the world into developed and developing countries and the belief that the developed world was obligated to help the rest of the world progress economically and socially.

Electra complex: in Freudian theory, the dynamic where girls and eventually women come to perceive other women as competitors in the ultimate goal of possessing their own penis.

Ethnography: a social science research technique that involves time spent in the field, living among the group you're studying to develop an insider's perspective on its culture and society.

Formal economy: the sector of the economy that is regulated by the government or legal systems.

Homophobia: fear of homosexuality.

Homosociality: the continuum of cooperative relations between people of the same gender.

Informal economy: the sector of the economy that is not regulated by the government or legal systems.

Man the hunter: the assumption that the hunters were, in fact, all men.

Matrilineal: social systems in which both kinship and property are passed down based on one's mother.

Meta-analysis: using the results from many published studies on sex difference to obtain an average difference based on a larger and more diverse sample size.

Meta-narrative: a story that attempts a comprehensive and universal explanation of some phenomenon.

Modernization theory: the belief that in the postcolonial world, the global South would become like the countries of the developed world if it followed their lead economically, politically, and socially.

Nongovernmental organizations: organizations that work on various problems that might be considered the role of government to solve—social, political, economic, or environmental problems—but that have no government affiliation and are privately funded entities.

Oedipus complex: Freud's articulation of how men develop a sense of masculine identity through boys' desire to have sex with their mothers and destroy their fathers.

Patrilineal: social systems where descent is through the father and the male line.

Penis envy: in Freudian theory, women's envy of men's possession of a penis.

Phallic symbols: any abstract representation of the penis.

Sex difference research: a body of research that identifies whether a wide range of traits, dispositions, and behaviors differ significantly between men and women.

Structural adjustment: demands on the part of the IMF and the World Bank that developing countries change the structure of their market and economy as a condition of receiving loans.

Subsistence agriculture: growing crops to feed or provide directly for yourself and your family.

Superego: in Freudian theory, the part of our personality responsible both for our own morality and for lining our morality up with societal standards.

Women in development: a perspective that emphasizes women's roles in development and the potentially adverse effects of development on women.

Women as gatherers: the assumption that the gatherers were, in fact, all women.

Part II

Chapter 4: How Do We Learn Gender?

Agents of socialization: the people, groups, and institutions who are doing the socializing.

Androcentrism: the belief that masculinity and what men do in our culture is superior to femininity and what women do.

Androgenization: adopting some of the qualities of the opposite gender.

Cognitive-development theory: a perspective on gender socialization that emphasizes the ways in which children acquire a sense of a gender identity and the ability to gender-type themselves and others.

Cumulative disadvantage: the ways in which inequalities that persist between men and women over the whole course of their lives become intensified in old age.

Ego boundaries: the sense of personal psychological division between ourselves and the world around us.

Enculturation: how culture comes to reside inside individuals.

Gender congruency: in cognitive-development theory, the need for children to create cognitive consistency by lining up their gender identity with their behaviors.

Gender constancy: the understanding that even changing the outward physical appearance of a person does not change his or her underlying sex category.

Gender identity: the way in which being feminine or masculine, a woman or a man, becomes an internalized part of the way we think about ourselves.

Gender norms: sets of rules for what is appropriate masculine and feminine behavior in a given culture.

Gender polarization: the way in which behaviors and attitudes that are viewed as appropriate for men are viewed as inappropriate for women and vice versa.

Gender schema: a cognitive structure that enables us to sort characteristics and behaviors into masculine and feminine categories and then create various other associations with those categories.

Gender schema theory: a theory of gender socialization that builds on the framework of both cognitive development and social learning theory to formulate an explanation that is specific to gender socialization, rather than to socialization as a more general process; influenced by feminism and emphasizes the content of gender schema.

Gender stability: when children know that their gender is permanent and that it is the gender they will be for the rest of their lives.

Gender socialization: the process through which individuals learn the gender norms of their society and come to develop an internal gender identity.

Gender transgression zone: any activities or behaviors that have the potential to be perceived as violating gender norms in some way.

Genital tubercle: undifferentiated embryonic tissue that develops into a penis in males and a clitoris in females.

Hegemonic masculinity: how the dominant ideas about what it means to be a man influence the behaviors of actual men in any given society.

Hermaphrodites: another term for intersexed individuals that originates in Greek mythology.

Identification: when a child copies whole patterns of behavior of the same sex parents without necessarily being trained or rewarded for doing so.

Intersexed: individuals who, based on a variety of biological conditions, do not fit into the contemporary Anglo-European biological sex categories of male and female.

Language asymmetry: the way in which the structure and vocabulary of a language reflects and helps to re-create the social inequalities of the culture in which it exists.

Life-course perspective: Looking at how key experiences and social roles influence individuals' lives through biographies that transform through exposure to new social settings, unfolding in an interactive trajectory.

One-child policy: China's governmental policy to control population growth in which couples are limited to having only one child.

Primary group: the intimate, enduring, unspecialized relationships among small groups who generally spend a great deal of time together.

Primary socialization: the initial process of learning the ways of a society or groups that occurs in infancy and childhood and is transmitted through the primary groups to which we belong.

Psychoanalytic identification: the way in which a child modifies her own sense of self and order to incorporate some ability, attribute, or power she sees in others around her.

Psychoanalytic theory: a theory of gender socialization drawing on Freud and feminism that emphasizes the ways in which gender becomes deeply embedded in the psychic structure of our personalities.

Roleless role: a social status in society with unclear or minimal expectations attached to it so that knowing how to enact the role becomes difficult.

Schema: a cognitive structure and network of associations that helps to organize an individual's perception of the world.

Secondary groups: generally larger, more temporary, more impersonal, and more specialized groups that also engage in socialization.

Secondary socialization: the learning process that takes place each time we join one of these new secondary groups.

Sex-typed behaviors: a behavior that is more expected and therefore perceived as appropriate when performed by one sex, but less expected and therefore perceived as inappropriate when performed by the other sex.

Social learning theory: a theory of gender socialization based in behaviorism that posits that we learn norms through the rewarding or withholding of rewards for certain behaviors.

Socialization: the ways in which we learn to become a member of a group.

Suttee: a Hindu practice of ritual self-immolation of a widow on her husband's funeral pyre.

Target of socialization: the persons being socialized.

Chapter 5: How Does Gender Matter for Whom We Want and Desire?

Boston marriages: passionate attachments between middle- and upper-class white women, in the 18th and the 19th century, which often lasted through marriages and relationships, and were generally not looked on with disapproval.

Berdache: an Anglo-European word for individuals within Native American tribes who are born into one sex category but live as a different gender.

Compulsive heterosexuality: the way in which heterosexuality becomes institutionalized into the practice of daily life and therefore enforced as a way of regulating our behaviors and distributing power and privilege.

Discourse: the means by which institutions wield their power through a process of definition and exclusion.

Double standard: a cultural belief in Anglo-European society that the exact same sexual behaviors or feelings are OK for one gender but not for the other.

Egalitarian: a relationship that is more embedded in a belief in equality among all people.

Ethnocentrism: believing one's own culture to be better, more correct, or right relative to another culture.

Hegemonic curriculum: educational institutions that legitimize the dominant culture and marginalize or reject other cultures and forms of knowledge.

Heteronormativity: the way in which heterosexuality is viewed as the normal, natural way of being.

Hetero-privilege: the set of unearned rights that are given to heterosexuals in many societies.

Homosexual identity: considering being lesbian or gay to be a part of one's individual self-concept.

Homosocial: social relationships between those of the same sex.

Ideal love: submission to and adoration of an idealized other whom one would like to be like and from whom one wants confirmation and recognition.

One-sex model: a model of sex categories in which certain kinds of men (free, citizens) are perceived as at the top of the hierarchy while everyone else (women, slaves, eunuchs) are perceived as inferior versions of men, rather than as a completely different type of person.

Privilege: an unearned right that is attached to a social status.

Romance tourism: the ways in which many women as sex tourists are looking more to be swept away by men than to assert their strong control over their paid male counterparts.

Sex: any act that is identified as sexual.

Sexology: the scientific study of sex.

Sexual desire: a combination of objective physical responses and subjective psychological or emotional responses to some internal or external stimulus.

Sexual identity: the particular category into which people place themselves based on the current, Anglo-European division of the world into heterosexual, homosexual, bisexual, or asexual.

Sexual inversion: an early theory in sexology that attributes homosexuality to an inborn inversion of gender traits; i.e., gay men are internally women.

Sexual object: the passive recipient of sexual behavior and sexual desire, or the one who is sexually acted on and sexually desired, rather than the one doing the sexual acting and sexual desiring.

Sexual script: the learned guidelines for sexual expression that provide individuals with a sense of appropriate sexual behaviors and sexual desires for their particular culture.

Studying up: the need to study those at the top of any particular power structure or hierarchy.

Two-sex model: a model of sex categories in which women and men are believed to be two completely different types of people, and sex is viewed as a bounded category.

Chapter 6: How Does Gender Impact the People You Spend Your Time With?

Affective love: emotionally and affectionately connected through communication and the sharing of feelings.

Bridewealth: the money that a man pays the bride's family in the form of money or goods.

Calling: a courtship practice in which men come to call at the homes of their potential love interest.

Collectivistic: societies in which the needs of the group are more important than the needs of individuals.

Conjugality: the personal relationship between husband and wife.

Face-to-face: in sex difference research, a way of describing women's friendships as characterized by an intimate confidante with whom they can share their feelings.

Friendship: a voluntary, informal, and personal relationship.

Friends with benefits: having sex in a nonromantic friendship.

Going steady: a courtship practice in which the ideal was for a boy and girl to pair up and date each other exclusively for long periods of time.

Hookups: sexual encounter, usually lasting only one night between two people who were strangers or brief acquaintances.

Individualistic: in a society where the collective needs are generally perceived as less important than the needs of individual.

Instrumental love: doing things for the other person by providing material help and practical assistance.

Kin: people with whom you're related either by actual blood or through symbolic ties.

Nuclear family: a historically and culturally specific way of organizing a family life that consists of only two generations, parents and children.

Rating-dating-mating complex: a courtship practice in which girls and boys dated many different people at the same time.

Residential segregation: separation in terms of where people live.

Romantic love: an intense attraction in which the love object is idealized.

Sexual convergence: the trend for the norms and ideals surrounding women's sexuality to become increasingly similar to the norms and ideals of men's sexuality.

Side-by-side: in sex difference research, a way of describing men's friendships as characterized by someone with whom they can share activities and interests.

Social segregation: the separation of some realm of social life into different groups on the basis of some category.

Chapter 7: How Does Gender Matter for How We Think About Our Bodies?

"Be a Sturdy Oak": the idea that men should never show vulnerability or weakness, even when they are in a potentially vulnerable position; one of the rules of manhood.

Beauty myth: the belief in a quality called beauty that is real and universal and that women, as a result of biological, sexual, and evolutionary factors, should want to embody, while men should desire the women who embody that ideal of beauty.

Body dysmorphic disorder: a psychological disorder whose symptoms include frequent mirror checking, excessive grooming, face picking, reassurance seeking, applying excessive amounts of makeup, buying excessive amounts of hair products, as well as engaging in hair removal to excess.

Body image: the perception and evaluation of one's own bodily appearance.

Body image distortion syndrome: disturbance in body size awareness, or an indicator of anorexia both among clinicians and doctors responsible for making diagnoses and in popular culture depictions.

Commodity: something that can be bought and sold.

Eugenics: the study or practice of improving the human race through selective breeding and sometimes restrictive migration policies.

Face-off masculinity: the way in which men look directly into the camera and at the viewer in advertising images, conveying a sense of being powerful, armored, and emotionally impenetrable.

Gender attributions: to make guesses about who is male and who is female.

"Give 'em Hell": the idea that real men should exude an aura of manly daring and suggestion, encouraging men to go for it and take risks.; one of the rules of manhood.

Hysteria: a disease (headache, muscular aches, weakness, depression, menstrual difficulties, indigestion) that began to be diagnosed among primarily upper-class and sometimes famous and prominent, white women in the United States and Europe in the late 19th century.

Male gaze: in feminist film theory, the way in which women are generally perceived as objects of the gaze (of both the audience and characters in the film) because control of the camera is shaped by the assumption of a heterosexual, male audience.

Metrosexuals: heterosexual men who demonstrate a level of concern with their appearance that is considered outside the norm for masculinity.

Mind–body dualism: in Anglo-European society, the split between the physical body and the nonmaterial entity we call mind, where the mind is perceived as superior in many ways to the inferior body.

"No Sissy Stuff": real men don't do anything that carries with it the least suggestion of femininity; one of the rules of manhood.

Premenstrual syndrome: a feminine hormonal disorder sometimes diagnosed as occurring in women just before their menstrual cycle.

Psychology of the uterus: the belief that women's whole persona was dictated by her uterus.

Sex hormones: types of hormones that are thought of as being involved in sex characteristics and sexuality, e.g., estrogen and progesterone.

Part III

Chapter 8: How Does Gender Impact the People We Live Our Lives With?

Be a Big Wheel: emphasizes the way in which masculinity is measured by power, success, wealth, and status; one rule of manhood.

Cooperation theory of marriage: argument that marriage is not about protection or exploitation; rather, it is about cooperation and circulation.

Demographics: in social science research, a way to describe the basic population characteristics of a group, including distribution, the sex ratio, how long people live, the composition of different racial and ethnic groups.

Diaspora: the worldwide scattering of a nationality or ethnic group.

Doctrine of separate spheres: an ideology that evolved in Anglo-European societies with industrialization that separates a man and a woman's spheres of duties and responsibilities.

Dual-earner couples: a household in which both spouses are engaged in paid work outside the home.

Endogamy: the tendency to marry only others within the same social group.

Exploitation theory of marriage: argument that marriage is an institution that primarily involves the exploitation of women in a system of exchange.

Families of choice: forming families with people who are not necessarily related to one by blood.

Genteel patriarch: an 18th-century American model of masculinity that emphasized a man's role as head of the family and supervisor of his estate; examples include George Washington and Thomas Jefferson.

Good provider role: a man's primary obligation to his family is to work to meet his family's material needs.

Heroic artisan: an 18th-century American model of masculinity that emphasized physical strength, family duties, and dedication to country and budding ideals of democracy; examples include Paul Revere and yeoman farmers.

Household division of labor: the way in which the tasks necessary to the care and running of a household are distributed.

Love marriage: a type of marriage that evolved in Anglo-European societies in the last 150 years and emphasizes the importance of romantic attraction.

Market manhood: an ideal of American masculinity in which men derive their identity from their success in the capitalist economy in accumulating wealth, power, and status.

Marriage gradient: demographic pattern in which women tend to marry up in social status (men with a higher social status than themselves) while men marry down in social status.

Marriage squeeze: a shortage of one sex or the other in the age group in which marriage generally occurs.

Modern marriage: an institution freely chosen on the basis of love and compatibility and composed of a sole male breadwinner plus economically dependent wife and children.

Nuclear family: a family grouping that consists of a mother, father, and their children.

Othermothers: women who may or may not be biologically related to children but play a crucial role in assisting bloodmothers with the task of caring for and raising children.

Polyandry: a marriage that involves one woman and multiple men.

Polygamy: a marriage between one man and several women.

Postmodern family: the changing, unsettled, recombination of different family forms that some argue had taken the place of traditional family structures.

Private sphere: anything that's not public, primarily the home and family life that defined as the place where women belong with the separation of spheres.

Protection theory of marriage: argument that marriage is an institution that evolved to provide for the protection of women.

Public sphere: the world of market relations and productive behavior, a world that is labeled as distinctly masculine with the separation of spheres.

Quarter shifts: in dual-earner couples, when one partner does much of the work and childcare during the week, while the second partner does most of the housework and childcare on the weekends.

Reduction of needs: a method for addressing the household division of labor in which one partner who is unwilling or unable to maintain the home according to certain standards simply lowers household expectations by denying that work needs to be done.

Remittances: the money and income that immigrants send back to their families in their country of origins.

Second shift: the extra burden of childcare and housework added on for women who also do paid work outside of the home.

Sex ratio: the number of males to females in a given society.

Sexual division of labor: a sense that certain tasks are more appropriate to one sex category or the other.

Social institution: a group, organization, behavior, or pattern that serves a specific function in society.

Subsistence agriculture: farming that's done to feed one's self or one's family.

Transnational marriage: a marriage that crosses national borders.

Transnational motherhood: relationship between mother and children that crosses national borders.

Chapter 9: How Does Gender Affect the Type of Work We Do and the Rewards We Receive for Our Work?

Comparable worth: a policy that seeks to raise the wages of low-paying jobs occupied predominantly by women by evaluating the gendered ways in which jobs are socially constructed.

Compensatory masculinity: an exaggerated form of masculinity supposedly used by lower-class men involving drugs, alcohol, and sexual carousing that is used to demonstrate defiance and independence from both the control of their wives and the establishment.

Contributing family workers: a kind of own-account worker who works without pay in an establishment operated by a related person living in the same household.

Egalitarian masculinity: more civilized, refined, and closer to a situation of gender equity supposedly demonstrated among upper- and middle-class men.

Family wage: the idea that workers should be paid a sum necessary to sustain all family members.

Formal paid labor: an hourly wage or salary, or some kind of paycheck from which taxes were taken.

Gender wage gap: the ratio of women's average earnings in an occupation to men's average earnings in the same occupation.

Gendered organization: any social group where gender is built into the formal and informal rules in a way that is taken for granted as part of the normal functioning of the group.

Ghettoization: when lower paid women's jobs are separated from the better paid men's jobs within an occupation through informal gender typing and then paid less.

Glass ceiling: the fact that despite the progress women have made into many managerial positions in the business world, they are still far less likely than men to have jobs that involve exercising authority over people and resources.

Glass escalator: the invisible pressure that men in some occupations face to move upward in their professions.

Human capital: the skills workers may acquire through education, job training, and job tenure that affect their ability to be productive in their jobs.

Human capital theory: an explanation for sex segregation that argues that individual occupational choice results from the attempt to maximize the benefits and reduce the costs involved in any particular job.

Informal labor: where there is no legal contract for the work, not necessarily a set hourly wage or salary, and none of the income is reported to the government by the employer or the employee.

Job evaluation: determining how pay is assigned to jobs and evaluating those pay rates as fair or unfair.

Job-level sex segregation: sex segregation in the specific positions that workers hold within a specific establishment.

Labor markets: the particular pool of people in a given population from which employers hire or promote.

Maquiladoras: assembly plants largely owned by American and other foreign companies that are located along the Mexico–U.S. border to take advantage of the supply of cheap labor, reduced taxes, and more relaxed regulation across the border.

Occupational sex segregation: the concentration of women and men in different occupations.

Oppositional culture: a coherent set of values, beliefs, and practices that mitigates the effects of oppression and reaffirms that which is distinct from the majority culture by defining itself in opposition.

Own-account workers: those who are self-employed with no employees working for them.

Patriarchal dividend: the advantages men in general gain from the subordination of women.

Piece-meal work: paying on the basis of the number of products produced rather than on the hours worked.

Precarious work: employment that is uncertain, unpredictable, and risky from the point of view of the worker.

Protective labor legislation: laws that prohibit certain kinds of persons—usually women and children—from working in certain types of jobs or in certain kinds of situations based on the belief that these jobs or situations would be dangerous or harmful.

Putting-out industry: a model of work in which rather than workers traveling to one central building to labor, the raw materials are distributed to an individual household where individuals assemble them on their own schedule within their own homes.

Resegregation: when an entire occupation transitions from one gender to another, usually from a predominantly male to a predominantly female occupation.

Sex segregation: the concentration of women and men into different jobs, occupations, and firms.

Socialization theory: an explanation for sex segregation that argues that our experiences with gender socialization lead men and women to prefer different types of jobs.

Vulnerable employment: the informal jobs that are lower paid overall relative to jobs in the formal economy.

Chapter 10: How Does Gender Affect What You Watch, What You Read, and What You Play?

Adult leisure: leisure spent purely in the company of adults.

Audience power theory: a perspective that places the power to determine the popularity of any given media content squarely in the hands of the people who are making informed decisions about what they like and what they don't like.

Commodification: turning any project, idea, or behavior into something that can be bought and sold.

Contaminated leisure: leisure that involves other nonleisure activities.

Culture industry: the groups, organizations, individuals, and corporations involved in producing cultural products such as television, movies, magazines, books, Internet content, etc.

False consciousness: a concept from Marx that explains that some institutions can exert enough power over the way people think to mislead them about the true state of affairs, and usually about relations of power in a society.

Family leisure: leisure spent with children.

Formal organization: a group of people who interact on a regular basis and have a set of explicit, written rules.

Harried leisure: leisure that is opposed to the more relaxed and comfortable extended block of leisure time.

Leisure: the opposite of paid work, or what we do with our time when we're not working jobs in the formal economy.

Machismo: in Latin American cultures, a cultural tradition of male dominance, particularly as it related to matters of sexuality and family.

Marianismo: in Latin American cultures, a cultural tradition of female superiority and the ability to endure suffering and self-sacrifice.

Market segmentation: a sales approach in which corporations divide a large and diverse market for a product into smaller segments based on characteristics like gender, social class, and race.

Media power theory: a perspective that perceives the culture industry as a well-oiled machine producing entertainment products to make a profit and that places most of the power over content in the hands of the producers.

Midriff: in popular media, a young woman who intentionally uses her sexuality as a means of empowerment.

Overspill of meanings: in a cultural product, when the viewers create their own meaning that goes beyond the preferred meaning.

Preferred meaning: in a cultural product, the meaning that was intended by the producers of that cultural product.

Pure leisure: leisure that involves no nonleisure activities.

Relaxation: one explanation for boys' attraction to video games that perceives them as escape from the weight of adult demands and of the rules of social decorum.

Restoration: one explanation for boys' attraction to video games that perceives part of the appeal in the sense of control they give their players that is not available in the real world.

Revenge: one explanation for boys' attraction to video games that perceives them as a way to get revenge against those who have usurped what you thought was yours.

Reverse discrimination: the perception that affirmative action policies result in discrimination against dominant groups such as whites and men.

Telenovela: a type of Latin American soap opera equivalent with a finite number of episodes.

Telenovela de ruptura: a genre of *telenovela* that incorporates social and cultural issues in Latin America.

Telenovela rosa: a type of *telenovela* that portrays a fairly one-dimensional set of characters who are clearly good or evil.

White, working-class male buffoon: a gender, racial, and social class stereotype shown on television as dumb, immature, irresponsible, and lacking in common sense.

Chapter 11: How Does Gender Help Determine Who Has Power and Who Doesn't?

Authority: the power that comes from a position in an organization or institution that is widely regarded as legitimate.

Burka: a tent-like version of hijab that covers the woman's entire body, leaving a tiny transparent bit of cloth for women to see out of and negotiate the world.

Coercive power: the ability to impose one's will by force, threats, or deceit.

Cultural relativism: a perspective that encourages us to view other cultures through their own lens rather than through ours.

Deference: when individuals engage in behavior that seems to support a dominant groups' power over them.

Descriptive or numerical representation: the number of individuals in a particular group who hold office.

Emasculation: taking away or reducing masculinity whether by women or other men.

Equality through complementarity: women and men may not do exactly the same things in the culture, but the contributions of both genders are equally valued.

Ethic of care: the idea that women are more worried about not helping others when they could be helped in moral dilemmas.

Ethic of justice: the idea that men tend to be more concerned about people interfering with each other's rights in moral dilemmas.

Feminization of poverty: the economic disadvantages experienced by female-headed households relative to male-headed households.

Gender outlaws: those who seek to dismantle the existing categories of power by demonstrating their contradictions and flaws.

Geography of fear: the assessments of vulnerability and fear in certain spaces that women are forced to make as a result of the threat of potential violence, assault, or harassment.

Head and master laws: laws that gave husbands the final say regarding all household decisions and joint property in a marriage, meaning that a husband could make decisions about these matters without his wife's knowledge or consent.

Hegemonic definition of manhood: what it means to be masculine; or a man in power, a man with power, and a man of power.

Hijab: belief in certain cultures that women should demonstrate modesty in their dress; comes from the word for veil and can take the form of a full body covering or a head scarf.

Human trafficking: using coercive force to sell a person for exploitative uses.

Institutionalized power: power that derives from the strength of an institution.

Interaction routines: patterns or norms of speech or action that are followed with regularity to accomplish some task in an interaction.

Marital rape: nonconsensual sex occurring within a marriage.

Missile envy: using the language of sex and anatomy to describe the weapons and their affects.

Nation: akin to ethnicity in that it defines the cultural ties that unite, or attempt to unite, the group of people who may or may not reside within the borders of the state.

Nation-state: a concept that links the nation as a sense of identity and community with the state as a political and geographic entity.

Political efficacy: the sense that what a person does really matters in terms of making change.

Political elite: the small group of people who hold high-level positions of power and responsibility in a government.

Political socialization: the process by which people learn what is expected of them in their particular political system.

Power: the ability of some actors to influence the behavior of others, whether through the use of persuasion, authority, or coercion.

Rape culture: a certain set of ideas about the nature of women, men, sexuality, and consent that may lead to a higher or lower incidence of sexual assault.

Rape-free: cultures where because sexual aggression is socially disapproved of and severely punished, sexual assault is either infrequent or does not occur.

Rape myths: incorrect beliefs about sexual assault that shift the blame for these crimes from the perpetuator to the victim.

Rape prone: cultures where the incident of rape is high because rape is excused as a ceremonial expression of masculinity or used as an act by men to punish or threaten women.

Slavery: an economic and social relationship between two people characterized at its core by violence.

State: the set of institutions and laws that govern life within a set of geographical boundaries.

State feminism: all the activities of governments that are officially charged with furthering women's status and rights.

Substantive representation: the representation of group interests, or the extent to which the members of the particular group represent the interest of that group when they are elected.

Index

About the Author

Robyn Ryle is an associate professor of sociology at Hanover College in Hanover, Indiana, where she has been teaching sociology of gender and other courses for 10 years. She went to Millsaps College in Jackson, Mississippi, for her undergraduate degree in sociology and English with a concentration in women's studies. She received her Ph.D. in sociology from Indiana University. She is a member of the American Sociological Association and has served on the editorial board of the journal *Teaching Sociology*. She grew up in a small town in Northern Kentucky and now lives in another small town just down the Ohio River. When she's not teaching classes, she writes, gardens, plays the fiddle, and knits. She currently lives in a 170-year-old house in scenic Madison, Indiana, with her husband, stepdaughter, and two peculiar cats. You can find her on Twitter (@robynryle), on the Facebook page for *Questioning Gender* (https://www.facebook.com/questioninggender?ref = hl), or at her blog: you-think-too-much.com.